Applied Ordinal Logistic Regression Using Stata®

SAGE was founded in 1965 by Sara Miller McCune to support the dissemination of usable knowledge by publishing innovative and high-quality research and teaching content. Today, we publish more than 850 journals, including those of more than 300 learned societies, more than 800 new books per year, and a growing range of library products including archives, data, case studies, reports, and video. SAGE remains majority-owned by our founder, and after Sara's lifetime will become owned by a charitable trust that secures our continued independence.

Los Angeles | London | New Delhi | Singapore | Washington DC

Applied Ordinal Logistic Regression Using Stata®

From Single-Level to Multilevel Modeling

Xing Liu

Eastern Connecticut State University

Los Angeles | London | New Delhi
Singapore | Washington DC

Los Angeles | London | New Delhi
Singapore | Washington DC

FOR INFORMATION:

SAGE Publications, Inc.
2455 Teller Road
Thousand Oaks, California 91320
E-mail: order@sagepub.com

SAGE Publications Ltd.
1 Oliver's Yard
55 City Road
London EC1Y 1SP
United Kingdom

SAGE Publications India Pvt. Ltd.
B 1/I 1 Mohan Cooperative Industrial Area
Mathura Road, New Delhi 110 044
India

SAGE Publications Asia-Pacific Pte. Ltd.
3 Church Street
#10-04 Samsung Hub
Singapore 049483

Acquisitions Editor: Helen Salmon
Editorial Assistant: Anna Villarruel
eLearning Editor: Robert Higgins
Production Editor: Kelly DeRosa
Copy Editor: Sheree Van Vreede
Typesetter: C&M Digitals (P) Ltd.
Proofreader: Scott Oney
Indexer: Will Ragsdale
Cover Designer: Michael Dubowe
Marketing Manager: Nicole Elliott

Printed in the United States of America

Library of Congress Cataloging-in-Publication Data

Liu, Xing (Education professor)

Applied ordinal logistic regression using Stata : from single-level to multilevel modeling/Xing Liu, Eastern Connecticut State University.

pages cm
Includes bibliographical references.

ISBN 978-1-4833-1975-9 (pbk.)

1. Logistic regression analysis—Data processing.
2. Regression analysis—Data processing.
3. Stata. I. Title.

QA278.2.L585 2016
519.5'35—dc23   2015023615

This book is printed on acid-free paper.

15 16 17 18 19 10 9 8 7 6 5 4 3 2 1

Brief Contents

Detailed Contents

Acknowledgments

I am grateful to many people without whom this book would not have been possible. First and foremost, I would like to thank Dr. Ann A. O'Connell, who opened the door to the fascinating world of logistic regression and multilevel modeling for me at the University of Connecticut. Thank you for your help and encouragement! I also would like to thank my graduate students from fall 2014 at Eastern Connecticut State University, who proofread the initial drafts of most chapters. Special thanks goes to Samantha Baumgart, who proofread all of the chapters. I also would like to thank Lisa Ferrari for proofreading several chapters.

In addition, I would like to thank the following reviewers for their thoughtful reviews of the book. They provided valuable comments and suggestions on an earlier version of the manuscript, which strengthened the book: Jennifer Hayes Clark, University of Houston; Lisa M. Dilks, West Virginia University; Anirudh V. S. Ruhil, Ohio University; Lu Liu, University of La Verne; David Peterson, Iowa State University; and Jessica N. Terman, George Mason University.

Furthermore, I would like to thank SAGE's editor, Helen Salmon, who has been very helpful and supportive throughout my writing process. I am also thankful to the production editor, Kelly DeRosa, for her help during the production stages.

A special thanks goes to the authors, listed in alphabetical order, who wrote the following user-written Stata programs: Maarten Buis, author of `seqlogit`; Vincent Kang Fu, author of `gologit`; Joseph Hilbe, author of `aic`; Ben Jann, author of `esttab`; Sophia Rabe-Hesketh, Anders Skrondal, and Andrew Pickles, authors of `gllamm`; J. Scott Long and Jeremy Freese, authors of the `SPost` package; Roy Wada, author of `outreg2`; Richard Williams, author of `gologit2`; and Rory Wolfe, author of `ocratio`.

I also would like to thank my wonderful colleagues in the Education Department at Eastern Connecticut State University, where we have collaborated on various research projects. They include Jeffrey Trawick-Smith, Sudha Swaminathan, Catherine Tannahill, Jeanelle Day, and David Stoloff. And special thanks goes to Dr. Hari Koirala, with whom I have worked collaboratively on multiple research grants, papers, and presentations.

Finally, very special thanks go to my wife, Li He, for her encouragement and support, and to my children, Lucy, Aileen, and Jerry. Lucy and Aileen are always curious about whether this is a storybook they can read, and Jerry grows with my writing of this book. Thanks to all of you who brought me joy during this journey. This book is dedicated to you all.

Preface

Purpose

The purpose of this book is to provide readers with advanced techniques for analyzing ordinal response variables using the statistical software package Stata*, a powerful tool with extensive analytic capabilities for ordinal regression analysis. This book presents comprehensive coverage of modern ordinal regression techniques from single-level proportional odds models to more advanced multilevel models. Furthermore, it provides a unified framework for both single-level and multilevel modeling of ordinal categorical data in a single text. Emphasis is given to demonstrating how to conduct ordinal logistic regression analysis using Stata, how to interpret the models, and how to present the results for publication.

The main intentions of the book are twofold. First, to provide a variety of modern ordinal regression models by focusing on how to fit these models, interpret the results with step-by-step explanations, and present the results with practical guidelines. Second, to demonstrate the use of Stata as a powerful tool for analyzing ordinal response data.

My motivation for writing this book stems from my experience in statistical consulting and teaching. I have observed researchers and students who find advanced statistical models for categorical data analysis as obstacles. They are perplexed when deciding which statistical analysis is appropriate to address their research questions, and they have difficulty in conducting such analysis. My expectation is to have a nontechnical book on ordinal regression models, which will be readily accessible to graduate students, applied researchers, and practitioners in education and social and behavioral sciences.

This book is designed as a supplementary text for graduate-level quantitative methods courses on logistic regression models, ordinal regression models, categorical data analysis, or multilevel modeling in education, social sciences, and behavioral sciences. Such courses are generally required in graduate programs in education, psychology, sociology, political science, and other related fields. In addition, it may be of interest to researchers in the areas of epidemiology, medicine, and public health, where

---

*Stata® is a registered trademark of StataCorp LP.

categorical data are extensively analyzed. Furthermore, it is designed as a reference for self-study for graduate students, applied researchers, and practitioners who are interested in categorical data analysis, in particular, in analyzing ordinal response data.

Approach of the Book

This book focuses on how to conduct ordinal logistic regression analysis using Stata, how to interpret results from the Stata output, and how to present the results in scholarly writing. It is organized in a structured format, which will help readers understand ordinal logistic regression models better and choose an appropriate model for their own research. In each chapter, the concept of a new model with related literature is introduced first, followed by the research problem, questions, and variables. Stata is then used to fit models using real-world example data. The Stata commands are explained and the output is interpreted in detail. The model is interpreted in terms of odds ratios and estimated probabilities using the `margins` command. Finally, how to create publication-quality tables using Stata is introduced, and the guidelines for reporting results are discussed. Readers can replicate results by practicing the Stata commands, which are summarized at the end of each chapter. In addition, the end-of-chapter exercises help readers to understand the model further.

A specific pedagogical approach used in this book is learning by doing. I find that students understand statistical concepts more deeply when the underlying theories of the models are closely connected with the application of real-world data using statistical software. The organization of the models, research questions and data, Stata commands, examples of model fitting, and exercises in sequence in each chapter provide students with hands-on opportunities to practice Stata and later conduct their own research.

Features of the Book

This book makes a close connection between statistical models and the statistical package Stata and makes these advanced techniques readily accessible to readers. One of the features is that each model in the text is illustrated with a real-world example from national large-scale datasets, such as the Education Longitudinal Studies (ELS:2002) and the General Social Survey of 2012 (GSS 2012). Another feature is that detailed instructions are provided on how to estimate the model, interpret the model, and present the results. The ordinal logistic regression models are interpreted using odds ratios and estimated probabilities with the `margins` command when applicable; the results are presented in publication-quality tables using Stata and reported in text. This will help readers understand each of the ordinal regression models, learn how to choose appropriate techniques, and apply them appropriately to address different research questions.

This book is application oriented rather than theoretical. It provides practical workbook-type problems and techniques in applying ordinal regression analysis with Stata. It is nontechnical; the theory and mathematical computations are kept minimal. Although the theories of the ordinal regression models are not the focus, this book does cover basic methodologies, which build foundations for readers to understand these models before applying them with real examples. The equation for each statistical model used by Stata is presented since other statistical software packages may parameterize the model differently. Presenting the equations may clarify some misunderstandings existing in the literature. Readers should not shy away from these equations, although skipping them will not hinder their understanding of the model and their conducting analysis.

Readers are assumed to have basic knowledge of introductory statistics and to be able to conduct multiple regression analysis. To facilitate the understanding of ordinal regression models promptly, a chapter introducing logistic regression models for binary data is included. Readers are also assumed to have limited or no previous knowledge in using Stata.

What This Book Does

- Presents comprehensive coverage of modern ordinal regression techniques from proportional odds models to advanced multilevel models in a logical way.
- Provides a unified treatment of single-level and multilevel models for ordinal response variables in a single text.
- Offers step-by-step explanations of model fitting using Stata.
- Focuses on interpreting the models and presenting the results for publication.
- Applies a structured approach to presenting materials for ordinal regression models, including an introduction of a model, the research problem, questions, and variables; the model estimation with Stata; the interpretation and presentation of the results; a summary of the commands; and the end-of-chapter exercises.
- Uses real examples from national large-scale datasets, such as the Education Longitudinal Studies (ELS:2002) and the General Social Survey of 2012 (GSS 2012), which can be generalized to a wide range of fields.

Organization of the Book

This book features chapters devoted to introducing Stata basics (Chapter 1), a review of basic statistics (Chapter 2), logistic regression for binary data (Chapter 3), proportional odds models for ordinal response variables (Chapter 4), partial proportional odds models and generalized ordinal logistic regression models (Chapter 5), continuation ratio models (Chapter 6), adjacent categories logistic regression models (Chapter 7), stereotype logistic regression models (Chapter 8), ordinal regressions with complex survey sampling designs (Chapter 9), multilevel modeling for continuous and binary response variables (Chapter 10), multilevel modeling for ordinal response variables (Chapter 11), and beyond ordinal logistic regression models: ordinal probit regression models and multinomial logistic regression models (Chapter 12).

The presentation of the chapters moves from basic to advanced statistical techniques for ordinal response variables. The initial two chapters review the basics of Stata and basic statistics using Stata. Readers with advanced knowledge in Stata and statistics may skip these two chapters or only read two sections at the end of Chapter 2, which cover how to make publication-quality tables using Stata, and the general guidelines for reporting results. Readers who are familiar with binary logistic regression may even skip Chapter 3 and go directly to Chapter 4.

Chapters 4 through 9 present various techniques for analyzing ordinal response variables for single-level data. They start with the most commonly used model (the proportional odds model), then proceed with more complex models (such as partial proportional odds models and generalized ordinal logistic models, continuation ratio models, adjacent categories models, and stereotype logistic regression models), and end with ordinal regression models for complex survey data.

Chapters 10 and 11 focus on multilevel modeling, covering models for continuous, binary, and ordinal response variables. Chapter 10 starts with multilevel models for continuous response variables and then moves to binary response variables, and it is a prerequisite chapter for Chapter 11, which focuses on more advanced techniques for ordinal response variables. The last chapter, Chapter 12, covers ordinal probit regression models and multinomial logistic regression models. Readers who want to learn about probit regression models, rather than about logistic regression models, may find this chapter useful. In addition, the multinomial logistic regression model is used to analyze nominal response variables. It is an alternative to ordinal logistic regression models when the proportional odds assumption is not met.

Instructors may assign the first two chapters to students for review, or they may use the materials for an introductory class with a focus on a general introduction to Stata and data analysis. Chapter 3 can be assigned for a class on binary logistic regression models. Chapters 4 through 9 can be used as teaching materials for single-level ordinal logistic regression models, and Chapters 10 and 11 can be used when introducing multilevel ordinal regression models. Instructors may also assign the first section of Chapter 12, which focuses on ordinal probit models, to students after discussing proportional odds models in Chapter 4, and they may read the second section of Chapter 12 before the adjacent categories models (Chapter 7). Readers can go directly to any chapter if a particular ordinal model is the focus. It is up to the instructors and students to decide the ordering of the chapters.

The following is a brief synopsis of the chapters:

- Chapter 1 begins with an introduction to the basics of Stata. The first section shows novice Stata users the interface, menus, and Stata toolbar. It also introduces the structure and rules of Stata commands, official and user-written commands, do-files, comments in do-files, and a do-file template. The second section shows readers various data management techniques, such as how to create a new variable, recode an existing variable, label a variable, and label values. The third section shows how to draw histograms, bar charts, box plots, and scatter plots, and it shows how to use Graph Editor to edit graphs. This chapter will help readers get started with Stata and be familiar with basic statistical techniques that are necessary for the remaining chapters.
- Chapter 2 reviews descriptive statistics and various inferential statistics using Stata. When introducing each type of inferential statistics, an example of research design is provided followed by the research questions, Stata commands, and output. The Stata commands are explained and the output is interpreted in detail. In addition, a sample of reporting the results for each analysis is provided. Finally, the commands for creating publication-quality tables using Stata are introduced, and the guidelines for reporting results are discussed. This chapter will help readers become familiar with basic statistical techniques that are necessary for the remaining chapters.
- Chapter 3 presents logistic regression models for binary data. It introduces the concepts of odds, odds ratios, goodness-of-fit statistics of the model, how to test the significance of predictors, and how to interpret parameter estimates. After the description of the data, two logistic regression models using Stata are illustrated with step-by-step instructions. A summary of Stata commands used in this chapter and exercises are provided at the end. This is a prerequisite, warm-up chapter for the following various ordinal logistic regression models.
- Chapter 4 covers proportional odds models for ordinal response variables. It starts with an introduction of the proportional odds model and the Brant test of the proportional odds assumption, followed by a description of the data. Then it shows two examples on how to fit proportional odds models using Stata. Finally, the chapter discusses how to interpret the models in terms of odds ratios and estimated probabilities, present the results in publication-quality tables using Stata, and report the results with practical guidelines.
- Chapter 5 extends proportional odds models by presenting partial proportional odds models and generalized ordinal logistic regression models when the proportional odds assumption is untenable. It begins with an introduction of the models, odds and odds ratios, model fit statistics, and interpretations of parameter estimates. After the description of the research example and the data, two examples on how to fit these two models using Stata are illustrated. This

chapter focuses on fitting partial proportional odds models and generalized ordinal logistic regression models using Stata, interpreting the models, and presenting the results.

- Chapter 6 focuses on continuation ratio models. After the introduction of the models, the conditional probabilities, odds, and odds ratios are reviewed; model fit statistics and interpretations of parameter estimates are discussed. It shows how to conduct the analysis using Stata with step-by-step instructions. Stata commands are explained and output is interpreted in detail. The chapter also illustrates how the results are displayed in publication-quality tables using the Stata command and are reported in text.

- Chapter 7 deals with adjacent categories logistic regression models. After introducing the models, we review the odds, odds ratios, and model fit statistics, and we discuss how to interpret the parameter estimates. Next, the multinomial logistic regression model is briefly introduced, and the transformation from the multinomial logistic model to the adjacent categories model is discussed. After the description of the data, the chapter shows how to conduct the analysis using Stata with step-by-step instructions. It also illustrates how the results are displayed in publication-quality tables using Stata and are reported in text.

- Chapter 8 discusses the stereotype logistic regression model. It first introduces the advantages of this model, ordinality constraints, odds and odds ratios, measures of goodness of fit, and interpretation of parameter estimates. After the description of the data, it illustrates two examples of stereotype logistic regression models using Stata with detailed instructions, and it compares these models with the proportional odds model. The chapter also illustrates how the results are displayed in publication-quality tables using the Stata command and are reported in text.

- Chapter 9 presents ordinal logistic regression models with complex survey sampling designs. It starts with an introduction of the features of complex sampling designs, followed by discussions of variance estimation in complex survey sampling. After a description of the research example and the data, various types of models are built, including the proportional odds model without weights, the model with weights, and the model with complex survey sampling designs. In addition, methods are discussed to deal with the issue of single sampling units with strata. The chapter also illustrates how the results are displayed in publication-quality tables using the Stata command and are reported in text.

- Chapter 10 introduces multilevel modeling. To facilitate the understanding of multilevel models for ordinal response variables in the next chapter, this chapter presents multilevel modeling for continuous outcome variables and binary response variables. It starts with an introduction to multilevel modeling, followed by an illustration of these two models with two research examples using

Stata. In each example, it discusses when the multilevel model can be employed and how to specify a two-level model, followed by the description of the research questions and data. Then several models, from the unconditional (null) model to the random-intercept model and random-coefficient model to the contextual models are illustrated using Stata with step-by-step instructions. The chapter also illustrates how the results are displayed in publication-quality tables using the Stata command and are reported in text.

- Chapter 11 introduces multilevel modeling for ordinal response variables. It starts with an introduction of the model, followed by a discussion of the model specification, the odds and odds ratios in the model, the log likelihood ratio test, and the description of the research questions and data. Then several models are illustrated using Stata with step-by-step instructions, including the unconditional (null) model, the random-intercept model, the random-coefficient model with a level 1 variable, the contextual model with both level 1 and level 2 variables, and the contextual model with cross-level interactions. The chapter also illustrates how the results are displayed in publication-quality tables using the Stata command and are reported in text.

- Chapter 12 discusses ordinal probit regression models and multinomial logistic regression models. The first section starts with an introduction of the probit regression model, followed by a discussion of how to interpret parameter estimates. Next is a description of a research example, the data, and the sample, and then a multiple-predictor ordinal probit regression model using Stata is illustrated with step-by-step instructions. The second section starts with an introduction of the multinomial logistic regression model, followed by a discussion of the odds and odds ratios or relative risk ratios in the model. After the description of a research example, the data, and the sample, a multiple-predictor multinomial logistic regression model is fitted with Stata.

After reading each chapter, readers will be able to identify when an ordinal regression model is used, conduct data analysis using Stata, develop models using the likelihood ratio test and other fit statistics, interpret the output, present the results in publication-quality tables using Stata, and write the results for publication.

Topics Not Covered

Due to space limitations, some important topics are omitted. For example, longitudinal data analysis of ordinal response variables is not covered. This type of analysis can be conducted within the framework of multilevel modeling by specifying the level 1 time variable. Another important topic is the residual diagnostics for ordinal logistic regression models since the analytic techniques have not been well developed. The current approach (Hosmer & Lemeshow, 2002; Hosmer, Lemeshow, & Sturdivant, 2013; Long & Freese, 2006, 2014) is to dichotomize the ordinal response variable and apply the residual diagnostics strategies developed for binary logistic models outlined by Pregibon (1981). Refer to O'Connell and Liu (2011) on the illustration of model diagnostics for ordinal logistic regression models using this approach.

Software and Versions

Why Stata? Stata is a powerful tool with extensive analytic capabilities for ordinal regression analysis. It can be used to fit various ordinal regression models covered in this book. Students do not need to switch between software packages when conducting ordinal regression analysis.

I assume that readers are using either Stata 13 or the current version, Stata 14, although most programs in the book will work in previous versions up to at least Stata 9. The new functions of the `margins` and `marginsplot` commands are introduced for both Stata 13 and 14, and the differences between these two versions are highlighted. When a program does not work in previous versions, it is noted. For example, the program `meologit` was introduced in Stata 13 and does not work in versions earlier than that. However, an alternative user-written program, `gllamm`, works even in Stata 7. Both programs are explained when they first appear. When a user-written program is introduced, its installation information is provided. For example, Section 5.3.1 introduces how to install the `gologit` and `gologit2` programs. I also assume readers are using Stata on a Windows PC. If other operating systems, such as Macintosh (Mac) or Unix System, are used, readers may notice minor differences when running programs.

Notation

First, I use `typewriter font` (i.e., `Courier New`) for Stata commands, output, and variable names. For example, `logit` is the command for logistic regression models; `educ` and `age` are two variables. Second, a Stata command shown in the output is followed by a period. Readers should not include the period to rerun the command when replicating the results. Third, when copying and pasting the selected output into a Microsoft Word document, `Courier New` font with the font size of 9 or smaller should be used. Otherwise, the output will not be shown properly. Finally, although the `xi` prefix command and the new factor variables notation (`i.` prefix) are both introduced for categorical predictor variables, the `xi` prefix is used throughout the book since the factor variables coding does not work with some of the user-written programs.

User-Written Stata Programs

As introduced in the first chapter, one strength of Stata is the free user-written programs. The following is a complete list of the user-written programs used in this book.

- aic (Hardin & Hilbe, 2012; Hilbe, 2009)
- brant (Part of the SPost package) (Long & Freese, 2014)
- esttab (Jann, 2005, 2007)
- fitstat (Part of the SPost package) (Long & Freese, 2014)
- gllamm (Rabe-Hesketh, Skrondal, & Pickles, 2004)
- gologit (Fu, 1998)
- gologit2 (Williams, 2006)
- listcoef (Part of the SPost package) (Long & Freese, 2014)
- mlogtest (Part of the SPost package) (Long & Freese, 2014)
- mtable (Part of the SPost package) (Long & Freese, 2014)
- ocratio (Wolfe, 1998)
- outreg2 (Wada, 2008)
- seqlogit (Buis, 2007)

Book Website

There is a companion website for this book located at https://study.sagepub.com/liu-aolr. This website contains datasets, the Stata commands used in this book, and solutions to the end-of-chapter exercises for students and instructors.

About the Author

Xing Liu, Ph.D., is an associate professor of educational research and assessment at Eastern Connecticut State University. He received his Ph.D. in measurement, evaluation, and assessment in the field of educational psychology from the University of Connecticut, Storrs. His interests include categorical data analysis, multilevel modeling, longitudinal data analysis, structural equation modeling, and educational assessment. His major publications focus on advanced statistical models. His articles have been recognized among the most popular papers published in the *Journal of Modern Applied Statistical Methods (JMASM)*. Dr. Liu is the recipient of the Excellence Award in Creativity/Scholarship at Eastern Connecticut State University.

CHAPTER 1

Stata Basics

Objectives of This Chapter

This chapter begins with an introduction to the basics of Stata. The first section shows novice Stata users the interface, menus, and Stata toolbar. It also introduces the structure and rules of Stata commands, official and user-written commands, do-files, comments in do-files, and a do-file template. The second section shows readers various data management techniques, such as how to create a new variable, recode an existing variable, label a variable, and label values. The third section shows how to draw histograms, bar charts, box plots, and scatter plots, as well as how to use Graph Editor to edit graphs. After reading this chapter, you should be able to

- Start and exit Stata using different approaches.
- Enter data using Stata data-editor, and save a new data file.
- Open existing data files.
- Enter commands, create do-files, and save output.
- Create new variables, recode and label variables, and label values for categorical variables.
- Draw different types of graphs.

1.1 Introduction to Stata

Stata is a general-purpose statistical software package for data management, data analysis, and graphing. It can be run on almost all operating systems across Windows, Mac, Linux, and Unix. Stata is relatively small; yet it is powerful. It does not require much hard drive space. The space needed for version 8 or earlier is less than 30 MB. For versions 9 to 10, 100 MB of space should be good enough for the installation. With the inclusion of Stata manuals in the installation folder, versions 11 and 12 need about 250 MB and 500 MB of disk space, respectively; Stata 13 needs about 700 MB of hard drive space, and Stata 14 needs 900 MB. Excluding other files, the Stata executable and ado-files (i.e., the programs) in versions 13 and 14 only take about 150 MB.

Stata is fast. It starts fast! You will not need to sit there waiting. It also runs fast since it loads data directly into your memory when you work on your data and conduct analysis. Before Stata 12, if you had to work on a large dataset that exceeded the allowed amount of memory, you needed to set the memory size by using the set memory command. Starting with Stata 12, however, there are no longer memory constraints and you do not need to set it manually. It automatically adjusts your memory based on your needs via the new automatic memory manager.

Stata is a powerful tool for data management, graphics, and data analyses. As a general-purpose statistical software package, it is meant for conducting various statistical analyses. These analyses include, but are not limited to, the following techniques: basic statistical analyses, such as descriptive statistics, t test, analysis of variance (ANOVA), correlation, regression, and nonparametric tests; categorical data analyses for binary, ordinal, multinomial, and count outcomes; multivariate analyses, such as multivariate analysis of variance, cluster analysis, discriminant analysis, factor and principle component analysis, multidimensional scaling, correspondence analysis, and biplot; time series; survival analysis; propensity score analysis; structural equation modeling; longitudinal/panel data analysis; power analysis; survey data analysis; and multilevel modeling for continuous and categorical outcomes. The latest Stata 14 includes new statistical models for Bayesian analysis and item response theory (IRT).

Stata has nice programming capabilities. One advantage of Stata is that users can write programs with new features and functions that can be shared with other users in the Stata community. With the contributions of experts from various fields, new statistical techniques can be quickly implemented in Stata. For example, the user-written program gllamm for generalized linear latent and mixed modeling was developed by Rabe-Hesketh, Skrondal, and Pickles (2004) before the official release of the xtmixed program. We can search and install these user-written programs within Stata using the search or findit commands. After installation, these programs can be executed in the same way as the official Stata programs. So, when functions of interest are unavailable in the official programs, users can see whether they have been developed by other users.

Stata has complete documentation and is readily available to users. Starting with version 11, PDF manuals were installed within Stata so that users can now directly access them from the help menu. Stata 13 has 20 manuals with more than 10,000 pages, and the newest version, Stata 14, has 23 manuals with more than 12,000 pages. Stata also has a user-friendly help system. To get help for a particular command, you can either select it from the menu or type the `help`, `search` or `findit` commands on the command line. The help file you find is linked directly to its related entry in the PDF manuals.

Stata is mainly a command-driven statistical package. To execute a command, you need to enter it on the command line and press the **Enter** key. This task may seem daunting to a novice Stata user who may then turn around and look for a more user-friendly program. But wait a minute! Stata also has a graphic user interface (GUI), which was introduced in Stata 8. This point-and-click menu system helps new users to learn about the program. It makes it easier to familiarize yourself with the features of data management, graphics, and statistical tools. When you perform an analysis using the GUI, Stata displays the corresponding command in the **Preview** window and the output in the **Results** window. You can then save the command for future use.

Which flavor of Stata do you need? Stata offers several different flavors (or versions): Small Stata, Stata/IC (intercool), Stata/SE (special edition), and Stata/MP (multiprocessor). Among them, Stata/SE might be the most commonly used. All flavors have the same statistical functions but differ in how they handle the number of variables and observations, as well as the speed of computation. The Small Stata stands for "Stata for small computers" and can only handle up to 99 variables and 1,200 observations, so it is for smaller datasets. The other three flavors are the professional versions of Stata, and they almost do not have any limitation on observations as long as your computer memory allows for it. For example, Stata/IC and Stata/SE allow 2 billion observations and Stata MP allows up to 20 billion observations, but you need to have a computer with large memory. Stata/IC can handle a maximum of 2,047 variables, which is good enough for normal data analysis. Stata/SE is for a single-process computer, whereas Stata/MP is for a computer with a multicore processor. Both of them can handle 32,767 variables, and you rarely see a dataset with more variables than that.

1.1.1 Do You Still Need to Use Commands?

Since Stata has a well-built GUI system, why do we still need text commands in a non-DOS era? There is nothing wrong with using the GUI pull-down menus, and you should never feel inferior because you are using the functions via the GUI instead of via text commands. As long as you know how to use it, it will fulfill your needs. As you become an experienced Stata user, however, you will find it will be more efficient to type commands. There are three reasons for this efficiency. First, it saves you time. For example, to run a simple regression analysis with a dependent variable y and an independent variable x, you simply type:

```
regress y x
```

Second, you can save all your commands to a do-file so that you can replicate your analysis easily. In addition, you can edit your do-file for other analyses. For example, if we have three independent variables in a linear regression analysis, then we can modify the previous command as follows:

```
regress y x1 x2 x3
```

Third, you know what analysis you are doing by typing commands interactively. The GUI environment will allow you to point and click to any menus and produce results without knowing the underlying statistical techniques. Statisticians have always been concerned about abusing the use of statistical software when users have limited knowledge of the models. Stata helps alleviate this concern.

1.1.2 Stata at First Sight: Interface, Menus, and Toolbar

Stata can be started in three ways. First, you can run Stata by clicking the icon on the desktop. Second, you can start it by clicking the **Start** Menu on Windows, **All Programs**, and then **Stata**. Third, you can open it by double clicking a Stata data file.

To exit Stata, you can either type the command `exit, clear` and press **Enter** or use the pull-down menu. To use the menu, go to **File** and then click on **Exit**.

Stata Windows

Once you start Stata, you will see five windows (Figure 1.1): **Command**, **Results**, **Review**, **Variable**, and **Properties**.

Command window: In this window, you need to type a command and press the **Enter** key to execute it. You can copy a command from a Notepad or a Word document and paste it here. You can also edit a command before execution. You can only execute one command at a time in this window, whereas you can run a series of commands via the do-file. Normally you will also need to open a data file before running commands for data analysis.

Results window: This window displays the output after you execute a command. All the commands you entered in the Command window are echoed in the Results

Figure 1.1   Stata Windows

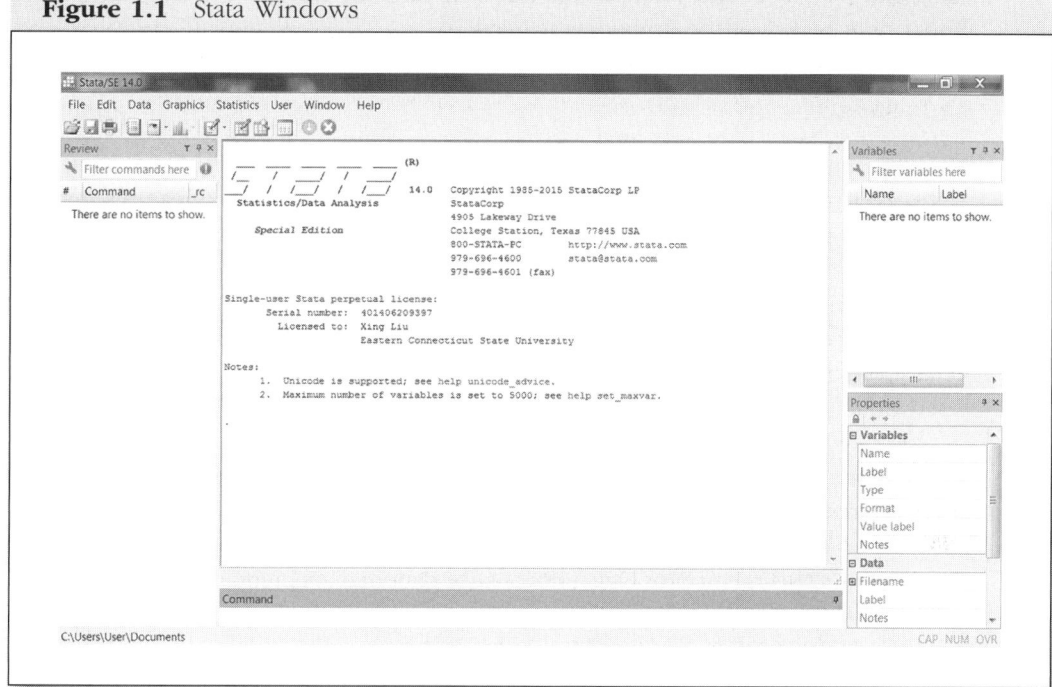

window, followed by the output. If you see –more– at the bottom of the Results window, it means that the screen is full and that you need to click on it so that you can see more results. To see all the output without showing this message, you can type the command set more off. The output can be copied and pasted into a text file or a Word document. It can also be saved into a log file, which will be explained in more detail next.

Review window: This window displays the commands that have been executed from the Command window. A nice feature of this window is that you can click on any command and bring it back to the Command window without retyping it. Then, you can modify it for the following analyses. You can also select any commands in the Review window and save them into a do-file.

Variable window: This window contains variable names in the dataset. Instead of typing variable names in a command, clicking any variable in this window will bring it to the Command window, which will be helpful to save time and reduce typos.

Properties window: This function was introduced in Stata 12. This window shows the properties of variables and the dataset. If you select a variable in the Variable window, the Properties window will display the variable name, label, type, format, value label, and notes. To modify these properties, you need to unlock the function first. To see the properties of the next variable, click on the arrow key at the upper right corner of the Properties window.

Stata Menus

Stata has eight pull-down main menus, including File, Edit, Data, Graphics, Statistics, User, Windows, and Help. These menus provide tools and features that can be used by pointing and clicking.

File menu: The purpose of this menu is to help you with files. Options in this menu help you open existing Stata datasets, graphs, and recent files; save data and graphs; open do-files; create and close log files; import and export data files; print results; open example datasets; and exit Stata.

Edit menu: The Edit menu helps you copy texts and tables from the output and paste them in another location so that you can save them.

Data menu: This menu provides functions for data management. By using this menu, you can describe data; edit data using the data editor; browse data using the data browser; create and change variables; sort variables; combine datasets; label datasets, variables, and values; add or delete notes to datasets; rename variables; add or drop variables or observations; identify duplicate observations; move the position of variables; create matrixes and work with them; and use other utilities.

Graphics menu: You can create and manage graphics by using various options on this menu.

Statistics menu: This menu provides the statistical analysis tools Stata offers. The list of Statistics menu options has grown as Stata has incorporated new statistical procedures into the package.

User menu: This menu is seldom used. Clicking on it brings you three empty submenus. It is useful only if you would like to add your own menu items for Data, Graphics, and Statistics menus.

Window menu: This menu contains a series of submenus, such as follows: Command, Results, Review, Variable, Properties, Graph, Viewer, Data Editor, Do-file Editor, and Variables Manager.

Help menu: This menu provides help and search facilities, such as having access to the PDF documents, getting advice, learning Stata via the contents, searching Stata commands, getting help with a command, finding resources for Stata learning, finding and installing user-written programs, and checking for updates.

Stata Toolbar

The toolbar, located below the main menus, comprises a set of icons. It helps you quickly access the most frequently used features. Familiarizing yourself with these icons will make the use of Stata more efficient.

These icons include Open, Save, Print Results, Log, Viewer, Graph, Do-File Editor, Data Editor (Edit), Data Editor (Browse), Variables Manager, Clear –More– Condition, and Break. Table 1.1 shows the icons of the toolbar, their titles, and their functions.

1.1.3 Creating a File and Entering Data

To create a new dataset, you can use the Stata Data Editor either by clicking on the **Data Editor (Edit)** icon on the Toolbar or by going to the **Data** Menu and selecting **Data Editor**. Figure 1.2 shows the window of Data Editor (Edit). Once the Data Editor window is opened, you can enter data just as you would in a spreadsheet. Stata automatically names your variables to var1, var2, var3, ... from the left column to the right. Since the release of Stata 12, you can add/edit variable names, types, formats, variable labels, value labels, and notes in the Properties window at the bottom right corner. If you use an earlier version, you can change the variable name by double clicking on it. To save the data you have entered, click on the **Save** icon on the Toolbar or go to **File** and then **Save As**. You can also enter the command save on the command line with a file name, for example, save data1. The extension of the Stata dataset is .dta.

For Stata 13 users, although Stata 13 can read data saved by previous versions, older versions may not be able to read data created by newer versions. The command saveold can save a dataset in Stata 12 format. To create a dataset compatible with versions older than 12, use the user-written command use13.

For Stata 14 users, the improved saveold command can create a dataset readable by versions 11 to 13. Since Stata 14 now supports Unicode, whereas older versions do not, you may see that variable names in a dataset created by Stata 14 are not properly displayed in older versions. The command use13 currently does not work in Stata 14. It is hoped that the issues of data transfer will be solved once use14 is available. Be careful with data conversions. Another option is to use Stat/Transfer, a program for converting datasets among various formats. The latest Stat/Transfer 13 supports the data formats for all versions of Stata.

Table 1.1 Icons of the Toolbar, Their Titles, and Their Functions

Icon	Icon Title	Functions
	Open	Opens an existing Stata file (e.g., .dta, .gph, .do)
	Save	Saves the current dataset to your computer
	Print	Prints your Stata output
	Log	Begins, closes, suspends, or resumes your log file
	Viewer	Opens the Viewer to get help
	Graph	Brings a graph window to the front
	Do-File Editor	Opens the do-file editor so that you can make or edit a do-file
	Data Editor (Edit)	Opens the data editor so that you can edit the current data file
	Data Editor (Browse)	Opens the data editor so that you can browse rather than edit the current data file
	Variables Manager	Opens the variables manager so that you can change variable properties, such as the variable name, label, type, format, and value label
	Clear –More– Condition	Tells Stata to show more output
	Break	Stops executing the current program

Figure 1.2   Data Editor (Edit) Window

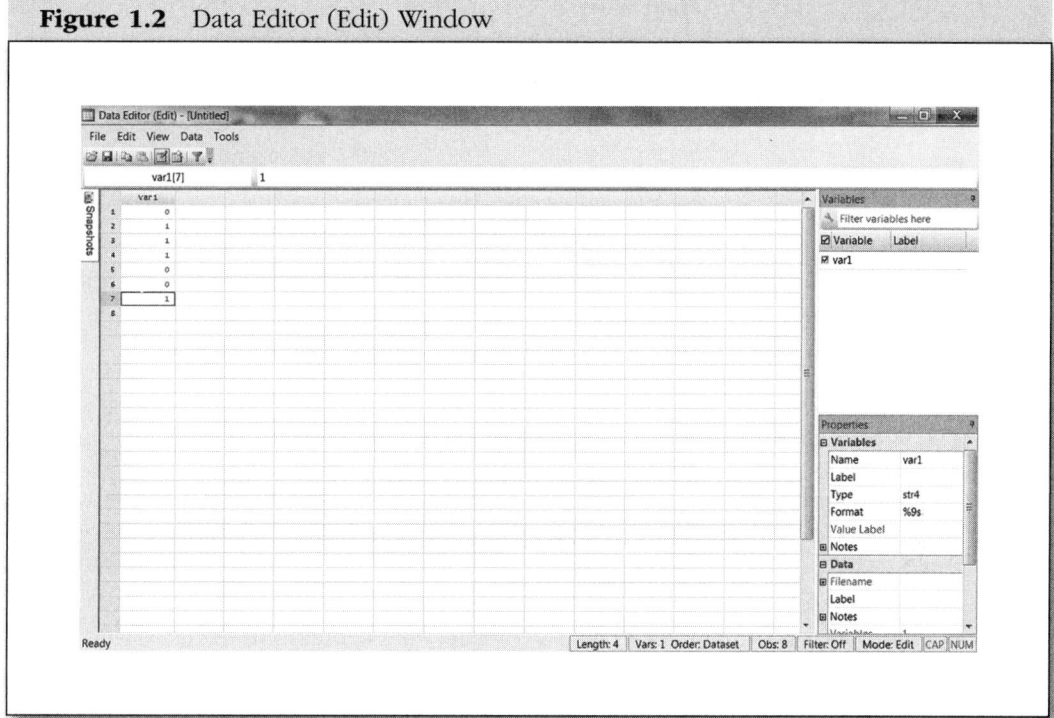

1.1.4 How to Open an Existing Data File

To open an existing Stata .dta file, you can use one of the following four ways:

- First, **double click** the Stata data file, which starts the Stata program, and Stata will read the data.
- Second, **enter** the command use followed by a file path on the command line. If a Stata data file is saved in the current working directory, you only need to type use followed by the file name. For example, **type** use Chapter1.dta and **press** the **Enter** key. If the file is saved elsewhere on the computer, you need to type the full path. The use command can also open a data file on the web followed by the web link. To open a dataset for Stata manuals, type the webuse command followed by the filename. In addition, the sysuse command can be used to open datasets stored in the system directories. For example, to use a system data file auto.dta, simply **type** sysuse auto in the Command window and press **Enter**.

- Third, click the **Open icon** on the Toolbar. Once you click on the icon, it will open the dialog. Select the dataset and then click on **Open**.
- Fourth, go to **File** on the main menu, and then click on **Open**. Select the data file and then click on **Open**.

Other Data Formats

Besides its own data format, Stata can read a variety of other data formats. For example, ASCII files, Microsoft Excel files (both .xls and .xlsx formats), and SAS transport format (SAS XPORT). Since the release of version 12, Stata can directly read Excel files using the `import excel` command or via the pull-down **File** menu.

1.1.5 The Structure of Stata Commands

Stata commands are concise, so do not be afraid of learning the command syntax. As a new user of Stata, you may ask, "Is it difficult to learn how to use Stata, particularly Stata commands?" The answer is "no." You just need to understand the structure of Stata syntax and get some practice on your computer. Now, let us look at several examples of Stata commands. Then, I will explain its generic structure.

The simplest structure of a command is the command name itself. For example, to get the descriptive statistics of all variables in a dataset, you just type the command `summarize`. If we want to see the descriptive statistics of a particular variable, type the variable name after the command `summarize`. For example, `summarize var1`. Entering more variable names after `summarize` will generate descriptive statistics for those variables.

To get a frequency of a categorical variable, type the command `tabulate` plus the variable name. For example, `tabulate var2`. If you get frequency tables for more than one variable, use the `tab1` command. For example, `tab1 gender class`.

Suppose you want to run descriptive statistical analysis on a subset of the data. To do this, you can use the `if` qualifier. For example, to see the descriptive statistics of students' reading scores in class 1 of two classes, enter the command `summarize reading if class == 1`.

If you want to see the descriptive statistics of a variable in the first 50 observations, you can use the `in` qualifier. For example, `summarize reading in 1/50`.

From the previous examples, we can induce a simple structure of Stata commands:

command [varlist] [if] [in]

The `command` stands for any Stata command. The `varlist` means variable list, which can be one variable or a list of variables. `If` stands for the "if qualifier," and `in` stands for the "in qualifier." In the command syntax, the command name is required and the elements in brackets are optional. When we enter a command, we should not enter any brackets, which are only used to explain the optional components in the command syntax here.

This simple structure of Stata commands can be expanded to a more generic one. We can add weight and command options to the syntax. The options are specified at the end and separated from the command by a comma. A prefix can also be added before the command. So the generic Stata command syntax you see in the manuals or in the command help dialog looks like this:

```
command [varlist] [if] [in] [weight] [, options]
```

A prefix can also be added before a Stata command. The `by` prefix is used to repeat a command for each category or group of the data. The `svy` prefix is used for the complex sampling survey data. The `xi` prefix creates dummy variables in an analysis.

Rules of Stata Commands

1. Stata commands should be error-free. In other words, you need to type them exactly.

2. Stata commands are case sensitive, so the lowercase and uppercase letters are different. Be careful when the variable names are mixed with lowercase and uppercase letters. When you name a variable, it is always a good idea to use lowercase.

3. Stata commands are typically in lowercase. It is rare to see a capital letter in Stata commands.

Official Commands and User-Written Commands

The official commands are those developed and released by Stata. They come directly with the Stata installation disk or files. Once Stata is installed, you can immediately execute these commands. In the Stata community, a wide range of Stata commands are developed by users. These commands do not come with your installation disk or files. They complement the official command by extending the capabilities of Stata. They are free and can easily be installed to your Stata. Most of the user-written commands were published in the *Stata Journal* or its predecessor, the *Stata Technical Bulletin*. Most of them are stored in the Statistical Software Component (SSC) Archive at Boston College (http://ideas.repec.org/s/boc/bocode.html) and can be accessed freely.

If you want to install a user-written program that you know the name of, just type the command `ssc install` followed by the program name. For example, to install a program named `ocratio`, which is used for continuation ratio models, type ssc install ocratio, and Stata will install it for you. You can also use the command findit to search a user-written program or to see whether the program has been installed. As a result, Stata will search official help files, FAQs, examples, and web resources, and it will provide a list of related links for this command. You can browse these links and decide which you want to install. Using the findit command gives you more control than directly installing it with the ssc install command.

If you want to check the version of a user-written program you have installed, type the which command followed by the program name. For example, if the outreg2 program has been installed, type which outreg2 and the version of the program will be displayed.

If you want to uninstall a user-written program previously installed from SSC, type ssc uninstall followed by the program name. If a program is not installed from SSC and you want to uninstall it, type the ado, dir command. After seeing the listed program in the Results window, type ado uninstall followed by the program name.

Different from the official Stata programs, the user-written programs cannot be updated automatically. If you want to update a user-written program to the latest version, type the adoupdate command. Stata will check the status of all the installed user-written programs and list the numbered programs that need to be updated. By typing the adoupdate, update command, all recommended programs will be updated. If you only want to update one of the listed programs, specify the program name after the adoupdate command.

1.1.6 Do-Files

If you use the Command window to enter commands, you can enter one piece at a time. Do-files can help you put a list of commands in one file and execute them together as a batch. Using do-files helps you to organize commands, keep a record, and understand what you have done when you need them in the future. It is also helpful when you collaborate with other researchers on a research project. Your colleagues can simply replicate the analyses using the do-files you provide, and they can modify them for new analyses. It will save you time when you need to replicate your statistical analyses. If you are an instructor, these do-files are also helpful when preparing your instruction and for grading students' assignments.

To create a do-file, you can use the Do-file Editor by clicking the **Do-file Editor** icon on the toolbar, or you can just type the command doedit. This will open up the Do-file Editor. Figure 1.3a shows the window of the Do-file Editor. After entering a list of commands, you can save it from the **Review** window as a do-file with a .do extension, for example, chapter1.do. To execute this do-file, type do chapter1.do.

You should save the do-file to the current working directory. Otherwise, you will need to type the whole path with the file name when you execute it. To check your current working directory, type the command pwd or cd.

Figure 1.3a Window of the Do-file Editor

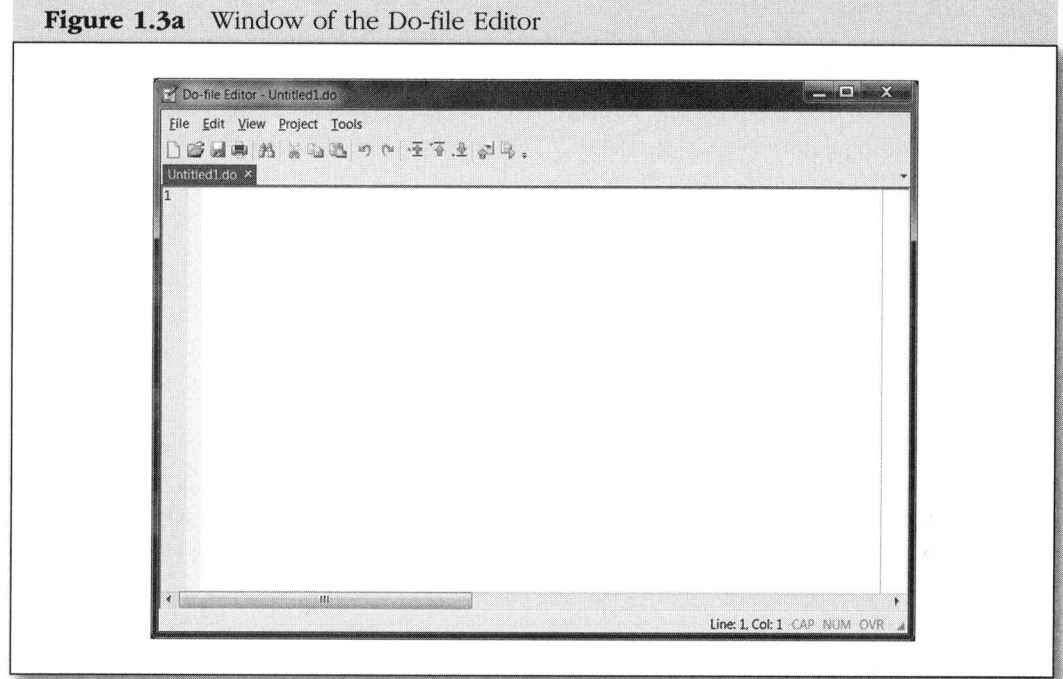

A good habit when you create a do-file is to have comments in the file. The comments may include time, project name, who wrote it, and for what purposes. The clearer your comments are, the better your documentation will be. The comments in the do-files will not be executed but will be displayed in your output. There are several ways to write comments. A comment would start with a * after a double slash, //, or after a triple slash, ///. For a long comment, you can use a combination of /* and */. Stata recognizes it as a comment if you follow these four formats. Please note that the triple slash (///) is often used as a line-join indicator for long command lines in do-files. It tells Stata that the following command line is a continuation of the current line and both should be executed together. Users may feel confused with the two functions of the triple slash. In addition, a comment with //, ///, and /* */ can only be used in do-files or ado-files, and it should not be entered in the Command window.

Long (2009) provided detailed suggestions on how to write do-files. He suggested that the do-files be "robust and legible" (p. 50). They are robust when researchers can produce the same results when rerunning the same command or conduct the same analysis using a different computer. They also need to be legible so that researchers can understand what analyses can be done using these do-files. Specific suggestions for making robust do-files by Long (2009) mainly include making the do-files independent of each other, using version control so that the do-files work with previous versions, and excluding directory locations so that the do-files can be run in the current working directory. Another suggestion is to set the seeds when you generate random numbers so that the same results can be produced.

To make the do-files legible, Long (2009) also suggested using various comments, employing alignment and indentation, setting line size, limiting the use of abbreviations, and being consistent in the way of creating do-files.

How to Create a Do-File Template

A do-file template is a template you can use to create do-files for any statistical analysis. It includes some common commands that ease your work. Figure 1.3b is an example of a do-file template with comments.

- Line 1: A comment with the content information and the date.
- Line 2: A comment for the version control.
- Line 3: The command for version control. This command is useful when you need to run commands in older versions. The command version 13 tells us that the do-file will work in this version.

Figure 1.3b   Example of a Do-File Template

```
Do-file Editor - dotemplate*

File  Edit  View  Project  Tools

dotemplate* ×  Untitled.do ×
 1  *do-file template for chapter1-1/10/2014
 2  *version control
 3  version 13
 4  set more off   /*running do-file without interruption*/
 5  capture log close
 6  //creating a log file
 7  log using chapter1, replace
 8  /*the following 4 commands run descriptive statistics for age*/
 9  use GSS2012.dta, clear
10  codebook age
11  summarize age
12  summarize age, detail
13  log close
14  exit

                                              Line: 14, Col: 5  CAP  NUM  OVR
```

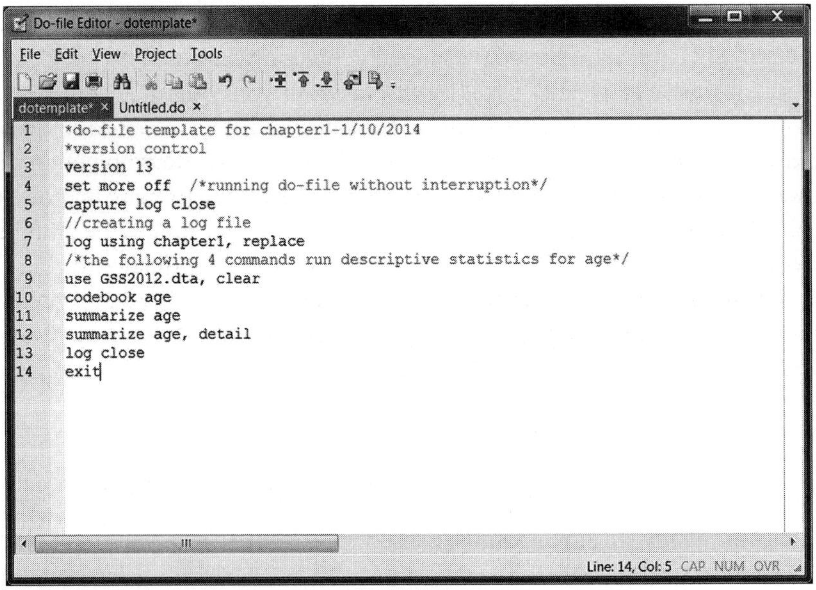

- Line 4: The set more off command is used so that your do-file can be run without interruption. A comment follows the command.
- Line 5: If you would like to create log files so that the output will be automatically saved, use the capture log close command before creating the log files. This command tells Stata that no other log files are open or in use.
- Line 6: A comment for creating a log file using a command below.
- Line 7: The command log using chapter1 creates a log file to save the output.
- Line 8: A comment for the following four commands.
- Lines 9–12: The core commands for descriptive statistics.
- Line 13: The log close command closes the log file after the analysis is completed.
- Line 14: The exit command closes Stata. This command is optional in the template. It can be omitted if you would like to continue your data analysis.

1.1.7 How to Save Stata Results

When you execute a command from the Command window or from a pull-down menu, Stata will display the output in the Results window. There are two ways to save the output. First, a simple way is to copy and paste it. You can copy the selected output as text and paste it into a Word document or a text file. If you paste it into a Word document, then you need to specify the font as Courier New and set the font size to 9 or smaller. Otherwise, it will not be shown properly since the table will be misaligned. You can also copy and paste it as a table or as a picture. One disadvantage of this method is that you need to do it manually. It will be tedious and time-consuming to use the copy-and-paste method if your output is long or if you save different results to separate folders.

The second and most common way to save results is to create a log file, which is more efficient than the copy-and-paste method. By using a log file, you can save all of your results at the beginning of your Stata session. The log file can be created in either of the following two formats. One is a plain text format with the extension .log, and the other is a Stata Markup and Control Language format (.smcl), which is the default log format. The plain text format log is preferred since it can be easily opened and edited by other programs, whereas the log in .smcl extension can only be opened by Stata itself. To create or start a log file, type:

```
log using filename, text
```

The "filename" is the log file you name. The "text" means that the format is plain text. If you want to replace a previous log with the same name, you can add "replace" to the command. So the modified command is:

```
log using filename, replace text
```

If you want to add the current log file to an existing log file, you can add the append option as follows:

```
log using filename, append text
```

To suspend your logging, type:

```
log off
```

You can resume your logging by typing:

```
log on
```

To close your log file after you complete an analysis, simply type `log close`.

If you use the point-and-click menu to create a log, go to **File**, select **Log**, and click on **Begin**. Provide a name for your new log file, select either the .txt or .smcl format type, and then click **Save**.

A log file will be saved in the current working directory. If you are uncertain of your working directory, type `pwd` in the **Command** window. You can use the command `cd` to change the working directory.

1.1.8 What If I Have a Question? How Do I Get Help?

We have many questions when we use Stata. Thankfully, Stata has a complete, convenient help system, which provides rich resources for users in different ways.

First, Stata itself provides help. Stata has online help and search facilities, which provide a wide range of help. If you have a question for a particular command, you can either click on **Help** from the menu or type `help`, `search` or `findit` commands with the name of the command on the command line. For example, if you are looking for the syntax of the `ologit` command for the proportional odds model for ordinal outcomes, you could type `help ologit`. The screenshot shown in Figure 1.4 shows the syntax for `ologit` with descriptions. Examples are also provided in the help file, which is truncated in the screenshot.

The same screenshot can be reached if you use the pull-down menu. Go to **Help** on the menu, click on **Stata Command**, and then enter `ologit` in the command box. You can see the **help ologit** after clicking on **OK**.

The help command only focuses on a particular command. The other three options for broader help are the `search` command, the `hsearch` command, and the `help contents` command.

The `search` command provides a keyword search. For example, `search ologit` provides a keyword search for the `ologit` command, which is displayed in Figure 1.5.

Figure 1.4 Screenshot of the `help ologit` Command

Figure 1.5 Screenshot of the `search ologit` Command

The `hsearch` command searches the contents of a command on the Internet. For example, typing the `hsearch ologit` command brings you to the screenshot shown in Figure 1.6.

The `help contents` command helps users learn about Stata in a general way. It is not for a particular command, so it is not followed by any Stata command names. If you type `help contents`, you will get the screenshot displayed in Figure 1.7. Topics are organized into the following categories: getting started, data manipulation and management, utilities, graphics, statistics, matrix commands, programming, and interface features. Clicking on any category brings you to the PDF documentation, help, and search files related to that topic.

Second, Stata has provided complete PDF documentation since version 11. Users have access to these manuals directly from the menu or the installation folder within Stata. Since the help file is linked directly to its related entry in the PDF manuals, users can focus on a specific command in the manuals via the help file.

Third, you can watch Stata tutorials on YouTube. Stata has an official YouTube channel for users to learn Stata under the username "StataCorp". Stata periodically uploads short videos regarding various topics that show users how to use Stata. They are free to watch at any time.

Fourth, you can join the Stata listserver. The free listserver is organized for Stata users. After signing up, you can post your questions and get answers from the Stata user community. Even if you are not a member of the listserver, you can still benefit

Figure 1.6 Screenshot of the `hsearch ologit` Command

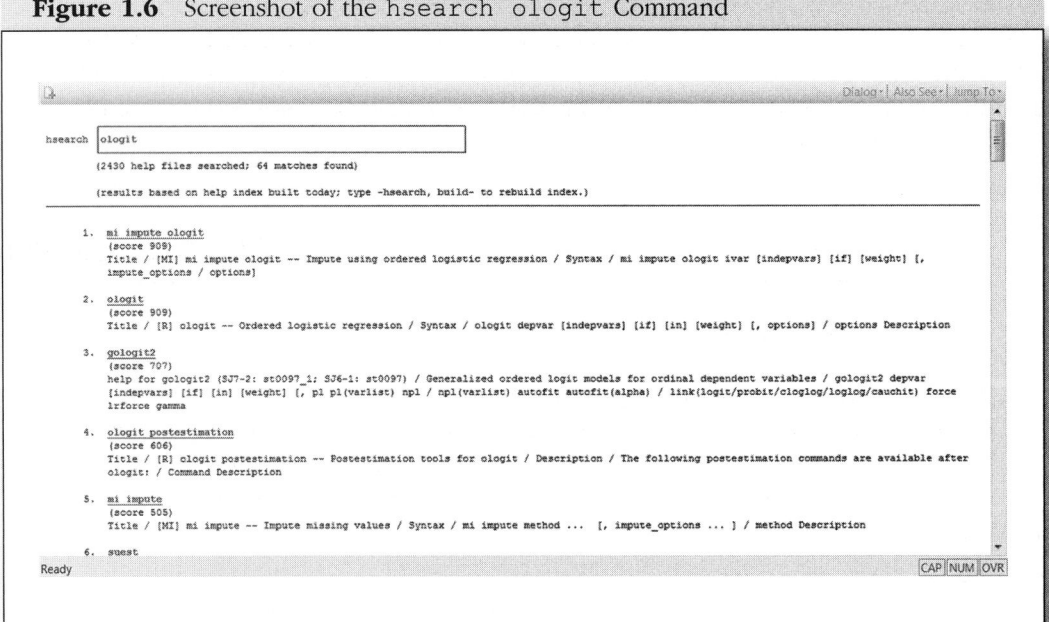

Figure 1.7 Screenshot of the help contents Command

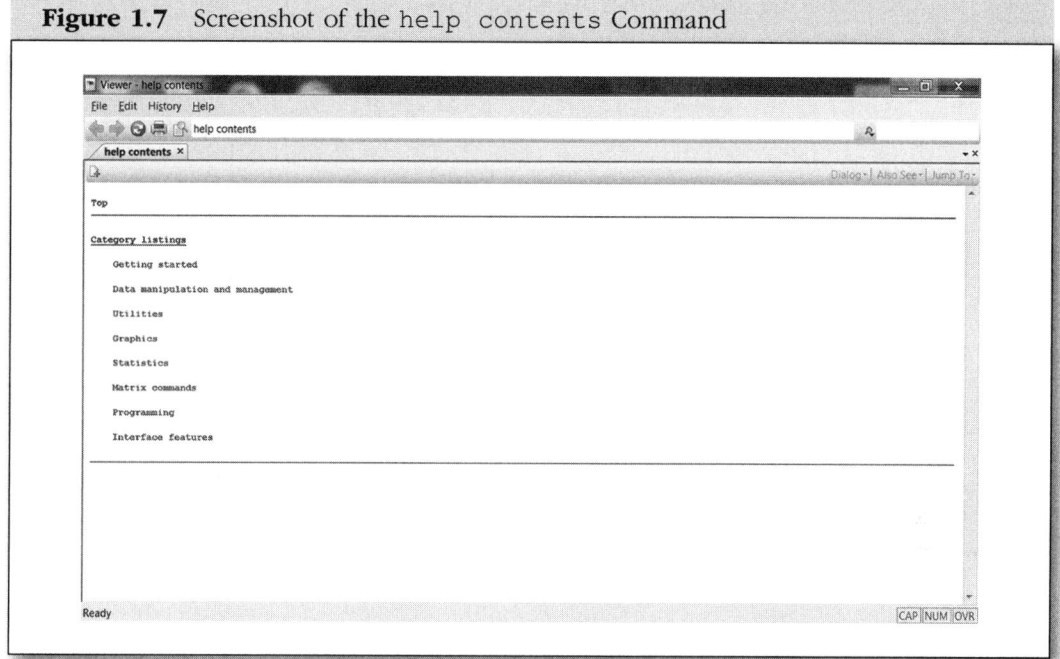

from it. After searching your question online, you may find that it has been answered by other users via the listserver.

Fifth, you can get help from the Stata website (www.stata.com). Stata provides FAQs, technical support, onsite training, third-party workshops or short courses, and NetCourses. Stata also publishes books and journals (the *Stata Journal* and its predecessor, the *Stata Technical Bulletin*).

1.2 Data Management

1.2.1 Creating a New Variable

When you work on a research project, you often need to create new variables, recode an existing variable to a new variable, or combine several variables into one variable. There are also situations when you need to label a variable or label values for a categorical variable. In this section, we will briefly introduce commands to fulfill these tasks.

The generate command is used to create a new variable.

```
generate var1 = 1
```

This command creates a variable `var1` with a value of 1 for all observations.

```
generate var2 = var1
```

This command creates a variable `var2`, which is the same as the original variable `var1`.

```
generate var3 = var1+var2
```

This command creates a variable `var3`, which is the sum of the two variables `var1` and `var2`.

Now, let's see an example using the General Social Survey 2012 (GSS 2012) dataset. I would like to create a new variable `help`, which is the sum of two variables `helppoor` and `helpsick`.

```
. generate help = helppoor + helpsick
(707 missing values generated)
```

The `replace` command is used to change the values or contents of an existing variable. To protect your data, you should avoid making changes to existing variables. Any time you change or modify a variable, it is always a good habit to create a new variable that is the same as the original one and then make changes to the new one. Therefore, the two commands `generate` and `replace` are often used together. For example, suppose you have a variable `sex` with a value of 1 for males and a value of 2 for females. If you want to recode the value of males to be 1 and females to be 0, type:

```
generate male = 1 if sex ==1
replace male = 0 if sex ==2
```

The following output is displayed.

```
. generate male = 1 if sex ==1
(1088 missing values generated)

. replace male = 0 if sex ==2
(1088 real changes made)
```

The double equal sign "==" is different from the single equal sign "=". The former sign is one of the logical operations, and it works with the `if` qualifier, whereas the single equal sign does not work with the `if` qualifier. Instead, it is used when we generate or recode a variable or replace the value of a variable.

You can also create a dummy or binary variable with a combination of the generate and replace commands. For example, let us create a variable educa-tion with a value of 1 for respondents' highest year of school completed greater than 13.53 years, and a value of 0 otherwise. You type:

```
generate education = 0
replace education = 1 if educ > 13.53
```

```
. generate education = 0

. replace education =1 if educ > 13.53
(953 real changes made)
```

Instead of using two commands, a shortcut for doing this is to type:

```
generate education = educ > 13.53
```

Let us see a more complex example. If we would like to create a new categorical variable SES, according to the family income realinc, then we enter the following five commands:

```
generate SES = realinc
replace SES = 1 if realinc >= 0 & realinc <= 10412
replace SES = 2 if realinc >= 10412.5 & realinc <= 22050
replace SES = 3 if realinc >= 22051 & realinc <= 40425
replace SES = 4 if realinc >= 40426
```

The first command creates a new SES, which is the same as the variable realinc. The second command creates the level 1 for SES if the value of realinc is equal to or greater than 0 and less than 10,412. The third command creates the level 2 for SES if the value of realinc is equal to or greater than 10,412.5 and less than 22,050. The fourth command creates the level 3 for SES if the value of realinc is equal to or greater than 22,051 and less than 40,425. The fifth command creates the level 4 for SES if the value of realinc is equal to or greater than 40,426.

The following is the Stata output. As we can see, there is no error message so all the commands are executed successfully.

```
. generate SES = realinc
(216 missing values generated)
```

```
. replace SES = 1 if realinc >= 0 & realinc <= 10412
(415 real changes made)

. replace SES = 2 if realinc >= 10412.5 & realinc <= 22050
(560 real changes made)

. replace SES = 3 if realinc >= 22051 & realinc <= 40425
(422 real changes made)

. replace SES = 4 if realinc >= 40426
(577 real changes made)
```

1.2.2 Recoding a Variable

The recode command is used to recode values of an existing variable. For example:

```
recode educ (min/13.53 = 0) (13.54/max = 1)
```

This command tells Stata to recode the values of the variable educ from the minimum to 13.53 to 0, and from 13.54 to the maximum to 1.

Since you do not want to overwrite the original variable, generate a new variable when recoding its values:

```
recode educ (min/13.53 = 0) (13.54/max = 1),
generate(education)
```

This command tells Stata to generate a new variable education by recoding the values of the variable educ from the minimum to 13.53 to 0 and from 13.54 to the maximum to 1. The following is the Stata output.

```
. recode educ (min/13.53 = 0) (13.54/max = 1), generate (education)
(1969 differences between educ and education)
```

Let us see another example. We would like to recode the continuous variable income into a new categorical variable inclevel, which has three categories. The values between the minimum and 9 of the variable income will be coded to 1 for the new variable, and the values from 10 to 11 and from 12 to the maximum will be coded to 2 and 3, respectively. We enter the command:

```
recode income(min/9 = 1) (10/11 = 2) (12/max = 3),
generate(inclevel)
```

The following is the Stata output.

```
. recode income (min/9 = 1) (10/11 = 2) (12/max = 3), generate (inclevel)
(1733 differences between income and inclevel)
```

You can use the `recode` command to reverse a variable. For example, a survey item uses a Likert scale of 1 to 5, with 1 = strongly agree and 5 = strongly disagree. You want to reverse the order and define strongly disagree to be 1 and strongly agree to be 5. The following is the syntax:

```
recode goodlife (1=5)(2=4)(3=3)(4=2)(5=1), generate
(gliferev)
```

The command tells Stata to reverse the values of the variable `goodlife` and create a new variable `gliferev`.

The following is the Stata output without any error message.

```
. recode goodlife (1=5) (2=4) (3=3) (4=2) (5=1), generate (gliferev)
(1108 differences between goodlife and gliferev)
```

1.2.3 Labeling a Variable

Labeling shows the meaning of a variable. To label a variable, use the `label variable` command as follows:

```
label variable varname "label text"
```

For example, if you want to label a variable `efficacy` with the text `mathematics self-efficacy`, enter:

```
label variable efficacy "mathematics self-efficacy"
```

Another example:

```
label variable watchtv "number of hours in watching TV"
```

Now, using the GSS 2012 dataset, let us label the new variable `gliferev` with the text `standard life of you will improve (recoded)`. You would enter:

```
label variable gliferev "standard life of you will improve
(recoded)"
```

1.2.4 Labeling Values

To analyze a categorical variable, first you need to code the text data into numeric values. Next, you need to make sense of these values by labeling them. The purpose of labeling values is to define the numeric values of a categorical variable. This will make your analysis easier and interpretation clearer. For example, say a variable gender is coded as 1 and 2 in your data. Without proper labeling of these two values, people will feel confused when reading your data and output. Once you label values of a categorical variable, the label will appear in your output when you conduct an analysis.

In Stata, labeling values involves two steps:

- First, you use the label define command to define a label, which connects numeric values to meaningful descriptions of categories.
- Second, you use the label values command to assign the label to that variable. For example, suppose we want to label values of a variable gender with 1 for female and 0 for male. The following are the commands:

```
label define gendlabel 1 "female" 0 "male"
label values gender gendlabel
```

The first command tells Stata to define a label named gendlabel with 1 for female and 0 for male. The second command assigns the label gendlabel to the variable gender.

Let us take a look at an example. We would like to add value labels to the newly created variable gliferev. The scale is from 1 to 5, with 1 = strongly disagree, 2 = disagree, 3 = neither, 4 = agree, and 5 = strongly agree.

First, we use the command label define to define a label named glifelabel. The following is the command:

```
label define glifelabel 1 "strongly disagree" 2 "disagree"
3 "neither" 4 "agree" 5 "strongly agree"
```

Then, we attach the value label glifelabel to the variable gliferev. The command is as follows:

```
label values gliferev glifelabel
```

The following is the Stata output. Using the command codebook gliferev, we can see the labels are correctly listed.

```
. label define glifelabel 1 "strongly disagree" 2 "disagree" 3 "neither" 4 "agr
> ee" 5 "strongly agree"

. label values gliferev glifelabel

. codebook gliferev

-------------------------------------------------------------------------------
gliferev                         standard life of you will improve (recoded)
-------------------------------------------------------------------------------

                  type:  numeric (byte)
                 label:  glifelabel

                 range:  [1,5]                       units:  1
         unique values:  5                      missing .:   0/1974
       unique mv codes:  2                      missing .*:  642/1974

            tabulation:  Freq.   Numeric  Label
                            71         1  strongly disagree
                           306         2  disagree
                           224         3  neither
                           549         4  agree
                           182         5  strongly agree
                             6        .c
                           636        .i
```

1.2.5 The **egen** Command

The egen command is an extension of the generate command. It has functions that are unavailable in the generate command or more efficient in tasks done by the generate command. For example, you can use this command to create a mean, standard deviation, minimum, maximum, median, or mode of a variable. You can also create a new variable, which is a summation of several items in a survey. For example, let us enter the following command:

```
egen var4 = rowtotal(var1 var2 var3)
```

It creates a variable with a total row score of var1, var2, and var3.
Another example is to create a variable with a row mean score of a set of variables:

```
egen mean = rowmean(var1 var2 var3)
```

If you use the generate command, you need to type and enter an equation like this:

```
generate mean = (var1 + var2 + var3)/3
```

The egen command computes more accurate results of row means than the generate command when there are missing values among the original variable. For example, if some observations are missing in one of the above three variables, the egen command computes the mean based on the values of two existing variables. However, the generate command creates a missing value if any of the three variables are missing.

1.2.6 How to Deal With Missing Values When Recoding Variables

Most missing values are coded as a period (.) in a dataset. In Stata, all missing values take on the largest possible values, which are greater than any nonmissing values. In other words, nonmissing values are less than any missing values. So an expression var1 < . means nonmissing values in var1 since nonmissing values in var1 are less than the system missing values.

For example, using the ELS:2002 data, we want to generate a binary variable proficiency and code it as 1 when students' math scores (bytxmirr) are larger than 38.1, and 0 otherwise.

```
. generate proficiency = 0

. replace proficiency = 1 if bytxmirr > = 38.1
(8280 real changes made)
```

Using this command, all the missing values (if there are any) will be coded to be 1 since they are larger than any nonmissing values. To exclude missing values, we need to add another "if" condition, if bytxmirr < .. The following is the Stata command.

```
. drop proficiency

. generate proficiency = 0

. replace proficiency = 1 if bytxmirr > = 38.1 & bytxmirr <.
(8004 real changes made)
```

To rerun the commands, we need to drop the originally created variable proficiency. With the new command, the missing values have been coded correctly.

In addition to being coded as `.`, different missing codes are allowed in Stata. For example, `.a`, `.b`, `.c`, … `.z`. If these missing codes appear in your data, check the meaning of them. Identify reasons for these missing codes, and decide whether you want to code the other types of missing codes to the default missing (`.`) for your analysis.

If there are different types of missing codes in your data, and you want to exclude missing values in some of the variables, use the following expression:

```
!missing(var)
```

1.2.7 Other Useful Data Management Commands

The following commands will be briefly introduced, but the examples using real data will be omitted here due to space limitations:

1. Combining data

   - The `append` command can be used to add cases to the existing variables. When we have two datasets containing the same variables with the same variable types, this command can be applied to combine different cases into one dataset.
   - The `merge` command can be used to add variables from two datasets that have a common unique identification variable. Please note that the command ID variable should be sorted before using the `merge` command.

2. Reshaping data

   The `reshape` command is useful when we reorganize data into different forms. The `reshape long` command reorganizes the dataset in the long form, whereas the `reshape wide` command transforms the dataset in the wide form. For example, in longitudinal data analysis, a person-level dataset is in the wide form, a multivariate layout with one record per individual; on the other hand, a person-period dataset is in the long form with multiple records for each individual, representing each time-point for data collection.

3. Converting variable types

   A variable can be coded in either a numeric or a string format. A numeric variable deals with numbers, whereas a string variable contains text data. The `encode` command can be used to convert a string variable into a numeric variable. For example, if a string variable `group` is coded as "small," "medium," and "large," we can convert it to a numeric variable `size` with the following command: `encode group, generate(size)`.

4. Using the `display` command as a calculator

   The `display` command can be used a calculator. After typing the command and the math expression, you can see the calculated results in the output.

```
. display -2*[-734.755 -(-701.147)]
67.216
```

1.3 Graphs

One advantage of Stata is that it can draw various publication-quality graphs. In this section, some basic functions in Stata Graphics will be introduced. I will focus on histograms, bar charts, box plots, and scatter plots. For each type of graph, I will start from the Stata command to the pull-down menu. Commands for Stata graphics can be either simple or complex. They are simple because the basic types of graphs can be easily created using the graph command, which is a combination of graphs and the type of graph you would like to draw. For example, the graph bar is the command for bar graphs, graph pie is for pie charts, and graph twoway scatter is for scatter plots. The one-word histogram command, as its name suggests, handles histograms.

For each type of graph, Stata offers rich options, which may seem complicated. You may make graphs either by entering the command or via the pull-down menus of the GUI. The GUI is helpful when you draw a complex graph that normally needs a long syntax. When you use the GUI to draw graphs, it automatically generates Stata commands that are shown in the output in the **Results** window. You may save the commands into a do-file so that the graphs can be easily reproduced or edited. When you make any changes or modifications to the existing graphs, Stata displays corresponding commands.

In many situations, you will need to modify your graphs. Although you can do it by entering commands or using the menus, thanks to the **Graph Editor**, you can directly edit your graphs based on your needs. The **Graph Editor** is a function that was added to Stata in version 10. Instead of typing long, complete commands on the command line, you may create a simple graph first, and then explore various options you need by using the **Graph Editor**.

In the following sections, I will show you how to create basic graphs, such as histograms, bar charts, box plots, and scatter plots. At the end, I will introduce the Graph Editor for when you need to modify graphs.

1.3.1 Histograms

The histogram is one of the most frequently used graphs, and it is used to present in a visual manner a frequency distribution of data. In a histogram, scores appear on a horizontal scale and frequency counts are displayed on a vertical scale. Histograms are normally used for continuous variables. They can also be used for ordinal variables if their underlying traits are continuous.

The Stata command for histograms is histogram. For example, to see the distribution of the variable price using the auto.dta data file, simply enter the histogram price command. The histogram is shown in Figure 1.8.

```
. histogram price
(bin=8, start=3291, width=1576.875)
```

In the histogram in Figure 1.8, the distribution is shown as density, which is the default. If you would like it to be shown as frequencies, fractions, or percentages, you would add the options of frequency, fraction, and percent. For another example, if you wanted to draw a histogram of students' math scores (Figure 1.9) using the ELS:2002 data, you would enter the following command with the option frequency:

```
. histogram BYTXMIRR_nomiss, frequency
(bin=42, start=12.523, width=1.3618096)
```

1.3.2 Bar Charts

There are two different types of bar charts in Stata. The first type is similar to histograms since bars in both types of graphs display frequency counts. However, in a

Figure 1.8 Histogram of Price

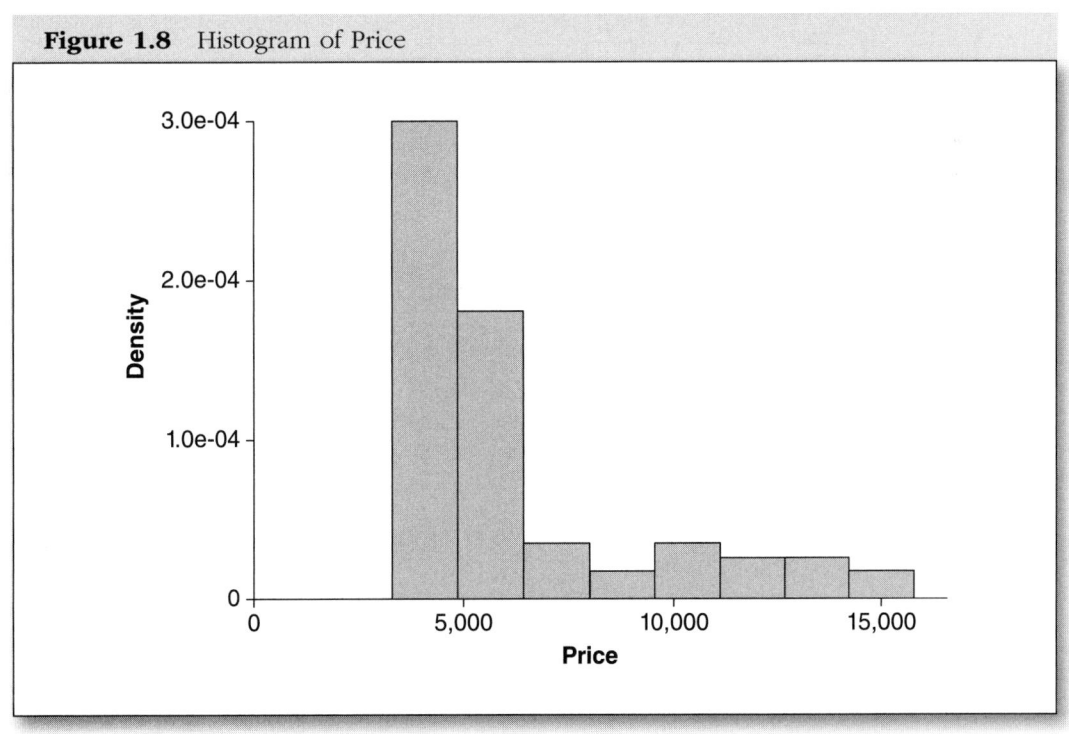

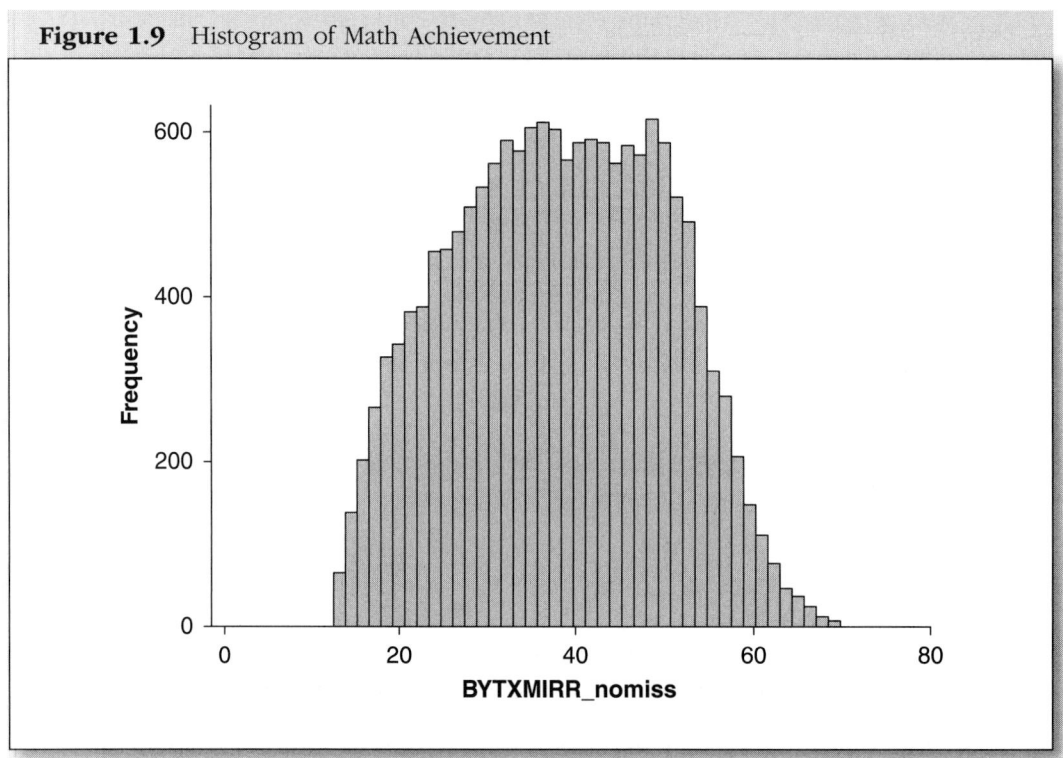

Figure 1.9   Histogram of Math Achievement

histogram, each bar is separated, whereas there is no space between bars in a bar chart. Bar charts are normally used to display frequencies for categorical variables. The command for this type of bar chart is the same as that for histograms. For example, to draw a bar chart for a categorical variable BYINCOME, type the following command:

```
histogram BYINCOME, frequency
```

You will then see the following output and the graph shown in Figure 1.10:

```
. histogram BYINCOME, frequency
(bin=42, start=1, width=.28571429)
```

If the variable is not coded as categorical in the dataset, you need to add the option discrete to draw a bar chart. The same bar graph can be drawn if you type the following command:

```
histogram BYINCOME, frequency discrete
```

The second type of bar chart is for a continuous variable across categories. The graph bar command does not produce the first type of bar charts we introduced

Figure 1.10 Bar Chart of Family Income

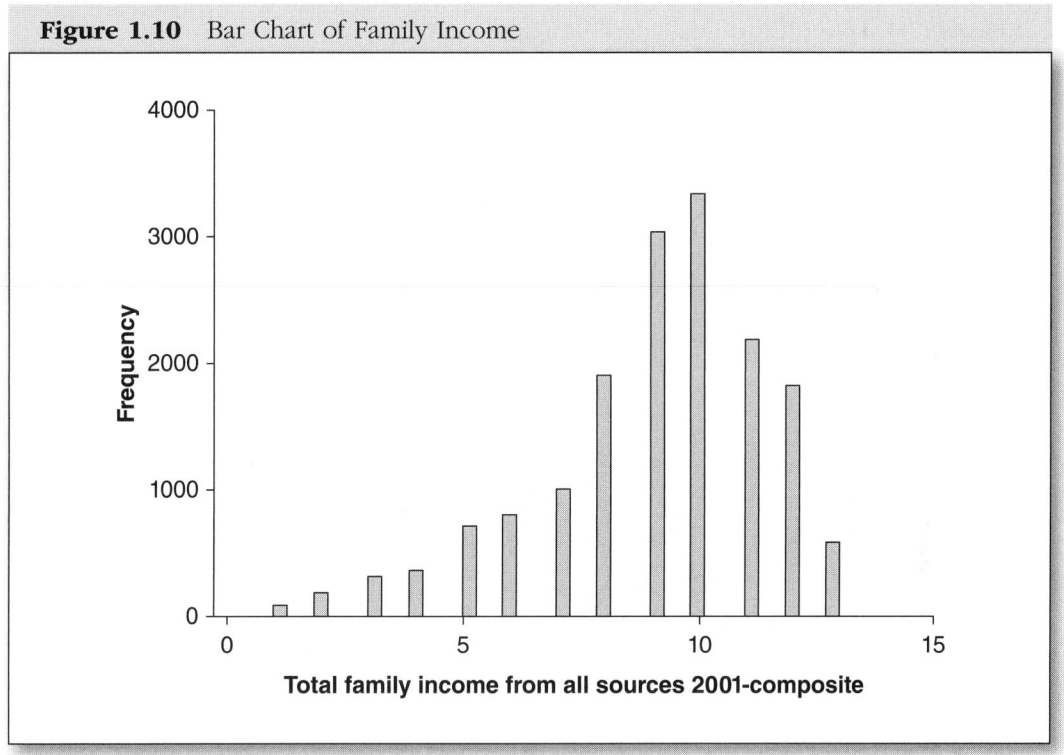

earlier. Instead, it shows a two-way vertical bar chart, with the y axis for the continuous variable and the x axis for the categorical variable. Statistics of the continuous variable, such as mean, median, sum, count, minimum, maximum, or percentiles from 1 to 99 can be shown on the y axis, where mean is the default. A basic bar chart can be obtained if you enter the following command:

```
graph bar varname
```

This command only shows one bar with the mean of the variable on the y axis. This might not be the graph you want. If you would like to draw bar charts for a continuous variable across a categorical variable, you need to use the right command with the over option. For example, the command `graph bar yvar, over(xvar)` tells Stata to draw a bar chart for a continuous variable `yvar` over the categorical variable `xvar`.

The following `graph bar` command draws bar charts (Figure 1.11) for the math achievement score (BYTXMIRR_nomiss) over the categorical variable BYGENDER:

```
. graph bar BYTXMIRR_nomiss, over(BYGENDER)
```

Figure 1.11   Bar Chart of Math Achievement Over Gender

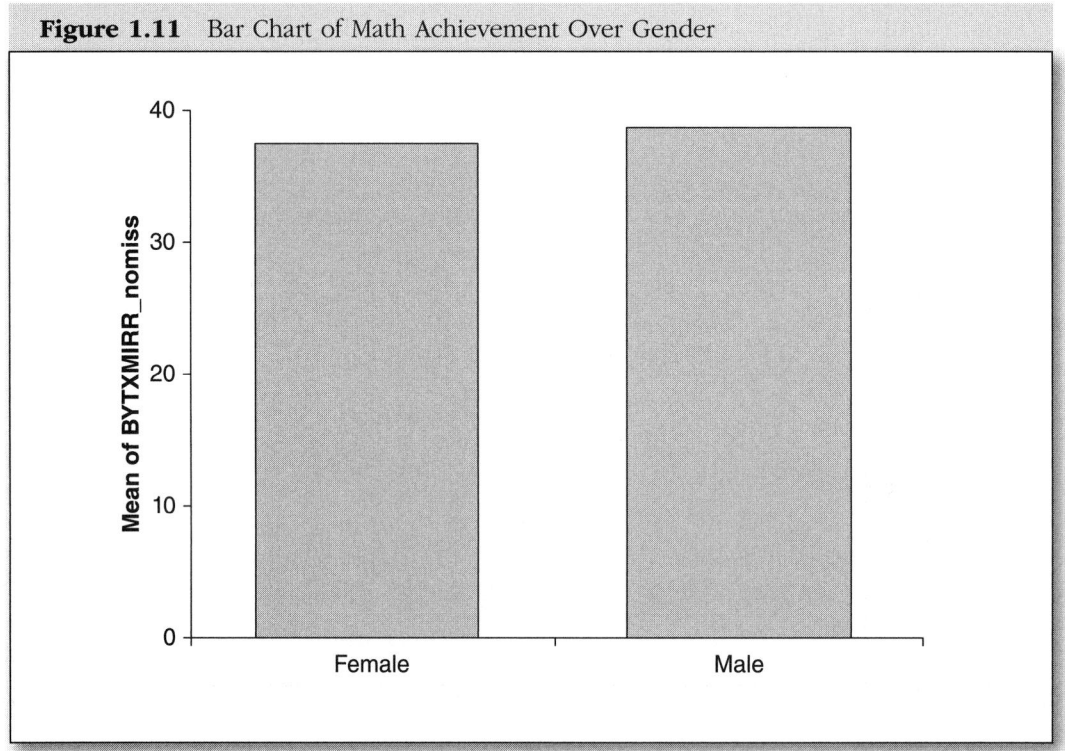

1.3.3 Box Plots

Box plots are useful for displaying a distribution and identifying outliers for a continuous variable. They display the 25th and 75th percentile, median, whiskers, and outliers. In a box plot, the lower and upper ends of the box indicate the 25th and 75th percentile, respectively. The width of the box indicates the interquartile range. The vertical line in the box is the median, which is the 50th percentile. The vertical lines below and above the box are whiskers, which indicate the spread of your data. Observations beyond the whiskers are shown as dots, which are outliers.

To get a basic box plot, enter the following command:

```
graph box var1
```

This command tells Stata to draw a box plot for the variable var1. Let us draw a box plot for the variable BYINCOME using the ELS:2002 data (Figure 1.12):

```
. graph box BYINCOME
```

Figure 1.12 Box Plot of Family Income

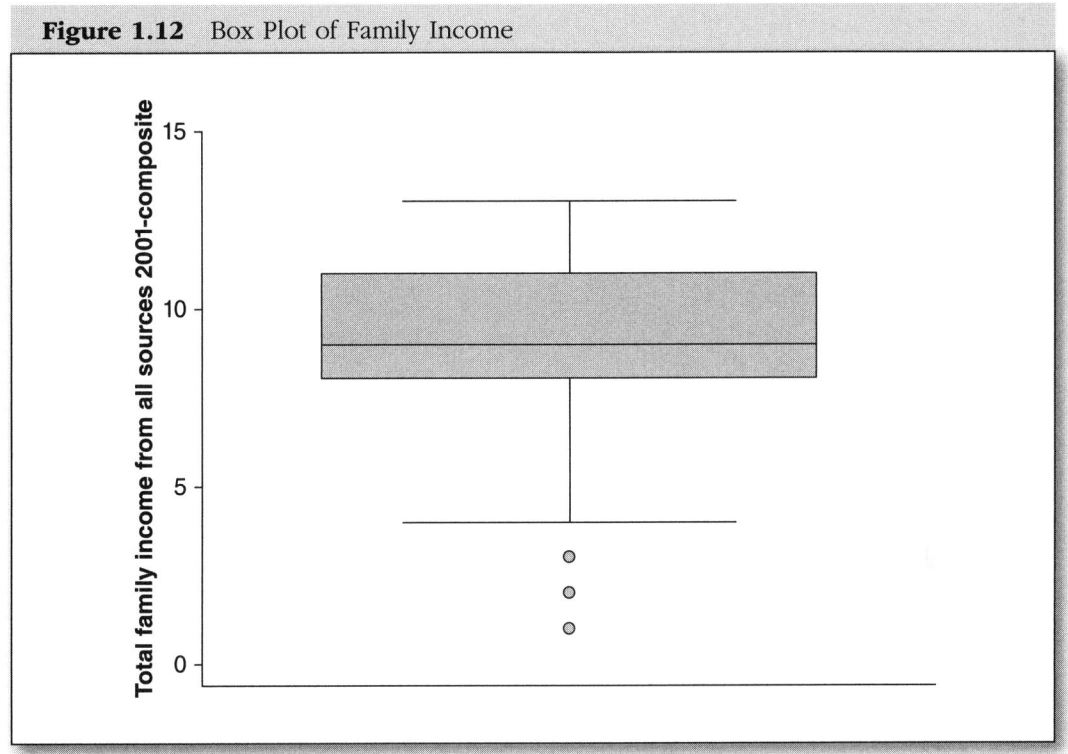

To display several box plots in one graph, enter the command with several variables:

```
graph box var1 var2
```

For example, the following command draws box plots for the two variables BYINCOME and BYSES2 in one graph (Figure 1.13):

```
. graph box BYINCOME BYSES2
```

1.3.4 Scatter Plots

Scatter plots are used to show a relationship between two variables. It is a two-dimensional graph, with the x axis displaying values of one variable and the y axis displaying values for the other variable (Figure 1.14).

To see a scatter plot of two variables var1 and var2, type the following command:

```
twoway scatter var2 var1
```

Figure 1.13 Box Plots of Family Income and Socioeconomic Status

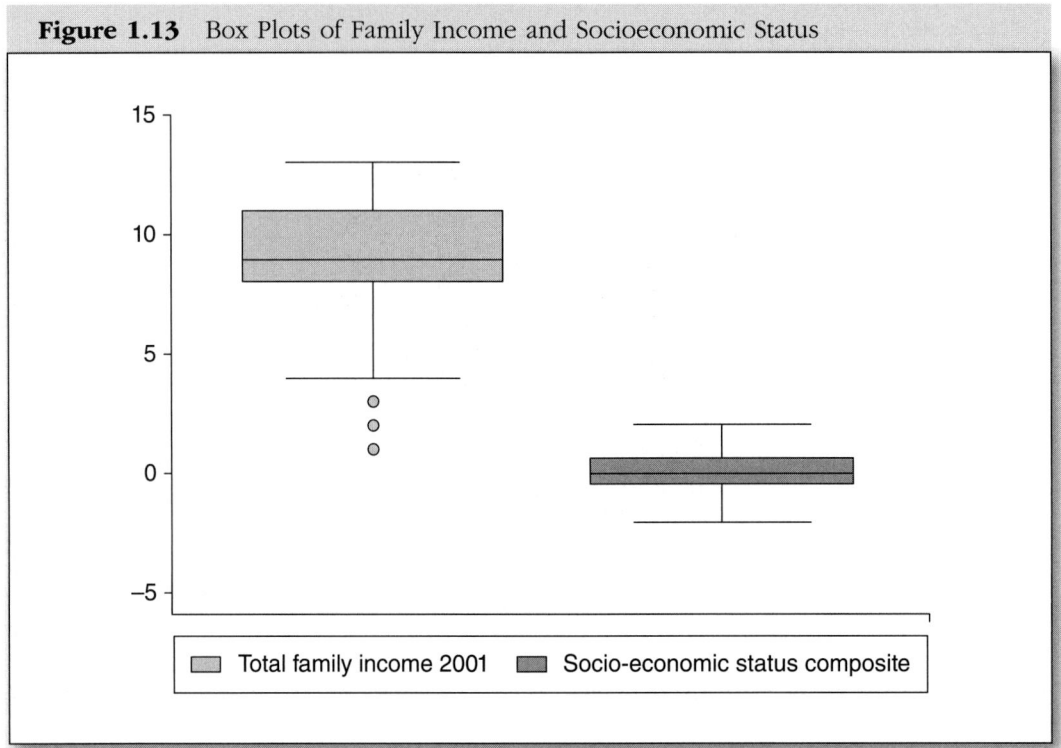

This command tells Stata to draw a two-way scatter plot for var1 and var2. For example:

```
. twoway scatter BYSES2 BYTXMIRR_nomiss
```

For illustration purposes, we may focus on a subsample of 200 cases so that we can get a clearer picture. You still use the same `twoway scatter` command with the `in` qualifier:

```
twoway scatter BYSES2 BYTXMIRR_nomiss in 1/200
```

Figure 1.15 shows the scatterplot of math scores and SES for the subsample.

1.3.5 How to Save Graphs

Stata does not automatically save the graphs you create. The graphs are independent from the log output files; therefore, they are not saved into log files. To save your graph, enter the `graph save` command. For example:

```
graph save graphname
```

Figure 1.14 Scatterplot of Math Achievement and Socioeconomic Status

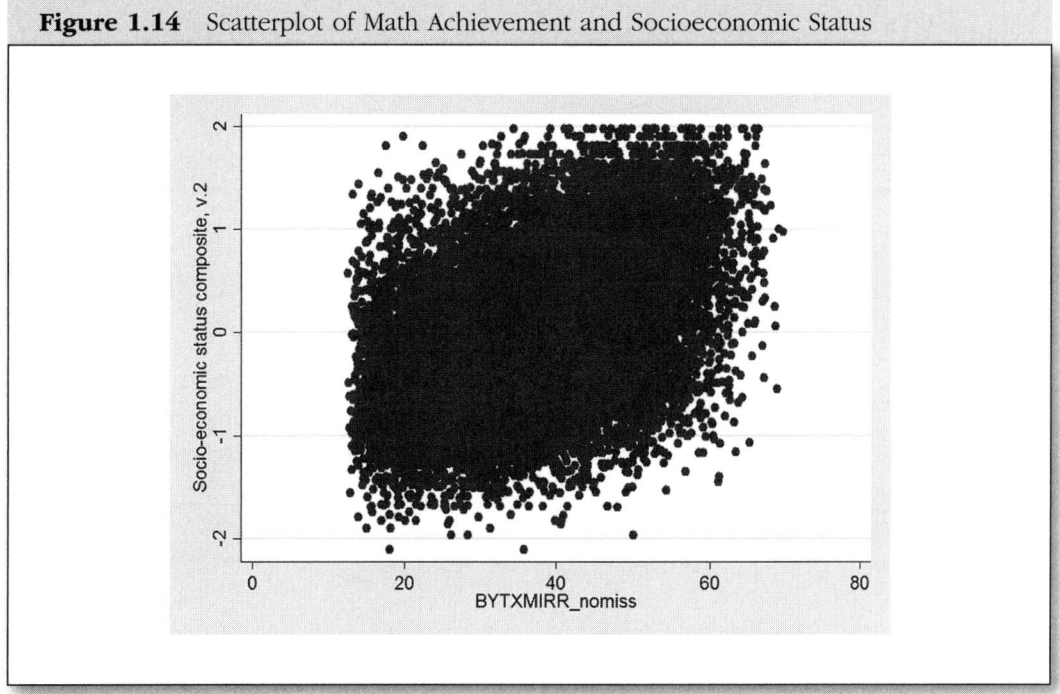

The graph will be saved in the .gph format, which can only be opened by Stata. Using the `graph export` command, Stata can also save graphs into different formats that can be easily opened by other programs. These formats include Encapsulated PostScript (.eps), PDF (.pdf), Portable Network Graphics (.png), PostScript (.ps), and Windows Metafile (.wmf). For example, to save the graph in the .wmf format, enter the following command:

```
graph export graphname.wmf
```

The created graph can be copied and pasted into a Word document, but this method would be tedious if you needed to save a lot of graphs all at one time.

To open an existing graph for editing, enter the `graph use` command:

```
graph use filename
```

1.3.6 Stata Graph Editor

The Stata Graph Editor is a good helper if you need to make edits to your graph. Please keep in mind that it is an editor, so it cannot create a new graph. You first need to have a graph at hand to make edits.

Figure 1.15   Scatterplot of Math Achievement and Socioeconomic Status: Subsample of 200

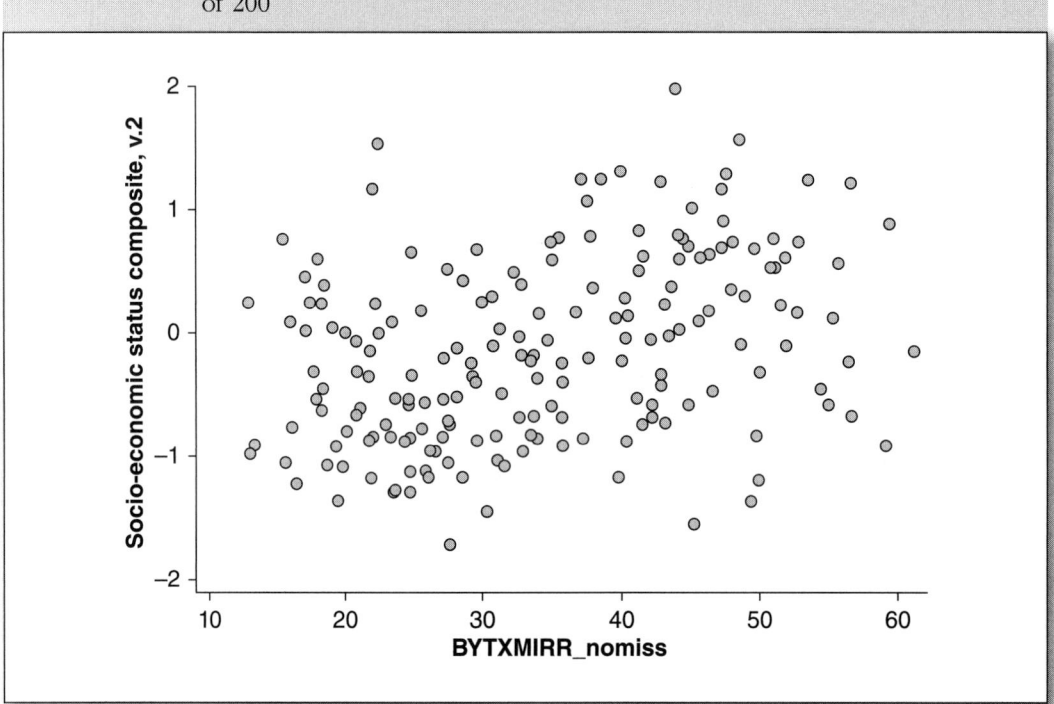

What can the Graph Editor do? It is powerful and convenient. By using the Editor, you can select and make edits on objects, as well as add texts for title, subtitle, caption, note, x axis, y axis, and the legend. You can also add lines and arrows, add markers, and edit the graph grid.

There are three ways to start the Graph Editor. First, after the file is open, go to **File**. Then click **Start Graph Editor**. Another option is to click on the Graph Editor icon on the toolbar. The third option is to right click on the graph, and then select **Start Graph Editor**. When it starts, it displays as shown in Figure 1.16.

You can see five tool buttons on the left of the graph and the Object Browser on the right. The five tools in sequence are a pointer tool for selecting objects; three tools for adding texts, lines, and markers; and a tool for editing the grid. The Object Browser shows you all the objects or contents of a graph, which include the plot region, y axis, x axis, legend, and positional titles. If you want to make edits on any one of the objects, you just need to click on it, and it will be selected. We will see that it is highlighted in the Object Browser and in the graph if it exists. It also opens the Contextual Toolbar, which shows the properties of the object.

Figure 1.16   Graph Editor

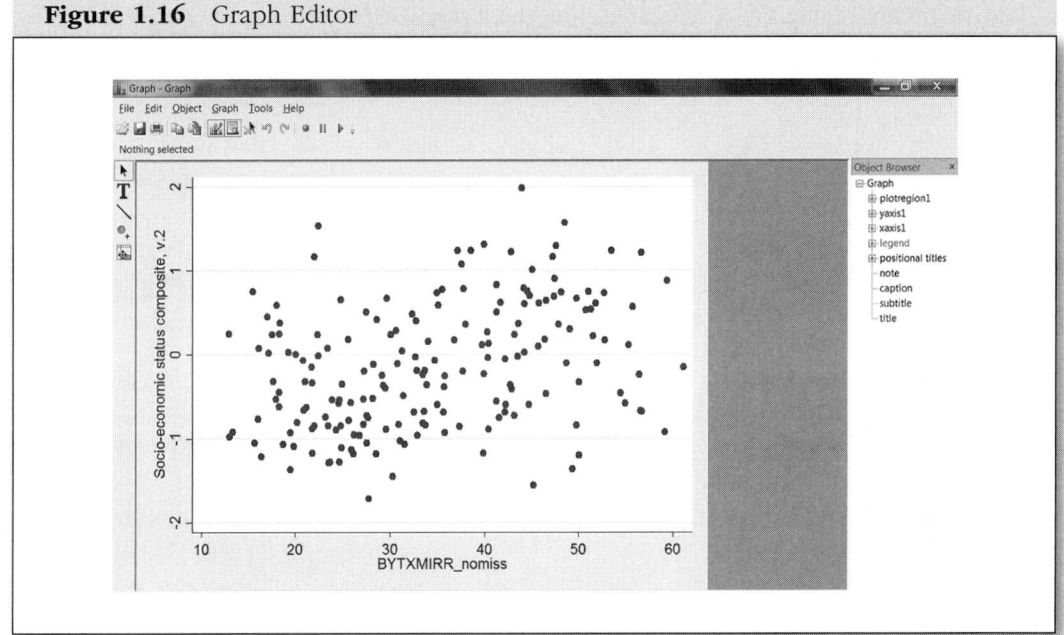

In Figure 1.17, I will show you how to change the title of the x axis.

First, let us click on the title under the object `xaxis1`. Once it is selected, you will see the title of the x axis, `BYTXMIRR_nomiss`, is also highlighted in Figure 1.17.

Next, double click or right click on the title. You will see a dialogue box for the textbox properties displayed in Figure 1.18. In the text box, type the new title for the x axis: "Math Scores of High School Students". Click **Apply**, and you will see the new title appear in the graph.

As a second example, I will add a title for the graph. At the bottom of the **Object Browser**, next to **subtitle**, select **title**. Since there is no title in the graph, double click **title**. It will open up a dialog box for **textbox properties** (Figure 1.19). **Enter** the title "Relationship between Math Achievement and SES" in the text box, and then click on **Apply**.

You can see that the new title appears in Figure 1.20. In addition to entering the title name in the textbox properties, a shortcut is to enter it in the Contextual Toolbar, which is located just above the graph.

For more information about data management and graphics, see Acock (2014), Baum (2006), Cameron and Trivedi (2010), Hamilton (2012), Juul (2014), Kohler and Kreuter (2012), Longest (2015), and Mitchell (2010, 2012a). Refer to Baum (2009) and Long (2009) for advanced topics, such as programming in data management.

Figure 1.17   Editing the X-Axis Title With the Graph Editor

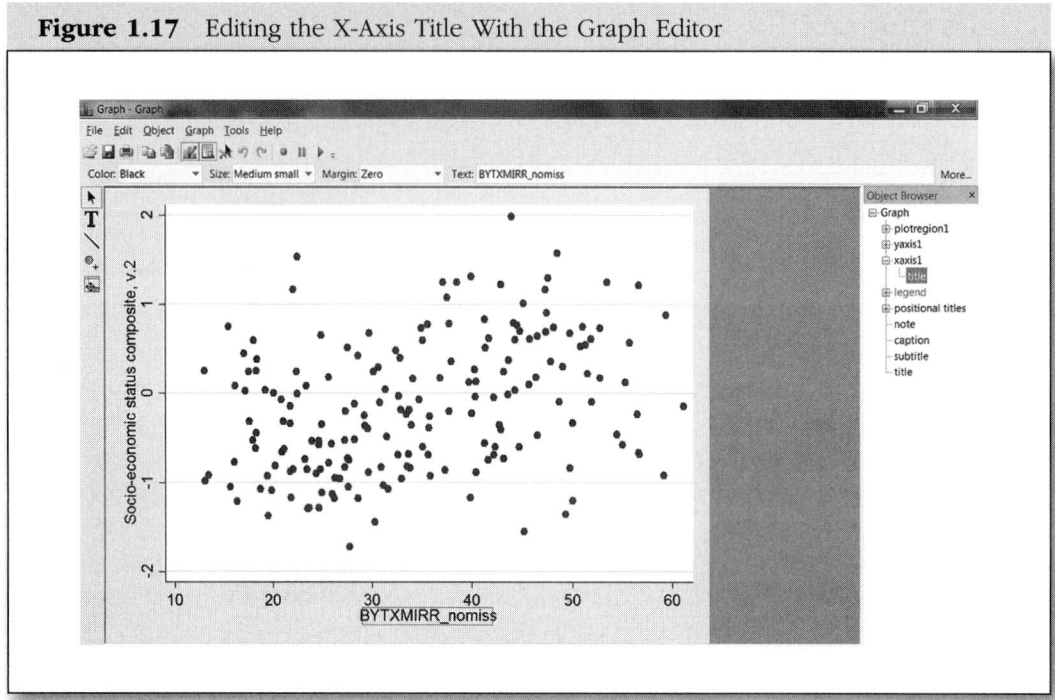

Figure 1.18   Dialogue Box for the Textbox Properties

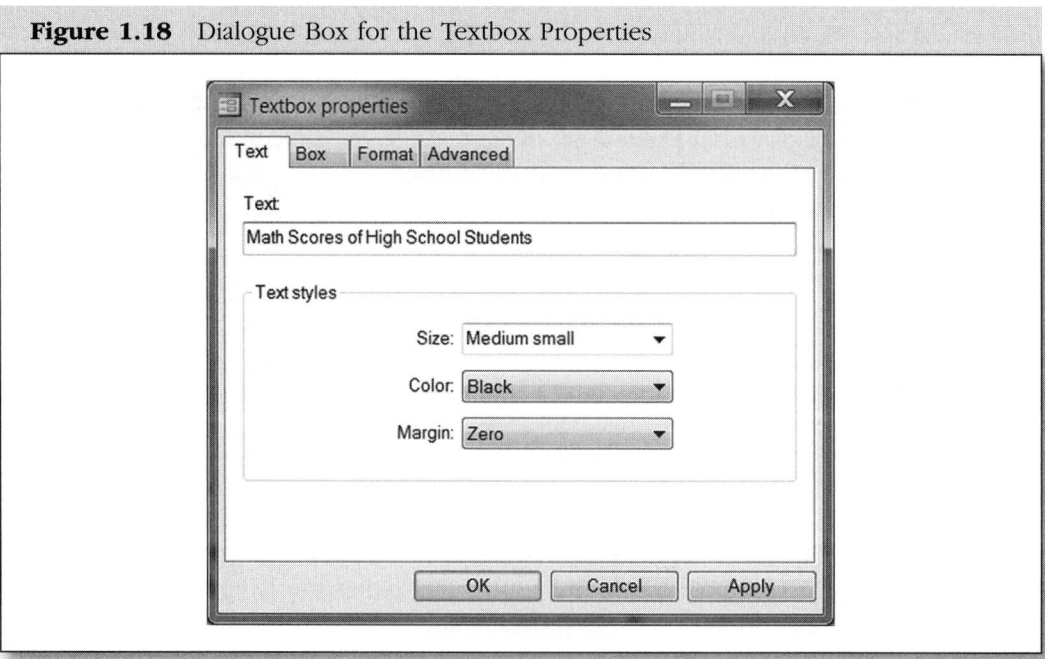

Figure 1.19 Adding a Graph Title Using the Textbox Properties

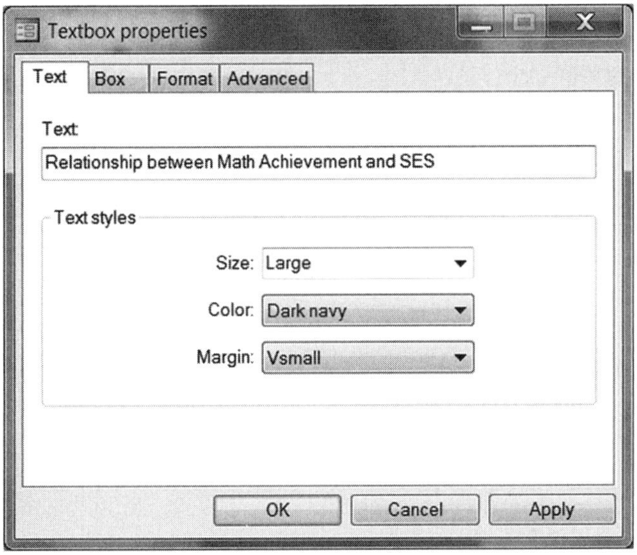

Figure 1.20 A New Title in the Graph

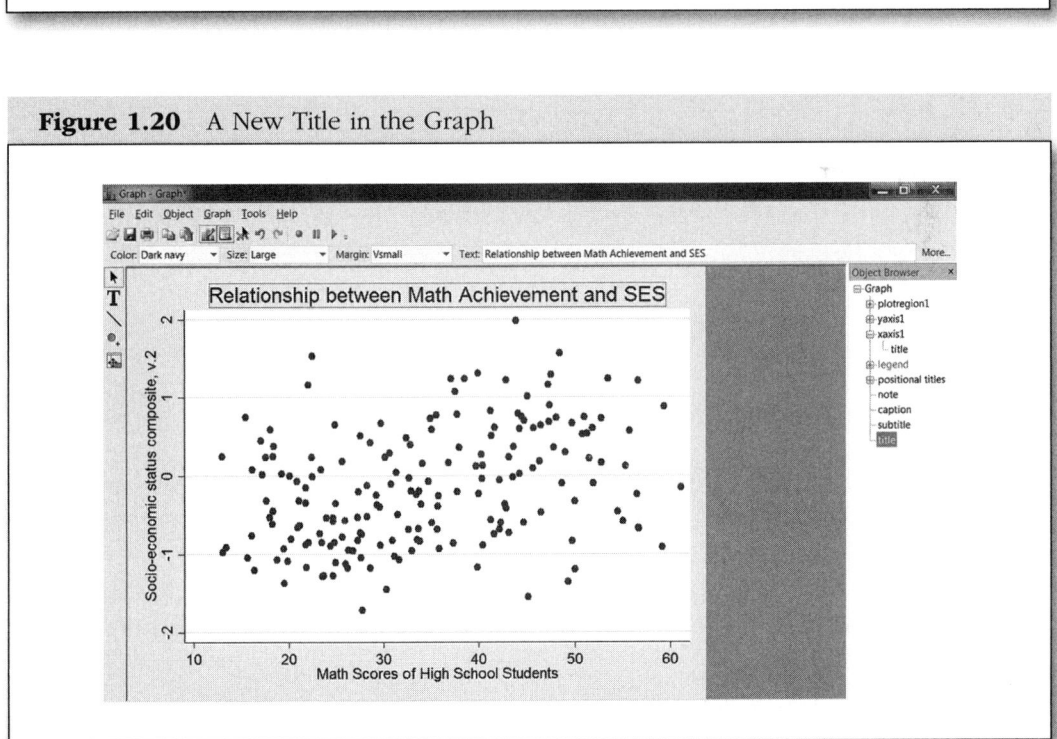

1.4 Summary of Stata Commands in This Chapter

```
*1.2 Data Management

*1.2.1 Creating a new variable
use chap1-gss2012, clear
generate help = helppoor+helpsick

*Recoding the values of an existing variable
generate male = 1 if sex ==1
replace male = 0 if sex ==2

*Creating a dummy variable
generate education = 0
replace education =1 if educ > 13.53

*Another way to create a dummy variable

*drop education first since it has already existed
drop education
generate education = educ > 13.53

*A more complex example of creating a new categorical variable
generate SES = realinc
replace SES = 1 if realinc >= 0 & realinc <= 10412
replace SES = 2 if realinc >= 10412.5 & realinc <= 22050
replace SES = 3 if realinc >= 22051 & realinc <= 40425
replace SES = 4 if realinc >= 40426

*1.2.2 Recoding a variable

*Recoding values of an existing variable: Overwriting the original variable
recode educ (min/13.53 = 0) (13.54/max = 1)

*Recoding values of an existing variable: Generating a new variable
drop education
recode educ (min/13.53 = 0) (13.54/max = 1), generate(education)

*Another example
recode income(min/9 = 1)(10/11 = 2)(12/max = 3), generate(inclevel)

*Reversing the order of a variable
recode goodlife (1=5)(2=4)(3=3)(4=2)(5=1), generate(gliferev)

*1.2.3 Labeling a variable

*label variable varname "label text"
*label variable efficacy "mathematics self-efficacy"
*label variable watchtv "number of hours in watching TV"
*label variable gliferev "standard life of you will improve (recoded)"

*1.2.4 Labeling values

*Example 1

*label define gendlabel 1 "female" 0 "male"
*label values gender gendlabel
```

```
*label values gender gendlabel

*Example 2
label define glifelabel 1 "strongly disagree" 2 "disagree" ///
3 "neither" 4 "agree" 5 "strongly agree"
label values gliferev glifelabel
codebook gliferev

*1.2.5 egen

*egen var4 = rowtotal(var1 var2 var3)
*egen mean = rowmean(var1 var2 var3)
*generate mean = (var1 + var2 + var3)/3

*1.2.6 Dealing with missing values
use chap1-gss2012, clear

*Example 1
generate proficiency = 0
replace proficiency = 1 if bytxmirr > 38.1

*Example 2
drop proficiency
generate proficiency = 0
replace proficiency = 1 if bytxmirr > = 38.1 & bytxmirr <.

*1.2.7 Other useful commands: display
display -2*[-734.755 -(-701.147)]
```

*1.3 Graphs

```
*1.3.1 Histograms
sysuse auto, clear
histogram price
use chap1-els2002, clear
histogram BYTXMIRR_nomiss, frequency
histogram BYINCOME, frequency
histogram BYINCOME, frequency discrete

*1.3.2 Bar charts
graph bar BYTXMIRR_nomiss, over(BYGENDER)

*1.3.3 Box plots
graph box BYINCOME
graph box BYINCOME BYSES2

*1.3.4 Scatter plots
twoway scatter BYSES2 BYTXMIRR_nomiss
twoway scatter BYSES2 BYTXMIRR_nomiss in 1/200

*1.3.5 How to save graphs

*graph save graphname
*graph export graphname.wmf
*graph use filename
```

1.5 EXERCISES

Use the GSS 2012 data for the following problems.

1. Find the variable happy, and recode it to a new variable happyrev. Recode the values of 1, 2, and 3 in happy into the values of 3, 2, and 1, respectively, for the new variable.

2. Label the new variable happyrev, happiness, and then label its values 1 "not too happy", 2 "pretty happy", and 3 "very happy".

3. Produce a histogram for educ.

4. Draw a scatter plot to explore the relationship between satfam7 and satfin.

5. Save the graph with a name.

CHAPTER 2

Review of Basic Statistics

Objectives of This Chapter

This chapter reviews descriptive statistics and various inferential statistics using Stata. When introducing each type of inferential statistics, an example of research design is provided followed by the research questions, Stata command, and output. The Stata commands are explained and the output is interpreted in detail. In addition, a sample of reporting the results for each analysis is provided. Finally, the commands for creating publication-quality tables using Stata are introduced, and the guidelines for reporting results are discussed. This chapter focuses on conducting basic statistical analyses using Stata, as well as on interpreting and presenting the results. After reading this chapter, you should be able to

- Conduct analysis of descriptive statistics for continuous and categorical variables.
- Perform the independent samples t test, the paired-samples t test, and the analysis of variance (ANOVA).
- Conduct correlational analysis, simple linear regression, and multiple regression.
- Conduct the chi-square test.
- Interpret Stata output for these analyses.
- Make publication-quality tables using Stata.
- Writing results for research reports.

This section reviews the basic statistics covered in most introductory statistics courses. It shows you how to use Stata to perform basic statistics analysis in descriptive statistics, independent samples t tests, paired-samples t tests, analysis of variance (ANOVA), simple linear regression, multiple regression, and the chi-square test.

2.1 Understand Your Data Using Descriptive Statistics

Descriptive statistics should never be overlooked. Whenever you conduct statistical analysis, you first need to understand your data. Why? First, you need to check for any possible errors. For example, are the data entered correctly? Are the missing values coded correctly and do they make sense? Are the variables properly labeled? Are the types of variables properly specified? Second, statistical models may require various assumptions. For example, in multiple regression analysis, it assumes that the relationships between independent variables and the dependent variables are linear. Finally, you need to choose sound statistical methods for different types of variables. For example, it is inappropriate to use a linear regression model to estimate a binary response variable.

The descriptive statistics analysis helps you describe your data by showing you types of data, graphs of the distribution of variables, and various statistical indices, such as the central tendency and variability of your data. For continuous variables, you can draw graphs, such as histograms, box plots, and stem-and-leaf plots to show frequency distributions of your data. You can also create a scatter plot to check the relationship between two variables.

The central tendency of a variable tells you the central values of a distribution. Common measures of central tendency include the mean, median, and mode. The mean of a variable is the arithmetic average of scores, the median is the middle score when all values of a variable are ordered, and the mode is the score that is most frequently occurring. In addition to the central tendency of a distribution, we also need to understand its variability. Measures of variability show the spread of a variable. They include the range, variance, and standard deviation. The range is simply the difference between the highest and lowest scores. Variance is the average summed square of each score from the mean. To compute the variance of a set of scores, we need to follow three steps. First, subtract each score from the mean. Second, square each deviation score from the first step and sum them together to get the total sum of squares. Third, get the average sum of squares by dividing the total sum of squares by the number of scores. The standard deviation is just the square root of the variance.

2.2 Descriptive Statistics for Continuous Variables Using Stata

The command `summarize` can be used for descriptive statistics analysis. It provides the basic descriptive statistics, such as the mean, the standard deviation, the minimum, and the maximum values. Before running a descriptive analysis, you may want to know whether the missing values are properly coded and how they are coded since Stata handles different types of missing data. Although you may browse the data, you may not be able to identify all types of missing values. A better way is to take a look at the codebook so that you get the variable type, label, range, and missing values. If you just type the command codebook, you will get a codebook describing the contents for all the variables in the dataset, which is useful when your data only have a few variables. If you have many variables in your dataset, the Stata output will be very long. This may not be what you expect! So if you just want to see a codebook for one variable or several variables, you just enter the command codebook and the variable name(s). Let us take a look at the codebook of the variable age, which is the age of a respondent to the General Social Survey 2012 (GSS 2012), and run an analysis of descriptive statistics.

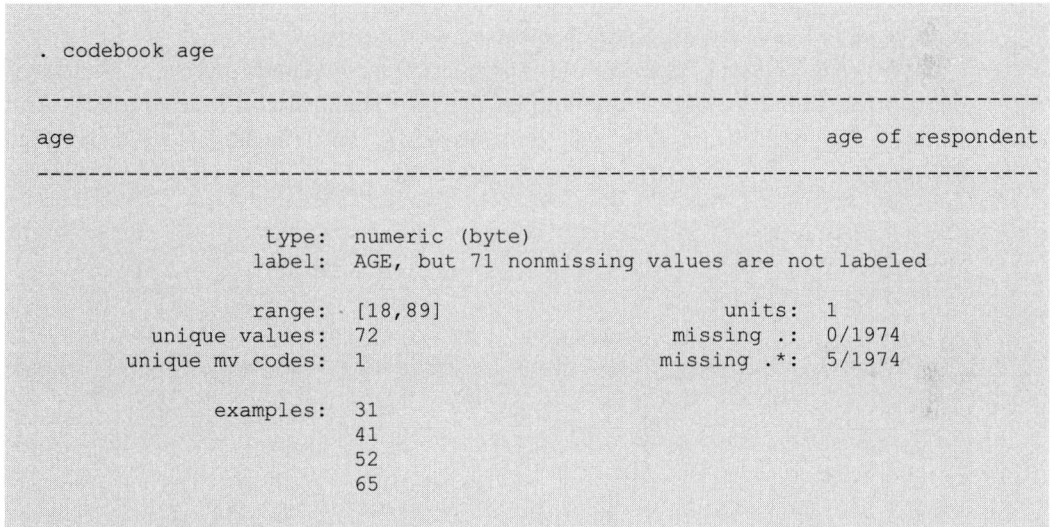

```
. codebook age

-----------------------------------------------------------------------------
age                                                          age of respondent
-----------------------------------------------------------------------------

                 type:  numeric (byte)
                label:  AGE, but 71 nonmissing values are not labeled

                range:  [18,89]                   units:  1
        unique values:  72                      missing .:  0/1974
       unique mv codes:  1                      missing .*:  5/1974

             examples:  31
                        41
                        52
                        65
```

The Stata output of the codebook command shows the variable name, type, range, and unique values, the number of unique codes for the missing values (mv), how many missing values coded to be "." and coded as other types of missing, and examples of the variable. In this example, the variable name is age, which is the age of the respondent. It is numeric with a range from 18 to 89. There is one unique code for missing values. Of 1,974 cases, there is no missing value coded as ".", but there are five missing values coded as the other type of missing. The codebook also lists four examples of the variable age.

The command summarize provides descriptive statistics for all the variables in the dataset if you do not specify variables after it. If you want to get descriptive statistics for a particular variable or several variables, you just enter the variable name(s) after the command. In this example, we would like to see the basic descriptive statistics for the variable age.

```
. summarize age

    Variable |        Obs        Mean    Std. Dev.        Min        Max
-------------+--------------------------------------------------------
         age |       1969     48.1935     17.68711         18         89
```

The Stata output displays the entered command and a table showing the variable name, number of observations, mean, standard deviation, minimum, and maximum of the variable. In this example, we can see that of 1,969 respondents in the dataset, the mean age is 48.194 years and the standard deviation is 17.687 years. The minimum age is 18 years old, and the maximum is 89 years old.

With the detail option, the summarize command provides additional descriptive statistics, such as percentiles, the four smallest and four largest values, variance, skewness, and kurtosis of a variable. The results are displayed as follows.

```
. summarize age, detail

                         age of respondent
-------------------------------------------------------------
      Percentiles      Smallest
 1%            19            18
 5%            22            18
10%            25            18       Obs              1969
25%            33            18       Sum of Wgt.      1969

50%            47                     Mean          48.1935
                      Largest         Std. Dev.     17.68711
75%            61            89
90%            73            89       Variance      312.834
95%            79            89       Skewness     .2946918
99%            88            89       Kurtosis     2.204701
```

In this example, we see the various percentiles from 1% to 99%. The four smallest values are 18, and the four largest values are 89. In addition to the mean and standard deviation, the variance is 312.834. The skewness is .295, and the kurtosis is 2.205.

In Stata, we do not need to split the data file if you are only interested in descriptive statistics of a variable for a subsample. Instead, we can use the `if` qualifier. In this example, we would like to see the descriptive statistics for male and female respondents, respectively. Since the male is coded to be 1 and the female to be 2 for the variable sex, the `if` qualifier, `if sex==1`, is for the subsample male and `if sex==2` is for the subsample female.

```
. summarize age if sex==1

    Variable |        Obs        Mean    Std. Dev.       Min        Max
-------------+--------------------------------------------------------
         age |        886    47.75395    17.62755        18         89

. summarize age if sex==2

    Variable |        Obs        Mean    Std. Dev.       Min        Max
-------------+--------------------------------------------------------
         age |       1083    48.55309    17.73573        18         89
```

The Stata output shows the descriptive statistics of age for males and females separately. The command `summarize age if sex==1` tells Stata to compute descriptive statistics for the variable age if the variable sex is male. Of 886 male respondents, the mean of the age is 47.754 years and the standard deviation is 17.628 years. The command `summarize age if sex==2` tells Stata to compute separate descriptive statistics for the variable age if the variable sex is female. Of 1,083 female respondents, the average age is 48.553 years and the standard deviation is 17.736 years.

Another way to get the same result is to use the `bysort` prefix command. It tells Stata to repeat a command for subsets of the data. For example, the command `bysort sex: summarize age` tells Stata to sort the data by the variable sex and repeat the `summarize age` command for each category of sex. In other words, Stata first sorts the variable sex and then runs descriptive statistics analysis of age for two categories of the variable sex. The following is the output.

```
. bysort sex: summarize age

-----------------------------------------------------------------------
-> sex = male
```

```
-------------+----------------------------------------------------------
        age |        886    47.75395    17.62755          18          89

-------------------------------------------------------------------------

-> sex = female

    Variable |        Obs        Mean    Std. Dev.         Min         Max
-------------+----------------------------------------------------------
        age |       1083    48.55309    17.73573          18          89
```

The previous examples focus on the summarize command with one variable. If you would like to see the descriptive statistics for more than one variable, simply list them after the command summarize. For example, the command summarize age educ computes descriptive statistics for the two variables age and educ. The following results are displayed.

```
. summarize age educ

    Variable |        Obs        Mean    Std. Dev.         Min         Max
-------------+----------------------------------------------------------
        age |       1969     48.1935    17.68711          18          89
       educ |       1972    13.52789    3.126576           0          20
```

With the detail option, we can request additional descriptive statistics in addition to the mean and standard deviation of these two variables.

```
. summarize age educ, detail

                            age of respondent
-------------------------------------------------------------------------
              Percentiles      Smallest
 1%               19               18
 5%               22               18
10%               25               18        Obs                1969
25%               33               18        Sum of Wgt.        1969
50%               47                           Mean            48.1935
                               Largest         Std. Dev.      17.68711
75%               61               89
90%               73               89        Variance         312.834
```

| 95% | 79 | 89 | Skewness | .2946918 |
| 99% | 88 | 89 | Kurtosis | 2.204701 |

highest year of school completed
```
----------------------------------------------------------------
      Percentiles      Smallest
1%         4              0
5%         8              0
10%       10              0        Obs                    1972
25%       12              1        Sum of Wgt.            1972
50%       13                       Mean               13.52789
                     Largest       Std. Dev.          3.126576
75%       16             20
90%       18             20        Variance           9.775477
95%       19             20        Skewness          -.3976432
99%       20             20        Kurtosis           4.163187
```

The variable educ is the highest year of school completed. The smallest values are 0 and 1, whereas the largest value is 20. The 25, 50, and 75 percentiles are 12, 13, and 16 years old, respectively. Of 1,972 respondents, the average highest year of school completed is 13.528 years old with the standard deviation of 3.127. In addition, the skewness is −.398, and the kurtosis is 4.163.

This example displays descriptive statistics for each of the two variables separately. If we would like to display the summary statistics for several variables in a table, you can use the command tabstat. For example, the command tabstat age educ tells Stata to display the means for variables age and educ in a table.

```
. tabstat age educ

    stats |       age      educ
----------+--------------------
     mean |   48.1935  13.52789
----------------------------------
```

The Stata output shows that the means of age and educ are 48.194 and 13.528, respectively. If we would like to see the mean of these two continuous variables across the categories of a third variable, we can add the by(varname) option. The following example displays each mean of age and educ for males and females separately.

```
. tabstat age educ, by (sex)

Summary statistics: mean
  by categories of: sex (respondents sex)

    sex |       age      educ
--------+--------------------
   male |  47.75395  13.50566
 female |  48.55309  13.54596
--------+--------------------
  Total |   48.1935  13.52789
----------------------------
```

The Stata output shows the means of two variables by two categories of sex. The last row of the table labeled `Total` shows the overall mean of the two variables.

The default summary statistic is the mean, which has been demonstrated in the previous examples. We can request more descriptive statistics by adding the `statistics()` option. For example, the option `statistics (mean sd min max)` tells Stata to report the mean, standard deviation, minimum, and maximum. The `statistics` option can be combined with the `by` option. The following command `tabstat age educ, by (sex) statistics (mean sd min max)` tells Stata to provide separate summary statistics, including the mean, standard deviation, minimum, and maximum for the variables `age` and `educ` across males and females, and the overall descriptive statistics.

```
. tabstat age educ, by (sex) statistics (mean sd min max)

Summary statistics: mean, sd, min, max
  by categories of: sex (respondents sex)

    sex |       age      educ
--------+--------------------
   male |  47.75395  13.50566
        |  17.62755  3.172106
        |        18         0
        |        89        20
--------+--------------------
 female |  48.55309  13.54596
        |  17.73573  3.090434
        |        18         0
        |        89        20
--------+--------------------
  Total |   48.1935  13.52789
        |  17.68711  3.126576
        |        18         0
        |        89        20
----------------------------
```

In the Stata output, we can easily see that the mean, standard deviation, minimum, and maximum of the variable age for males are 47.754, 17.628, 18, and 89, respectively. The last row in the table shows the overall summary statistics for the two variables.

The statistics() option can also provide other descriptive statistics, such as the number of observations (n), kurtosis, skewness, variance, coefficient of variation (cv), range, and various percentiles (e.g., p1, p25, p75, and p95).

2.3 Frequency Distribution for Categorical Variables Using Stata

The summarize command is normally used for continuous variables. For categorical variables, such as gender or ethnicity, it does not make much sense to get summary statistics, such as the mean and the standard deviation. Instead we normally do a frequency analysis to get the frequency of each value for categorical variables. The Stata command tabulate is used for creating frequency tables. The following example shows you how to create a frequency table for the nominal variable degree from the GSS 2012 dataset. Before we start, let us let a look at the codebook of this variable first.

The codebook Command

```
. codebook degree

------------------------------------------------------------------------------
degree                                                       rs highest degree
------------------------------------------------------------------------------

              type:  numeric (byte)
             label:  LABAA

             range:  [0,4]                          units:  1
     unique values:  5                          missing .:  0/1974

       tabulation:  Freq.   Numeric  Label
                      288         0  lt high school
                      976         1  high school
                      151         2  junior college
                      354         3  bachelor
                      205         4  graduate
```

Interpreting Stata Output

The codebook degree command provides the contents of the variable degree. In the Stata output, the variable degree is labeled as the respondent's highest degree.

It has five categories (unique values) ranging from 0 to 4. Of 1,974 cases, there are no missing values. The bottom of the output shows the number of observations (Freq.) for each category (Numeric) and its corresponding label (Label).

The tabulate Command for a Single Categorical Variable

Next, let us see the frequency table of the variable degree using the command tabulate degree.

```
. tabulate degree

    rs highest |
        degree |      Freq.      Percent         Cum.
---------------+-----------------------------------
lt high school |        288        14.59        14.59
   high school |        976        49.44        64.03
junior college |        151         7.65        71.68
      bachelor |        354        17.93        89.61
      graduate |        205        10.39       100.00
---------------+-----------------------------------
         Total |      1,974       100.00
```

Interpreting Stata Output

In the Stata output, the first column shows the label of the degree, the labels of five categories, and the total number of respondents. The Frequency column (labeled "Freq.") shows the number of respondents who reported their highest degrees. The Percent column shows that 14.59% of respondents have a degree less than high school, 49.44% have a high-school degree, 7.65% have a junior college degree, 17.93% have a bachelor's degree, and 10.39% have a graduate degree. The last column labeled Cum. provides the cumulative percent for each category. For example, the cumulative percent of having a junior college degree or less is 71.68%, which equals the percent of respondents having less than a high-school degree (14.59%) plus the percent having a high-school degree (49.44%), plus the percent having a junior college degree (7.65%).

The tab1 Command

If there is more than one categorical variable, you can use the command tab1 to get frequency tables. The command tab1 marital wrkstat tells Stata to produce a frequency table for each of the two variables marital, which is the marital status, and wrkstat, which is the labor force status.

```
. tabl marital wrkstat

-> tabulation of marital

     marital |
      status |      Freq.      Percent        Cum.
-------------+-----------------------------------
     married |        900        45.59       45.59
     widowed |        163         8.26       53.85
    divorced |        317        16.06       69.91
   separated |         68         3.44       73.35
never married |       526        26.65      100.00
-------------+-----------------------------------
       Total |      1,974       100.00

-> tabulation of wrkstat

    labor force |
         status |      Freq.      Percent        Cum.
----------------+-----------------------------------
working fulltime |       912        46.22       46.22
working parttime |       226        11.45       57.68
temp not working |        40         2.03       59.71
unempl, laid off |       104         5.27       64.98
         retired |       357        18.09       83.07
          school |        70         3.55       86.62
   keeping house |       210        10.64       97.26
           other |        54         2.74      100.00
----------------+-----------------------------------
          Total |      1,973       100.00
```

Interpreting Stata Output

In the Stata output, two frequency tables are created. One is for the variable `mar-ital`, and the other is for the variable `wrkstat`. The first table is titled `tabulation of marital`. Of 1,974 subjects, 45.59% were married, 8.26% were widowed, 16.06% were divorced, 3.44% were separated, and 26.65% were never married. The `Freq.` column provides the number of respondents for each category of the marital status, and the last column gives the cumulative percent for each category or below. The second table is titled `tabulation of wrkstat`. The first column lists eight categories of the working status. The other three columns give the number of subjects for each category, the percent of subjects for each category, and the cumulative percent. The last row (labeled `Total`) gives the total number of subjects, which is the sum of the frequency across all eight categories.

The tabulate Command for a Two-Way Table

If we want to get a two-way cross-tabulation table for two categorical variables, we use the tabulate command followed by the two variables. For example, the

```
. tabulate degree race

   rs highest |           race of respondent
       degree |    white       black       other |      Total
--------------+---------------------------------+----------
lt high school |      176          59          53 |        288
  high school |      741         160          75 |        976
junior college |     114          22          15 |        151
     bachelor |      290          36          28 |        354
     graduate |      156          24          25 |        205
--------------+---------------------------------+----------
        Total |    1,477         301         196 |      1,974
```

command tabulate degree race tells Stata to create a two-way table of frequency counts for two nominal variables degree and race.

Interpreting Stata Output

The Stata output displays a two-way cross-tabulation table, where the first column lists five categories for the variable degree and the first row lists three categories for the variable race. The variable degree is the row variable since its five categories are across the rows of the table. The variable race is the column variable since its three categories are across the top of the table. Each cell shows a relative frequency of subjects in each subgroup. For example, the first cell tells us there were 176 White respondents who did not have a high-school degree. The last column is the row total. It shows the total frequency for each of the five degree levels. The last row is the column total. It provides the total frequency for each of the three categories of the variable race. The row and column totals are also called marginal totals or frequencies.

The tabulate Command With the row and column Options

To get the relative frequency of each cell within its row and column, you need to add the row and column options. For example, the command tabulate degree race, row column tells Stata to provide row percentage and column percentage in each cell.

```
. tabulate degree race, row column

+-------------------+
| Key               |
|-------------------|
|     frequency     |
|   row percentage  |
| column percentage |
+-------------------+
```

rs highest degree	race of respondent white	black	other	Total
lt high school	176	59	53	288
	61.11	20.49	18.40	100.00
	11.92	19.60	27.04	14.59
high school	741	160	75	976
	75.92	16.39	7.68	100.00
	50.17	53.16	38.27	49.44
junior college	114	22	15	151
	75.50	14.57	9.93	100.00
	7.72	7.31	7.65	7.65
bachelor	290	36	28	354
	81.92	10.17	7.91	100.00
	19.63	11.96	14.29	17.93
graduate	156	24	25	205
	76.10	11.71	12.20	100.00
	10.56	7.97	12.76	10.39
Total	1,477	301	196	1,974
	74.82	15.25	9.93	100.00
	100.00	100.00	100.00	100.00

Interpreting Stata Output

In the Stata output, each category of the first variable degree takes one row, and each category of the second variable race takes one column. The row percentage displays the relative frequency of a cell within each row (i.e., each degree level), and the column percentage displays the relative frequency of a cell with each category of race. For example, 741 White respondents had a high-school degree, which was 75.92% of the 976 respondents who had a high-school degree (741/976 = 75.92%). Its column percentage was 50.17%. This means 50.17% of the 1,477 White respondents had a high-school degree (741/1,477 = 50.17%).

2.4 Independent Samples *t* Test Using Stata

The independent samples *t* test compares the mean difference between two indepen-dent groups on a continuous dependent variable. The test computes a *t* statistic, which is the difference between the mean of two samples divided by the standard error of the mean difference.

In a research design, we have participants randomly assigned to two indepen-dent groups, and then we measure the dependent variable of interest. After data are collected, we have an independent variable, which indicates the group membership, and one dependent variable, which has scores for all participants from each of the two groups. We would like to investigate whether the mean of the dependent vari-able for one group is significantly different from the one for the other group. The independent samples *t* test assumes that the variances of the dependent variable from two populations are equal and they are normally distributed within each population.

In the following example, researchers are interested in comparing the annual income between males and females. The dependent variable is realrinc, which is the annual income, and the independent variable is sex, which is gender with males coded as 1 and females as 2.

Research Question: Is there a significant mean difference in the annual income between males and females?

Null hypothesis: There is not a significant mean difference in the annual income between males and females.

Alternative hypothesis: There is a significant mean difference in the annual income between males and females.

Stata Command

The generic Stata command for the independent sample *t* test with equal variance is as follows:

```
ttest y, by(x)
```

In this syntax, the command ttest is the *t* test, y is the dependent variable, and x is the independent variable, which is a categorical variable with two levels. In this example, the dependent variable is realrinc and the independent variable is sex. The command ttest realrinc, by(sex) tells Stata to run an independent sample *t* test on realrinc between males and females.

```
. ttest realrinc, by(sex)

Two-sample t test with equal variances
-----------------------------------------------------------------------------
   Group |     Obs        Mean    Std. Err.   Std. Dev.   [95% Conf. Interval]
---------+-------------------------------------------------------------------
    male |     561    37065.47     3070.89    72735.39     31033.59    43097.34
  female |     585    18498.33    1456.561     35229.5      15637.6    21359.07
---------+-------------------------------------------------------------------
combined |    1146    27587.48    1698.657    57503.99     24254.65    30920.31
---------+-------------------------------------------------------------------
    diff |             18567.13     3354.93                 11984.63    25149.64
-----------------------------------------------------------------------------
    diff = mean(male) - mean(female)                            t =    5.5343
Ho: diff = 0                                      degrees of freedom =     1144

    Ha: diff < 0                  Ha: diff != 0                  Ha: diff > 0
 Pr(T < t) = 1.0000      Pr(|T| > |t|) = 0.0000         Pr(T > t) = 0.0000
```

Interpreting Stata Output

In the Stata output, we can easily see the observations, the means, the standard errors, the standard deviations, and the 95% confidence intervals for males, females, and the overall data (the combined data). We also see the mean difference in the annual income between males and females (18567.13), its standard error (3354.93), and the 95% confidence interval [11984.63, 25149.64]. t = 5.5343, which is the ratio of mean difference between males and females (18,567.13) and its standard error (3,354.93). The degrees of freedom (df) = 1,144, which equals the total number of observations, *n*, minus 2.

The bottom of the table shows three hypothesis tests. The first one tests whether the mean difference between two groups is less than 0, the second one tests whether the mean difference is not equal to 0, and the third one tests whether the mean difference is larger than 0. In this example, we are interested in the second hypothesis test: whether there is a significant mean difference in the annual income between males and females. Pr (|T| > |t|) = 0 tells us the two-tailed probability of having a *t* value larger than the absolute value of the observed *t* value of 5.534, if the null hypothesis is true. Since $p = 0.000$ (< .001), we conclude that we reject the null hypothesis and are in favor of the alternative hypothesis that there is a significant mean difference in the annual income between males and females.

The sdtest command is used to test whether the standard deviations are the same across groups. The command sdtest realrinc, by(sex) tells Stata to test whether the standard deviations of the annual income are the same between males and females.

```
. sdtest realrinc, by(sex)

Variance ratio test
------------------------------------------------------------------------------
   Group |     Obs        Mean    Std. Err.   Std. Dev.   [95% Conf. Interval]
---------+--------------------------------------------------------------------
    male |     561    37065.47     3070.89    72735.39    31033.59    43097.34
  female |     585    18498.33    1456.561     35229.5     15637.6    21359.07
---------+--------------------------------------------------------------------
combined |    1146    27587.48    1698.657    57503.99    24254.65    30920.31
------------------------------------------------------------------------------
    ratio = sd(male) / sd(female)                             f =    4.2626
Ho: ratio = 1                                 degrees of freedom = 560, 584

    Ha: ratio < 1               Ha: ratio != 1                 Ha: ratio > 1
 Pr(F < f) = 1.0000        2*Pr(F > f) = 0.0000          Pr(F > f) = 0.0000
```

The *F* test statistic = 4.263, and the degrees of freedom are 560 for males and 584 for females. 2*Pr(F > f) = 0.0000 tells us to reject the null hypothesis that the standard deviations are the same across the two groups. Therefore, we can conclude that the assumption of the equal variances across the two groups is violated.

Since the variances are unequal across groups, we add the unequal option to the ttest command. For example, we enter:

```
ttest realrinc, by(sex) unequal
```

```
. ttest realrinc, by(sex) unequal

Two-sample t test with unequal variances
------------------------------------------------------------------------------
   Group |     Obs        Mean    Std. Err.   Std. Dev.   [95% Conf. Interval]
---------+--------------------------------------------------------------------
    male |     561    37065.47     3070.89    72735.39    31033.59    43097.34
  female |     585    18498.33    1456.561     35229.5     15637.6    21359.07
---------+--------------------------------------------------------------------
combined |    1146    27587.48    1698.657    57503.99    24254.65    30920.31
---------+--------------------------------------------------------------------
    diff |             18567.13    3398.814                11895.51    25238.76
------------------------------------------------------------------------------
    diff = mean(male) - mean(female)                           t =    5.4628
Ho: diff = 0                      Satterthwaite's degrees of freedom =  801.417

    Ha: diff < 0                 Ha: diff != 0                  Ha: diff > 0
 Pr(T < t) = 1.0000        Pr(|T| > |t|) = 0.0000          Pr(T > t) = 0.0000
```

A modified *t* test is used by adjusting the degree of freedom and the standard error of the mean difference. The mean difference is still 18,567.13, but the standard error is 3,398.814, which is different from the one when the group variances are the same. The *t* value = 5.4628, and the adjusted degree of freedom is 801.417.

The command esize twosample realrinc, by(sex) is used to calculate the measures of the effect size for the independent samples *t* test discussed earlier. The esize command is for effect size and was new in Stata 13. An alternative user-written program is cohend, which also works in older versions of Stata. Type ssc install cohend to install this program.

```
. esize twosample realrinc, by(sex)

Effect size based on mean comparison

                                    Obs per group:
                                         male =     886
                                       female =    1088
----------------------------------------------------------
        Effect Size |    Estimate     [95% Conf. Interval]
--------------------+-------------------------------------
          Cohen's d |    .3270351    .2103728    .4435563
         Hedges's g |    .3268207    .2102348    .4432655
----------------------------------------------------------
```

In the output, Cohen's d = .327, which is the standard mean difference in the annual income between males and females. It is interpreted as follows: The annual income for males is .327 standard deviations higher than that for females. Its 95% confidence interval is [.210 .444]. The Hedges *g* is an adjustment for Cohen's *d*. Both results are almost the same.

Reporting the Results for the Independent Samples t Test

One major focus of the book is to present results in a professional way. A good report of the results will be needed when you submit your research paper for a conference presentation or journal publication. Knowing what should be reported and how to write the results clearly is important since they impact the reviewers and/or journal editors' decision on the acceptance of your research articles. In addition, a good presentation of the results will help readers to understand your analyses and the findings of your study. Guidelines and examples on reporting the results are provided throughout the book. See more details in Section 2.12 on the general guidelines for presenting the results. The following is an example of reporting the results for the independent samples *t* test.

> This study investigated whether males and females were different in their annual income using the GSS 2012 data. An independent samples t test was used to address the research question. The result found that there was a statistically significant mean difference in the annual income between males and females, $t(801.417) = 5.463, p < .001$. Of 1,146 subjects, the average annual income for males was \$37,065.47, and for females, it was \$18,498.33. Male income was higher than female income with a mean difference of \$18,567.13. The effect size, Cohen's $d = .327$.

2.5 Paired-Samples t Test

The paired-samples t test is used when we compare scores of two related samples. For example, we compare the mean of the difference between two variables in matched-subject designs or compare scores of pre- and post- tests. In a matched-subject design, one group of participants is matched with the other group of participants in pairs, and then they are assigned to different conditions within each pair. In the dataset, we have ID numbers indicating the number of matched samples and scores of two variables for two different treatments. In a repeated-measures design at two time points (pre- and post- tests), one group of participants take both tests on two occasions. In the dataset, we have ID numbers of a group of participants and scores of two variables (i.e., repeated measures).

In the following example, researchers are interested in comparing the mathematics scores of high-school students on two occasions, the scores of the base-year (2002) and the first follow-up (2004), using the Educational Longitudinal Studies (ELS:2002–2004) dataset (Ingels, Pratt, Roger, Siegel, & Stutts, 2004, 2005). The first variable is BYACH, which is the math achievement for the base year, and the second variable is F1ACH, which is the math achievement for the first follow-up year.

> **Research Question**: Is there a significant mean difference in the math achievement of high-school students between the base-year and the first follow-up year?

> **Null hypothesis**: There is not a significant mean difference in the math achievement of high-school students between the base-year and the first follow-up year.

> **Alternative hypothesis**: There is a significant mean difference in in the math achievement of high-school students between the base-year and the first follow-up year.

Stata Command

The generic Stata command for the paired-samples t test is as follows:

```
ttest x1 = x2
```

In this syntax, the command `ttest` is the *t* test. `x1` is the first variable, and `x2` is the second variable, with `x1` and `x2` being paired variables. In the following example, the command `ttest BYACH = F1ACH` tells Stata to run a paired-samples *t* test on the mean difference between the two variables BYACH and F1ACH.

```
. ttest BYACH=F1ACH

Paired t test
------------------------------------------------------------------------------
Variable |     Obs        Mean    Std. Err.   Std. Dev.   [95% Conf. Interval]
---------+--------------------------------------------------------------------
   BYACH |   13448    39.13401     .10099     11.71135    38.93605    39.33196
   F1ACH |   13448    49.68324    .1303751    15.11902    49.42769     49.9388
---------+--------------------------------------------------------------------
    diff |   13448   -10.54924    .0607146    7.040799   -10.66824   -10.43023
------------------------------------------------------------------------------
     mean(diff) = mean(BYACH - F1ACH)                          t =   -1.7e+02
 Ho: mean(diff) = 0                           degrees of freedom =      13447

 Ha: mean(diff) < 0            Ha: mean(diff) != 0            Ha: mean(diff) > 0
 Pr(T < t) = 0.0000      Pr(|T| > |t|) = 0.0000            Pr(T > t) = 1.0000
```

Interpreting Stata Output

In the Stata output, we can see the observations, means, standard errors, standard deviations, and 95% confidence intervals for the base-year math scores (BYACH) and the first follow-up math scores (F1ACH). We also see the mean difference in the math scores between the base-year and the first follow-up year (−10.54924), its standard error (.0607146), and the 95% confidence interval [−10.66824, −10.43023].

The *t* value is the ratio of mean difference in the match scores on two occasions (-10.54924) and its standard error (.0607146). t = −1.7e+02, which is an exponential notation. It means that $t = -1.7 \times 10^2 = -170$. The degrees of freedom (df) = 13,447, which equals the total number of observations, *n*, minus 1.

The bottom of the table shows three hypothesis tests. The first one tests whether the mean difference between two paired samples is less than 0, the second one tests whether the mean difference is not equal to 0, and the third one tests whether the mean difference is larger than 0. The first and the third tests are one-tailed, and the second one is two-tailed. Let us look at the second hypothesis test: whether there is a significant mean difference between the base-year and the first follow-up math scores. Pr (|T| > |t|) = 0 tells us the two-tailed probability of observing a *t* value larger than the absolute value of −170 if the null hypothesis is true. Since p = 0.000 (<.001), we conclude that we can reject the null hypothesis and are in favor of the alternative hypothesis that there is a significant mean difference in the math scores between two occasions.

Next, let us take a look at the first test (Ha: mean(diff) < 0). This tests whether the mean difference between two paired samples is less than 0. In other words, it tests whether the math scores for the base-year are smaller than those for the first follow-up. Since p <.001, we can conclude that the math scores for the base-year are significantly smaller than those for the first follow-up year.

Reporting the Results for the Paired-Samples t *Test*

In this study, a paired-samples *t* test was used to investigate whether the mathematics scores of high-school students are different on two occasions, the scores of the base-year (2002) and the first follow-up (2004), using the Educational Longitudinal Studies (ELS: 2002–2004) dataset. The results indicated that there was a statistically significant mean difference in the math scores between the two occasions. $t(13,447) = -170$, $p < .001$. Of 13,448 subjects, the mean math score for the base-year was 39.13 and 49.68 for the first follow-up year. The score for the first follow-up year was higher than the base-year score with a mean difference of 10.55. The effect size, Cohen's $d = 10.55/7.04 = 1.50$, which is large.

2.6 Analysis of Variance (ANOVA)

ANOVA can be seen as an extension of an independent samples *t* test when we compare the means of two or more groups on a continuous dependent variable. In the ANOVA, the total variability (total sum of squares) is partitioned into two components, that is, the between-group variability (between-group sum of squares) and the within-group variability (within-group sum of squares). The *F* statistic is used for ANOVA rather than for the *t* test, which is a ratio of the between-group mean squares and the within-group mean squares. When using the ANOVA to compare the means between two groups, the results are equivalent to those of the independent samples *t* test. When there is only one independent variable with multiple groups, it is referred to as one-way ANOVA.

In the following example, researchers are interested in comparing the annual family income among different ethnic groups. The dependent variable is realinc1, which is the annual family income on the scale of $10,000, and the independent variable is race, which is the ethnic group, with White coded as 1, Black coded as 2, and all others as 3.

Research Question: Is there a significant mean difference in the annual family income among three ethnic groups?

Null hypothesis: There is not a significant mean difference in the annual family income among three ethnic groups.

Alternative hypothesis: There is a significant mean difference in the annual family income among three ethnic groups. In other words, at least the mean difference between any two groups is significant.

The Stata Command anova

The Stata generic command for the one-way ANOVA is:

```
anova y x
```

In this syntax, anova is the command for analysis of variance, y is the dependent variable, and x is the independent variable. In the one-way ANOVA, the dependent variable is continuous and the independent variable is categorical. It is named the one-way ANOVA since it only includes one independent variable or factor.

We first create a new variable for the annual family income realinc1 from the original variable realinc so that the income is on the $10,000 scale. By doing this, the results of the sum of squares can be properly shown in the ANOVA table. The command generate realinc1 = realinc/10000 tells Stata to create a new variable realinc1, which equals realinc divided by 10,000. Next, the command anova realinc1 race tells Stata to run a one-way ANOVA. The dependent variable is realinc1, and the independent variable is race.

```
. generate realinc1=realinc/10000
(216 missing values generated)

. anova realinc1 race

                      Number of obs =    1758    R-squared      =  0.0192
                      Root MSE      =  3.9355    Adj R-squared =  0.0180
          Source |   Partial SS     df      MS               F     Prob > F
        ---------+----------------------------------------------------------
           Model |   531.185181      2   265.592591          17.15     0.0000
                 |
            race |   531.185181      2   265.592591          17.15     0.0000
                 |
        Residual |   27181.7233   1755   15.4881614
        ---------+----------------------------------------------------------
           Total |   27712.9085   1757   15.7728563
```

Interpreting Stata Output

In the Stata output, the source of variance lists Model, race, which is the same as the Model, Residual, and Total. The total sum of squares of the variance (SST) is partitioned into the model sum of squares (SSM) and the residual sum of squares (SSR). SST = SSM + SSR = 531.185 + 27,181.723 = 277,121.908.

The df column lists the degrees of freedom for their corresponding variances. The total degrees of freedom = $n - 1 = 1,758 - 1 = 1,757$. The degrees of freedom for the model = $k - 1 = 3 - 1 = 2$. The degrees of freedom for the residual = $n - k = 1,758 - 3 = 1,755$.

The MS column lists the mean squares of the variance, which is the ratio of the sum of squares and the corresponding degrees of freedom. For example, MS (model) = $531.285/2 = 265.593$.

The F statistic equals the ratio of the mean squares of variance for the Model over the mean squares of variance for the Residual. $F = \text{MSM}/\text{MSR} = 265.593/15.488 = 17.15$.

The last column labeled Prob > F shows the probability associated with the F statistic, $p < .001$, which indicates that there is a statistically significant difference in annual family income among ethnic groups.

The command estat esize is used to estimate the effect size for the analysis.

```
. estat esize

Effect sizes for linear models
-------------------------------------------------------------------
          Source |   Eta-Squared      df    [95% Conf. Interval]
-----------------+-------------------------------------------------
           Model |    .0191674         2     .0080602     .0330434
                 |
            race |    .0191674         2     .0080602     .0330434
-------------------------------------------------------------------
```

In the output, the η^2 for the overall model is the same as that for race since we only have one independent variable. The model $\eta^2 = .019$, which indicates the overall model with one independent variable explains 1.9% of the variability in the annual family income.

The Stata Command oneway

Another more convenient way to use the one-way ANOVA is to use the oneway command. It provides a summary descriptive statistics table and conducts multiple comparison tests after the one-way ANOVA.

For example, the command oneway realinc1 race, tabulate bonferroni tells Stata to conduct the one-way ANOVA on the dependent variable realinc1 across the categories of the independent variable race. With the options of tabulate and bonferroni, Stata also provides a summary descriptive statistics table and conducts post hoc comparisons with the bonferroni adjustment.

```
. oneway realinc1 race, tabulate bonferroni

      race of |        Summary of realinc1
   respondent |       Mean    Std. Dev.        Freq.
--------------+-----------------------------------------
        white |   3.7167461    4.170336         1312
        black |   2.1919367    2.6360744         269
        other |   3.1808741    3.7831109         177
--------------+-----------------------------------------
        Total |   3.4294747    3.9715055        1758


                       Analysis of Variance
      Source              SS        df        MS            F      Prob > F
---------------------------------------------------------------------------
Between groups       531.185181     2    265.592591       17.15    0.0000
Within groups        27181.7233   1755   15.4881614
---------------------------------------------------------------------------
      Total          27712.9085   1757   15.7728563


Bartlett's test for equal variances:   chi2(2) =   76.6112  Prob>chi2 = 0.000

                   Comparison of realinc1 by race of respondent
                                  (Bonferroni)
Row Mean-|
Col Mean |      white        black
---------+----------------------------
   black |    -1.52481
         |      0.000
```

Interpreting Stata Output

In the Stata output, the first table displays the means, standard deviations, and frequencies of family income across three ethnic groups (i.e., White, Black, and other).

The second table is the Analysis of Variance table. It provides the sum of squares (SS), the degrees of freedom (df), and the means of squares (MS) for the between groups, the within groups, and the total. It also provides the F statistic and the associated p value. The SS for the between groups in the table is the same as the SS for the model, and the SS for the within groups is the same as the SS for the residual in the previous ANOVA table using the anova command. $F(2, 1,755) = 17.15, p = 0.0000$, which is less than the significant level of .001. Therefore, there is a significant difference in the mean scores of family income (realinc1) across three ethnic groups.

Next, Bartlett's test for equal variances examines whether the variance in the family income is the same across three ethnic groups: $\chi^2_{(2)} = 76.611, p < .001$. This indicates that homogeneity of variance assumption is violated.

The last table is for the multiple comparisons using the bonferroni adjustment. They are the post hoc pairwise comparisons after a significant mean difference among groups was identified from the ANOVA table. These comparisons tell us where the mean differences on family income among the three groups are. In the table, two groups, Black and other, are shown on rows, and two groups, White and Black, are listed on columns. The label `Row Mean-Col Mean` tells us the mean difference between two groups in each comparison. The mean difference on family income between the Black and the White is −1.526, and this difference is statistically significant ($p < .001$), which indicates that the family income for the White Americans is greater than that for the Black Americans. The mean difference between the other and the White is −.536, but these two groups are not statistically significant from each other ($p = .268$). The third comparison is between the other and the Black groups. The mean difference = .989, $p =$ .029, which is less than the .05 level. Therefore, there is a statistically significant difference in the means between these two groups (i.e., other vs. Black), with family income for the other group being greater than that for the Black group.

Reporting the Results for the One-Way ANOVA

> This study investigated whether three ethnic groups (White Americans, Black Americans, and other races) differed in their annual family income using the GSS 2012 data. A one-way ANOVA was used to address the research question. The dependent variable was the annual family income on a scale of $10,000, and the independent variable was race, which was an ethnic group with three categories: White Americans, Black Americans, and all other races. The study found that there was a statistically significant mean difference in the annual family income among three ethnic groups, $F(2, 1,755) = 17.15$, $p < .001$.
>
> Post hoc pairwise comparisons with the bonferroni method were conducted to identify mean differences among groups. There were significant differences in the means between White Americans (M = 3.72, SD = 4.17) and Black Americans (M = 2.19, standard deviation [SD] = 2.64), as well as in the means between Black Americans (M = 2.19, SD = 2.64) and the other races (M = 3.18, SD = 3.78). However, no significant difference in the means between White Americans and the other races was identified. The effect size $\eta^2 = .019$.

2.7 Correlation

The Pearson correlation coefficient is used to evaluate whether two continuous variables are linearly correlated. It is a ratio between covariance and standard deviations of two variables, and it can be considered a standardized measure of covariance. Pearson's *r* has a range of −1 to 1, which indicates both the direction and the strength.

A positive sign of the coefficient indicates the relationship between two variables is positive, whereas a negative sign indicates a negative relationship. If the values of one variable increase and the corresponding values of the other variable also increase, these two variables have a positive relationship. Conversely, if the values of one variable increase while the values of the other variable decrease, the two variables have a negative relationship. When the coefficient r equals 0, it means that there is no linear relationship between two variables. The closer the correlation coefficients move from 0 to 1, the stronger the relationships between variables. When r equals 1 or -1, the relationship between variables is perfect. When comparing the magnitude of two correlation coefficients, we look at their absolute values. For example, a correlation coefficient of $-.80$ indicates a stronger relationship than a correlation coefficient of .30. Although the former indicates a negative relationship, its absolute value is larger than that of the latter.

In the following example, researchers are interested in investigating the relationships among annual family income, years of education, and age. All three variables are continuous.

Research Questions:

1. Is there a significant relationship between annual family income and years of education completed?

2. Is there a significant relationship between annual family income and age?

3. Is there a significant relationship between age and years of education completed?

Or, in other words, are there significant relationships among the three variables, annual family income, years of education completed, and age?

Null hypothesis: There is not a significant relationship between any of the two variables.

Alternative hypothesis: There is a significant relationship between at least two of the variables.

The Stata Command `correlate`

The generic Stata syntax for the correlational analysis is:

```
correlate x1 x2
```

The command `correlate` is followed by a list of variables. In the previous syntax, we are telling Stata to correlate the two variables X1 and X2. The number of a list of variables is not limited to two variables.

In this example, the command `correlate realinc1 educ age` produces a correlation matrix for the three variables.

```
. correlate realinc1 educ age
(obs=1755)

             | realinc1      educ       age
-------------+---------------------------------
    realinc1 |   1.0000
        educ |   0.3929    1.0000
         age |  -0.0228   -0.0443    1.0000
```

Interpreting Stata Output

The correlation matrix table displays the Pearson correlation coefficients. There are a total of 1,755 observations for the analysis, which means that listwise deletion is used for the missing values and all three variables have the same 1,755 cases. The listwise deletion removes all cases from the analysis when there is a missing value.

The correlation coefficient for annual family income and years of education completed is .393, the correlation coefficient for annual family income and age is −.023, and the correlation coefficient for years of education completed and age is −.044.

```
. correlate realrinc educ age, means
(obs=1144)

    Variable |        Mean   Std. Dev.         Min         Max
-------------+---------------------------------------------------
    realrinc |       27622    57547.85         245    341672.4
        educ |    13.97028    3.033639           1          20
         age |    43.14773     13.9444          18          86


             | realrinc      educ       age
-------------+---------------------------------
    realrinc |   1.0000
        educ |   0.2705    1.0000
         age |   0.1315    0.0524    1.0000
```

The Stata Command `pwcorr`

Although the correlation command provides the correlation coefficients, we do not know whether these correlation coefficients are significant. The command `pwcorr` has this function. It computes the pairwise correlation coefficient for each pair of the variables and reports the significance level for each correlation coefficient. Since it computes the correlation coefficient for each pair of two variables, the sample size may be different due to different combinations.

The command `pwcorr realinc1 educ age, sig obs` tells Stata to compute pairwise correlation coefficients between each pair of the three variables `realinc1`, `educ`, and `age`. The two options `sig` and `obs` request the significance level and the number of observations used for the correlation coefficient. You can also request other options, such as `star (.05)`, which puts the stars for the correlation coefficient significant at the .05 level, and `bonferroni`, which used the bonferroni adjustment for significance levels.

```
. pwcorr  realinc1 educ age, sig obs

             | realinc1     educ       age
-------------+------------------------------
   realinc1 |    1.0000
             |
             |     1758
             |
       educ |    0.3930    1.0000
             |    0.0000
             |     1758     1972
             |
        age |   -0.0228   -0.0529    1.0000
             |    0.3405    0.0190
             |     1755     1967     1969
             |
```

Interpreting Stata Output

The correlation matrix table displays the Pearson correlation coefficients, significance levels, and observations. To interpret the correlation coefficient, we first examine both the direction and the magnitude. Then, we look at the significance level to test whether the relationship between variables is statistically significant. We also check the number of observations, which is used for computing the correlation coefficient.

In the table, the correlation coefficient for annual family income and years of education completed is .393, $p = .000$, $n = 1,758$. It tells us that there is a positive relationship between annual family income and years of education completed, and that the relationship is moderate. The significance level, $p < .001$, indicates that the relationship between the two variables is significant.

The correlation coefficient for annual family income and age is $-.023$, $p = .341$, $n = 1,755$, which indicates that the relationship between annual family income and age is negative and small. However, it is not significant!

The correlation coefficient for years of education completed and age is −.053, $p = .019$, $n = 1,967$, which indicates that there is a significant negative relationship between these two variables, but it is small.

Reporting the Results for the Correlational Analysis

A correlational analysis was conducted to investigate the relationships among the three continuous variables, annual family income, years of education, and age. The Pearson correlation coefficients were calculated. The correlation coefficient for annual family income and years of education completed, $r(1,758) = .393$, $p < .001$, which indicates there was a medium, positive relationship between the two variables. The correlation coefficient for years of education completed and age, $r(1,967) = −.053$, $p = .019$, which indicates that there was a significant negative relationship between these two variables, but the relationship was small. In addition, no significant relationship between annual family income and age was identified, $r(1,755) = −.023$, $p = .341$.

2.8 Simple Linear Regression

Regression is used when we predict a dependent variable from an independent variable or multiple independent variables. When there is only one independent variable, the regression is simple linear regression; when there are two or more independent variables, it is called multiple linear regression. The independent variable is the predictor variable, whereas the dependent variable is the outcome variable, the one we are interested in predicting from the predictor variable. In linear regression, the dependent variable is continuous and the independent variable(s) can be either continuous or categorical. The simple linear regression can be expressed as follows:

$$Y = \beta_0 + \beta_1 X_1 + e \qquad (2.1)$$

where Y is the observed value of the dependent variable, X_1 is the value of the independent variable, β_0 is the intercept, β_1 is the regression coefficient that is the slope of the regression line, and e is the error term, also known as the residual. The error term e is the difference between the observed value and predicted value of the dependent variable. It is assumed to have a normal distribution with a mean of 0 and a constant variance at every value of the independent variable. The values of e are mutually independent from each other.

The simple linear regression can also be expressed as a simple equation, which is the predicted regression equation:

$$\hat{Y} = b_0 + b_1 X_1 \qquad\qquad (2.2)$$

where $\hat{Y}$ is the predicted or fitted value of the dependent variable, b_0 is the intercept, and b_1 is the regression coefficient.

To estimate the intercept β_0 and the regression coefficient β_1, the method of ordinary least squares (OLS) is used, which minimizes the sum of squared residuals. Since the residual e is the difference between the observed value Y and predicted value $\hat{Y}$, of the dependent variable, the estimated intercept and regression coefficients are the values when the total amount of the squared errors is as small as possible.

In the following example, we are interested in how well annual family income can be predicted by a respondent's individual income using the GSS 2012 dataset. The dependent variable is annual family income, and the independent variable is the respondent's annual income.

Research Questions:

1. Can an individual's income significantly predict the family income?

2. How well can the annual family income be predicted from an individual's income?

The Stata Command `regress`

The Stata command `regress` is used for linear regression analysis. In simple linear regression, there is only one dependent variable and one independent variable, so the syntax is the command `regress` followed by the dependent variable and the independent variable. For example, the command `regress y x` tells Stata to run a simple regression analysis predicting the dependent variable y with an independent variable x.

In the following example, the command `regress realinc1 realrinc1` tells Stata to predict the dependent variable `realinc1` from the independent variable `realrinc1`. The results are displayed as follows.

Interpreting Stata Output

Two new variables, `realinc1` and `realrinc1`, are created so that the family income and the respondent's income are on a scale of $10,000. With this transformation, the Stata output would show the results properly. In the output for the regression analysis, you can see two tables. At the top is the ANOVA table, which was discussed in the one-way ANOVA section. At the bottom is the regression model table. In addition to these two tables, we can see the summary statistics at the upper right corner. We will interpret each part as follows:

```
. generate realinc1=realinc/10000
(216 missing values generated)

. generate realrinc1=realrinc/10000
(828 missing values generated)

. regress realinc1 realrinc1

      Source |       SS       df       MS              Number of obs =     1113
-------------+------------------------------           F(  1,  1111) =   664.50
       Model |  6882.82217     1  6882.82217           Prob > F      =   0.0000
    Residual |  11507.6552  1111  10.3579255           R-squared     =   0.3743
-------------+------------------------------           Adj R-squared =   0.3737
       Total |  18390.4774  1112  16.5381991           Root MSE      =   3.2184


    realinc1 |      Coef.   Std. Err.      t    P>|t|     [95% Conf. Interval]
-------------+----------------------------------------------------------------
   realrinc1 |   .4271563   .0165707    25.78   0.000     .394643    .4596696
       _cons |   2.691405   .1071445    25.12   0.000    2.481177    2.901634
```

First, let us take a look at the ANOVA table. The source of variance includes three components: Model, Residual, and Total. The total sum of squares of the variance (SST) is partitioned into the model sum of squares (SSM) and the residual sum of squares (SSR):

$$\text{SST} = \text{SSM} + \text{SSR} = 6{,}882.822 + 11{,}507.655 = 18{,}390.477$$

The df column lists the degrees of freedom for their respective variances. The total degrees of freedom $= n - 1 = 1{,}113 - 1 = 1{,}112$. The degrees of freedom for the model equal the number of predictors $k = 1$. The degrees of freedom for the residual $= n - 1 - k = 1{,}113 - 2 = 1{,}111$.

The MS column lists the mean squares of the variance, which is the ratio of the sum of squares and the corresponding degrees of freedom. For example, MS (model) $= 6{,}882.822/1 = 6{,}882.822$.

The F statistic equals the ratio of the mean squares of variance for the Model to the mean squares of variance for the Residual. $F = \text{MSM}/\text{MSR} = 6{,}882.822/10.358 = 664.50$, which is the one shown in the summary model statistics at the upper right corner, $F(1, 1{,}111) = 664.50$.

The F statistic tests whether the overall model with one predictor in this example can significantly predict the dependent variable.

Null hypothesis: The overall model with one predictor in this example can significantly predict the dependent variable.

Alternative hypothesis: The overall model with one predictor in this example cannot significantly predict the dependent variable.

$F(1, 1, 111) = 664.50$, $p < .001$, which indicates that the model with one predictor, respondent's income, is significantly different from zero.

R^2 is called the coefficient of determination, telling us the strength of the prediction. It is the proportion of error variance explained by the predictor. $R^2 = .374$, which indicates that 37.4% of the variance in the family income is explained by the respondent's income. The adjusted R^2 takes the sample size and the number of predictors into consideration, and it is a less biased estimate of the population R^2. Adjusted $R^2 = .374$. Root MSE = 3.218, which is the square root of the mean squares of variance for the Residual.

Next, let us take a look at the regression model table at the bottom. This table provides the regression coefficients, standard errors, t values, probabilities, and the 95% confidence intervals for the constant and predictors. In this model, we have one independent variable, or the predictor, and the constant.

The regression coefficient $\beta_1 = .427$. The t statistic tests whether the regression coefficient of the independent variable, respondent's income, is significantly different from zero. In other words, it tests whether the effect of the independent variable on the dependent variable is significant.

The t value equals the ratio of the regression coefficient to its standard error: $t = .427/.017 = 25.78$, P>|t|= 0.000, which means that the probability of having a t value larger than the absolute value of the observed t value of 25.78 is zero if the null hypothesis is true. Since $p < .001$, we can conclude that the regression coefficient of the independent variable, respondent's income, is significantly different from zero. Therefore, there is a significant effect of respondent's income on the dependent variable, family income.

In the model, the constant (β_0), shown as _cons, = 2.691. It is the mean family income when the value of the independent variable equals zero.

The regression coefficient $\beta_1 = .427$. It can be interpreted as follows: For a one-unit increase in the respondent's annual income, the dependent variable, the family income, increases by a value of .427.

Substituting the values of the constant and regression coefficient into the equation of the estimated regression model:

$$\hat{Y} = b_0 + b_1 X_1$$

We get:

$$\hat{Y} = 2.691 + .427 X_1$$

When $X_1 = 0$, the predicted outcome variable = 2.691, which is the constant.

Reporting the Results

> A simple regression analysis was conducted to investigate whether an individual's annual income was a significant predictor of the family income and how accurate the prediction was: $F(1, 1,111) = 664.50$, $p < .001$, which indicated that the model with one predictor, respondent's income, was significantly different from zero. The regression coefficient $\beta = .427$, $p < .001$, which indicates that there was a significant effect of respondent's income on the dependent variable, family income. For a one-unit increase in the respondent's annual income, the family income increased by a value of .427.

2.9 Multiple Linear Regression

Multiple linear regression is simply an extension of the simple linear regression when there are two or more independent variables. It is used to predict a continuous dependent variable from a combination of predictors that can be either continuous or categorical variables. Similar to simple regression, multiple linear regression can be expressed as follows:

$$Y = \beta_0 + \beta_1 X_1 + \beta_2 X_2 + \dots + \beta_k X_k + e$$

where Y is the continuous dependent variable; X_1, X_2, ..., and X_k are a set of independent variables; β_0 is the intercept; and β_1, β_2, and β_k are the regression coefficients for predictors. Just as with simple linear regression, the intercept and regression coefficients are estimated using the OLS method.

In the following example, we are interested in how well the family income can be predicted by a combination of three independent variables using the GSS 2012 dataset. The dependent variable is the annual family income, and the independent variables are the respondent's annual income, the highest years of education completed, and age.

Research Question: How accurately can the family income be predicted from a set of three independent variables, the respondent's annual income, the highest years of education completed, and age?

The Stata Command `regress`

Just as with simple regression, the syntax for multiple regression is still the same command, `regress`, followed by the dependent variable and a list of independent variables. In the following example, the command `regress realinc1 realrinc1 educ  age` tells Stata to predict the dependent variable `realinc1` from the three independent variables `realrinc1`, `educ`, and `age`.

```
. regress realincl realrincl edu age

      Source |       SS         df         MS                  Number of obs =      1111
-------------+------------------------------              F(  3,   1107) =    274.73
       Model |  7842.71902        3   2614.23967            Prob > F       =    0.0000
    Residual |  10533.8025     1107   9.51563011            R-squared      =    0.4268
-------------+------------------------------              Adj R-squared  =    0.4252
       Total |  18376.5216     1110   16.5554248            Root MSE       =    3.0847


------------------------------------------------------------------------------
    realincl |      Coef.   Std. Err.      t    P>|t|     [95% Conf. Interval]
-------------+----------------------------------------------------------------
   realrincl |   .3798311   .0166127    22.86   0.000     .3472352    .412427
        educ |   .3181111   .0319029     9.97   0.000     .2555141    .3807081
         age |   .0092051   .0067389     1.37   0.172    -.0040173    .0224275
       _cons |  -2.030146   .5274237    -3.85   0.000    -3.065009   -.9952834
------------------------------------------------------------------------------
```

Interpreting Stata Output

Just as explained for simple regression, in the ANOVA table, the SS column displays the sum of squares for the model (also called the model sum of squares), the sum of squares for the residual (also called the sum of squared errors), and the total sum of squares. The df column displays the respective degrees of freedom related to each type of sum of squares. The MS column shows the mean squares of the variance, which is the ratio of the sum of squares to the corresponding degrees of freedom.

The F statistic in the multiple regression tests whether the overall model with all the predictors can significantly predict the dependent variable.

Null hypothesis: The regression coefficients of all three independent variables are equal to zero ($\beta_1 = \beta_2 = \beta_3 = 0$).

In other words, all three independent variables are not significant predictors of the dependent variable, family income.

Alternative hypothesis: At least one of the three regression coefficients of the independent variables is different from zero, controlling for the others. (At least one β_j is not equal to 0.)

In other words, at least one of the three independent variables is a significant predictor of the dependent variable, family income.

$F(3, 1,107) = 274.73$, $p < .001$, which indicates that the overall model with three predictors, respondent's income, the highest years of education completed, and age, is significant. In other words, at least one independent variable significantly predicts the dependent variable.

R^2 is the coefficient of multiple determination, which tells us the strength of the prediction with a set of predictors. It is the proportion of error variance explained by the model with all the predictors. $R^2 = .427$, which indicates that 42.7% of the variance in the family income is explained using a fitted model with the three predictors overall. The adjusted R^2 takes the sample size and the number of predictors into consideration, and it is a less biased estimate of the population R^2. Adjusted $R^2 = .425$.

Root MSE = 3.218, which is the square root of the mean squares for the residual (i.e., mean square error).

In the regression model table, we can see the estimated regression coefficients and the associated statistics, such as standard errors, t tests, probabilities, and 95% confident intervals for the constant and the three predictors.

The t statistic in the regression table tests whether each regression coefficient of the three independent variables is significantly different from zero, controlling for the other independent variables. It is the ratio of the estimated regression coefficient to its standard error.

The regression coefficient for the first predictor, respondent's annual income (realrinc1), $\beta = .380$. The t value equals the ratio of the regression coefficient to its standard error: $t = .3798311/.0166127 = 22.86$.

Under the heading, P>|t|, $p = 0.000$. It means that the probability of having a t value larger than the absolute value of 22.86 is zero if the null hypothesis is true. Since $p < .001$, we can conclude that the regression coefficient of the independent variable, respondent's income, is significantly different from zero when holding the other two predictors constant. Therefore, the respondent's income is a significant predictor of the dependent variable, family income.

The regression coefficient $\beta_1 = .380$. It can be interpreted as follows: For a one-unit increase in the respondent's annual income, the dependent variable, the family income, increases by a value of .380 when holding the other two predictors constant.

The 95% confidence interval for the predictor realrinc1 is [.3472352, .412427]. It can be interpreted as follows: For a one-unit change in the respondent's annual income, we are 95% confident that the change in the dependent variable, the family income, is between .347 and .412 when controlling for the other two variables.

The regression coefficient for the second predictor, the highest years of education completed (educ), $\beta_2 = .318$: $t = .3181111/.0319029 = 9.97$.

Under the heading, P>|t|, $p = 0.000$. This means that the probability of having a t value larger than the absolute value of 9.97 is zero if the null hypothesis is true. Since $p < .001$, we can conclude that the regression coefficient of the independent variable, the highest years of education completed, is significantly different from zero when holding the other two predictors constant. Therefore, the predictor, the highest years of education completed, is a significant predictor of the dependent variable, family income.

The regression coefficient $\beta_2 = .318$, which means that for a one-unit increase in the highest years of education completed, the dependent variable, the family income, increases by a value of .318 when holding the other two predictors constant.

The 95% confidence interval for the predictor educ is [.2555141, .3807081]. It means that for a one-unit change in the highest years of education completed, we are 95% confident that the change in the dependent variable, the family income, is between .256 and .381 when holding the other two predictors constant.

The regression coefficient for the third predictor age (age) β_3 = .009: t = .0092051/.0067389 = 1.37.

Under the heading, P>|t|, p = 0.172. It means that the probability of having a t value larger than the absolute value of 9.97 is 0.172 if the null hypothesis is true. Since $p > .05$, we can conclude that the regression coefficient of the independent variable, age, is not significantly different from zero when controlling for the other two predictors. Therefore, the predictor, age, is not a significant predictor of the dependent variable, family income.

In the model, the constant (β_0), shown as _cons, = −2.030. It is also known as the intercept of the model and is the mean family income when the values of the independent variables equal zero.

Substituting the values of the constant and regression coefficients into the equation of the estimated regression model:

$$\hat{Y} = \beta_0 + \beta_1 X_1 + \beta_2 X_2 + \beta_3 X_3$$

We get:

$$\hat{Y} = -2.030 + .380X_1 + .318X_2 + .009X_3$$

The command estat esize is used to estimate the effect size for the multiple regression analysis following the regress command.

```
. estat esize

Effect sizes for linear models

------------------------------------------------------------------
        Source |   Eta-Squared     df     [95% Conf. Interval]
---------------+--------------------------------------------------
         Model |    .4267793        3       .38498      .463268
               |
      realrinc1 |   .3207581        1      .2783181     .3611523
          educ |    .0824132        1      .0541179     .1144591
           age |    .0016827        1         0         .009879
------------------------------------------------------------------
```

In the output, the overall model η^2 = .427, which is the same as R^2 in the regression output. It indicates that 42.7% of the variance in the family income is explained by the overall model. The partial η^2 for respondent's annual income (realreinc1) is .321, which indicates that 32.1% of the variance in the family income is explained by this predictor variable. The partial eta squares (η^2) for the other two predictor variables educ and age can be interpreted in the same way.

The margins command, introduced in Stata 11, can be used to compute the estimated outcome variable if we know the specified values of a predictor variable or multiple predictor variables. This command is particularly useful for nonlinear models, such as ordinal logistic regression models that will be introduced in the following chapters. For more information about this command, type help margins. The following command margins, atmeans produces the predicted margins or the predicted mean of the outcome variable at the means of all three predictor variables.

```
. margins, atmeans

Adjusted predictions                          Number of obs   =       1111
Model VCE     : OLS

Expression    : Linear prediction, predict()
at            : realrinc1     =     2.817199  (mean)
                educ          =      14.0171  (mean)
                age           =     43.35284  (mean)

-------------------------------------------------------------------------------
             |            Delta-method
             |    Margin   Std. Err.      t    P>|t|     [95% Conf. Interval]
-------------+-----------------------------------------------------------------
       _cons |  3.897977   .0925469    42.12   0.000     3.71639    4.079564
-------------------------------------------------------------------------------
```

In the output, when the three predicator variables realreinc1, educ, and age are held at their means, 2.817, 14.017, and 443.353, respectively, then the estimated margin or the estimated mean outcome is 3.898. The same value can be obtained if we substitute the means of the three predictor variables into the multiple regression equation provided earlier.

If we are interested in the estimated means or margins for particular values of one predictor variable while holding other predictors constant at their means, we can use the at() option. In the following example, we estimate the margins when the predictor variable educ is specified at the values of 8, 13, and 16 and the other two predictors are held at their means.

```
. margins, at(educ = (8 13 16)) atmeans vsquish

Adjusted predictions                          Number of obs   =       1111
Model VCE     : OLS
```

```
1._at          : realrinc1       =     2.817199  (mean)
                 educ            =            8
                 age             =    43.35284  (mean)
2._at          : realrinc1       =     2.817199  (mean)
                 educ            =           13
                 age             =    43.35284  (mean)
3._at          : realrinc1       =     2.817199  (mean)
                 educ            =           16
                 age             =    43.35284  (mean)


                  |                Delta-method
                  |     Margin    Std. Err.       t     P>|t|      [95% Conf. Interval]
            ------+---------------------------------------------------------------------
              _at |
                1 |    1.98387    .2131073      9.31    0.000      1.56573      2.40201
                2 |    3.574426   .0980705     36.45    0.000      3.382001     3.766851
                3 |    4.528759   .1121016     40.40    0.000      4.308804     4.748715
            ----------------------------------------------------------------------------
```

In the output, for the first combination (labeled 1._at), when educ = 8, and the other two predictor variables are held at their means (realrinc1 = 2.817 and age = 43.353), the estimated margin or the mean expected outcome ($\hat{Y}$) is 1.984.

For the second combination (labeled 2._at), when educ = 13, and the other two predictor variables are held at their means, the estimated margin ($\hat{Y}$) is 3.574.

For the third combination (labeled 3._at), when educ = 16, and the other two predictor variables are held at their means, the estimated margin ($\hat{Y}$) is 4.529.

In the margins table at the bottom, in addition to margins, standard errors, t tests, their associated p values, and the 95% confidence intervals are also reported. It shows that all three predicted margins are significantly different from 0 since $p < .001$.

Reporting the Results

A multiple regression analysis was conducted to predict the annual family income using three predictors, the respondent's annual income, the highest years of education completed, and age. $F(3, 1,107) = 274.73$, $p < .001$, which indicated that the overall model with the three predictors was significant. $R^2 = .427$, which indicated that 42.7% of the variance in the family income was explained by the fitted model with the three predictors overall. Adjusted $R^2 = .425$.

The respondent's income was a significant predictor of the dependent variable, family income ($\beta = .38$, $t = 22.86$, $p < .001$). For a one-unit increase in the respondent's annual income, the dependent variable, the family income, increased by a value of .38 when holding the other two predictors constant.

The predictor, the highest years of education completed, was also a significant predictor of the dependent variable, family income ($\beta = .32$, $t = 9.97$, $p < .001$). For a one-unit increase in the highest years of education completed, the dependent variable, the family income, increased by a value of .32 when holding the other two predictors constant.

However, age was not a significant predictor of the dependent variable, family income ($\beta = .01$, $t = 1.37$, $p = .172$).

2.10 Chi-Square Test

The chi-square test of independence is used to investigate the relationship between two categorical variables. Each categorical variable has two or more levels/categories. A two-way contingency table can be constructed with one variable as the row variable and the other as the column variable. The rows list different levels of the row variable, and the columns represent categories of the column variable. For example, a 4×5 contingency table shows frequencies for a row variable with 4 levels and a column variable with 5 levels.

A simple case for the chi-square test is a two-by-two frequency table, which includes two categorical variables with two levels for each category. In it the rows represent two categories of one variable and the columns represent two categories of the other. Each cell where a row and column intersects tells the frequency number of participants that fall into each subgroup.

In the following example, we are interested in whether two categorical variables, health status and marital status, are related using the GSS 2012 dataset. One variable, health status, has four levels, including poor, fair, good, and excellent. The other variable, marital status, has five levels: married, widowed, divorced, separated, and never married.

Research Question: Is there a significant relationship between the two categorical variables, health status and marital status? Or, in other words, are health status and marital status independent of each other?

Null hypothesis: There is a significant relationship between the two categorical variables, health status and marital status.

Alternative hypothesis: There is no significant relationship between the two categorical variables, health status and marital status.

The Stata Command `tabulate` *With the* `chi` *Option*

Remember that we use the command `tabulate` to create a two-way cross-tabulation table in the frequency distribution for the categories variables. The Stata syntax for the chi-square test is still the `tabulate` command along with the addition of the option `chi2`. For example, the syntax `tabulate health marital, chi2 V` tells Stata to create a two-way table for two categorical variables `health` and `marital`, conduct a chi-square test, and provide Cramér's V for the test.

```
. tabulate health marital, chi2 V

condition |                    marital status
of health |   married    widowed   divorced  separated  never mar |     Total
----------+-------------------------------------------------------+----------
excellent |       200         13         37         10         90 |       350
     good |       274         46        104         13        161 |       598
     fair |       107         29         49         14         76 |       275
     poor |        30         14         18          8         13 |        83
----------+-------------------------------------------------------+----------
    Total |       611        102        208         45        340 |     1,306

          Pearson chi2(12) =  60.5894   Pr = 0.000
             Cramér's V =     0.1244
```

Interpreting Stata Output

The Stata output provides a two-way contingency table at the top and the result of the chi-square test at the bottom. The row variable, health status, has four levels, and the column variable, marital status, has five levels. The last row shows the column totals (611, 102, 208, 45, and 340), and the last column displays the row totals (350, 598, 275, and 83). The row and column totals are also called the marginal totals according to the places where they are located. The total number of observations is shown at the lower right bottom of the table ($n = 1,306$).

The Pearson $\chi^2 = 60.589$ with the degrees of freedom of 12. The number of degrees of freedom in the χ^2 test $= (r - 1) \times (c - 1) = 3 \times 4 = 12$ since r is the number of rows and c is the number of columns.

$\chi^2_{(12)} = 60.589$, $p < .001$, which indicates that there is a significant relationship between health status and marital status.

Cramér's V indicates the strength of association between two categorical variables. Its size lies between 0 and 1. In the output, Cramér's $V = .124$, which indicates that the relationship between two variables is weak.

We can also request the row percentage and column percentage by adding the options of `row` and `column` to the syntax.

```
. tabulate health marital, row column chi2 V

+-------------------+
| Key               |
|-------------------|
|      frequency    |
|   row percentage  |
|  column percentage|
+-------------------+
```

condition of health	married	widowed	marital status divorced	separated	never mar	Total
excellent	200	13	37	10	90	350
	57.14	3.71	10.57	2.86	25.71	100.00
	32.73	12.75	17.79	22.22	26.47	26.80
good	274	46	104	13	161	598
	45.82	7.69	17.39	2.17	26.92	100.00
	44.84	45.10	50.00	28.89	47.35	45.79
fair	107	29	49	14	76	275
	38.91	10.55	17.82	5.09	27.64	100.00
	17.51	28.43	23.56	31.11	22.35	21.06
poor	30	14	18	8	13	83
	36.14	16.87	21.69	9.64	15.66	100.00
	4.91	13.73	8.65	17.78	3.82	6.36
Total	611	102	208	45	340	1,306
	46.78	7.81	15.93	3.45	26.03	100.00
	100.00	100.00	100.00	100.00	100.00	100.00

```
         Pearson chi2(12) =  60.5894   Pr = 0.000
            Cramér's V =     0.1244
```

In the output, we can easily see that each cell displays the row and column percentages in addition to the frequency. The row percentage equals the number of frequency in a cell divided by its corresponding row total at the margin (i.e., the marginal total).

The Pearson χ^2 test discussed earlier indicated that there is a significant relationship between health status and marital status. It is an omnibus test for the overall model. If we are interested in the relationship between the variables for subcategories, then we can conduct follow-up tests. For example, we can examine the health status between married and widowed. The research question would be as follows: Do the married and the widowed differ among the levels of health status?

Since we know the frequency distribution from the earlier output, we can use the command tabi to display a contingency table without splitting the data. Each row is

```
. tabi 200 13 \ 274 46\ 107 29 \ 30 14 , chi2 V row column


+-------------------+
| Key               |
|-------------------|
|      frequency    |
|   row percentage  |
| column percentage |
+-------------------+

           |          col
       row |         1          2 |      Total
-----------+----------------------+----------
         1 |       200        13  |        213
           |     93.90       6.10 |     100.00
           |     32.73      12.75 |      29.87
-----------+----------------------+----------
         2 |       274        46  |        320
           |     85.63      14.37 |     100.00
           |     44.84      45.10 |      44.88
-----------+----------------------+----------
         3 |       107        29  |        136
           |     78.68      21.32 |     100.00
           |     17.51      28.43 |      19.07
-----------+----------------------+----------
         4 |        30        14  |         44
           |     68.18      31.82 |     100.00
           |      4.91      13.73 |       6.17
-----------+----------------------+----------
     Total |       611       102  |        713
           |     85.69      14.31 |     100.00
           |    100.00     100.00 |     100.00

           Pearson chi2(3) =  28.1619   Pr = 0.000
               Cramér's V =   0.1987
```

separated by "\". The option `chi2 V row column` tells Stata to provide the row percentages and the column percentages, conduct a chi-square test, and provide Cramér's V.

$\chi^2_{(3)}$ = 28.162, p < .001, which indicates that there is a significant relationship between health status and marital status and that the married have better health status levels than the widowed.

Cramér's V = .199, which indicates that the relationship between two variables is still weak but that it is larger than the one in the overall model.

Other follow-up comparisons can be done in a similar way. To control the Type I error due to multiple comparisons, we need to adjust the significance level. To do this, we can use the bonnferroni adjustment.

Reporting the Results

A Pearson chi-square test was conducted to investigate the relationship between two categorical variables, health status and marital status, using the GSS 2012 dataset. $\chi^2_{(12)} = 60.59$, $p < .001$, which indicated that there was a significant relationship between health status and marital status. Cramér's V was used to indicate the strength of association between two categorical variables. Cramér's $V = .124$, which indicated that the relationship between two variables was weak.

2.11 Making Publication-Quality Tables Using Stata

Once you have conducted statistical analyses and interpreted results, the final step of the research process is to report the results and submit your manuscript for publication. While writing the manuscript, you may find that you need to summarize your research findings and display them in tables. You will likely find it time-consuming and tedious to look for a particular number from a stack of printed output and then manually enter it into the table. Here is a common scenario for researchers: You are conducting a series of nested regression analysis models where the reduced model contains a subset of predictors of the full model. Next, you would like to make a table containing the regression coefficients and model fit statistics for all the competing models. What you normally do is look for those numbers from each output of the model and enter them into the table. Once the entry is done, you double check whether you have made any typos. You might wonder whether there are any tools to reduce your workload, or automatically combine results from the fitted models and generate a single publication-style regression table that follows a particular format, such as American Psychological Association (APA) style.

The answer is yes! All the tedious, error-prone typing work can be easily handled by a simple command. Stata has several programs to accomplish this job with no hassle. These programs include user-written programs, such as `estout` (Jann, 2005, 2007), mktab (Winter, 2005), `outreg` (Gallup, 1998, 1999, 2000), outreg2 (Wada, 2008), `tabout` (Watson, 2007), `xml_tab` (Lokshin & Sajaia, 2008), and Stata official programs, such as the combination of `estimates store` and `estimates table` programs and the new `putexcel` program, which was introduced in Stata 13 and improved in Stata 14.

To install these user-written programs, type `ssc install` with the program name or type `findit` followed by the program name. For example, if you would like to install outreg2, type `ssc install outreg2` or type `findit outreg2` and then install the program. To uninstall a user-written program, type `ado dir` and you will see a list of previous installed programs. Find the program you would like to uninstall, which has a number at the beginning, and then type `ado uninstall` followed by the number. For example, if you want to uninstall the program `outreg2`, which is numbered 11 in the directory on my computer, type `ado uninstall [11]`, and the program will be uninstalled.

When you have time, explore the various functions of these programs. You will be amazed to see what each program can accomplish in making publication-style tables. But you do not need to be familiar with all the programs to make publication-quality tables. Users have different preferences. You only need to find one that meets your needs. A good suggestion is that you start from a simple table. Once you are familiar with the simple structure of the syntax, you may then build more complex tables. Keep in mind that you do not need to memorize all the options. Actually, this is impossible since there are so many options. Do not be intimidated by the long syntax with many options. Once you understand them, you can use the syntax as a template by saving it as a do-file and then modifying it for future uses.

The `outreg2` command is used throughout this book because of its ability to produce regression tables for all the models covered here. Other programs will also be introduced if necessary. One strength of the `outreg2` command is that it works with user-written programs. Once you are familiar with `outreg2`, the other programs can be learned easily. The help files of these programs provide various examples with explanations on how to use them. The following is an example of using the `outreg2` command to make a table for the results of two regression models.

```
1.  . quietly regress realinc1 realrinc1
2.  . est store mod1
3.  . quietly regress realinc1 realrinc1 edu age
4.  . est store mod2
5.  . outreg2 [mod1 mod2] using regout, dec(3) word replace
```

The first command estimates the simple regression model without showing the output. The second command stores the estimates named `mod1`. The next two commands estimate the multiple regression model and save the estimates with a name of `mod2`, respectively. The fifth command, which is the last one, produces a table containing the estimates from the two regression models and saves it to a Word document named "regout". It is assumed that you have Microsoft Word installed on your computer. Table 2.1 shows the results of two regression models using the `outreg2` command.

A similar table can be created if you use the `estout` or `esttab` command (Jann, 2005, 2007). The `esttab` command is a wrapper for `estout` with simpler

Table 2.1   Results of Two Regression Models: An Example

Variables	(1) mod1 realinc1	(2) mod2 realinc1
realinc1	0.427***	0.380***
	(0.017)	(0.017)
Educ		0.318***
		(0.032)
Age		0.009
		(0.007)
Constant	2.691***	−2.030***
	(0.107)	(0.527)
Observations	1,113	1,111
R^2	0.374	0.427

Note: Standard errors in parentheses.

***$p < .01$.

syntax. You need to use `ssc install estout` to install these two programs. The following box provides an example of using the `esttab` commands to generate a table similar to Table 2.1. The `eststo` command is used to store estimates for two regression models without naming them. To save space, the created table is omitted here.

```
1.  . eststo: quietly regress realinc1 realrinc1
2.  . eststo: quietly regress realinc1 realrinc1 edu age
3.  . esttab using regout2.rtf, se ar2
```

The `outreg2` command can be a good helper when you create a regression table since at least you do not need to start from scratch. This command is particularly useful when you fit a series of models and need to summarize parameter estimates from each model. Please note that different fields or journals have different requirements for the table format. Although the table produced by `outreg2` has close-to-publication quality, it may not be the final table in the manuscript submitted for publication. Sometimes you may still need to do a lot of editing to the created table.

2.12 General Guidelines for Reporting Results

Once the data analysis is complete, the next step is to present and interpret the results, which normally are included in the results section of a research report. When reporting the results of your statistical analyses, a general rule is to provide sufficient information for readers to understand your analyses and the findings of your study. What are the major elements that should be included in the results section? The answer varies since different disciplines and journals may have their own reporting requirements, and one statistic commonly reported in one field may not be needed by another field. We provide the following general guidelines for reporting results, but it is up to you to decide what should be reported.

First, describe the analyses you have conducted, explain the variables with descriptive statistics, and state what research questions have been addressed.

Second, when reporting the results of a statistical test, provide the value of the test, the degrees of freedom, and the associated p value, followed with an explanation of the meaning of your findings. You may also need to form a conclusion about whether the test is significant. If the test is significant, provide the effect size if available. The reporting of the effect sizes was recommended in APA (2010), and various measures of effect size for continuous and categorical outcome variables and their estimation methods were introduced in Kline (2013).

For example, when reporting the results of a multiple regression model, provide the F statistic for the model, its degrees of freedom, and the associated p value. Explain whether the overall model is significant. The coefficient of determination R^2, which is a measure of the effect size, may also be reported and interpreted in plain English. For each predictor variable in the model, provide the value of regression coefficient β, its t statistic, and the associated p value. Since the degree of freedom for each t statistic is 1, it can be omitted. Explain whether each predictor is significant. If there are categorical predictor variables in the model, then explain how they are coded and report the regression coefficients of the coded variables, t statistics, and the associated p values. The results of the coded categorical predictor variables also need to be interpreted. When there are multiple predictor variables in the model, construct a table containing all the regression coefficients, values of t tests, and p values. The 95% confidence intervals of the coefficients may also need to be reported.

According to APA style (2010), a general rule is to report the numbers up to two decimal places, but more decimals may be reported when there is a need (e.g., statistical precision). The p values can be reported up to three decimal places. When $p = .000$ in the output, report it as $p < .001$. In addition, there are different ways to present the results. They can be summarized in numbers, tables, figures, and text or statements. The APA (2010) has provided examples on what information needs to be included.

Third, when summarizing numerical information in tables, make sure they can be easily interpreted. The labels, categories, and numbers in the tables should be concise and clear so that readers can understand them without much effort. Complex tables may confuse readers with excessive information. When creating a table, imagine yourself as the reader. A table should be informative and self-evident. Ask the following questions: Do we need a table? Are the labels and categories clear? Is the numerical information accurate? Is there anything missing?

Fourth, tables and written text should be complementary to each other. If the results can be summarized in a sentence, you do not need a table. On the other hand, if you have many categorical variables and you need report frequencies for all categories, it will be tedious to report them in paragraph after paragraph in written text, and it will be boring for readers to read your description. Instead, a table containing these frequencies is good enough to help readers quickly understand this information.

Fifth, after the results of statistical tests are presented, you also need to interpret them. Readers are more interested in the meaning of the statistical results than in the technical information related to the tests. The numerical information is important since it is the evidence supporting your conclusion. To help readers understand these statistics, you need to interpret them in a clear manner. You should have a concise statement to summarize the conclusions and explain your findings in standard English, which will help those who have a limited background in statistics to understand your results. Since a more general, detailed summary of the major findings will be presented in the discussion section, the summary of the results in the results section should be focused on the specific statistical tests.

Finally, when summarizing the results, also keep the audiences in mind. If you submit your article to a journal for publication, reviewers are interested in reading the statistics, tables, and graphs. You might receive feedback from them asking you to provide more statistics. It always happens! But if your readers are those who have limited background in statistics and hate all that jargon, you will need to explain your results in plain English in addition to the statistics you provide in the text. Those readers can skip over the statistics that they find complicated and directly read the text summarizing the results. So you need to keep a good balance between numbers and statements.

In summary, first, you should provide complete statistics without missing the important information; second, your explanation of the results should be clear and easily understood by readers even if they do not understand those statistics.

For more information about introductory statistics using Stata, see Acock (2014), Hamilton (2012), Juul (2014), Kohler and Kreuter (2012), and Longest (2015). Consult Baum (2006), Cameron and Trivedi (2010), Hardin and Hilbe (2012), Hilbe (2009), and Long and Freese (2014) for more advanced statistics using Stata. Refer to Cargill and O'Connor (2013), Dunn (2012), McInerney (2002), Sternberg and Sternberg (2010), Thody (2006), and Thyer (2008) for guidelines on reporting the results of research or writing research articles for publication.

2.13 Summary of Stata Commands in This Chapter

```
*Descriptive statistics for continuous variables
use chap2-gss2012, clear
codebook age
summarize age
summarize age, detail
summarize age if sex==1
summarize age if sex==2
bysort sex: summarize age
summarize age educ
summarize age educ, detail
tabstat age educ
tabstat age educ, by (sex)
tabstat age educ, by (sex) statistics (mean sd min max)

*Frequency analysis for categorical variables
codebook degree
tabulate degree
tab1 marital wrkstat
tabulate degree race
tabulate degree race, row column

*Independent samples t test
ttest realrinc, by(sex)
sdtest realrinc, by(sex)
ttest realrinc, by(sex) unequal
esize twosample realrinc, by(sex)

*Paired samples t test

*Using a different dataset
use els_math, clear
ttest BYACH = F1ACH

*ANOVA
use chap2-gss2012, clear
generate realinc1=realinc/10000
anova realinc1 race
estat esize
oneway realinc1 race, tabulate bonferroni

*Correlational analysis
correlate realinc1 educ age
correlate realrinc educ age, means
pwcorr  realinc1 educ age, sig obs

*Simple linear regression
drop realinc1
generate realinc1=realinc/10000
generate realrinc1=realrinc/10000
regress realinc1 realrinc1
```

```
*Multiple regression
regress realinc1 realrinc1 edu age
estat esize
margins, atmeans
margins, at(educ = (8 13 16)) atmeans vsquish

*Chi-square test
tabulate health marital, chi2 V
tabulate health marital, row column chi2 V
tabi 200 13 \ 274 46\ 107 29 \ 30 14 , chi2 V row column

*Making publication-quality tables using Stata

*Using outreg2
quietly regress realinc1 realrinc1
est store mod1
quietly regress realinc1 realrinc1 edu age
est store mod2
outreg2 [mod1 mod2] using regout, dec(3) word replace

*Using esttab
eststo: quietly regress realinc1 realrinc1
eststo: quietly regress realinc1 realrinc1 edu age
esttab using regout2.rtf, se ar2
```

2.14 EXERCISES

Use the GSS 2012 data for the following problems.

1. Run a descriptive statistics analysis for coninc, and interpret the results.

2. Conduct a frequency analysis for happy7. What percentages of respondents are very happy?

3. Make a two-way table for degree and class.

4. Perform a correlational analysis for satfam7 and satfin.

5. Run a chi-square test to investigate the relationship between degree and class.

6. Conduct a multiple regression analysis to estimate tvhours from three predictor variables sex, educ, and age.

   a. Write a research question.

   b. Find the F statistic from the output, and interpret whether the overall model is statistically significant.

c. Which predictor variables are significant? Interpret the coefficients for all three predictor variables.

d. Compute effect sizes using the `estat esize` command. List the effect sizes for the model and three predictor variables.

e. Produce a table for the regression output using `outreg2`.

f. Write a concise report to summarize the results.

CHAPTER **3**

Logistic Regression for Binary Data

Objectives of This Chapter

This chapter introduces logistic regression models for binary data. It will introduce concepts of odds, odds ratio, and goodness-of-fit statistics of the model; describe how to test significance of predictors; and show how to interpret parameter estimates. Following the description of data, two logistic regression models using Stata will be illustrated with step-by-step instructions. Stata commands and output will be explained in detail. The focus of this chapter is on fitting binary logistic regression models using Stata, as well as on interpreting and presenting the results. After reading this chapter, you should be able to

- Determine when a logistic regression model is used.
- Conduct logistic regression using Stata.
- Interpret the output.
- Interpret the model in terms of odds ratios.
- Compute and plot the estimated probabilities using the `margins` and `marginsplot` commands, respectively.
- Compare models using the likelihood ratio test and other fit statistics.
- Present results in publication-quality tables.
- Write the results for publication.

3.1 Logistic Regression Models: An Introduction

In multiple linear regression, a set of independent variables is used to predict a continuous outcome variable. But when the dependent variable is a dichotomous categorical variable with two categories, can we still use multiple regression? If not, then what statistical method can we use to model this type of binary data? In this chapter, logistic regression models will be introduced for the analysis of dichotomous response variables.

Let us look at several research examples:

1. Researchers in education would like to conduct a study on college student retention in their freshmen year. They are interested in what factors are associated with student retention or dropout. They plan to collect student data, such as SAT scores, first-year GPA, the number of courses the student has taken, whether a student received financial aid, whether he or she lives on campus or off campus, gender, ethnicity, and socioeconomic status (SES).

2. Researchers in education would like to investigate factors related to faculty tenure. They would like to know how well faculty's tenure status would be predicted by a set of variables, such as the number of publications, number of grants, teaching evaluations, previous teaching experience, gender, and ethnicity.

3. Researchers in medical science are interested in identifying risk factors for certain kinds of diseases, for example, coronary heart disease (CHD). The variables of interest include age, weight, cholesterol level, blood pressure, smoking, family history, gender, and ethnicity.

4. Researchers in political science are interested in investigating factors related to voting behavior (voting or not voting). Variables of interest include party identification, religion, education, socioeconomic status, attitude toward policy issues, gender, and ethnicity.

5. Researchers in psychology are interested in identifying risk factors related to depression. They look at variables such as self-esteem, age, income, relationship with family members, family history, health status, and gender; investigators would like to know how well these variables would predict the likelihood of having depression.

6. Researchers in sociology are interested in identifying factors related to employment status (employed or unemployed). They would like to predict whether people are employed by using a set of variables, such as previous working experience, highest degree they received, age, gender, and ethnicity.

In the preceding examples, the outcome variables of interest, retention or not, tenured or not, having CHD or not, voting or not, having depression or not, and being employed or not, are all dichotomous. These binary response variables have two categories: having an event or not having an event. They are normally coded as 1 for having an event and 0 for not having an event.

Since the response variable only has two values (i.e., 1 and 0), it is inappropriate to use multiple linear regression to estimate the outcome variable. Why? Using multiple regression to estimate a binary response variable violates the assumptions of normality, homoscedasticity, and the linear relationship between the dependent variable and independent variables. In multiple regression, we assume that the dependent variable has a normal distribution. When the dependent variable is binary, it has a Bernoulli distribution, which is not a normal distribution. The homoscedasticity assumption suggests that the error variances of the dependent variable are the same across each value of the independent variables. This assumption is violated when the dependent variable is dichotomous. Multiple regression also assumes that the relationship between the dependent variable and independent variables are linear so that the predicted values fall on the regression line. When the dependent variable is dichotomous, the relationship between this variable and a set of independent variables is nonlinear, which can be easily seen if you draw a scatter plot showing the relationship between a dichotomous dependent variable and each independent variable.

What will happen if you do use multiple regression to estimate the binary data even though you are cautioned not to do so? If you do use it, you will see the predicted values of the dependent variable are out of the range of 0 and 1. In other words, you will see the estimated values of the outcome variable are negative (i.e., smaller than 0) or larger than 1. These values do not make sense since they are the predicted probabilities from the independent variables.

3.1.1 Why Do We Need a Logistic Transformation?

As explained previously, if we use linear regression to estimate binary data, the predicted value will have values less than 0 or larger than 1. To estimate the probability of success for having an event, a logistic transformation $\text{logit}(\pi)$ or a probit transformation probit (π) can be used. A regression model with the logistic transformation is called the logistic regression model, whereas a model with the probit transformation is called the probit regression model. Both models produce similar results, and use of either model is just a choice by researchers. This chapter focuses on the binary logistic regression models only. Readers interested in the probit models should refer to the last chapter (Chapter 12) for details.

The form of the simple logistic regression model can be expressed as follows:

$$\text{logit}(\pi) = \alpha + \beta X$$

where π is the probability when the outcome variable equals 1, $P(Y = 1)$; logit(π) is the logistic transformation of the probability of success or of an event occurrence; and on the right side, α is the intercept and β is the logit regression coefficient.

This equation looks similar to that for simple linear regression. The noticeable difference is the logistic transformation on the left side of logistic regression. Instead of directly estimating the dependent variable, we estimate the logistic transformation (i.e., logit) of the probability of a success, which is also known as the logarithm of the odds or "log odds." Odds are the ratio of the probability of success to the probability of failure. The transformation between probabilities and odds will be introduced in detail in the next section. So in simple logistic regression, we estimate the relationship between an independent variable and the binary outcome variable on a scale of the logit or log odds. In other words, the relationship between the predictor variable X and the logit of the outcome (i.e., the logit transformation of the probability when the outcome variable $Y = 1$) is linear. Therefore, to estimate the relationship between a predictor variable X and the binary outcome variable Y, we cannot use a linear regression model, which assumes the relationship between the predictor variable and the outcome variable is linear. Instead, we transform the outcome variable into the logit of the outcome and assume that there is a linear relationship between the predictor and the logit. The logit of the outcome logit(π) can be easily transformed back to the probability of the outcome $P(Y = 1)$ since the logit is the natural logarithm of the odds or log odds.

Since logit(π) is ln(odds), which is expressed as $\ln \frac{\pi}{1-\pi}$, the form of the simple logistic regression can also be rewritten as:

$$\ln \frac{\pi}{1-\pi} = \alpha + \beta X \tag{3.1}$$

where ln is the natural logarithm. For simplicity, we read "ln(odds)" log odds or log of the odds.

When the probability π varies from 0 to 1, the log odds or logit will vary from negative infinity to positive infinity.

In multiple logistic regression, we have more than one predictor variable. The following is the form of the multiple logistic regression model:

$$\ln \left(\frac{\pi(x)}{1-\pi(x)} \right) = \alpha + \beta_1 X_1 + \beta_2 X_2 + \dots + \beta_p X_p \tag{3.2}$$

where $X_1, X_2, \dots, X_p$ are the predictor variables and $\beta_1, \beta_2, \dots, \beta_p$ are the logit coefficients of these predictors. This equation can be also expressed as:

$$\text{logit}[\pi(x)] = \alpha + \beta_1 X_1 + \beta_2 X_2 + \dots + \beta_p X_p \tag{3.3}$$

3.1.2 Probabilities, Odds, and Odds Ratios

Just as the mean and the standard deviation are the cornerstone for the descriptive statistics and linear models, probabilities, odds, and odds ratios are the key concepts for logistic regression models. Probability is the chance of success or of having an event occur. It can be expressed as a proportion. It equals the frequency of success or of having an event divided by the total number of observations or events. Probabilities range from 0 to 1. When probability is 0, it means that the chance of success is 0 or that there is no chance an event will occur. When probability is 1, it means that the chance of success is certain.

If p is the probability of success or having an event, then the probability of failure or of not having an event is $1 - p$ since these two probabilities are complementary. The sum of all possible probabilities equals 1.

The odds are the ratio of two probabilities, the probability of success or of having an event (p) to the probability of failure or of not having an event ($1 - p$):

$$\text{Odds} = \frac{p}{1-p}$$

Since the probability p varies from 0 to 1, the odds vary from 0 to positive infinity. With the increase of the probability of success or of having an event, the odds also get larger. Let us have a look at several examples:

When $p = 0.1$, odds $= 0.1/(1 - 0.1) = 0.11$.
When $p = 0.5$, odds $= 0.5/(1 - 0.5) = 1.00$.
When $p = 0.8$, odds $= 0.8/(1 - 0.8) = 4.00$.
When $p = 0.9$, odds $= 0.9/(1 - 0.9) = 9.00$.
When $p = 0.99$, odds $= 0.99/(1 - 0.99) = 99$.
When $p = 0.999$, odds $= 0.999/(1 - 0.999) = 999$.
...

From the preceding examples, we can easily see that when the probability of success or of having an event is less than the probability of failure or of not having an event, then the odds are less than 1. When the probability of having an event is equal to the probability of not having an event (0.5 vs. 0.5), the odds are equal to 1. When the probability of having an event is larger than the probability of the event not occurring, the odds are greater than 1. When the probability of having an event moves from 0.9 to 1, the odds start from 9, become larger and larger, and finally reach positive infinity. Odds $= 99$ means that the probability of having an event is 99 times the probability of not having an event. Knowing whether the odds are less than, greater than, or equal to 1 helps us interpret regression coefficients in the following logistic regression models.

The preceding examples focus on the odds of having an event. What are the odds of not having that event? When the odds of having an event = 99, the odds of not having an event = 1/99 = 0.01 since they are just the reciprocal of the odds of having an event. This is an important property of odds. It is useful when we compare different categories for the odds in various ordinal logistic regression models in the following chapters.

The examples show us how to compute odds if we know the probability of success or of having an event. But if we know the odds of success or of having an event, can we compute the probability of success or of having an event? This backward transformation is also easy. Since Odds = $\frac{p}{1-p}$, with a simple transformation, we get $p = \frac{\text{odds}}{1+\text{odds}}$. In other words, the probability of success or of having an event equals the odds divided by (1 + odds).

For example, the command tab healths marital produces a two-way cross-tabulation table of two binary variables, the health status and marital status. The health status has two categories with 1 = good health and 0 = not good health; the marital status is coded as currently married and not married.

```
. tab healths maritals


RECODE of |
   health |     RECODE of marital
(condition |     (marital status)
of health) |        0           1 |      Total
-----------+----------------------+----------
        0 |      221         137 |        358
        1 |      474         474 |        948
-----------+----------------------+----------
    Total |      695         611 |      1,306
```

Since the probability of success or of having an event is a ratio of the frequency of success or of having an event to the total observations, the probability of having good health p(health = 1) = 948/1,306 = .726. The probability of not having good health p(health = 0) = 358/1,306 = .274. Therefore, overall, the odds of having good health = p(health = 1)/p(health = 0) = .726/.274 = 2.650.

Next, let us compute the odds of having good health for the unmarried.

To compute the probability of having good health for people who are currently not married, we look at the ratio of the frequency of people with good health who are not married to the total frequency of people who are not married: p(health = 1 | unmarried) = 474/695 = .682.

The probability of having poor health for unmarried people p(health = 0 | unmarried) = 221/695 = .318.

For the unmarried people, the odds of having good health = p(health = 1 | unmarried)/p(health = 0 | unmarried) = .682/.318 = 2.145.

For married people, the odds of having good health = p(health = 1 | married)/p(health = 0 | married). Since p(health = 1 | married) = 474/611 = .776 and p(health = 0 | married) = 137/611 = .224, the odds of having good health for the married people = .776/(1 − .776) = 3.460.

3.1.3 Transformation Among Probabilities, Odds, and Log Odds in Logistic Regression

Forward Transformation

If we know the probability of success or of having an event, the odds of success are just the probability of success divided by the probability of failure. By taking the natural logarithm of odds, we get log odds or logit. Table 3.1 presents the forward transformation from probabilities to odds to log odds.

Backward Transformation

If we know the logit, since it is the natural logarithm of the odds or log odds, then the odds are the antilogarithm. In other words, the odds of success can be obtained by taking the exponential of the logit. To transform the odds back to the probability of success, we divide the odds by one plus the odds. The backward transformation is displayed in Table 3.2.

Table 3.1 Forward Transformation From Probabilities to Odds to Log Odds

Transformation	Probabilities	Odds	Logit or Log Odds
Forward	p	$\frac{p}{1-p}$	$\ln \frac{p}{1-p}$

Table 3.2 Backward Transformation From Log Odds to Odds to Probabilities

Transformation	Logit or Log Odds	Odds	Probabilities
Backward	$\text{logit}(p)$ or $\ln \frac{p}{1-p}$	Odds = exp(logit)	$p = \frac{\text{odds}}{1+\text{odds}}$

Odds Ratio

What is an odds ratio (OR)? It is just a ratio of two odds. Since the odds of having good health for the married are 3.464 and the odds of having good health for the unmarried are 2.145, the ratio of the odds for the married to the odds for the unmarried = 3.460/2.145 = 1.613. In other words, the odds of having good health for the married are 1.613 times the odds for the unmarried.

When OR > 1, the odds of success or of having an event for one group are larger than the odds for the other group. For example, OR = 2 indicates the odds of success for one group are two times the odds for the other group.

When OR < 1, the odds in one group are less than the odds in the other group. For example, OR = 0.5 indicates that the odds for one group are 0.5 times the odds for the other group. In other words, the odds for the second group are two times the odds for the first group. When OR is less than 1, we can take the inverse of it and make it more interpretable.

When OR = 1, the odds for one group are the same as the ones for the other group.

3.1.4 Goodness-of-Fit Statistics

To assess whether a model fits the data well, we normally look at several measures of fit statistics rather than at a single measure. The deviance, log likelihood ratio test, pseudo R^2, AIC, and BIC statistics are introduced next.

Deviance

Deviance is one of the goodness-of-fit statistics used in the logistic regression. It compares the currently fitted model and the saturated model. It is defined as −2 times the difference in log likelihood between these two models, which can be expressed as −2(log likelihood of the current model − log likelihood of the saturated model). The saturated model is the model that fits the data perfectly.

In binary logistic regression, the saturated model estimates one parameter for each observation so the number of parameters equals the number of sample size. Its likelihood of fitting the data is 1 so the log likelihood of the saturated model equals ln1, which is 0. The expression of deviance can be simplified to −2 log likelihood of the current model. It is often abbreviated as −2LL.

Deviance shows how well a model fits the data compared with a perfect model. Since the saturated model fits the data perfectly, we can also ask this question: How poorly does a model fit the data? If the discrepancy in log likelihood between the fitted model and the saturated model is small, then this model has a good fit. On the contrary, if the deviance is large, then the fitted model has a poor fit. Therefore, smaller deviance means a better fit. In a linear regression model, we minimize the error variance and would like to see the sum of squared residuals as small as possible. Similarly, in a logistic regression model, we would like to minimize the deviance and would like to see the discrepancy in log likelihood between the fitted model and the saturated model as small as possible.

Model Comparisons Using the Deviance Difference or Log Likelihood Ratio Test

Deviance is often used to compare nested models. Models are nested when more constraints can be put on parameters in one model than in the other. One model is called the reduced model, which contains fewer parameters, and the other is called the full model, within which the reduced model is nested but which has more parameters. In the logistic regression, the reduced model has fewer variables than the full model, and the former model is a subset of the latter. The difference in deviance between nested models follows a chi-square distribution. The degrees of freedom of the distribution equal the difference in the number of parameters between these two models. The difference in deviance is often expressed as G = Deviance for the reduced model − Deviance for the full model or as $D_{Reduced} - D_{Full}$. This test is also known as the likelihood ratio test since the difference in deviance is the difference in −2LL, which can be expressed as a ratio of likelihood in logarithm.

In simple logistic regression with only one independent variable, the likelihood ratio test compares the deviance between the null model with only the intercept and the model with one independent variable. If the likelihood ratio chi-square test is significant, then we reject the null hypothesis and conclude that the model with one independent variable fits the data better than the model with only the intercept (null model).

Predictor Selections Using the Likelihood Ratio Test

The likelihood ratio chi-square test can be used to test whether a predictor variable contributes to the model by comparing the models with and without the variable. A significant test means that the added variable contributes to the model. It is also useful for model developing. We can build a series of nested models from a simple model with one predictor to more complex models with multiple predictors. The likelihood ratio chi-square test can be used to decide which model fits the data better and whether the predictor variables should be kept in or removed from the model.

Pseudo R^2

In linear regression models, the coefficient of determination, R^2, is the index for the model fit. It is a ratio of the variance explained by the model to the total variance. It indicates how much variance in the dependent variable is accounted for by an independent variable or a set of independent variables. Analogous to R^2 in the linear regression, several pseudo R^2 measures are used in logistic regression (Hardin & Hilbe, 2012; Long & Freese, 2014; Menard, 2010). But the interpretation of these pseudo R^2 measures in logistic regression are different from those in the linear regression. Table 3.3 displays three major pseudo R^2 measures and their formulas.

1. The Likelihood Ratio R^2

The likelihood ratio R^2, written as R^2_L, is also known as the McFadden's R^2. It is the reduction in deviance from the fitted model (D_m) to the null model that only contains the intercept (D_0). It is expressed as:

$$R^2_L = 1 - \frac{D_m}{D_0} = 1 - \frac{-2LL_m}{-2LL_0} \tag{3.4}$$

When model deviance equals −2 times log likelihood ($D = -2LL$), the likelihood ratio R^2 can also be the reduction in log likelihood between the fitted model and the null model. It is expressed as:

$$R^2_L = 1 - \frac{-2LL_m}{-2LL_0} = 1 - \frac{LL_m}{LL_0} \tag{3.5}$$

2. Cox and Snell R^2

Cox and Snell's R^2, written as R^2_{ML}, is also known as the maximum likelihood R^2. It is based on the likelihood function of the fitted model (L_m) and the null model, which only contains the intercept (L_0). It is expressed as:

$$R^2_{ML} = 1 - \left(\frac{L_0}{L_m}\right)^{2/n} \tag{3.6}$$

where n is the total number of observations.

3. Nagelkerke R^2

This is also called Cragg and Uhler's R^2. It is an adjustment to Cox and Snell's R^2 by dividing the maximum value of Cox and Snell's R^2. It is expressed as:

$$R^2_N = R^2_{ML}/\text{maximum } R^2_{ML} \tag{3.7}$$

4. Other Pseudo R^2

Other pseudo R^2 include the adjusted McFadden's R^2, McKelvey and Zavoina's R^2, Efron's R^2, Tjur's R^2, and the Count and adjusted Count R^2. Refer to Hardin and Hilbe (2012) and Long and Freese (2014) for a detailed introduction to pseudo R^2 measures.

Table 3.3 Three Major Pseudo R^2 Measures

Pseudo R^2	Formulas
Likelihood ratio R^2 (McFadden's R^2)	$R^2_L = 1 - \dfrac{D_m}{D_0} = 1 - \dfrac{-2LL_m}{-2LL_0}$
Cox and Snell R^2 (maximum likelihood R^2)	$R^2_{ML} = 1 - \left(\dfrac{L_0}{L_m}\right)^{2/n}$
Nagelkerke R^2 (Cragg and Uhler's R^2)	$R^2_N = R^2_{ML}/\text{maximum } R^2_{ML} = \{1 - \left(\dfrac{L_0}{L_m}\right)^{2/n}\}/(1 - L_0)^{2/n}$

Information Criteria Indices: AIC and BIC

The likelihood ratio tests are used for comparisons between nested models. But how do you compare models if they are not nested (e.g., if two models contain different sets of independent variables or if one model has more missing values so the sample size is different between the two models)? To compare non-nested models, the Akaike information criterion (AIC) (Akaike, 1974) and the Bayesian information criterion (BIC) (Schwarz, 1978) tests are commonly used. Both tests are based on the deviance statistics and can be seen as an adjustment to the deviance. Table 3.4 presents formulas for different versions of the AIC and BIC statistics.

The AIC adjusts the deviance by the number of predictors and the sample size:

$$\text{AIC} = -2(LL_m - k)/n = (D_m + 2k)/n \tag{3.8}$$

where k is the number of parameters (the number of independent variables plus the intercept), n is the sample size, and $-2LL_m$ or D_m is the deviance of the fitted model.

Another version of the AIC is AIC $\times$ n, which equals $(D_m + 2k)$. The AIC $\times$ n adjusts the deviance by the number of parameters only. Remember that we would like to see smaller deviance; this is still true for the AIC statistic. A smaller AIC means a better fit of the model.

The BIC adjusts the deviance by its degrees of freedom and the sample size:

$$\text{BIC} = D_m - \text{df} \times \ln(n) \tag{3.9}$$

where df is the degrees of freedom associated with the deviance, n is the sample size, and D_m is the deviance of the fitted model. The degrees of freedom associated with the deviance are the degrees of freedom of the model, which equals the number of observations minus the number of independent variables and intercept:

$$\text{Another version of BIC} = -2LL_m + \ln(n) \times k = D_m + \ln(n) \times k \tag{3.10}$$

where k is the number of parameters (the number of independent variables plus the intercept). The third version of BIC is BIC', which is based on the log likelihood ratio chi-square test statistic:

$$\text{BIC'} = -LR \text{ chi-square} + \text{the number of predictors} \times \ln(n) \qquad (3.11)$$

Similar to the AIC statistic, we also want a smaller BIC statistic.

Table 3.4 Information Criteria Indices: AIC and BIC Statistics

Information Criteria Indices	Formulas
AIC	$-2(\text{LL}_m - k)/n = (D_m + 2k)/n$
AIC $\times$ n	$(D_m + 2k)$
AIC used by Stata	$2(\text{LL}_m - k)/n = (D_m + 2k)/n$
BIC	$D_m - \text{df} \times \ln(n)$
BIC used by Stata $= D_m + \ln(n) \times k$	$2\text{LL}_m + \ln(n) \times k = D_m + \ln(n) \times k$
BIC'	$-LR \text{ chi-square} + \text{the number of predictors} \times \ln(n)$

3.1.5 Testing Significance of Predictors

To test the statistical significance of each predictor, the Wald test is used. It is computed as a ratio of the parameter estimate for each predictor variable in the model to its corresponding standard error. Statistical software packages report either the univariate Wald test statistic, which follows a standard normal distribution, or a squared Wald chi-square test statistic, which follows a chi-square distribution. The null hypothesis is that the coefficient of each predictor variable is zero. Rejection of the null hypothesis indicates that the effect of a particular predictor variable is significant. The Wald test can also be used to test the effects of multiple predictors simultaneously.

Another way to test the significance of predictors is the likelihood ratio test, which compares -2LL of different models. With only one additional predictor added to the existing model, the difference in deviance between the nested models has a chi-square distribution with one degree of freedom (i.e., the degree of freedom equals the difference between the number of predictors). A significant likelihood ratio test means that the variable is a significant predictor in the model. For a univariate test of a single predictor, the results of the Wald test and the likelihood ratio test are equivalent. The likelihood ratio test can also be used to test the contribution of two or more variables when they are added to the current model. A significant likelihood ratio test for the nested models means that a set of variables overall makes a significant contribution to the model.

3.1.6 Interpretation of Model Parameter Estimates in Logistic Regression

Probabilities, Odds, and Odds Ratios in Logistic Regression

In 3.1.2, the transformation from probabilities to odds and to odds ratios was introduced. Let us take a look at their transformation in logistic regression. In the simple logistic regression, $\ln \frac{\pi}{1-\pi} = \alpha + \beta X$, where π is the probability when $Y = 1$. The estimated coefficient is the logit coefficient, which is the coefficient on the scale of logit or log odds. In the simple logistic regression, the outcome variable is dichotomous with values of 1 and 0. We estimate the relationship between the predictor variable and the logit function of the probability that $Y = 1$, the log odds.

Exponentiating both sides of the equation, we get the odds of success or of having an event:

$$\frac{\pi}{1-\pi} = \exp(\alpha + \beta X) \qquad (3.12)$$

It can be rewritten as:

$$\text{odds } (Y = 1) = \exp(\alpha + \beta X)$$

If the independent variable X is a categorical variable with the values of 0 and 1, what are the odds of success or of having an event?

When $X = 0$, odds $(Y = 1) = \exp(\alpha)$, which is the exponentiated intercept.

When $X = 1$, odds $(Y = 1) = \exp(\alpha + \beta)$, which is the exponentiated sum of intercept and logit coefficient.

The odds ratio of the group 1 $(X = 1)$ to the group 2 $(X = 0)$:

$$\text{OR} = \frac{\exp(\alpha+\beta)}{\exp(\alpha)} = \frac{\exp(\alpha) \times \exp(\beta)}{\exp(\alpha)} = \exp(\beta) \qquad (3.13)$$

This is the simple case for a one-unit increase in an independent variable (e.g., from 0 to 1 in the previous example); the change in the odds is the odds ratio, which is the exponentiated logit coefficient. When the independent variable is continuous, for a one-unit increase from any value of x to the value of $(x + 1)$, the change in the odds is still the exponentiated logit coefficient.

In simple linear regression, we interpret the regression coefficient as the change in the outcome variable for a one-unit increase in the predictor variable. Although the logit coefficient can be interpreted as the change in the predicted logit or the log odds for a one-unit increase in the predictor variable, it is difficult for people to understand what the change in the logit or log odds means.

By exponentiating the logit coefficient, we get the OR, which is more interpretable. The OR in logistic regression is also known as the exponentiated logit coefficient.

When the logit coefficient is positive, it indicates the relationship between the predictor variable and the logit function of the probability is positive. By exponentiating the logit coefficient, we get the OR, which is larger than 1. This means that the odds of success or of having an event increases for a one-unit increase in the predictor variable.

When the logit coefficient is negative, it indicates that the relationship between the predictor variable and the logit function is negative. The exponentiated coefficient, the OR, is less than 1. It means that the odds of success or of having an event decreases for a one-unit increase in the predictor variable.

When the logit coefficient equals 0, OR equals 1. It indicates that there is no relationship between the predictor and the odds of success.

Interpreting an Odds Ratio as a Percentage Change in Odds

Another way of interpreting odds ratios is the percentage change in odds. It can be calculated by using (Odds ratio − 1) × 100%. A positive percentage change in odds indicates there is an increase in the odds, whereas a negative percentage change corresponds to a decrease in the odds. A zero percentage change indicates no change in odds at all. In other words, the predictor variable does not influence the odds of success.

For example, if an OR for a predictor variable equals 1.2, then the percentage change in odds can be computed as follows: (1.2 − 1) × 100% = 20%. This indicates that each one-unit increase in the predictor variable corresponds to an increase of 20% in the odds of success.

In another example, if OR = .80, then the percentage change in odds is (.80 − 1) × 100% = −20%. Since the percentage change is negative, it indicates that for each one-unit increase in the predictor variable, there is a decrease of 20% in the odds of success.

Interpreting Coefficients in Terms of Estimated Probabilities

In addition to odds ratios, we can interpret logistic regression models in terms of estimated probabilities. Recall that the logit is the natural logarithm of the odds, or log odds, and the odds of success is obtained by taking the exponential of the logit. By using the backward transformation method introduced previously, we can easily transform the odds back to the probability of success. The estimated probability can be calculated at specified values of a set of predictor variables. When a combination of the values of the predictor variable changes, the estimated probabilities are different. The `margins` command, which is introduced in Stata 11, makes the cumbersome calculations of the predicted probabilities much easier.

Why Are Odds Ratios Preferred?

Odds ratios are more popular than the probabilities when we interpret the logistic regression coefficients (i.e., the logit coefficient) because the former is more interpretable. In logistic regression, when the values of an independent variable change, so do the probabilities. In other words, the change in the probabilities varies according to the values of the independent variable. Nevertheless, for any one-unit change in the dependent variable, the change in the odds is the exponentiated logit coefficient, which is constant.

3.2 Research Example and Description of the Data and Sample

Research Problem and Questions: In this example, we are interested in investigating whether the binary response variable, health status, can be predicted by four predictor variables, marital status, years of education, age, and gender. The research question is as follows: Which predictor variables are associated with the likelihood of having good health? In other words, can health status be significantly predicted by the four preceding variables?

Description of the Data and Sample: The data for the following analyses were the General Social Survey 2012 (GSS 2012). The following are the variables:

- healths: the recoded variable of health (health status) with 1 = good health and 0 = not good health
- maritals: the recoded variable of marital (marital status) with 1 = currently married and 0 = not currently married
- educ: the highest education
- age: respondent's age
- male: recoded variable of sex with 1 = male and 0 = female

3.3 Logistic Regression With Stata: Commands and Output

3.3.1 Simple Logistic Regression Using Stata

The Stata Command logit

Three commands can be used for binary logistic regression models, including logit, logistic, and glm. The first two commands are almost identical except that the output produced by the logit command displays logit coefficients, whereas the logistic command produces odds ratios. The third command glm is normally used to fit generalized linear models where a binary logistic regression model belongs.

In simple logistic regression, there is only one binary dependent variable with values of 1 and 0 and one independent variable, which can be either categorical or continuous. The syntax for logistic regression is the command `logit` followed by the dependent variable and the independent variable. For example, the command `logit y x` tells Stata to run a simple logistic regression analysis predicting the dependent variable y with an independent variable x. If we would like to display odds ratios, then we can use the command `logistic y x`. If we use the `glm` command to fit the same model, then we need to specify the exponential family of the distribution and the link function. For example, the command `glm y x, fam(binomial) link(logit)` fits the same binary logistic regression model by specifying the binomial family and the logit link function. Type `help logit`, `help logistic`, and `help glm`, respectively, for more details on how to use these three commands.

In the following example, the command `logit healths maritals` tells Stata to predict the dependent variable `healths` from the independent variable `maritals`. The output is shown in the following box.

```
. logit healths maritals
Iteration 0:   log likelihood = -765.10363
Iteration 1:   log likelihood = -757.76447
Iteration 2:   log likelihood = -757.74215
Iteration 3:   log likelihood = -757.74215


Logistic regression                              Number of obs    =       1300
                                                 LR chi2(1)       =      14.72
                                                 Prob > chi2      =     0.0001
Log likelihood = -757.74215                      Pseudo R2        =     0.0096


------------------------------------------------------------------------------
     healths |      Coef.   Std. Err.      z    P>|z|     [95% Conf. Interval]
-------------+----------------------------------------------------------------
    maritals |   .4824281   .1267693     3.81   0.000     .2339648    .7308914
       _cons |    .75457    .0815631     9.25   0.000     .5947092    .9144308
------------------------------------------------------------------------------
```

Interpreting Stata Output

In the Stata output for simple logistic regression, a series of iterations of the log likelihood estimation is displayed at the top since the maximum likelihood estimation involves an iterative process. The log likelihood for the initial iteration is −765.104, and it reaches the maximum at the value of −757.742, which is the maximum log likelihood.

The top right of the logistic regression table provides the number of observations in the dataset, the log likelihood ratio test statistic and the associated p value, and the pseudo R^2.

LR chi2 (1) = 14.72, which is the log likelihood ratio chi-square test statistic. It is the difference in the −2 log likelihood (−2LL) between the current model, which contains the one predictor and the intercept, and the null or empty model, which contains only the intercept. The log likelihood of the current model is −757.742 and that of the null model is −765.104 at Iteration 0 in the output. If we run an empty model with only the dependent variable with the command logit healths, then we will also get the same log likelihood for the empty model. The difference in the −2 log likelihood statistic is 14.72, with 1 degree of freedom, which is the difference in the predictors between these two models.

The null hypothesis of the log likelihood ratio chi-square test is that the logit coefficient of the predictor variable healths is not significant or that the predictor variable does not contribute to the model.

The alternative hypothesis is that the coefficient of the predictor is significant in the model or that it significantly contributes to the model.

The associated p value with the log likelihood ratio chi-square test Prob > chi2 = 0.0001 indicates that the null hypothesis is rejected. Therefore, the overall model with one predictor is significant.

Pseudo $R^2 = 0.0096$. It is the likelihood ratio R^2 or McFadden's R^2, which is normally written as R^2_L:

$$R^2_L = 1 - \frac{-2LL_m}{-2LL_0} = 1 - \frac{LL_m}{LL_0} = 1 - (-757.742/-765.104) = .0096$$

Next, the logistic regression table displays the parameter estimates for the predictor variable and intercept, their standard errors, the Wald z statistics, the associated p values, and the 95% confident intervals of the parameter estimates for the constant and predictors. The null hypothesis for the Wald test is that the coefficient of the predictor variable is zero, and the alternative hypothesis is that the coefficient of the predictor variable is significantly different from zero.

The Wald z statistic equals the parameter estimate divided by its standard error. For the predictor variable maritals, Wald $z = .482/.127 = 3.81$. Some other statistical software packages report the Wald chi-square test statistic, which is the squared Wald z statistic. The associated p value, P>|z|=.000, so we reject the null hypothesis. The rejection of the null hypothesis indicates that the predictor variable marital status is a significant predictor of the dependent variable health status.

Since the estimated intercept and coefficient are .755 and .482, respectively, the simple logistic regression model could be expressed as:

$$\text{logit}(\pi) = .755 + .482X$$

or:

$$\ln \frac{\pi}{1-\pi} = .755 + .482X$$

The regression coefficient of the predictor `maritals` is .482, which is also called the logit coefficient since we estimate the relationship between the predictor variable and the logit of the probability when the outcome variable `healths` takes the value of 1.

When $X = 0$, the estimated logit(π) = .755. Since logit is log odds, exponentiating .755 gives us the odds of having good health. The odds of having good health for people who were not currently married are exp(.755) = 2.128.

When $X = 1$, the estimated logit(π) = .755 + .482 = 1.237. So the odds of having good health for people who were married are exp(1.237) = 3.445.

Odds ratio = 3.445/2.128 = 1.619, which is the ratio of odds of having good health for the married to the odds for the unmarried.

If we directly exponetiate the logit coefficient exp(.482) = 1.619, then we get the same odds ratio as earlier. Stata reports odds ratios if you add the option `or` to the original `logit` command. You get the following output displaying the odds ratio if you run the following command:

```
logit healths maritals, or
```

```
. logit healths maritals, or

Iteration 0:   log likelihood = -765.10363
Iteration 1:   log likelihood = -757.76447
Iteration 2:   log likelihood = -757.74215
Iteration 3:   log likelihood = -757.74215


Logistic regression                             Number of obs   =       1300
                                                LR chi2(1)      =      14.72
                                                Prob > chi2     =     0.0001
Log likelihood = -757.74215                     Pseudo R2       =     0.0096


------------------------------------------------------------------------------
     healths | Odds Ratio   Std. Err.      z    P>|z|     [95% Conf. Interval]
-------------+----------------------------------------------------------------
    maritals |   1.620003   .2053667     3.81   0.000      1.2636    2.076931
       _cons |   2.126697   .1734601     9.25   0.000     1.812504    2.495355
------------------------------------------------------------------------------
```

The logistic regression table displays the odds ratio for the predictor, its standard error, the Wald z statistic, associated p value, and the 95% confidence interval of the odds ratio, with all others the same as those in the last output.

Odds ratio = 1.620. It is interpreted as follows: For each one-unit increase in the predictor variable, the odds of having good health increase by a factor of 1.620. Since the predictor variable marital status is also a binary variable, it can also be interpreted as the odds of having good health for the married are 1.620 times the odds of having good health for the unmarried.

The fitstat command, which is part of the SPost package by Long and Freese (2014), provides various fit statistics. For Stata 13 and 14 users, you need to install this user-written package first by typing net install SPost13_ado before running the fitstat command. Another way to install the package is to type the findit spost command, find the link for the package, and then follow the steps to install it. If you have the previous version SPost9 on your computer, then uninstall it by typing ado uninstall spost9_ado before installing SPost13. The output produced by the fitstat command is displayed as follows.

```
. fitstat


                            |        logit
----------------------------+-------------
Log-likelihood              |
                    Model   |     -757.742
            Intercept-only  |     -765.104
----------------------------+-------------
Chi-square                  |
        Deviance (df=1298)  |     1515.484
               LR (df=1)    |       14.723
                  p-value   |        0.000
----------------------------+-------------
R2                          |
                 McFadden   |        0.010
       McFadden (adjusted)  |        0.007
       McKelvey & Zavoina   |        0.017
              Cox-Snell/ML  |        0.011
   Cragg-Uhler/Nagelkerke   |        0.016
                    Efron   |        0.011
                 Tjur's D   |        0.011
                    Count   |        0.725
          Count (adjusted)  |        0.000
----------------------------+-------------
IC                          |
                      AIC   |     1519.484
           AIC divided by N |        1.169
               BIC (df=2)   |     1529.825
----------------------------+-------------
Variance of                 |
                        e   |        3.290
                   y-star   |        3.348
```

These fit statistics include the log likelihood for the null model with only intercept, the log likelihood for the full model, the deviance statistic, the log likelihood ratio test statistic and the associated p value, nine different types of pseudo R^2, and the AIC and BIC statistics.

The AIC and BIC statistics adjust the deviance by penalizing for the number of predictors and the sample size. They have various forms. The Stata fitstat command in SPost9 (Long & Freese, 2006) reports AIC, AIC*n, and AIC used by Stata. The fitstat command in the latest SPost13 (Long & Freese, 2014) only computes AIC and AIC divided by N.

In the preceding output, AIC divided by N = 1.169. The same value can be easily derived if we use the following equation:

$$\text{AIC divided by } N = -2(\text{LL}_m - k)/n = (D_m + 2k)/n$$

Since D_m, the deviance of the model, is 1,515.484, k, the number of parameters including the intercept and the predictor, equals 2 and n is the number of observations. By substituting them into the AIC equation, we get AIC divided by N = (1,515.484 + 2 × 2)/1,300 = 1.169.

AIC × n = 1.169 × 1,300 = 1,520, which is reported as AIC in the output.

BIC in the output is 15,29.825. It can be calculated using this equation:

BIC used by Stata = D_m + ln(n) × k = 1,515.574 + ln(1,300) × 2 = 1,529.825

The fitstat command in SPost9 (Long & Freese, 2006) reports this statistic labeled as BIC used by Stata.

BIC has another version reported by the old SPost9. BIC = D_m −df × ln(n). The current SPost13 in Stata 13 and 14 does not report this anymore. BIC' is the third version of BIC, which is based on the log likelihood ratio chi-square test statistic. BIC' = −LR chi-square + the number of predictors × ln(n). This statistic, originally reported by fitstat in SPost9, is not reported by the current SPost13 version. The AIC and BIC statistics themselves do not tell whether the model fits the data well. They are useful for model comparison purposes, particularly when comparing non-nested models. More details will be provided in the multiple logistic regression section.

Reporting the Results

When writing the results of logistic regression models, you should describe the statistical method you use for data analysis, the dependent variable and the independent variables in the models, and your research hypothesis or the purpose of your study. Also, you should report the likelihood ratio chi-square test statistic, the degrees of freedom, and the associated p value. Based on the log likelihood ratio test result, you should discuss whether the fitted model fits better than the null model containing only the intercept. If more than one model is fitted, then the deviance statistics for each model should be presented, and the results of the log likelihood ratio test comparing deviance statistics between nested models should be provided.

In the body of the text, you should report the parameter estimates for the predictor variables, the Wald z statistics, the degrees of freedom, and the associated p values. You might also need to interpret the odds ratios of the estimates.

You could have a table containing all the parameter estimates, their standard errors, p values, and odds ratios. A summary of major fit statistics also should be included in the table, which could include the log likelihood ratio chi-square test statistic with its degrees of freedom; a couple of pseudo R^2 values, such as the likelihood ratio R^2 and Cox and Snell R^2; the deviance statistic; and the AIC and BIC statistics. If more than one model is fitted, then the results of all the competing models from the simple model to the full model can be presented in the table. For detailed guidelines and recommendations on reporting the results of logistic regression, see Peng, Lee, and Ingersoll (2002). In addition, O'Connell and Amico (2010) provided a list of key elements of logistic regression that should be addressed when judging a manuscript from a reviewer's perspective. Huck (2012) also provided examples on reporting the results of logistic regression from published articles. The following is an example of summarizing results from the simple logistic regression model.

A simple logistic regression analysis was conducted to estimate the probability of having good health based on people's marital status. The dependent variable was having good health or not, and the single independent variable was marital status, which was also a dichotomous variable (married or unmarried). The log likelihood ratio chi-square test statistic $LR\ \chi^2_{(1)} =$ 14.72, $p < .001$, which indicated that the overall model with one predictor, `marital status`, was significantly different from zero. In other words, the predictor `marital status` contributed significantly to the overall model. The logit regression coefficient $\beta = .482$, Wald $z = 3.81$, $p < .001$, which also indicated that the coefficient of the predictor was significantly different from zero. Odds ratio = 1.620, which indicated that the odds of having good health for the married were 1.620 times the odds of having good health for the unmarried.

3.3.2 Multiple Logistic Regression

The multiple logistic regression model is simply an extension of the simple logistic regression model when there are two or more predictor variables in the model. The following equation is for the multiple logistic regression model:

$$\ln\left(\frac{\pi(x)}{1-\pi(x)}\right) = \alpha + \beta_1 X_1 + \beta_2 X_2 + \dots + \beta_p X_p \qquad (3.14)$$

where X_1, X_2, ..., X_p are the predictor variables and β_1, β_2, ..., β_p are the logit coefficients of these predictors.

The multiple logistic regression estimates the logit of the probability or log odds when the outcome variable equals 1 given a set of predictors, which can be either categorical or continuous.

Interpretation of Model Parameter Estimates and Odds Ratios in Multiple Logistic Regression

In multiple logistic regression:

$$\ln\left(\frac{\pi(x)}{1-\pi(x)}\right) = \alpha + \beta_1 X_1 + \beta_2 X_2 + \ldots + \beta_p X_p \tag{3.15}$$

where $\pi(x)$ is the probability when $Y = 1$ given a set of predictor variables. So the odds of success or of having an event in multiple regression is the ratio of probability of success to the probability of failure when given a set of predictor variables.

The estimated regression coefficients are the logit coefficients, which are the coefficients on the scale of logit or log odds. The logit coefficient can be interpreted as the change in the predicted logit or the log odds for a one-unit increase in the predictor variable when holding other predictors constant or controlling for the effects of other predictors.

The exponentiated logit coefficients are the odds ratios. The odds ratio of each predictor can be interpreted as the change in the odds for a one-unit change in the dependent variable when holding other predictors constant.

When an odds ratio is larger than 1, the odds of success are of having an event increase for a one-unit increase in the predictor variable when controlling for other predictors.

When an odds ratio is less than 1, the odds of success are of having an event decrease for a one-unit increase in the predictor variable when controlling for other predictors.

When an odds ratio equals 1, there is no relationship between the predictor and the odds of success when holding other variables constant.

Model Fitting Based on the Log Likelihood Ratio Test and Information Criteria Tests

Just like linear regression models, models in logistic regression can be fitted from a simple logistic regression model to a more complex model to the full model. A baseline model can be the starting point for model building. A series of nested models can be compared by using the log likelihood ratio test or the deviance difference test. Models are nested when one model is a special case of the other. As explained in the Goodness-of-Fit Statistics section, the difference in deviance between nested models has an approximately chi-square distribution with the degrees of freedom equal to the difference in the number of parameters between these two models. It is often expressed as G = Deviance for the reduced model – Deviance for the full model or as $D_{\text{Reduced}} - D_{\text{Full}}$. If the likelihood ratio chi-square test or the deviance test is significant, then we reject the null hypothesis and conclude that one model fits the data better than the other model.

The result of the log likelihood ratio test is valid when the competing models are fitted on the sample data. In real data analysis, it is common to see that the sample size varies among the nested models. If there is a great difference between sample sizes between fitted models, for example, if some variables have more missing values, then the results are questionable. Under this circumstance, two information criteria measures, the AIC and BIC statistics, are more appropriate for model comparisons. They are commonly used to compare non-nested models. The smaller the AIC and BIC statistics, the better the fit of the model.

In the following multiple logistic regression examples, a two-predictor model will be fitted first, and then the full model with all five predictors will be fitted.

The Stata Command `logit` for Multiple Logistic Regression Models

The command for multiple logistic regression is same as that for single logistic regression. The command `logit` is followed by the dependent variable and a list of independent variables. Let us start with a two-predictor logistic regression model. The outcome variable is still `healths`, and the two predictors are `marital` and `age`. The first predictor is `marital`, which is a categorical variable with five levels, and the second predictor `age` is a continuous variable. Please note that if the categorical predictor is not correctly coded, then Stata will treat it as a continuous variable and produce misleading results.

Stata has several different ways to handle categorical predictors. The first way is to use the `generate` command to create dummy variables, which have a value of 1 and 0. If there are k levels of the categorical variable, the number of dummy variables that needs to be created is $k - 1$ since we have a baseline or reference category that is omitted from the dummy variables. For example, the predictor `marital` has five levels, so four dummy variables need to be created with the lowest level as the reference group.

The second way is to use the `tabulate` command with the `generate` option, which is a better way than the first one. For example, `tabulate marital, generate(mari)` will create four dummy variables for `marital` with the variable names `mar1`, `mar2`, `mar3`, and `mar4`.

The disadvantage of these two ways is that you need to create dummy variables before you conduct any analyses. The third way is to use the `xi` prefix command, which automates the process of creating dummy variables for your analysis. All the created dummy variables are included in the analysis, and the results are shown in the output.

The fourth way is to use the factor variables coding, which was first introduced in Stata 11. You simply need to indicate that you would like to create factor variables for a categorical variable. For example, i.marital tells Stata that factor variables need to be created for the variable marital. This method is more convenient that the first three methods. In addition, it easily handles all kinds of interactions between predictor variables, whether they are categorical or continuous. The only caveat is that it does not work for versions earlier than Stata 11.

In the following example, the command xi: logit healths i.marital age tells Stata to run the logistic regression model by estimating the dependent variable healths from the two independent variables marital and age. The prefix xi command is used to create dummy variables for the categorical variable marital, and i.marital indicates that indicator variables or dummy variables will be created for this categorical variable. Please note that the prefix xi can be omitted for Stata 13 and 14 users since the new coding for factor variables is used. This prefix is still used throughout the book since user-written programs may not work with factor variables. The following output is displayed.

```
. xi: logit healths i.marital age
i.marital            _Imarital_1-5          (naturally coded; _Imarital_1 omitted)

Iteration 0:    log likelihood = -765.10363
Iteration 1:    log likelihood = -735.17599
Iteration 2:    log likelihood = -734.75481
Iteration 3:    log likelihood = -734.75476


Logistic regression                                 Number of obs   =        1300
                                                    LR chi2(5)      =       60.70
                                                    Prob > chi2     =      0.0000
Log likelihood = -734.75476                          Pseudo R2       =      0.0397


------------------------------------------------------------------------------
    healths |      Coef.   Std. Err.      z    P>|z|     [95% Conf. Interval]
------------+-----------------------------------------------------------------
 _Imarital_2 |  -.3936535   .2454783    -1.60   0.109    -.8747821     .087475
 _Imarital_3 |  -.4062934   .1799989    -2.26   0.024    -.7590848    -.053502
 _Imarital_4 |  -1.214882   .3207134    -3.79   0.000    -1.843468    -.5862947
 _Imarital_5 |   -.582633   .1745688    -3.34   0.001    -.9247816    -.2404844
        age |  -.0242829    .004494    -5.40   0.000     -.033091    -.0154748
       _cons |   2.474729   .2551222     9.70   0.000     1.974699     2.97476
------------------------------------------------------------------------------
```

Interpreting the Output

Just below the command syntax, there is a note for i.marital. It shows the indicator variables named _Imarital_1-5, and the first dummy variable is omitted from the model (naturally coded; _Imarital_1 omitted). In the logistic regression table, the first column lists the intercept and the predictors. We can see four dummy variables, including _Imarital_2, _Imarital_3, _Imarital_4, and Imarital_5, which all begin with "_I". The reference level for the categorical variable marital is 1, so the dummy variable _Imarital_1 is omitted from the model.

The maximum log likelihood value for the model with all predictors is −734.755. The log likelihood ratio chi-square test statistic LR chi2(5) = 60.7, and the associated p value Prob > chi2 = 0.0001. The log likelihood ratio test statistic is the difference in the −2 log likelihood (−2LL) between the current model that contains all five predictors and the intercept, as well as the null model that contains only the intercept. It follows a chi-square distribution with the degrees of freedom equal to the difference in the predictors between these two models. In the output, the difference in the log likelihood statistic was 60.70 with the degrees of freedom equal to 5.

The null hypothesis of the log likelihood ratio chi-square test is that the overall model with all predictor variables is not significantly better than the model with only the intercept.

The alternative hypothesis is that the overall model is significantly better than the model with only the intercept.

LR $\chi^2_{(5)}$ = 60.70, p < .001, which indicates that the null hypothesis is rejected. Therefore, the overall model with all five predictors is significant.

Pseudo R^2 = 0.0397, which is the likelihood ratio R^2 or McFadden's R^2:

$$R^2_L = 1 - \frac{-2LL_m}{-2LL_0} = 1 - \frac{LL_m}{LL_0} = 1 - (-734.75476/-765.10363) = .0397.$$

Please note that a pseudo R^2 measure should not be interpreted as variance explained by the predictors in the model as its analog in multiple linear regression. When it is used for model comparisons, instead of depending on this single statistic, it should be used together with other fit statistics.

The omnibus test indicates that the overall model with five predictors is significant. Next, we are interested in whether each predictor significantly contributes to the model. We can find the results from the logistic regression table, which displays the parameter estimates for the predictor variables and intercept, their standard errors, the Wald z statistics, the associated p values, and the 95% confident intervals of the parameter estimates for the constant and predictors.

The null hypothesis for the Wald test for each predictor is that the coefficient of the predictor variable is zero, and the alternative hypothesis is that the coefficient of the predictor variable is significantly different from zero.

The Wald z statistic equals the parameter estimate divided by its standard error. For the age predictor, the Wald z = −.0242829/.004494 = −5.40. The associated p value P>|z|=.000, so we reject the null hypothesis. Therefore, age is a significant predictor of the dependent variable health status.

Next, let us take a look at the dummy variables. For the categorical variable marital, four dummy variables are automatically created in the model. The first dummy variable _Imarital_1 is omitted from the model since the first level is the reference level. For _Imarital_2, the Wald $z = -1.60$, $p = .109$, so the effect of this dummy variable is not significant. The Wald z statistics for the other three dummy variables are -2.26, -3.79, and -3.34, respectively. The associated p values to these z statistics are less than .05, which indicates that these three predictors are significant.

By substituting the estimated intercept and coefficients into the multiple logistic regression equation, we can write the model as:

$$\text{logit}(\pi) = 2.475 - .394 \times \_Imarital\_2 - .406 \times \_Imarital\_3 - 1.215 \times \_Imarital\_4 - .583 \times \_Imarital\_5 - .024 \times age$$

The regression coefficients of the predictors in multiple logistic regression are the logit coefficients since they are on the scale of the logit or log odds. They are the partial effects of the predictor variables on the logit or log odds of the outcome variable (having good health). Each logit coefficient can be interpreted as the change in the logit or log odds of having good health for each one-unit increase in the predictor when controlling for the effects of other variables. A positive logit coefficient indicates there is an increase in the logit coefficient for a one-unit increase in the predictor, whereas a negative logit coefficient indicates that there is a decrease in the logit coefficient for a one-unit increase in the predictor when controlling for other predictors.

Interpretations using the odds ratios are more popular than logit coefficients. Adding the or option to the original command gives us the exponentiated logits, the odds ratios. The following is the output displaying the odds ratio.

```
. xi: logit healths i.marital age, or
i.marital          _Imarital_1-5        (naturally coded; _Imarital_1 omitted)

Iteration 0:    log likelihood = -765.10363
Iteration 1:    log likelihood = -735.17599
Iteration 2:    log likelihood = -734.75481
Iteration 3:    log likelihood = -734.75476


Logistic regression                             Number of obs   =        1300
                                                LR chi2(5)      =       60.70
                                                Prob > chi2     =      0.0000
Log likelihood = -734.75476                     Pseudo R2       =      0.0397


------------------------------------------------------------------------------
     healths | Odds Ratio   Std. Err.      z    P>|z|     [95% Conf. Interval]
-------------+----------------------------------------------------------------
  _Imarital_2 |   .6745877   .1655966    -1.60   0.109     .4169529    1.091415
  _Imarital_3 |   .6661147   .1198999    -2.26   0.024     .4680946     .947904
  _Imarital_4 |   .2967452   .0951702    -3.79   0.000     .1582676     .556385
  _Imarital_5 |   .5584261   .0974838    -3.34   0.001      .396618    .7862469
         age |   .9760095   .0043862    -5.40   0.000     .9674505    .9846443
------------------------------------------------------------------------------
```

Interpretation of Odds Ratios

For the dummy variable _Imarital_2, odds ratio = .675. Since the reference level is level 1 (married), the odds ratio compares level 2 with level 1, which is the ratio of the odds for level 2 to the odds for level 1. It can be interpreted that the odds of having good health for the widowed are .675 times as large as the odds for the married when holding other predictors constant. However, this odds ratio is not statistically significant.

For the dummy variable _Imarital_3, odds ratio = .666. Since it is the ratio of the odds for level 3 to the odds for level 1, it indicates that the odds of having good health for the divorced are .675 times the odds for the married when holding other predictors constant. It can also be interpreted in the opposite way: The odds of having good health for the married are 1.482 times the odds for the divorced if we take the inverse of the odds ratio of .666.

For the dummy variable _Imarital_4, odds ratio = .297. Since it is the ratio of the odds for level 4 to the odds for level 1, it can be interpreted as follows: The odds of having good health for the separated are .297 times the odds for the married when holding other predictors constant. In others words, the odds of having good health for the married are 3.370 times the odds for the divorced if we take the inverse of the odds ratio of .297.

For the dummy variable _Imarital_5, odds ratio = .558. Since it is the ratio of the odds for level 5 to the odds for level 1, it indicates that the odds of having good health for those who never married are .558 times the odds for the married when holding other predictors constant. It can also be interpreted that the odds of having good health for the married are 1.791 times as large as the odds for those who never married if we take the inverse of the odds ratio of .558.

The predictor age is a continuous variable. Odds ratio = .976, which is less than 1. Recall that when an odds ratio is less than 1, it is interpreted as a decrease in the odds of success for each one-unit increase in the predictor when holding other variables constant. The odds ratio for age can be interpreted that for each one-unit increase in age, the odds of having good health decrease by .976.

The fit statistics of the multiple logistic model can also be computed using the fitstat command. The output is displayed as follows.

```
. fitstat

                          |        logit
--------------------------+--------------
Log-likelihood            |
                  Model   |    -734.755
          Intercept-only  |    -765.104
--------------------------+--------------
Chi-square                |
      Deviance (df=1294)  |    1469.510
```

```
            LR (df=5) |       60.698
              p-value |        0.000
--------------------------+-------------
R2                        |
             McFadden |        0.040
   McFadden (adjusted) |        0.032
     McKelvey & Zavoina |      0.066
           Cox-Snell/ML |       0.046
  Cragg-Uhler/Nagelkerke |      0.066
                Efron |        0.046
             Tjur's D |        0.047
                Count |        0.727
        Count (adjusted) |      0.008
--------------------------+-------------
IC                        |
                  AIC |     1481.510
        AIC divided by N |      1.140
            BIC (df=6) |     1512.530
--------------------------+-------------
Variance of               |
                    e |        3.290
               y-star |        3.524
```

Classification Table

The command `estat class` is a postestimation tool for logistic regression. It provides the classification table and summary statistics, such as sensitivity and specificity of the model. The output created by the command is displayed as follows.

```
. estat class
Logistic model for healths

                -------- True --------
Classified |        D            ~D  |      Total
-----------+----------------------------+-----------
      +    |      922           335  |       1257
      -    |       20            23  |         43
-----------+----------------------------+-----------
   Total   |      942           358  |       1300

Classified + if predicted Pr(D) >= .5
True D defined as healths != 0
--------------------------------------------------------
Sensitivity                      Pr( +| D)    97.88%
Specificity                      Pr( -|~D)     6.42%
Positive predictive value        Pr( D| +)    73.35%
Negative predictive value        Pr(~D| -)    53.49%
--------------------------------------------------------
```

```
False + rate for true ~D        Pr( +|~D)    93.58%
False - rate for true D         Pr( -| D)     2.12%
False + rate for classified +   Pr(~D| +)    26.65%
False - rate for classified -   Pr( D| -)    46.51%
-----------------------------------------------------
Correctly classified                         72.69%
```

In the classification table, "D" means event occurring or when the outcome variable takes the value of 1 (i.e., $Y = 1$), whereas "~D" means no event occurring or when the outcome variable takes the value of 0 (i.e., $Y = 0$). "Classified +" means the model correctly predicts the category when the predicted probability is larger than or equal to .5 (predicted Pr(D) >= .5), whereas "Classified -" means that the model predicts the wrong category if the predicted probability is less than .5. The overall percentage of correctly classified = 922/1300 = 72.69%, which is shown at the bottom of the output. This indicates when the predictor variable marital status is included in the model; 72.69% of the cases are correctly predicted by the model. The sensitivity of the model is the conditional probability of the cases that are correctly classified given an event occurs. In this example, it is the probability of the cases that are correctly classified (97.88%) for the people having good health ($Y = 1$). The specificity is the percentage of the cases that were incorrectly classified for those having no good health ($Y = 0$). In this example, the specificity = 6.42%.

Hosmer–Lemeshow Goodness-of-Fit Test

The command estat gof is a postestimation command for the Hosmer–Lemeshow goodness-of-fit test statistic. The group(10) option requests that the sample be divided into 10 groups according to the size of the predicted probabilities. Although 10 groups are normally chosen, you can specify fewer groups. Without this option, the sample will be grouped by the covariate patterns, and you might see long output with more groups, which might be undesirable. The table option requests that the results are displayed in a table. The results are displayed as follows.

```
. estat gof,group(10) table
Logistic model for healths, goodness-of-fit test

(Table collapsed on quantiles of estimated probabilities)
+------------------------------------------------------------+
| Group |   Prob | Obs_1 | Exp_1 | Obs_0 | Exp_0 | Total |
|-------+--------+-------+-------+-------+-------+-------|
|     1 | 0.5845 |    70 |  67.4 |    60 |  62.6 |   130 |
|     2 | 0.6468 |    91 |  81.5 |    41 |  50.5 |   132 |
|     3 | 0.6846 |    75 |  87.3 |    56 |  43.7 |   131 |
```

```
|      4 | 0.7201 |     88 |  90.9 |    41 |  38.1 |    129 |
|      5 | 0.7491 |     90 |  94.1 |    38 |  33.9 |    128 |
|--------+--------+--------+-------+-------+-------+--------|
|      6 | 0.7707 |    103 | 105.9 |    36 |  33.1 |    139 |
|      7 | 0.7874 |     98 |  97.7 |    27 |  27.3 |    125 |
|      8 | 0.8070 |    113 | 106.1 |    20 |  26.9 |    133 |
|      9 | 0.8321 |    108 | 105.6 |    21 |  23.4 |    129 |
|     10 | 0.8771 |    106 | 105.4 |    18 |  18.6 |    124 |
+---------------------------------------------------------+

        number of observations =      1300
             number of groups =        10
      Hosmer-Lemeshow chi2(8) =        12.14
                 Prob > chi2 =         0.1451
```

The first column lists 10 groups. The second column contains the estimated probabilities for these groups, which are ordered in sequence. The third column Obs_1 lists the number of observations for the outcome variable with the value of 1, and the fourth column Exp_1 is the expected number of frequencies for the event occurring. The last two columns are Obs_0 and Exp_0, respectively, which are the observed and the expected frequencies of cases that have no event. Within each group, the frequencies of the observed cases are compared with the expected frequencies, and we would like to see small discrepancies between them. This test follows a chi-square distribution with the degrees of freedom equal to the number of groups minus 2. A nonsignificant p value indicates that the model fits the data well since there is no significant difference between the observed and expected data. In this example, the Hosmer–Lemeshow chi-square test has a value of 12.14 with the degrees of freedom equal to 8. The associated p value is .1451, which is not significant. Therefore, the model fits the data well.

Next, we fit the full model by adding two more predictors, educ and male. The command is:

```
xi: logit healths i.marital age educ male
```

```
. xi: logit healths i.marital age educ male
i.marital           _Imarital_1-5        (naturally coded; _Imarital_1 omitted)

Iteration 0:   log likelihood = -765.10363
Iteration 1:   log likelihood = -702.48399
Iteration 2:   log likelihood = -701.14964
Iteration 3:   log likelihood = -701.14743

Logistic regression                          Number of obs   =       1300
                                             LR chi2(7)      =     127.91
                                             Prob > chi2     =     0.0000
Log likelihood = -701.14743                  Pseudo R2       =     0.0836
```

```
------------------------------------------------------------------------
    healths |     Coef.    Std. Err.      z    P>|z|    [95% Conf. Interval]
------------+-----------------------------------------------------------
 _Imarital_2 |  -.2708026   .2548324   -1.06   0.288   -.770265    .2286598
 _Imarital_3 |   -.386311   .1843546   -2.10   0.036  -.7476395   -.0249825
 _Imarital_4 |  -1.142387   .3375442   -3.38   0.001  -1.803962   -.4808128
 _Imarital_5 |   -.534184   .1796845   -2.97   0.003  -.8863591   -.1820088
        age |  -.0231949   .0045906   -5.05   0.000  -.0321924   -.0141975
       educ |   .1718877   .0219504    7.83   0.000   .1288657    .2149097
       male |   .0059245   .1325964    0.04   0.964  -.2539596    .2658087
       _cons |   .1240643   .3904797    0.32   0.751  -.6412619    .8893906
------------------------------------------------------------------------
```

$LR \ \chi^2_{(7)} = 127.91$, $p < .001$, which indicates the overall model with all seven predictors is significant. The likelihood ratio R^2, listed as the Pseudo $R^2 = .0836$, is larger than that in the previous model (Pseudo $R^2 = .040$ for the two-predictor model). It seems that the full model fits better than the two-predictor model, but we also need to look at other fit statistics.

The Wald tests for the four dummy variables and age have similar results as those in the two-predictor model.

For the new predictor variable educ, the Wald $z = .1718877/.0219504 = 7.83$. The associated p value, P>|z|=.000, so we reject the null hypothesis. Therefore, the predictor variable educ is a significant predictor of the dependent variable health status.

For the other new predictor variable male, the Wald $z = .0059245/.1325964 = .04$. The associated p value P>|z|=.964, so we fail to reject the null hypothesis and conclude that there is no significant effect of male on the outcome variable. In other words, whether a person is a male or female does not significantly predict whether that person has good health.

To get the odds ratio for all the predictor variables, we run the following command with the or option:

```
xi: logit healths i.marital age educ male, or
```

```
. xi: logit healths i.marital age educ male, or
i.marital          _Imarital_1-5         (naturally coded; _Imarital_1 omitted)

Iteration 0:   log likelihood = -765.10363
Iteration 1:   log likelihood = -702.48399
Iteration 2:   log likelihood = -701.14964
Iteration 3:   log likelihood = -701.14743

Logistic regression                             Number of obs   =       1300
                                                LR chi2(7)      =     127.91
                                                Prob > chi2     =     0.0000
Log likelihood = -701.14743                     Pseudo R2       =     0.0836
```

```
----------------------------------------------------------------------------
   healths | Odds Ratio   Std. Err.      z    P>|z|     [95% Conf. Interval]
-----------+----------------------------------------------------------------
_Imarital_2 |   .7627671   .1943778   -1.06   0.288     .4628904   1.256914
_Imarital_3 |   .6795592   .1252799   -2.10   0.036     .4734829   .9753269
_Imarital_4 |   .3190565   .1076957   -3.38   0.001     .1646453   .6182807
_Imarital_5 |   .5861474   .1053216   -2.97   0.003     .4121536   .833594
       age |    .977072   .0044854   -5.05   0.000     .9683203   .9859028
      educ |   1.187544   .0260671    7.83   0.000     1.137537   1.23975
      male |   1.005942   .1333843    0.04   0.964     .7757231   1.304485
----------------------------------------------------------------------------
```

The odds ratios for the four dummy variables and age look similar to those in the two-predictor model, so they can be interpreted in a similar way.

The new predictor educ is a continuous variable. Odds ratio = 1.188, which is larger than 1. This indicates that for each one-unit increase in education, the odds of having good health increase by 1.188. Another interpretation is the percentage change in odds. It can be calculated by using (Odds ratio − 1) × 100%. In this example, (1.188 − 1) × 100% = 18.8%. So for each one-unit increase in education, there is an increase of 18.8% in the odds of having good health.

The other new predictor male is a dichotomous variable. Odds ratio = 1.006, p = .964, which is not significant. When an odds ratio takes a value of 1, it means that there is no significant effect of the predictor variable on the odds of success when holding other variables constant. Therefore, being a male or not does not significantly influence whether that person has good health.

The estat class command provides the classification table, which is displayed as follows.

```
. estat class
Logistic model for healths

              -------- True --------
Classified |         D             ~D  |      Total
-----------+--------------------------+-----------
     +     |        901           302  |       1203
     -     |         41            56  |         97
-----------+--------------------------+-----------
   Total   |        942           358  |       1300

Classified + if predicted Pr(D) >= .5
True D defined as healths != 0
--------------------------------------------------
Sensitivity                     Pr( +| D)   95.65%
Specificity                     Pr( -|~D)   15.64%
Positive predictive value       Pr( D| +)   74.90%
Negative predictive value       Pr(~D| -)   57.73%
--------------------------------------------------
```

```
False + rate for true ~D          Pr( +|~D)    84.36%
False - rate for true D           Pr( -| D)     4.35%
False + rate for classified +     Pr(~D| +)    25.10%
False - rate for classified -     Pr( D| -)    42.27%
-------------------------------------------------------
Correctly classified                           73.62%
-------------------------------------------------------
```

After adding two more predictors, the overall percentage of correctly classified is 73.62, which is slightly higher than that for the previous model (72.69%). The sensitivity of the model shows that the probability of the cases that are correctly classified for the people having good health ($Y = 1$) is 95.65%. The specificity shows that the percentage of the cases that were incorrectly classified for those having no good health ($Y = 0$) is 15.64%.

The command estat gof requests the Hosmer–Lemeshow goodness-of-fit test statistic. The output is displayed as follows.

```
. estat gof, group(10) table
Logistic model for healths, goodness-of-fit test

(Table collapsed on quantiles of estimated probabilities)
+-----------------------------------------------------------+
| Group |  Prob  | Obs_1 | Exp_1 | Obs_0 | Exp_0 | Total |
|-------+--------+-------+-------+-------+-------+-------|
|     1 | 0.5438 |    51 |  54.8 |    79 |  75.2 |   130 |
|     2 | 0.6208 |    83 |  77.5 |    49 |  54.5 |   132 |
|     3 | 0.6727 |    80 |  82.8 |    48 |  45.2 |   128 |
|     4 | 0.7194 |    82 |  90.4 |    48 |  39.6 |   130 |
|     5 | 0.7497 |    96 |  95.7 |    34 |  34.3 |   130 |
|-------+--------+-------+-------+-------+-------+-------|
|     6 | 0.7805 |    94 | 100.4 |    37 |  30.6 |   131 |
|     7 | 0.8146 |   114 | 102.9 |    15 |  26.1 |   129 |
|     8 | 0.8450 |   113 | 107.9 |    17 |  22.1 |   130 |
|     9 | 0.8786 |   111 | 112.0 |    19 |  18.0 |   130 |
|    10 | 0.9453 |   118 | 117.6 |    12 |  12.4 |   130 |
+-----------------------------------------------------------+

         number of observations =     1300
              number of groups =       10
    Hosmer-Lemeshow chi2(8) =         13.41
               Prob > chi2 =          0.0985
```

In the output, the Hosmer–Lemeshow chi-square test has a value of 13.41 with the degrees of freedom equal to 8. The associated p value = .099, which is not significant. Therefore, the model fits the data well.

To identify which model fits the data better, the log likelihood ratio test or the deviance difference test can be used. Recall that this test compares the reduced model, which contains fewer parameters, and the full model, which contains all parameters. The difference in deviance is often expressed as G = Deviance for the reduced model − Deviance for the full model or as $D_{\text{Reduced}} - D_{\text{Full}}$. The difference in deviance between nested models has a chi-square distribution with the degrees of freedom equal to the difference in the number of parameters between these two models.

Likelihood Ratio Test

The `lrtest` command is used for the log likelihood ratio test or the deviance difference test. Before using this command, we need to use the `estimates store` command to store the results of the fitted models.

```
1. . quietly xi: logit healths i.marital age
2. . estimates store model1
3. . quietly xi: logit healths i.marital age educ male
4. . estimates store fullmodel
5. . lrtest model1 fullmodel
Likelihood-ratio test                              LR chi2(2)  =     67.21
(Assumption: model1 nested in fullmodel)           Prob > chi2 =    0.0000
```

The first command `quietly xi: logit healths i.marital age` tells Stata to fit the two-predictor model quietly without showing the output. The command `estimates store model1` tells Stata to store the estimated log likelihood statistic for the preceding model with a name of "model1". Next, the command `quietly xi: logit healths i.marital age educ male` tells Stata to fit the full model quietly without providing the output. By using the command `estimates store fullmodel`, we store the estimated log likelihood statistics for the full model and name it "fullmodel". Finally, the command `lrtest model1 fullmodel` performs the log likelihood ratio test comparing the two models. If you do not specify `fullmodel` in the command, the latest estimation results will be used for the comparison, so the command `lrtest model1` compares model1 with the latest model.

The log likelihood ratio chi-square test $\chi^2_{(2)} = 67.21$, $p < .001$. The same result can be obtained if you compute it using the following equation:

$$G = D_{\text{Reduced}} - D_{\text{Full}} = -2 \times [-734.755 - (-701.147)] = 67.216, \text{ df} = 7 - 5 = 2$$

We can also use the `display` command to ease your computation since it can be used as a calculator.

```
. display -2*[-734.755 -(-701.147)]
67.216
```

Recall that the `fitstat` command provides more fit statistics of the model. We can also compare them between the nested models. The procedure is to fit the reduced model first, save the model fit statistics using the `fitstat` command, and then fit the full model. Finally, compare the fit statistics between the two models.

The command `quietly xi: logit healths i.marital age` quietly fits the first model without showing the output. The command `quietly fitstat, saving(mod1)` requests the fit statistics and saves the results with the name "mod1". Keep in mind the name cannot be more than five characters, so give it a short name. The next command `quietly xi: logit healths i.marital age educ male` fits the full model without showing the output. The fourth command `quietly fitstat, saving(full)` saves the results with a name of "full". The command actually can be omitted, but it is good to have it so you know what analysis you are doing. The final command `fitstat, using(mod1)` compares the final model (full) with the reduced model (mod1).

```
. quietly xi: logit healths i.marital age

. quietly fitstat, saving(mod1)

. quietly xi: logit healths i.marital age educ male

. quietly fitstat, saving(full)

. fitstat, using(mod1)
```

	Current	Saved	Difference
Log-likelihood			
Model	-701.147	-734.755	33.607
Intercept-only	-765.104	-765.104	0.000
Chi-square			
D (df=1292/1294/-2)	1402.295	1469.510	-67.215
LR (df=7/5/2)	127.912	60.698	67.215
p-value	0.000	0.000	0.000
R2			
McFadden	0.084	0.040	0.044
McFadden (adjusted)	0.073	0.032	0.041
McKelvey & Zavoina	0.143	0.066	0.077
Cox-Snell/ML	0.094	0.046	0.048

```
Cragg-Uhler/Nagelkerke |      0.135          0.066          0.070
              Efron |      0.099          0.046          0.052
           Tjur's D |      0.099          0.047          0.052
              Count |      0.736          0.727          0.009
     Count (adjusted) |    0.042          0.008          0.034
---------------------+-------------------------------------------
IC                   |
                AIC |   1418.295       1481.510        -63.215
       AIC divided by N |     1.091          1.140         -0.049
        BIC (df=8/6/2) |   1459.656       1512.530        -52.874
---------------------+-------------------------------------------
Variance of          |
                  e |      3.290          3.290          0.000
             y-star |      3.838          3.524          0.315

Note: Likelihood-ratio test assumes saved model nested in current model.

Difference of   52.874 in BIC provides very strong support for current model.
```

The first column of the table in the preceding box displays the names of various model fit statistics, the second column displays the values of model fit statistics for the final model (labeled "Current"), the third column shows those for the two-predictor model (labeled "Saved"), and the last column displays the differences in the fit statistics between these two models (labeled "Difference").

The deviance difference (labeled "D") equals 67.215 with df = 2. The difference between the log likelihood ratio test (labeled "LR") reports the same result as the value of deviance difference. All pseudo R^2 values improve from the two-predictor model to the full model. For example, the likelihood ratio R^2 (labeled McFadden's R^2) = .04 for the one-predictor model and .084 for the full model. The difference is .044, which indicates the full model has a better fit than the two-predictor model. The AIC and BIC statistics decrease by −63.215 and −52.874, respectively. Other versions of AIC and BIC statistics also show decreases from the two-predictor model to the full model. Since smaller AIC and BIC statistics mean a better model fit, all these results support that the full model fits the data better.

Interpreting the Estimated Probabilities With the margins Command

The margins command computes the margins or the predicted probabilities when the outcome variable $Y = 1$ for predictor variables at specified values. This command only works in Stata 11 or later versions. Please also note that it currently does not work with the xi prefix command in either Stata 13 or 14. An error message will be displayed when the margins command is used following a regression model with the xi prefix. The correct way is to use the factor variable notation with the i. prefix when there are category variables in the regression model.

The command margins, at(educ = (8 13 16)) atmeans vsquish tells Stata to compute the estimated probabilities when the binary outcome variable $Y = 1$ for the predictor variable educ at the values of 8, 13, and 16 while holding the other predictor variables at their means. The at(educ = (8 13 16)) option specifies the values for the predictor variable educ. The atmeans option specifies the values of other predictor variables at their means. The vsquish option removes the blank space between factor variables so that the output looks compact.

```
. quietly logit healths i.marital age educ male

. margins, at(educ = (8 13 16)) atmeans vsquish

Adjusted predictions                          Number of obs    =        1300
Model VCE     : OIM

Expression    : Pr(healths), predict()
1._at         : 1.marital         =      .4684615 (mean)
                2.marital         =      .0776923 (mean)
                3.marital         =      .1592308 (mean)
                4.marital         =      .0338462 (mean)
                5.marital         =      .2607692 (mean)
                age               =          48.2 (mean)
                educ              =             8
                male              =      .4438462 (mean)
2._at         : 1.marital         =      .4684615 (mean)
                2.marital         =      .0776923 (mean)
                3.marital         =      .1592308 (mean)
                4.marital         =      .0338462 (mean)
                5.marital         =      .2607692 (mean)
                age               =          48.2 (mean)
                educ              =            13
                male              =      .4438462 (mean)
3._at         : 1.marital         =      .4684615 (mean)
                2.marital         =      .0776923 (mean)
                3.marital         =      .1592308 (mean)
                4.marital         =      .0338462 (mean)
                5.marital         =      .2607692 (mean)
                age               =          48.2 (mean)
                educ              =            16
                male              =      .4438462 (mean)

------------------------------------------------------------------------------
             |            Delta-method
             |     Margin   Std. Err.      z    P>|z|     [95% Conf. Interval]
-------------+----------------------------------------------------------------
         _at |
          1  |   .5307819   .0315467    16.83   0.000     .4689514    .5926124
          2  |    .727648   .0131862    55.18   0.000     .7018034    .7534925
          3  |   .8173335   .0141801    57.64   0.000     .7895411    .8451259
------------------------------------------------------------------------------
```

The first section of the output displays three combinations of the values for all predictor variables. Among the first combination (labeled `1._at`), the predictor `marital` is a factor variable, which creates indicator variables, so all five categories are displayed. `1.marital = .4684615 (mean)`, which indicates that 47% of the respondents were married and is coded as category 1. The average age is 48.2 (`age = 48.2 (mean)`), `educ = 8`, and 44% of the respondents were male (`male = .4438462 (mean)`). The second and third combinations can be interpreted in a similar way.

The second section of the output located at the bottom displays the estimated margins, their Delta-method standard errors, z statistics and the associated p values, and the 95% confidence intervals.

- When `educ = 8`, and other predictor variables are held at their means (`1.marital = .468`, `2.marital = .078`, `3.marital = .159`, `4.marital =.034`, `5.marital = .261`, `age = 48.2`, and `male =.444`), the estimated margin or the probability when $Y = 1$ is .531.
- When `educ = 13`, and other predictor variables are held at their means, the estimated probability when $Y = 1$ is .728.
- When `educ = 16`, and other predictor variables are held at their means, the estimated probability when $Y = 1$ is .817.

These estimated probabilities can be plotted (Figure 3.1) using the `marginsplot` command as follows:

```
.marginsplot
```

The `marginsplot` command is used to plot the estimated results from the preceding `margins` command. For more functions of this command, type `help marginsplot`, or refer to Mitchell (2012a, 2012b) for various examples.

Figure 3.1 shows that with the increase of education, the probabilities of having good health ($Y =1$) increase.

3.4 Making Publication-Quality Tables

```
1.  .quietly xi:logit healths i.marital age
2.  .outreg2 using chap3out, e(ll df_m chi2) addstat(Pseudo R-squared,
    `e(r2_p)')word replace
3.  .outreg2 using chap3out, eform word append
4.  .quietly xi: logit healths i.marital age educ i.male
5.  .outreg2 using chap3out, e(ll df_m chi2) addstat(Pseudo R-squared, `e(r2_p)')
    word append
6.  .outreg2 using chap3out, eform word append
7.  .seeout
```

Figure 3.1 Estimated Probabilities When $Y = 1$ for Educ at 8, 13, and 16 With Others Fixed at Their Means

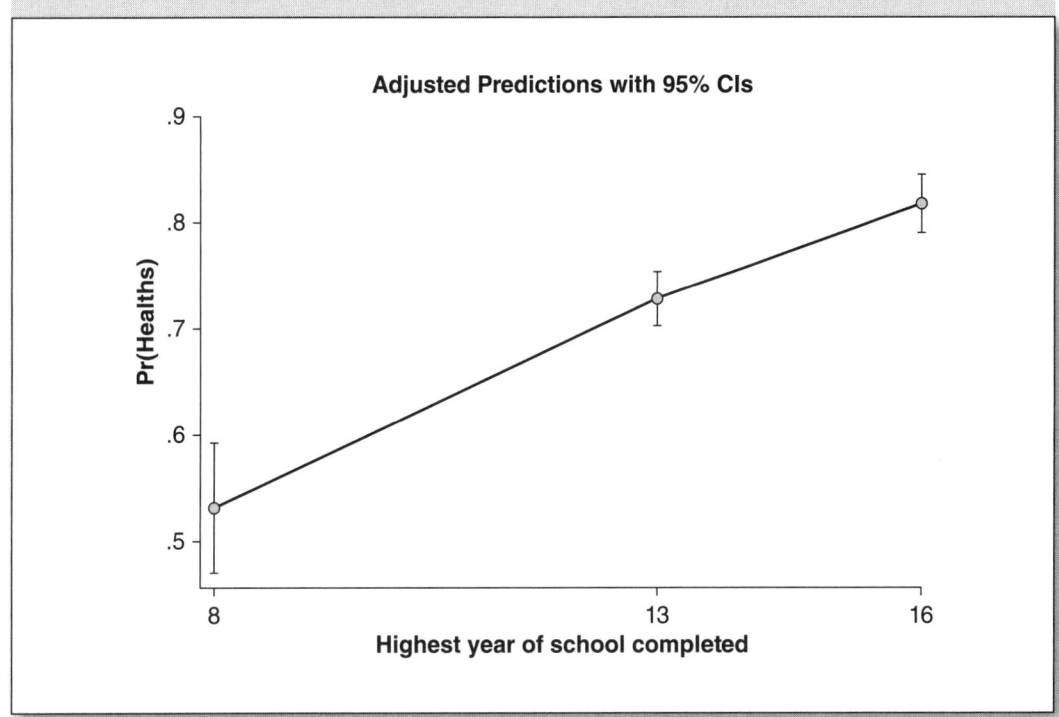

The first command `quietly xi:logit healths i.marital age` estimates the model without providing the output.

The second command `outreg2 using chap3out, e(ll df_m chi2)` `addstat(Pseudo R-squared, `e(r2_p)')word replace` tells Stata to create a regression table for the estimated logistic regression model. In the syntax, the command `outreg2` is used to create the table; the command `using chap3out` creates a file named "chap3out" containing the table. This file is saved in the current working folder where the dataset is located. With the combination of the `word` option, the table is created in the Microsoft Word document with the extension .rtf. Other formats such as .txt file, Microsoft Excel files, or LaTex files can also be requested. The `(ll df_m chi2)` option requests that the estimated log likelihood value, the degrees of freedom, and the chi-square test be listed in the table. The `addstat` option asks Stata to add the pseudo $R^2$ to the table. The `replace` option asks Stata to replace any files with the same name as chap3out.

The third command `outreg2 using chap3out, eform word append` tells Stata to add the estimated odds ratios to the table as a separate column by using the `eform` option. Please note that when requesting the odds ratios, the `or` option does not work here. The `append` option appends the results to the original table without replacing it.

The fourth command `quietly xi: logit healths i.marital age educ i.male` estimates the full model with four predictors without showing the output.

The fifth command `outreg2 using chap3out, e(ll df_m chi2) addstat (Pseudo R-squared, `e(r2_p)')` word append` creates a regression table for the full model and appends it to the original table.

The sixth command `outreg2 using chap3out, eform word append` reports the odds ratios of the estimates for the full model and appends the results to the original table by having an extra column.

The seventh command `seeout`, which is the last one, opens the final table in the data browser. By clicking the blue link in the Stata results window, you can see the created table, which is saved in the current working directory. The default is the tab delimited text file with the .txt extension.

These commands automatically produce Table 3.5, as shown here in its original format, presenting the results of both the two-predictor and the full logistic regression model.

With some edits to the column titles, names of the model fit statistics, and decimals of the numbers, Table 3.6 is the final table.

Table 3.5 Results of the Logistic Regression Models: Two-Predictor and Full Models (Shown in Original Format Generated by Stata)

Variables	(1) healths	(2) healths	(3) healths	(4) healths
_Imarital_2	−0.394	0.675	−0.271	0.763
	(0.245)	(0.166)	(0.255)	(0.194)
_Imarital_3	−0.406**	0.666**	−0.386**	0.680**
	(0.180)	(0.120)	(0.184)	(0.125)
_Imarital_4	−1.215***	0.297***	−1.142***	0.319***
	(0.321)	(0.0952)	(0.338)	(0.108)
_Imarital_5	−0.583***	0.558***	−0.534***	0.586***
	(0.175)	(0.0975)	(0.180)	(0.105)
Age	−0.0243***	0.976***	−0.0232***	0.977***
	(0.00449)	(0.00439)	(0.00459)	(0.00449)
Educ			0.172***	1.188***
			(0.0220)	(0.0261)

_Imale_1			0.00592	1.006
			(0.133)	(0.133)
Constant	2.475***	11.88***	0.124	1.132
	(0.255)	(3.030)	(0.390)	(0.442)
Observations	1,300	1,300	1,300	1,300
Pseudo R-squared	0.0397		0.0836	
Ll	−734.8		−701.1	
df_m	5		7	
chi2	60.70		127.9	

Note: Standard errors in parentheses.

***$p < 0.01$, **$p < 0.05$, *$p < 0.1$

Table 3.6 Results of the Logistic Regression Models: Two-Predictor and Full Models (Edited)

	Two-Predictor Model		Full Model	
Variables	**b (SE(b))**	**OR**	**b (SE(b))**	**OR**
_Imarital_2	−0.394	0.675	−0.271	0.763
	(0.245)	(0.166)	(0.255)	(0.194)
_Imarital_3	−0.406**	0.666**	−0.386**	0.680**
	(0.180)	(0.120)	(0.184)	(0.125)
_Imarital_4	−1.215***	0.297***	−1.142***	0.319***
	(0.321)	(0.095)	(0.338)	(0.108)
_Imarital_5	−0.583***	0.558***	−0.534***	0.586***
	(0.175)	(0.098)	(0.180)	(0.105)

(Continued)

Table 3.6   (Continued)

Variables	Two-Predictor Model		Full Model	
	b (SE(b))	OR	b (SE(b))	OR
Age	−0.024***	0.976***	−0.023***	0.977***
	(0.004)	(0.004)	(0.005)	(0.004)
Educ			0.172***	1.188***
			(0.022)	(0.026)
_Imale_1			0.006	1.006
			(0.133)	(0.133)
Constant	2.475***	11.88***	0.124	1.132
	(0.255)	(3.030)	(0.390)	(0.442)
Observations	1,300	1,300	1,300	1,300
LR R^2	0.040		0.084	
Log likelihood	−734.8		−701.1	
df_m	5		7	
LR χ^2	$\chi^2_5 = 60.70$		$\chi^2_7 = 127.9$	

Note: Standard errors are shown in parentheses.

**p < .05, ***p < .01.

3.5 Reporting the Results

The multiple logistic regression analysis was conducted to estimate the probability of having good health from four predictor variables. The dependent variable was having good health or not, and the independent variables were marital status, age, years of education, and gender. The marital status was a categorical variable with five levels, so dummy coding was used for the following models. A two-predictor model with marital status and age as the predictor variables were fitted first. The full model with all four predicted was fitted next. The log likelihood ratio chi-square test statistic for the two-predictor model LR $\chi^2_{(5)}$ = 60.70, p < .001, indicated that the overall model with all five predictors (including dummy variables for marital status) was significant. The log likelihood ratio chi-square test statistic for the full model LR $\chi^2_{(7)}$ = 127.91, p < .001, indicated that the overall model with all the predictors was significant. When comparing the two models, the log likelihood ratio chi-square test $\chi^2_{(2)}$ = 67.21, p < .001, indicated the full model had a better fit.

Table 3.6 presents the logit coefficients, standard errors, and odds ratios for the full model. The `marital` predictor was a categorical variable with five levels. In the model, four dummy variables were automatically created with level 1 (married) as the reference level. For the dummy variable `_Imarital_2`, odds ratio = .763, which indicated that the odds of having good health for the widowed were .763 times the odds for the married when holding other predictors constant.

For the dummy variable `_Imarital_3`, odds ratio = .680, which indicated that the odds of having good health for the divorced were .680 times the odds for the married when holding other predictors constant. In other words, the odds of having good health for the married were 1.471 times the odds for the divorced.

For the dummy variable `_Imarital_4`, odds ratio = .319, which indicated that the odds of having good health for the separated were .319 times the odds for the married when holding other predictors constant.

For the dummy variable `_Imarital_5`, odds ratio = .586, which indicated that the odds of having good health for those who are not married were .586 times the odds for the married when holding other predictors constant.

For the `age` predictor, odds ratio = .977, which is less than 1, indicating that for each one-unit increase in age, the odds of having good health decreased by .976.

For the `educ` predictor, odds ratio = 1.188, which is larger than 1, indicating that for each one-unit increase in education, the odds of having good health increased by 1.188.

For the predictor `male`, odds ratio = 1.006, p = .964, which was not significant, indicating that being a male or not did not significantly influence whether that person has good health.

For more information on logistic regression models, consult Agresti (2002, 2007); Allison (1999, 2012); Azen and Walker (2011); Collet (2003); Cramer (2003); Demaris (1992); Hosmer, Lemeshow, and Sturdivant (2013); Kleinbaum and Klein (2010); Long (1997); McCullagh and Nelder (1989); Menard (2002, 2010); O'Connell and Amico (2010); Powers and Xie (2000); and Stokes, Davis, and Koch (2000). In addition, refer to Dupont (2009); Hamilton (2012); Hilbe (2009); Long and Freese (2006, 2014); Rabe-Hesketh and Everitt (2007); and Vittinghoff, Shiboski, Glidden, and McCulloch (2011) on the illustration of logistic regression using Stata. Also see Nagler (1994) and Hilbe (2009) on the scobit or skewed logistic model, which is an alternative to logistic regression when the probability of success deviates from .5. For interpretation of logistic regression models, see Gould (2000) and Liao (1994). Finally, in addition to `margins` in Stata, for more information on computing marginal effects and estimated probabilities with user-written programs, refer to Long and Freese (2014) for m* programs in SPost13, and Tomz, King, and Wittenberg (2003) for the Clarify program.

3.6 Summary of Stata Commands in This Chapter

```
use chap3-gss2012, clear

*cross tabulation
tab healths maritals

*Simple logistic regression
logit healths maritals
logit healths maritals, or
fitstat

*Multiple logistic regression-model 1
xi: logit healths i.marital age
xi: logit healths i.marital age, or
fitstat
estat class
estat gof,group(10) table

*Multiple logistic regression-model 2
xi: logit healths i.marital age educ male
xi: logit healths i.marital age educ male, or
fitstat
estat class
estat gof,group(10) table

*Model comparison using the log likelihood ratio test
quietly xi: logit healths i.marital age
estimates store model1
quietly xi: logit healths i.marital age educ male
estimates store fullmodel
lrtest model1 fullmodel

*Model comparison using other fit statistics
quietly xi: logit healths i.marital age
quietly fitstat, saving(mod1)
quietly xi: logit healths i.marital age educ male
quietly fitstat, saving(full)
fitstat, using(mod1)

*margins and marginsplot
quietly logit healths i.marital age educ male
margins, at(educ = (8 13 16)) atmeans vsquish
marginsplot

*Making publication-quality tables
quietly xi:logit healths i.marital age
outreg2 using chap3out, e(ll df_m chi2) ///
addstat(Pseudo R-squared, `e(r2_p)')word replace
```

ف

```
outreg2 using chap3out, eform word append
quietly xi: logit healths i.marital age educ i.male
outreg2 using chap3out, e(ll df_m chi2) ///
addstat(Pseudo R-squared, `e(r2_p)') word append
outreg2 using chap3out, eform word append
seeout
```

3.7 EXERCISES

Use the GSS 2012 data for the following problems.

1. Conduct a logistic regression analysis to examine the relationship between the outcome variable gunlaw and three predictor variables from three predictor variables sex, educ, and age. Before conducting the analysis, recode gunlaw into a new variable named gun so that 1 = favoring gun permit and 0 = opposing gun permit.

2. Identify the likelihood ratio test of the model, and interpret it.

3. Compute the deviance statistic for the model.

4. List two measures of pseudo R^2 and the AIC and BIC statistics computed by the fitstat command.

5. Identify the logit coefficient, the Wald z test, and the 95% confidence interval for the predictor variable educ.

6. Interpret the odds ratio for educ and age.

7. Make a publication-quality table containing the estimated logit coefficients and odds ratios.

8. Write a report to summarize the results from the output.

CHAPTER 4

Proportional Odds Models for Ordinal Response Variables

This chapter introduces proportional odds models for ordinal response variables. It starts with an introduction of the model followed by a discussion of the odds and odds ratios in the model, goodness-of-fit statistics of the model, the Brant test of the proportional odds assumption, and how to interpret parameter estimates. After a description of the data, two proportional odds models using Stata are illustrated with step-by-step instructions. Stata commands and output are explained in detail. This chapter focuses on fitting proportional odds models using Stata, as well as on interpreting and presenting the results. After reading this chapter, you should be able to

- Identify when a proportional model is used.
- Conduct proportional odds models, and test the assumption using Stata.
- Interpret the output.
- Interpret the model in terms of odds ratios.
- Compute and plot the estimated probabilities using the `margins` and `marginsplot` commands, respectively.
- Compute the estimated probabilities using the `mtable` command.
- Compare models using the likelihood ratio test and other fit statistics.
- Present results in publication-quality tables using Stata.
- Write the results for publication.

4.1 Proportional Odds Models: An Introduction

In the last chapter, we focused on binary logistic regression models when the outcome variable is dichotomous with values of 1 and 0. In your research, you may often encounter ordinal outcome variables, which are categorical variables with ranks or orders, for example, students' socioeconomic status ordered from low to high; children's proficiency in early reading scored from level 0 to 5; and a response scale of a survey instrument with five levels, ordered from strongly disagree to strongly agree. Let us look at some research examples illustrating the estimation of ordinal response variables in the literature:

1. Agresti (2010) predicted the ordinal response variable, a three-level response scale for an item related to astrology in the 2006 General Social Survey (GSS) from a single-predictor variable: respondent's highest degree. The three categories of the response were ordered as very scientific, sort of scientific, and not at all scientific.

2. Borooah (2002) predicted the ordinal response variable: the deprivation level with three categories (not deprived, mildly deprived, or severely deprived) from various risk factors. The predictor variables in the analysis included age, higher education, middle education, retirement status, being economically inactive, unemployment status, number of persons in a household, being a single parent, living in a particular area or not, gender, and religion.

3. Heck, Thomas, and Tabata (2012) investigated factors related to the ordinal response variable, student persistence through high school with three levels (i.e., dropped out, still in school but behind peers, and persisted). The six predictor variables included student socioeconomic status, high-school absenteeism, percentage of students with high absences in schools, school expectation and educational processes, school enrollment, and a composite variable, student socioeconomic status, and language background in schools.

4. Hosmer and Lemeshow (2000) illustrated the estimation of the ordinal response variable, birthweight with four categories, from six risk factors. The predictor variables were the smoking status of the mother, mother's weight, ethnicity, hypertension, uterine irritability, and previous preterm delivery.

5. Liao (1994) replicated Greene's (1990) research example to estimate the ordinal response variable, new Navy recruits, classified as one of the three levels of skills from six predictor variables, including technical training guarantee, mother's education level, scores on Air Force Qualifying Test, years of education, marital status, and age.

6. Menard (2010) identified factors related to the ordinal response variable, drug user type, with four categories that used the National Youth Survey (NYS). The four predictor variables included exposure to delinquent friends, beliefs about how wrong it is to violate the law, gender, and ethnicity.

7. O'Connell (2006) estimated the ordinal response variable, children's literacy proficiency, with six categories, from 10 predictor variables using the first-grade sample from the Early Childhood Longitudinal Study—Kindergarten Cohort data. These predictor variables included gender, number of family risks, whether there is a family risk, frequency with which parents read books to their children, whether parents read books to children less than once or twice a week, half-day kindergarten, center-based daycare prior to kindergarten, whether they are a minority, family socioeconomic status, and age of children at kindergarten entry.

8. Powers and Xie (2000) estimated the level of agreement with the statement that a woman's place was "in the home not in the workplace" from seven predictor variables using the National Longitudinal Survey of Youth data. The ordinal response variable was on a 4-point Likert scale, ordered from strongly disagree to strongly agree. The predictor variables included age, nonintact family structure at age 14, mother's employment, mother's education, family income, number of siblings, and fundamentalist Protestant upbringing.

9. Rabe-Hesketh and Skrondal (2008, 2012) estimated the level of severity of illness with four categories from two predictor variables, week of assessment and whether being in the treatment group.

In the preceding examples, the outcome variables of interest, a three-level response to a survey item related to astrology: the deprivation level, student persistence through high school, the four-category birthweight, three levels of skills, type of drug use, literacy proficiency categories, level of agreement, and severity of illness are all ordinal with more than two categories. In the following chapters, we will focus on various logistic regression models for ordinal response variables, which are referred to as ordinal logistic regression models when they are broadly defined. The first model, which will be introduced in this chapter, is the proportional odds (PO) model. Do not be surprised when you see that the PO model is often called the ordinal logistic regression model in the literature since the PO model is the default model for ordinal response variables in most statistical software packages.

The PO model, which is also called the cumulative odds model (Agresti, 1996, 2002, 2010; Ananth & Kleinbaum, 1997; Armstrong & Sloan, 1989; Clogg & Shihadeh, 1994; Liu, 2009; Long, 1997; Long & Freese, 2006, 2014; McCullagh, 1980; McCullagh & Nelder, 1989; Menard, 2010; O'Connell, 2000, 2006; Powers & Xie, 2000; Tutz,

2012), is one of the most commonly used models for the analysis of ordinal categorical data, and it comes from the class of generalized linear models. It is a generalization of a binary logistic regression model when the response variable has more than two ordinal categories. The proportional odds model is used to estimate the odds of being at or below a particular level of the response variable. For example, if there are J levels of ordinal outcomes, then the model makes $J-1$ predictions, each estimating the cumulative probabilities at or below the jth level of the outcome variable. This model can estimate the odds of being beyond a particular level of the response variable as well because below and beyond a particular category are just two opposite directions.

The ordinal logistic regression model can be expressed in the logit form as follows:

$$\ln(Y_j') = \text{logit } [\pi(x)] = \ln\left(\frac{\pi_j(x)}{1-\pi_j(x)}\right) = \alpha_j + (-\beta_1 X_1 - \beta_2 X_2 - \ldots - \beta_p X_p) \tag{4.1}$$

where $\pi_j(x) = \pi(Y \leq j \mid x_1, x_2, \ldots, x_p)$, which is the probability of being at or below category j given a set of predictors, $j = 1, 2, \ldots, J-1$. α_j are the cut points, and $\beta_1, \beta_2 \ldots, \beta_p$ are the logit coefficients. This is the form of a PO model because the odds ratio of any predictor is assumed to be constant across all categories. Similar to logistic regression, in the proportional odds model, we work with the logit or the natural log of the odds. To estimate the ln(odds) of being at or below the jth category, the PO model can be rewritten as:

$$\text{logit}[\pi(Y \leq j \mid x_1, x_2, \ldots, x_p)] = \ln\left(\frac{\pi(Y \leq j \mid x_1, x_2, \ldots x_p)}{\pi(Y > j \mid x_1, x_2, \ldots x_p)}\right) = \alpha_j +$$
$$(-\beta_1 X_1 - \beta_2 X_2 - \ldots - \beta_p X_p) \tag{4.2}$$

Thus, this model predicts cumulative logits across $J-1$ response categories. By transforming the cumulative logits, we can obtain the estimated cumulative odds and the cumulative probabilities of being at or below the jth category.

To understand ordinal logistic regression, we can think of it as several binary logistic regression models that are estimated simultaneously. The outcome variables of these binary models are dichotomized from the ordinal outcome variable comparing outcomes at or below a category ($Y \leq$ cat. j) and above that category ($Y >$ cat. j). So, each binary logistic regression estimates odds of being at or below a category (coded as 1) versus above that category (coded as 0). The estimated logit coefficients are constrained to be equal; therefore, for each predictor variable, we only need to estimate one regression coefficient rather than multiple coefficients. This constraint is the proportional odds assumption or the parallel lines assumption.

4.1.1 Odds and Odds Ratios in PO Models

In binary logistic regression, the values of the outcome variable are either 1 or 0, and we model the odds of success or of having an event when the outcome variable takes the value of 1 ($Y = 1$). The odds of success are the probability of success (p) divided by the probability of failure ($1 - p$).

In proportional odds models, the outcome variable is ordered with multiple levels, and we estimate the odds of being at or below a particular category ($Y \leq j$). Similar to the odds in binary logistic regression, the odds of being at or below a category in ordinal logistic regression equals the probability of being at or below a category divided by the probability of being above that category:

$$\text{Odds } (Y \leq j) = \frac{p(Y \leq j)}{p(Y > j)}$$

Since the probability of being at or below a category and the probability of being above that category is complementary, $p(Y \leq j) + p(Y > j) = 1$, this equation can be rewritten as:

$$\text{Odds } (Y \leq j) = \frac{p(Y \leq j)}{1 - p(Y \leq j)}$$

It reads as follows: The odds of being at or below a category j in ordinal logistic regression equal the probability of being at or below a category divided by its complimentary probability, 1 minus the probability of being at or below that category.

The probability of being at or below a category $p(Y \leq j)$ is the cumulative probability since it equals the sum of the probabilities of all categories at or below that category:

$$p(Y \leq j) = p(Y = 1) + p(Y = 2) + \ldots, p(Y = j) \text{ when } j = 1, 2, \ldots, J$$

For example, an outcome variable, health status, is ordinal with four levels from 1 to 4, where $1 = $ poor, $2 = $ fair, $3 = $ good, and $4 = $ excellent:

$$p(Y \leq 4) = p(Y = 1) + p(Y = 2) + p(Y = 3) + p(Y = 4) = 1$$

$$p(Y \leq 3) = p(Y = 1) + p(Y = 2) + p(Y = 3)$$

$$p(Y \leq 2) = p(Y = 1) + p(Y = 2)$$

$$p(Y \leq 1) = p(Y = 1)$$

Since this variable has four outcomes, we can estimate the following cumulative odds: the odds of being at or below the category 1, the odds of being at or below the category 2, and the odds of being at or below category 3. The odds of being at or below a category in ordinal logistic regression are also called the cumulative odds.

Odds $(Y \leq 1)$ equal the ratio of probability of being at or below category 1 to the probability of being above this category. The probability, $p(Y > 1) = p(2) + p(3) + (p4)$, which is the sum of the probabilities when $Y = 2$, 3, and 4:

$$\text{Odds } (Y \leq 1) = \frac{p(Y \leq 1)}{1 - p(Y \leq 1)} = \frac{p(1)}{p(2) + p(3) + p(4)}$$

Odds $(Y \leq 2)$ equal the ratio of probability of being at or below category 2 to the probability of being above this category. Since $p(Y \leq 2) = p(1) + p(2)$, and $p(Y > 2) = p(3) + p(4)$, the odds of being at or below category 2 can be expressed as follows:

$$\text{Odds } (Y \leq 2) = \frac{p(Y \leq 2)}{1 - p(Y \leq 2)} = \frac{p(1) + p(2)}{p(3) + p(4)}$$

Odds $(Y \leq 3)$ equal the ratio of probability of being at or below category 3 to the probability of being above this category. Using the same method, we get the following equation:

$$\text{Odds } (Y \leq 3) = \frac{p(1) + p(2) + p(3)}{p(4)}$$

The odds of being at or below category 1 are the probability comparisons between category 1 and categories 2, 3, and 4; the odds of being at or below the category 2 comparing probabilities of 1 and 2 versus 3 and 4; and the odds of being at or below the category 3 comparing probabilities of categories 1, 2, and 3 versus 4. Therefore, the cumulative odds in ordinal logistic regression are basically comparisons between two complimentary probabilities [i.e., $p(Y \leq j)$ and $p(Y > j)$]. Table 4.1 presents the logits, odds, and category comparisons for the PO model for the health status with four levels.

Table 4.1   Category Comparisons for the Proportional Odds Model With Four Levels of Health Status ($j = 1, 2, 3, 4$)

Category	Logit $P(Y \leq j)$	Odds	Probability Comparisons
Level 1	logit $P(Y \leq 1)$	$\frac{P(Y \leq 1)}{P(Y > 1)}$	Category 1 vs. categories 2 through 4
Level 2	logit $P(Y \leq 2)$	$\frac{P(Y \leq 2)}{P(Y > 2)}$	Categories 1 and 2 vs. categories 3 and 4
Level 3	logit $P(Y \leq 3)$	$\frac{P(Y \leq 3)}{P(Y > 3)}$	Categories 1, 2, and 3 vs. category 4

Odds Ratios in Ordinal Logistic Regression

In binary logistic regression, the odds ratio is the ratio of two odds, the odds of success when the value of a predictor is $(x + 1)$ relative to the odds when the predictor has a value of x. In other words, it is the change in odds for a one-unit increase in the predictor variable. Similar to binary logistic regression, the odds ratio in ordinal logistic regression is the change in the odds (i.e., the odds of being above a particular category versus being at or below that category) for a one-unit increase from any value of x to the value of $(x + 1)$, and it is the exponentiated logit coefficient, $\exp(\beta)$. In contrast, the odds ratio of being at or below a particular category is the inverse of the odds of being above that category. It is the exponentiated logit coefficient with a negative sign before that [i.e., $\exp(-\beta)$].

4.1.2 Brant Test of the PO Assumption

PO Assumption

In the proportional odds models, we assume that each predictor has the same effects across the categories of the ordinal outcome variable. In other words, the logit regression coefficients for each predictor are the same across the ordinal categories. For example, if we predict the ordinal outcome variable, health status, from the predictor, marital status, we estimate the odds of being at or below a category of health status relative to being above that category, given that predictor variable. The estimated logits and the corresponding odds ratios of being at or below category 1, category 2, and category 3 for the predictor, marital status, are assumed to be the same. Although we assume that they are equal, how can we know whether the assumption holds?

Brant Test

To test whether the PO assumption is met, we can use the Brant test (Brant, 1990) to look at the logit coefficients of a series of underlying binary logistic regression models for the dichotomized ordinal outcome variable, comparing outcomes at or below a category versus beyond that category.

The Brant test of the PO assumption can be examined using the `brant` command of the Stata `SPost13` (Long & Freese, 2014) package. It provides both the univariate Brant test result for each predictor and the omnibus test for the overall model.

4.1.3 Goodness of Fit

Since ordinal logistic regression is an extension of binary logistic regression, all measures-of-fit statistics in binary logistic regression models, such as pseudo R^2 statistics, the deviance, the log likelihood ratio test, and Akaike's information criterion (AIC) and Bayesian information criterion (BIC), can also be applied to proportional odds models. The following is a brief summary.

Pseudo R^2

Among several pseudo R^2 statistics introduced in Chapter 3, three of them are commonly seen in the literature: the likelihood ratio R^2, Cox and Snell R^2, and Nagelkerke R^2. Since these are different from R^2 in linear regression models and there is no agreement as to which one is the most appropriate, they should be interpreted with caution.

1. The likelihood ratio R^2

The likelihood ratio R^2 is also known as McFadden's R^2. It is the reduction in deviance from the fitted model (D_m) to the null model, which only contains the intercept (D_0). Menard (2000, 2010) preferred this statistic since it was closed to the R^2 in linear regression models. It is expressed as:

$$R^2_L = 1 - \frac{D_m}{D_0} = 1 - \frac{-2LL_m}{-2LL_0}$$

2. Cox and Snell R^2

Cox and Snell's R^2 is also known as the maximum likelihood R^2. It is based on the likelihood function of the fitted model (L_m) and the null model, which only contains the intercept (L_0). It is expressed as:

$$R^2_{ML} = 1 - \left(\frac{L_0}{L_m}\right)^{2/n}$$

where n is the total number of observations.

3. Nagelkerke R^2

Nagelkerke's R^2 is also known as Cragg and Uhler's R^2. It involves adjusting Cox and Snell's R^2 by dividing its maximum values. It is expressed as:

$$R^2_N = R^2_{ML}/\text{maximum } R^2_{ML}$$

Model Comparisons Using the Deviance Difference or Log Likelihood Ratio Test

In the process of model fitting, we always ask the following question: Which model fits the data better? Or, which model do I need to choose? When fitting a series of nested models, a common practice is to use the deviance difference or log likelihood ratio test to compare these competing models.

As explained in Chapter 3, the deviance statistic in logistic regression models is defined as the log likelihood statistic of the current model multiplied by negative two. It is written as $-2LL$. The difference in $-2LL$ or deviance is expressed as $G = D_{\text{Reduced}} - D_{\text{Full}}$. It is the difference between the deviance for the reduced model and that for the full model. The likelihood ratio test statistic follows a chi-square distribution with the degrees of freedom of the distribution equal to the difference in the number of parameters between these two models. A significant likelihood ratio chi-square test indicates that the full model rather than the reduced model is preferred.

The log likelihood ratio test can also be used to test the significance of the currently fitted model by comparing the null model without predictor variables with the current model with one or several predictor variables. It is expressed as $G = D_{\text{null}} - D_{\text{current}}$. This log likelihood ratio test is known as the model chi-square test, which follows a chi-square distribution with the degrees of freedom of the distribution equal to the number of predictor variables in the current model.

Model Comparisons Using AIC and BIC Statistics

In addition to the log likelihood ratio test, the AIC and BIC statistics can be used to compare nested models. One advantage of these two statistics is that they can also be used to compare non-nested models.

The AIC adjusts the deviance by the number of predictors and the sample size:

$$\text{AIC} = -2(\text{LL}_m - k)/n = (D_m + 2k)/n$$

where k is the number of parameters (the number of independent variables plus the intercept), n is the sample size, and -2LL_m or D_m is the deviance of the fitted model.

Another version of the AIC statistic is AIC $\times$ n, which equals $(D_m + 2k)$. The AIC $\times$ n adjusts the deviance by the number of parameters only. Smaller AICs mean a better fit of the model. Stata reports this version of the AIC with the estat ic command. The fitstat command in SPost9 (Long & Freese, 2006) computes AIC, AIC $\times$ n, and AIC used by Stata. Among these three versions of AIC statistics, the first one includes the division by the sample size n and the last two are essentially identical. The fitstat command in the latest SPost13 (Long & Freese, 2014) computes AIC and AIC divided by N.

The BIC adjusts the deviance by its degrees of freedom and the sample size:

$$\text{BIC} = D_m - \text{df} \times \ln(n)$$

where df is the degrees of freedom associated with the deviance, n is the sample size, and D_m is the deviance of the fitted model. The degrees of freedom associated with the deviance are the degrees of freedom of the model, which equals the number of observations minus the number of independent variables and intercept. The fitstat command in `SPost9` (Long & Freese, 2006) reports this statistic labeled BIC in the output:

$$\text{Another version of BIC} = -2LL_m + \ln(n) \times k = D_m + \ln(n) \times k$$

where k is the number of parameters (the number of independent variables plus the intercept). This is the BIC used by Stata computed by the `fitstat` command in the previous `SPost9` or the `estat ic` command. The old `fitstat` command also reports the third version BIC, BIC′. This version is based on the log likelihood ratio chi-square test statistic. BIC′ = –LR chi-square + the number of predictors × $\ln(n)$. The `fitstat` in the current SPost13 only reports one version of the BIC (Long & Freese, 2014), which is the BIC used by Stata.

As with the AIC statistic, a smaller BIC statistic suggests a better fit of the model.

4.1.4 Interpretation of Model Parameter Estimates

The odds ratio in ordinal logistic regression can be interpreted in a similar way as that of the binary logistic regression. In binary logistic regression, we estimate the odds of success when the outcome takes the value of 1 (i.e., $Y = 1$), whereas in ordinal logistic regression, the odds are the ones when the outcomes are at or below a particular category (i.e., $Y \leq j$).

Recall that the signs before the logit coefficients in the equation of the ordinal logistic regression (Equation 4.1) are negative. To get the odds ratio (OR) of being at or below a category, we need to exponentiate the logit coefficient with a negative sign before that. This odds ratio can be interpreted as the change in the predicted logit or the log odds of being at or below a particular category for a one-unit increase in the predictor variable. By removing the negative sign and then exponentiating the logit coefficient, we get the OR of being beyond a category. In contrast, taking the inverse of the odds of being at or below a particular category also gives us the odds of being beyond that category. The odds ratio of being a particular category can be interpreted as the change in the predicted logit or the log odds of being above that particular category for a one-unit increase in the predictor variable.

When the logit coefficient itself is positive, it indicates the relationship between the predictor variable and the logit function of the probability is positive. In other words, a positive coefficient increases the probability of being above a category. By exponentiating the logit coefficient, you get the OR, which is greater than 1. This means that the odds of being beyond a particular category increases for a one-unit increase in the predictor variable.

When the logit coefficient itself is negative, it indicates that the relationship between the predictor variable and the logit function is negative. A negative coefficient decreases the probability of being above a category. The exponentiated coefficient, the

OR, is less than 1. It means that the odds of being beyond a particular category decrease for a one-unit increase in the predictor variable.

When the logit coefficient equals 0, the OR equals 1. It indicates that there is no relationship between the predictor and the odds, so there is no change in the odds when the values of the predictor variable change.

4.2 Research Example and Description of the Data and Sample

Research Problem and Questions: In this chapter, the purpose of the research example is to investigate the relationships between the ordinal response variable, health status, and four predictor variables: marital status, the highest education completed, age, and gender. The research question is as follows: Do the four predictor variables significantly predict the ordinal response variable, health status? Specifically, do the four predictor variables significantly predict the cumulative odds and then the cumulative probabilities of being at or below a particular level of health status, or the cumulative odds and then the cumulative probabilities of being above that health status level?

Description of the Data and Sample: The data for the following analyses were from the General Social Survey 2012 (GSS 2012). The following are the variables used for data analysis in this chapter:

- `healthre`: the recoded variable of health (health status) with four ordinal categories (1 = poor health, 2 = fair health, 3 = good health, and 4 = excellent health)
- `maritals`: the recoded variable of marital (marital status) with 1 = currently married and 0 = not currently married
- `educ`: the highest education completed
- `age`: respondent's age
- `male`: recoded variable of sex with 1 = male and 0 = female

4.3 Proportional Odds Models With Stata: Commands and Output

The Stata Command `ologit`

The Stata command `ologit` is used for the ordinal logistic regression analysis. The syntax for ordinal logistic regression is the command `ologit` followed by the dependent variable and the independent variable. For example, the command `ologit` y x tells Stata to run a simple ordinal logistic regression analysis predicting the ordinal dependent variable *y* with an independent variable *x*. For more details on how to use this command, type the `help ologit` command.

4.3.1 The PO Model: One-Predictor Model

The command `xi: ologit healthre i.maritals` tells Stata to conduct the ordinal logistic regression to estimate the ordinal outcome variable `healthre` using the predictor variable `maritals`. In the command syntax `xi` is the prefix command for the categorical variable `maritals`, and `i.maritals` indicates that the predictor variable `maritals` is categorical, which generates a dummy variable beginning with "_I" in the output. For Stata 13 and 14 users, the prefix `xi` command can be omitted since the new coding for factor variables is used. Please note that user-written programs may not work with factor variables. The following output is displayed.

```
. xi: ologit healthre i.maritals
i.maritals          _Imaritals_0-1          (naturally coded; _Imaritals_0 omitted)

Iteration 0:   log likelihood = -1579.4068
Iteration 1:   log likelihood = -1566.2237
Iteration 2:   log likelihood = -1566.1973

Ordered logistic regression                       Number of obs   =       1300
                                                  LR chi2(1)      =      26.42
                                                  Prob > chi2     =     0.0000
Log likelihood = -1566.1973                       Pseudo R2       =     0.0084

------------------------------------------------------------------------------
    healthre |      Coef.   Std. Err.      z    P>|z|     [95% Conf. Interval]
-------------+----------------------------------------------------------------
 _Imaritals_1 |   .5327629   .1041707     5.11   0.000     .328592    .7369337
-------------+----------------------------------------------------------------
       /cut1 |  -2.466528   .1205322                      -2.702767   -2.230289
       /cut2 |  -.7358156   .0761842                       -.8851338   -.5864974
       /cut3 |   1.270491   .0823931                       1.109004    1.431979
------------------------------------------------------------------------------
```

Interpreting the Output

Similar to the output in binary logistic regression models, at the beginning, there is a note explaining dummy variables for `i.maritals`. It shows the indicator variables named `_Imaritals_0-1`, and the first dummy variable is omitted from the model (naturally coded; _Imaritals_0 omitted). This tells us that the variable maritals is a binary variable with the values of 1 and 0.

The log likelihood for the initial iteration is –1579.407 (Iteration 0: log likelihood = -1579.4068), which is the log likelihood when the model contains no predictors. The maximum value at Iteration 2 is –1566.197 when the model contains the predictor variable (Iteration 2: log likelihood = -1566.1973).

On the top right of the logistic regression table, it provides the number of observations in the dataset, the log likelihood ratio test statistic and the associated p value, and the pseudo R^2. The dataset has 1,300 cases for the analysis. LR chi2 (1) = 26.42, which is the log likelihood ratio chi-square test statistic. It is the difference in the –2 log likelihood (–2LL) between the current model that contains the one predictor and the cut points or intercepts and the null model that contains only the cut points or the intercepts.

The null hypothesis of the log likelihood ratio chi-square test is that the predictor variable does not contribute to the model, or the one-predictor model is not better than the null model with no independent variables.

The alternative hypothesis is that the predictor variable significantly contributes to the model, or the one-predictor model is better than the null model with no independent variables.

The associated p value with the log likelihood ratio chi-square test Prob > chi2 = 0.0000, which indicates that the null hypothesis is rejected. Therefore, the one-predictor model provides a better fit than the null model with no independent variables in predicting the cumulative odds of being at or below a category of health status.

The Pseudo R² = .0084, which is the likelihood ratio R^2_L, or McFadden's R^2, suggests that the relationship between the response variable, health status, and the predictor, marital status, is small.

The ordinal logistic regression table displays the parameter estimates for the predictor variable and the cut points or intercepts, their standard errors, the Wald z statistics, the associated p values, and the 95% confident intervals of the parameter estimates for the constant and predictors.

The null hypothesis for the Wald test is that the coefficient of the predictor variable is zero, and the alternative hypothesis is that the coefficient of the predictor variable is significantly different from zero.

The Wald test statistic equals the parameter estimate divided by its standard error. For the predictor variable maritals, Wald z = .5327629/.1041707 = 5.11. Some statistical software packages report the Wald chi-square test statistic, which is the square of the Wald z statistic. The associated p value P>|z|=.000, so we reject the null hypothesis. Therefore, the predictor variable, marital status, is a significant predictor of the ordinal outcome variable, health status. The 95% confidence interval of the regression coefficient is [.328592, .7369337]. It does not contain 0, which indicates the coefficient is significantly different from 0.

The logit regression coefficient (labeled Coef.) β = .533 indicates that for a one-unit increase in the predictor variable, the change in the logit or log odds of being beyond a category is .533. The corresponding odds ratio, which offers a better interpretation, is shown as follows.

The results table reports three cut points: _cut1, _cut2, and_cut3. These are the estimated cut points on the latent variable Y^* used to differentiate the adjacent levels of categories of health status. When the response category is 1, the latent variable falls at or below the first cut point α_1. When the response category is 2, the latent variable falls between the first cut point α_1 and the second cut point α_2; when the response category reaches 3, the latent variable falls between the second α_2 and the third cut point α_3; and when the response category reaches 4, the latent variable is at or beyond the third cut point α_3.

These cut points are also called intercepts in other software packages. They can be thought of as the intercepts for three underlying binary logistic regression models if we dichotomize the ordinal outcome variable.

To request estimated odds ratios, we can add the option or to the original ologit command. With the command xi: ologit healthre i.maritals, or, we get the following output displaying the odds ratio.

```
. xi: ologit healthre i.maritals, or
i.maritals          _Imaritals_0-1          (naturally coded; _Imaritals_0 omitted)

Iteration 0:    log likelihood = -1579.4068
Iteration 1:    log likelihood = -1566.2237
Iteration 2:    log likelihood = -1566.1973

Ordered logistic regression                     Number of obs    =        1300
                                                LR chi2(1)       =       26.42
                                                Prob > chi2      =      0.0000
Log likelihood = -1566.1973                     Pseudo R2        =      0.0084

-------------------------------------------------------------------------------
    healthre | Odds Ratio   Std. Err.      z    P>|z|     [95% Conf. Interval]
-------------+-----------------------------------------------------------------
 _Imaritals_1 |   1.703633    .1774686    5.11   0.000     1.389011    2.089519
-------------+-----------------------------------------------------------------
       /cut1 |  -2.466528    .1205322                     -2.702767   -2.230289
       /cut2 |  -.7358156    .0761842                     -.8851338   -.5864974
       /cut3 |   1.270491    .0823931                      1.109004    1.431979
-------------------------------------------------------------------------------
```

With all other results the same, the ordinal logistic regression table provides odds ratios, their standard errors, the Wald z statistics, the associated p values, and the 95% confidence intervals of the odds ratios for the predictor and the cut points.

The odds ratio for the categorical variable, marital status, which is labeled as _Imaritals_1, is 1.703633. It equals the exponentiated regression coefficient exp(.5327629). The 95% confidence interval of the odds ratio is [1.389011, 2.089519].

Using the command listcoef, we can get the same odds ratio for the predictor, marital status.

```
. listcoef

ologit (N=1300): Factor change in odds

  Odds of: >m vs <=m

------------------------------------------------------------------------
             |        b        z     P>|z|      e^b    e^bStdX     SDofX
-------------+----------------------------------------------------------
_Imaritals_1 |   0.5328    5.114     0.000    1.704      1.305     0.499
------------------------------------------------------------------------
```

The estimated odds ratio for the predictor shown in this output is the change in the odds of being beyond a category versus being at or below that category (labeled Odds of:>m vs <=m). The table displays the logit regression coefficient (b), its Wald z test statistic (z), the associated p value (P>|z|), the odds ratio (e^b), the odds ratio for a 1 standard deviation change in the predictor (e^bStdX), and the standard deviation of the predictor.

More measures of fit can be obtained when using the fitstat command, which is part of the SPost13 package (Long & Freese, 2014). In addition to the deviance statistic and McFadden's R^2, several other types of R^2 statistics are reported. The information measures, AIC and BIC, are used to compare either nested or non-nested models. Smaller AIC and BIC statistics indicate the better fitting model. The output is shown as follows.

```
. fitstat

                          |        ologit
--------------------------+--------------
Log-likelihood            |
                  Model   |    -1566.197
          Intercept-only  |    -1579.407
--------------------------+--------------
Chi-square                |
        Deviance (df=1296)|     3132.395
              LR (df=1)   |       26.419
               p-value    |        0.000
--------------------------+--------------
```

```
R2                           |
             McFadden |        0.008
  McFadden (adjusted) |        0.006
  McKelvey & Zavoina |        0.021
          Cox-Snell/ML |        0.020
 Cragg-Uhler/Nagelkerke |      0.022
                 Count |        0.457
       Count (adjusted) |      0.000
------------------------------+--------------
IC                           |
                   AIC |     3140.395
     AIC divided by N |        2.416
            BIC (df=4) |     3161.075
------------------------------+--------------
Variance of                  |
                     e |        3.290
                y-star |        3.361
```

Interpreting the Odds Ratio of Being at or Below a Particular Category

The estimated logit regression coefficient $\beta = .533$, $z = 5.11$, $p < .001$, which indicates that marital status is a significant predictor of the ordinal response variable, health status. By substituting the value of the coefficient into Equation 4.2, logit $[\pi(Y \leq j \mid \texttt{maritals})] = \alpha_j + (-\beta_1 X_1)$, we calculated logit $[\pi(Y \leq j \mid \texttt{maritals})] = \alpha_j - .533$ (maritals). OR $= e^{(-.533)} = .587$, which indicates that the odds of being at or below any category of health status (i.e., less healthy) for the married are .587 times the odds for the unmarried. Another way to interpret it is that being married decreases the odds of being at or below a category of health status by .587.

To estimate the cumulative odds of being at or below a certain category j for marital status, let us take a look at the logit form of the proportional odds model, logit $[\pi(Y \leq j \mid \texttt{maritals})] = \alpha_j - .533$ (maritals). For example, when $Y \leq 1$, $\alpha_1 = -2.467$, which is the first cut point for the model. By substituting it into Equation 4.2, we get logit $[\pi(Y \leq j \mid \texttt{maritals})] = -2.467 - .533$ (maritals). For the married $(x = 1)$, logit $[\pi(Y \leq 1 \mid \texttt{maritals})] = -3$. By exponentiating the logit, we calculate the odds of being at or below category 1 (poor health) for the married, $e^{(-3)} = .050$. For the unmarried $(x = 0)$, logit $[\pi(Y \leq 1 \mid \texttt{maritals})] = -2.467 - .533 \times 0 = -2.467$, so the odds of being at or below category 1 (poor health) for the unmarried, $e^{-2.467} = .085$. The odds ratio of the married relative to the unmarried $= .050/.085 = .588$.

Interpreting the Odds Ratio of Being Beyond a Particular Category

The proportional odds model can also estimate the ln(odds) of being beyond a category j. Again, these ln(odds) can be transformed into the cumulative odds and cumulative probabilities. For example, we can estimate the cumulative probability of health status beyond category 3, $P(Y > 3)$; beyond category 2, $P(Y > 2)$; and beyond category 1, $P(Y > 1)$. The cumulative logit form can be expressed as logit $[\pi(Y > j \mid \texttt{maritals})] = -\alpha_j + (\beta_1 X_1)$. In Stata, when estimating the odds of being beyond category j, the sign of the cut points needs to be reversed and their magnitude remains unchanged since we estimate the cut points from the right to the left of the latent variable Y^*, that is, from the direction when $Y = 4$ approaches $Y = 1$. Therefore, three cut points from right to left turn to −1.270, .736, and 2.467.

When the predictor is dichotomous, a positive sign of the logit coefficient indicates that it is more likely for the group ($x = 1$) to be at or beyond a particular category than for the relative group ($x = 0$). When the predictor is continuous, a positive coefficient indicates that when the value of the predictor variable increases, the odds of being beyond a particular category increase.

Both the `ologit` command with the `or` option and the command `listcoef` provide the odds ratios of being beyond a particular category: OR = 1.704. It can be interpreted that the odds of being beyond a particular category of health status (better health status) for the married are 1.074 times the odds for the unmarried. In other words, the odds of being beyond a particular category of health status are 7.4% greater for the married than for the unmarried.

The command `brant` is used to test the proportional odds assumption. Brant (1990) proposed a test of proportional odds assumption for the ordinal logistic model by examining the separate fits to the underlying binary logistic models. A nonsignificant omnibus test indicates that the proportional odds assumption is not violated. It also provides tests for each independent variable. When there is only one independent variable in the model, the results of the omnibus test and individual test are the same. The results are shown as follows.

```
. brant, detail

Estimated coefficients from binary logits

-----------------------------------------------------
    Variable |   y_gt_1       y_gt_2       y_gt_3
-------------+---------------------------------------
_Imaritals_1 |    0.472        0.482        0.584
             |    2.00         3.81         4.61
       _cons |    2.488        0.755       -1.300
             |   17.40         9.25       -14.02
-----------------------------------------------------
                                    legend: b/t

Brant test of parallel regression assumption

              |      chi2       p>chi2        df
--------------+--------------------------------------
          All |      0.52       0.771          2
--------------+--------------------------------------
 _Imaritals_1 |      0.52       0.771          2

A significant test statistic provides evidence that the parallel
regression assumption has been violated.
```

The Brant test of parallel regression assumption yields $\chi^2_{(2)} = .52$, $p = .771$, which indicates that the proportional odds assumption for the model is upheld, suggesting that the effect of the explanatory variable, marital status, is constant across separate binary models fit to the cumulative cut points. The table also provides the estimated coefficient from $j - 1$ binary logistic regression models. Each logistic regression model estimates the probability of being beyond health status level j.

4.3.2 The PO Model: Multiple-Predictor Model

The command xi: ologit healthre i.maritals educ age i.male tells Stata to predict the ordinal response variable healthre from four predictor variables maritals, edu, age, and male using ordinal logistic regression. The output is shown as follows.

```
. xi: ologit healthre i.maritals educ age i.male
i.maritals           _Imaritals_0-1       (naturally coded; _Imaritals_0 omitted)
i.male               _Imale_0-1           (naturally coded; _Imale_0 omitted)

Iteration 0:    log likelihood = -1579.4068
Iteration 1:    log likelihood = -1513.1847
Iteration 2:    log likelihood = -1512.4793
Iteration 3:    log likelihood = -1512.4786
Iteration 4:    log likelihood = -1512.4786

Ordered logistic regression                      Number of obs   =      1300
                                                 LR chi2(4)      =    133.86
                                                 Prob > chi2     =    0.0000
Log likelihood = -1512.4786                      Pseudo R2       =    0.0424

------------------------------------------------------------------------------
   healthre |    Coef.    Std. Err.      z     P>|z|    [95% Conf. Interval]
------------+-----------------------------------------------------------------
_Imaritals_1 |  .5140058   .1058117     4.86   0.000    .3066187    .721393
       educ |  .1344542   .0172594     7.79   0.000    .1006265    .168282
        age | -.0193069   .0030489    -6.33   0.000   -.0252826   -.0133313
   _Imale_1 |  .0530645   .104808      0.51   0.613   -.1523553    .2584843
------------+-----------------------------------------------------------------
      /cut1 | -1.701386   .2976461                    -2.284762   -1.11801
      /cut2 |  .1246361   .2863089                     -.4365191   .6857913
      /cut3 | 2.257928    .2936829                     1.68232    2.833536
------------------------------------------------------------------------------
```

The log likelihood ratio chi-square test $LR\ \chi^2_{(4)} = 133.86$, $p < .001$, which indicates that the full model with four predictors provides a better fit than the null model with no independent variables in predicting the ordinal response variable. The likelihood ratio $R^2_L = .042$ is larger than that of the one-predictor model but is still small, suggesting that the relationship between the response variable, health status, and a set of four predictors is still small. Compared with the one-predictor model, all R^2 statistics of the full model show improvement.

With the command `xi: ologit healthre i.maritals educ age i.male, or`, we get the following output displaying the odds ratio of being beyond a category. The `xi` prefix command creates dummy variables for two categorical variables `maritals` and `male` with the use of `i.` before them. The following output is displayed.

```
. xi: ologit healthre i.maritals educ age i.male, or
i.maritals          _Imaritals_0-1      (naturally coded; _Imaritals_0 omitted)
i.male              _Imale_0-1          (naturally coded; _Imale_0 omitted)

Iteration 0:    log likelihood = -1579.4068
Iteration 1:    log likelihood = -1513.1847
Iteration 2:    log likelihood = -1512.4793
Iteration 3:    log likelihood = -1512.4786
Iteration 4:    log likelihood = -1512.4786

Ordered logistic regression                    Number of obs   =       1300
                                                LR chi2(4)      =     133.86
                                                Prob > chi2     =     0.0000
Log likelihood = -1512.4786                     Pseudo R2       =     0.0424


------------------------------------------------------------------------------
    healthre | Odds Ratio   Std. Err.      z    P>|z|     [95% Conf. Interval]
-------------+----------------------------------------------------------------
 _Imaritals_1 |   1.671975   .1769146     4.86   0.000     1.358823    2.057297
        educ |   1.143912   .0197432     7.79   0.000     1.105863     1.18327
         age |   .9808782   .0029906    -6.33   0.000     .9750343    .9867572
    _Imale_1 |   1.054498   .1105197     0.51   0.613     .8586831    1.294966
-------------+----------------------------------------------------------------
       /cut1 |  -1.701386   .2976461               -2.284762    -1.11801
       /cut2 |   .1246361   .2863089               -.4365191    .6857913
       /cut3 |   2.257928   .2936829                1.68232     2.833536
------------------------------------------------------------------------------
```

The `listcoef` command produces more detailed results of logit coefficients and odds ratios (exponentiated coefficients). For the proportional odds model, interpretation of odds ratios is independent on the ancillary parameters (cut points) because they are constant across all levels of the response variable. The output is shown as follows.

```
. listcoef

ologit (N=1300): Factor change in odds

  Odds of: >m vs <=m

----------------------------------------------------------------------------
             |        b        z     P>|z|       e^b    e^bStdX      SDofX
-------------+--------------------------------------------------------------
 _Imaritals_1 |   0.5140    4.858    0.000     1.672      1.293      0.499
        educ |   0.1345    7.790    0.000     1.144      1.527      3.151
         age |  -0.0193   -6.333    0.000     0.981      0.714     17.432
    _Imale_1 |   0.0531    0.506    0.613     1.054      1.027      0.497
----------------------------------------------------------------------------
```

Interpreting the Odds Ratios of Being Beyond a Particular Category

Both the `ologit` command with the `or` option and the command `listcoef` provide the odds ratios of being beyond a particular category. A positive logit regression coefficient corresponds to an odds ratio greater than 1. It indicates that we can expect an increase in the odds of being beyond a particular category for a one-unit increase in the predictor variable when holding all other predictors constant.

A negative logit coefficient corresponds to an odds ratio of being beyond a particular category less than 1. It indicates the odds of being beyond a category decrease for a one-unit increase in the predictor variable when holding all other predictors constant.

If a logit coefficient is not significantly different from zero, then its corresponding odds ratio equals 1. It indicates that there is no relationship between the predictor and the odds of being beyond a category when all the other variables are held fixed.

For the `maritals` predictor, β = .514, which is positive; OR = 1.672, which is greater than 1. It indicates that the odds of being beyond a particular category of health status (better health status) for the married are 1.672 times the odds for the unmarried when holding all the other predictors constant.

For the `educ` predictor, β = .134, which is positive; OR = 1.144, which is greater than 1. It indicates that the odds of being beyond a particular category of health status (better health status) increase by 1.144 for a one-unit increase in the predictor, education, when holding all the other predictors constant.

For the `age` predictor, β = −.019, which is negative; OR = .981, which is less than 1. It indicates that the odds of being beyond a particular category of health status (better health status) decrease by a factor of .981 for a one-unit increase in the predictor, age, when all the other predictors remain constant. In other words, for a one-unit increase in age, the odds of being healthier decrease by 1.9%.

For the `male` predictor, β = .053, p = .613, which is not significantly different from 0; OR = 1.055, which almost equals 1. It indicates that there is no relationship between being a male and the cumulative odds of being in better health status. In other words, there is no significant difference between the male and the female in better health status.

Interpreting the Odds Ratios of Being at or Below a Particular Category

In the preceding section, we interpreted the odds ratio of being beyond a category. We can also interpret how these predictor variables contribute to the odds of being at or below a particular category if we reverse the sign before the estimated logit coefficients and then compute the corresponding odds ratios.

Using the command `listcoef` with the `reverse` option, we get the odds ratios of being at or below a particular category relative to beyond that category. The `reverse` option tells Stata to reverse the odds of being beyond a category versus at or below that category to the odds of being at or below a category versus beyond a category. The following is the output produced by the command.

```
. listcoef, reverse

ologit (N=1300): Factor change in odds

  Odds of: <=m vs >m

---------------------------------------------------------------------
             |         b         z      P>|z|       e^b   e^bStdX    SDofX
-------------+-------------------------------------------------------
_Imaritals_1 |    0.5140     4.858     0.000     0.598     0.774     0.499
        educ |    0.1345     7.790     0.000     0.874     0.655     3.151
         age |   -0.0193    -6.333     0.000     1.019     1.400    17.432
    _Imale_1 |    0.0531     0.506     0.613     0.948     0.974     0.497
---------------------------------------------------------------------
```

By substituting the values of the four logit coefficients into Equation 4.2, we get logit $[\pi(Y \le j)] = \alpha_j + (-.514 \times \text{maritals} + .019 \times \text{age} - .135 \times \text{educ} - .053 \times \text{male})$. The exponentiated logit coefficients are the odds ratios of being at or below a particular category.

For the predictor `maritals`, OR = .598, which is less than 1. It indicates that the odds of being at or below a particular category of health status (worse health status) for the married are .598 times the odds for the unmarried when holding all the other predictors constant.

For the predictor `age`, OR = 1.020, which is greater than 1. It indicates that the odds of being at or below a particular category of health status (worse health status) increase by a factor of 1.020 for a one-unit increase in the predictor, age, when all the other predictors remain constant. In other words, for a one-unit increase in age, the odds of having poorer health increase by 2%.

For the predictor `educ`, OR = .874, which is less than 1. It indicates that the odds of being at or below a particular category of health status (poorer health status) decrease by .874 for a one-unit increase in the predictor education when holding all the other predictors constant. In other words, the odds of being at or below a particular category decrease by 12.6% for a one-unit increase in education, given the effects of other predictors are held constant in the model.

For the predictor `male`, OR = .948 (p =.613), which is close to 1. It indicates that there is no relationship between being a male and the cumulative of being in poorer health status. In other words, there is no significant difference between the male and the female in poorer health status.

The `fitstat` command computes various fit statistics for the multiple-predictor model. The output is displayed as follows. The results will be interpreted in the next section when we compare fit statistics between the one-predictor model and the full model.

```
. fitstat

                              |       ologit
------------------------------+-------------
Log-likelihood                |
                     Model    |    -1512.479
             Intercept-only   |    -1579.407
------------------------------+-------------
Chi-square                    |
         Deviance (df=1293)   |     3024.957
                  LR (df=4)   |      133.856
                   p-value    |        0.000
------------------------------+-------------
R2                            |
                  McFadden    |        0.042
       McFadden (adjusted)    |        0.038
       McKelvey & Zavoina     |        0.105
              Cox-Snell/ML    |        0.098
   Cragg-Uhler/Nagelkerke     |        0.107
                     Count    |        0.464
           Count (adjusted)   |        0.013
------------------------------+-------------
IC                            |
                       AIC    |     3038.957
            AIC divided by N  |        2.338
                 BIC (df=7)   |     3075.148
------------------------------+-------------
Variance of                   |
                         e    |        3.290
                    y-star    |        3.676
```

To examine the assumption of proportional odds, we use the command brant with the detail option. It provides the results of the Brant test for the overall model with four predictors and the univariate tests for individual variables. With the detail option, it also provides the estimated coefficient from $j - 1$ binary logistic regression models. Since the ordinal outcome variable has four categories, three separate binary logistic regression models are estimated. The data are dichotomized by comparing outcomes beyond a category and those at or below that category. So each binary logistic regression model estimates the probability of being beyond health status level j. The output is shown as follows.

```
. brant, detail

Estimated coefficients from binary logits

-----------------------------------------------------------
     Variable |  y_gt_1      y_gt_2      y_gt_3
--------------+--------------------------------------------
 _Imaritals_1 |   0.472       0.479       0.590
              |    1.96        3.60        4.53
```

```
         educ |     0.160          0.172          0.084
              |     4.74           7.87           3.97
          age |    -0.026         -0.021         -0.016
              |    -4.04          -5.70          -4.01
     _Imale_1 |    -0.224         -0.005          0.197
              |    -0.96          -0.04           1.53
         _cons |    1.882         -0.467         -1.812
              |     3.13          -1.32          -5.06
------------------------------------------------------
                                         legend: b/t


Brant test of parallel regression assumption

                    |     chi2      p>chi2      df
     ---------------+-------------------------------
               All  |    19.45      0.013        8
     ---------------+-------------------------------
      _Imaritals_1  |     0.55      0.758        2
              educ  |    12.98      0.002        2
               age  |     2.28      0.320        2
         _Imale_1   |     3.16      0.206        2


A significant test statistic provides evidence that the parallel
regression assumption has been violated.
```

Let us first take a look at the table for the Brant Test of Parallel Regression Assumption. For the omnibus Brant test $\chi^2_{(8)} = 19.45$, $p = .013$, which indicates that the proportional odds assumption for the full model is violated. By examining the Brant tests for each predictor variable, we find that the proportional odds assumption is upheld for maritals, age, and male, whereas it is violated for educ. For educ, the Brant test $\chi^2_{(2)} = 12.98$, $p = .002$, which is significant.

Next, let us check the estimated coefficients for each independent variable across three binary logistic regression models. The three logit regression coefficients for educ are .160, .172, and .084, respectively. The third coefficient is only half of the first two coefficients, which explains why the proportional odds assumption is violated. The logit coefficients for all the other predictor variables look similar across three binary logistic models, which supports the results of univariate Brant tests for these variables.

Interpreting the Estimated Probabilities
With the margins *Command in Stata 13*

With the margins command, we can estimate the margins or the predicted probabilities at each category of the ordinal response variable for predictor variables at specified values. The command margins, predict (outcome(1)) at(educ = (8 13 16)) atmeans vsquish tells Stata 13 to compute the estimated probabilities when the ordinal response variable $Y = 1$ for the predictor variable educ at the values

of 8, 13, and 16 when holding the other predictor variables at their means. The `predict`
(`outcome(1)`) `option`   estimates probability for the first category of the ordinal
response variable $Y = 1$. The `at(educ = (8 13 16))` option specifies the values
for the predictor variable `educ`. The `vsquish` option makes the output compact.

```
. margins, predict (outcome(1)) at(educ = (8 13 16)) atmeans vsquish

Adjusted predictions                            Number of obs    =        1300
Model VCE     : OIM

Expression    : Pr(healthre==1), predict(outcome(1))
1._at         : 0.maritals       =      .5315385  (mean)
                1.maritals       =      .4684615  (mean)
                educ             =             8
                age              =          48.2  (mean)
                0.male           =      .5561538  (mean)
                1.male           =      .4438462  (mean)
2._at         : 0.maritals       =      .5315385  (mean)
                1.maritals       =      .4684615  (mean)
                educ             =            13
                age              =          48.2  (mean)
                0.male           =      .5561538  (mean)
                1.male           =      .4438462  (mean)
3._at         : 0.maritals       =      .5315385  (mean)
                1.maritals       =      .4684615  (mean)
                educ             =            16
                age              =          48.2  (mean)
                0.male           =      .5561538  (mean)
                1.male           =      .4438462  (mean)

---------------------------------------------------------------------------
             |            Delta-method
             |    Margin   Std. Err.      z    P>|z|    [95% Conf. Interval]
-------------+-------------------------------------------------------------
        _at |
          1 |   .1080545    .013559    7.97   0.000    .0814794    .1346295
          2 |   .0582475   .0063735    9.14   0.000    .0457558    .0707393
          3 |   .0396806   .0049499    8.02   0.000     .029979    .0493822
---------------------------------------------------------------------------
```

In the output, for the first combination (labeled `1._at`), when `educ` = 8 and the
other predictor variables are held at their means (`1.maritals` = .468, age = 48.2,
and `1.male` = .444), the estimated margin or the probability when $Y = 1$ is .108.

For the second combination (labeled `2._at`), when `educ` = 13 and the other
three predictor variables are held at their means, the estimated probability when $Y = 1$
is .058.

For the third combination (labeled `3._at`), when `educ` = 16 and the other predictor variables are held at their means, the estimated probability when $Y = 1$ is .040.

These three predicted margins or probabilities are significantly different from 0
since $p < .001$.

The expected probabilities for other categories (i.e., 2, 3, and 4) of the ordinal response variable can be computed in a similar way. For example, the margins command with the option predict (outcome(2)) computes the estimated probability for the second category of the ordinal response variable. To save space, the output for the expected probabilities for categories 2 and 3 is omitted. The following output shows the results for the expected probability for category 4.

```
. margins, predict (outcome(4)) at(educ = (8 13 16)) atmeans vsquish

Adjusted predictions                                    Number of obs   =       1300
Model VCE      : OIM

Expression    : Pr(healthre==4), predict(outcome(4))
1._at         : 0.maritals       =      .5315385  (mean)
                1.maritals       =      .4684615  (mean)
                educ             =             8
                age              =          48.2  (mean)
                0.male           =      .5561538  (mean)
                1.male           =      .4438462  (mean)
2._at         : 0.maritals       =      .5315385  (mean)
                1.maritals       =      .4684615  (mean)
                educ             =            13
                age              =          48.2  (mean)
                0.male           =      .5561538  (mean)
                1.male           =      .4438462  (mean)
3._at         : 0.maritals       =      .5315385  (mean)
                1.maritals       =      .4684615  (mean)
                educ             =            16
                age              =          48.2  (mean)
                0.male           =      .5561538  (mean)
                1.male           =      .4438462  (mean)

------------------------------------------------------------------------------
             |            Delta-method
             |     Margin   Std. Err.      z    P>|z|     [95% Conf. Interval]
-------------+----------------------------------------------------------------
         _at |
          1  |   .1360439   .0144063     9.44   0.000     .107808     .1642798
          2  |    .235723   .0121122    19.46   0.000    .2119836    .2594623
          3  |   .3158499   .0158311    19.95   0.000    .2848216    .3468781
------------------------------------------------------------------------------
```

The bottom of the output displays the margins, their Delta-method standard errors, Wald z statistics and the associated p values, and 95% confidence intervals. When educ = 8, 13, and 16 and other predictor variables are held at their means, the margins or estimated probabilities for category 4 are .136, .236, and .316, respectively. They are all statistically different from zero. To visualize the results, the command margins-plot is used next. For more functions of this command and examples of using it, refer to Mitchell (2012a, 2012b). Figure 4.1 shows the estimated probabilities when $Y = 4$ (i.e., excellent health condition).

Figure 4.1   Estimated Probabilities When $Y = 4$ for educ at 8, 13, and 16

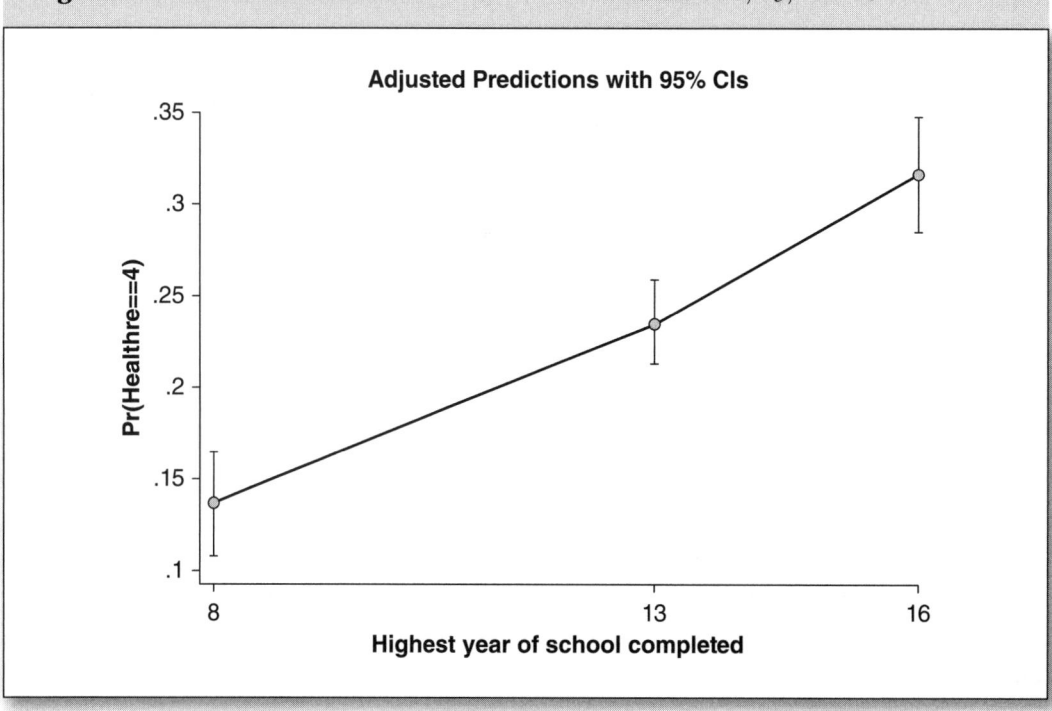

Note: CI = confidence interval.

Figure 4.1 created by marginsplot shows that with the increase of the highest year of school completed, the probabilities of being at category 4 increase.

After each of the four plots is created using marginsplot, we can combine all four graphs for the estimated probabilities of being in each category (i.e., $Y = 1, 2, 3,$ and 4) into a single graph (Figure 4.2) using the graph combine command. The detailed commands are as follows.

```
quietly ologit healthre i.maritals educ age i.male
margins, predict (outcome(1)) at(educ = (8 13 16)) atmeans vsquish
marginsplot, name(plot1)
margins, predict (outcome(2)) at(educ = (8 13 16)) atmeans vsquish
marginsplot, name(plot2)
margins, predict (outcome(3)) at(educ = (8 13 16)) atmeans vsquish
marginsplot, name(plot3)
margins, predict (outcome(4)) at(educ = (8 13 16)) atmeans vsquish
marginsplot, name(plot4)
graph combine plot1 plot2 plot3 plot4, ycommon
```

Figure 4.2 Estimated Probabilities of Being in Categories 1, 2, 3, and 4 for educ

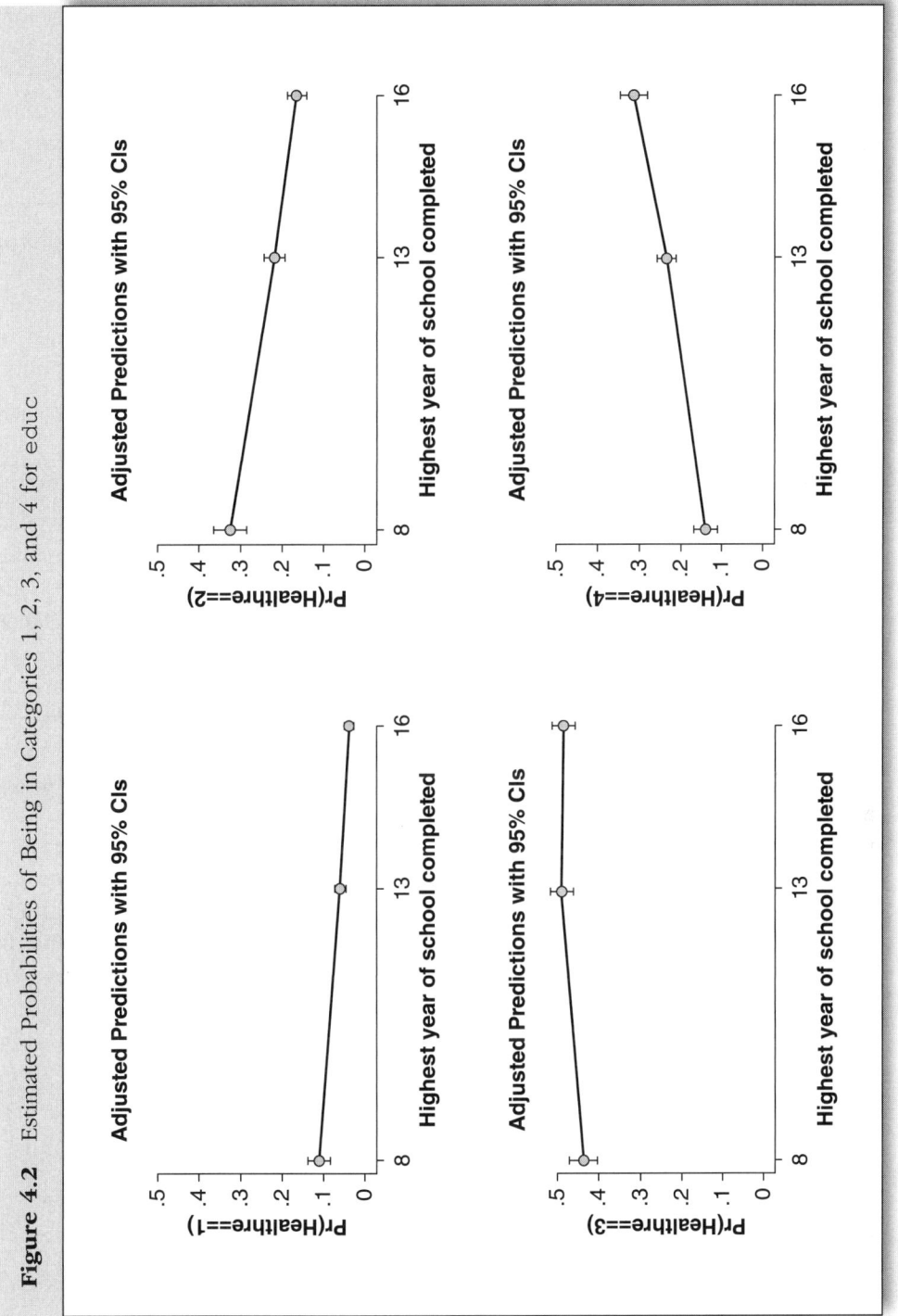

Note: CI = confidence interval.

The command `graph combine plot1 plot2 plot3 plot4, ycommon,` combines all four plots named plot1 to plot4, which are created by the `marginsplot` command.

As we can see, with the increase of the highest year of school completed, the probabilities of being in categories 1 and 2 decrease, whereas the probabilities of being in categories 3 and 4 increase.

Figure 4.2 demonstrates how the `margins` command can be used for the continuous variable `educ`. This command can also be easily used to estimate the probabilities for categorical predictor variables. A simple syntax is the command `margins` followed by a category predictor variable. For example, the command `margins maritals, predict (outcome(1))` computes the estimated probabilities of being in category 1 for the categorical variable `maritals`. To save space, the output of the results is omitted.

Estimated Probabilities With the Improved *margins* and *marginsplot* Commands in Stata 14

Both the `margins` and `marginsplot` commands for ordinal response variables in Stata 14 have been improved. Compared with the same commands in Stata 13, the updated commands in Stata 14 are more powerful and the command syntax is simpler. Specifically, in Stata 13, we need to compute the estimated probabilities for each category of an ordinal response variable separately. So if the ordinal response variable has four categories, we need to run the `margins` command four times with the `predict (outcome())` option. The command in Stata 14 now simultaneously provides the results for all categories of the ordinal response variable without requesting the `predict (outcome())` option.

To replicate the results in the preceding section, the command `margins, at(educ = (8 13 16)) atmeans vsquish` tells Stata 14 to compute the estimated probabilities for all four categories of the ordinal response variable when $Y = 1, 2, 3,$ and 4 for the predictor variable `educ` at the values of 8, 13, and 16 when holding the other predictor variables at their means. The output is shown as follows.

```
. margins, at(educ = (8 13 16)) atmeans vsquish

Adjusted predictions                          Number of obs     =      1,300
Model VCE    : OIM

1._predict   : Pr(healthre==1), predict(pr outcome(1))
2._predict   : Pr(healthre==2), predict(pr outcome(2))
3._predict   : Pr(healthre==3), predict(pr outcome(3))
4._predict   : Pr(healthre==4), predict(pr outcome(4))
1._at        : 0.maritals       =     .5315385 (mean)
```

```
               1.maritals     =     .4684615 (mean)
               educ           =            8
               age            =         48.2 (mean)
               0.male         =     .5561538 (mean)
               1.male         =     .4438462 (mean)
 2._at       : 0.maritals     =     .5315385 (mean)
               1.maritals     =     .4684615 (mean)
               educ           =           13
               age            =         48.2 (mean)
               0.male         =     .5561538 (mean)
               1.male         =     .4438462 (mean)
 3._at       : 0.maritals     =     .5315385 (mean)
               1.maritals     =     .4684615 (mean)
               educ           =           16
               age            =         48.2 (mean)
               0.male         =     .5561538 (mean)
               1.male         =     .4438462 (mean)
```

	Margin	Delta-method Std. Err.	z	P>|z|	[95% Conf. Interval]	
_predict#_at						
1 1	.1080545	.013559	7.97	0.000	.0814794	.1346295
1 2	.0582475	.0063735	9.14	0.000	.0457558	.0707393
1 3	.0396806	.0049499	8.02	0.000	.029979	.0493822
2 1	.3212358	.0209774	15.31	0.000	.2801209	.3623507
2 2	.219228	.0119511	18.34	0.000	.1958043	.2426516
2 3	.1644977	.0110984	14.82	0.000	.1427453	.1862501
3 1	.4346659	.0181304	23.97	0.000	.3991308	.4702009
3 2	.4868016	.0146464	33.24	0.000	.4580951	.515508
3 3	.4799718	.0146295	32.81	0.000	.4512985	.5086451
4 1	.1360439	.0144063	9.44	0.000	.107808	.1642798
4 2	.235723	.0121122	19.46	0.000	.2119836	.2594623
4 3	.3158499	.0158311	19.95	0.000	.2848216	.3468781

The first part of the output lists the number of the predicted probabilities for four ordinal categories (labeled from 1._predict to 4._predict) and three combinations of the predictor variables (labeled from 1._at to 3._at). The second part of the output is the table for the margins or estimated probabilities. The results are the same as those combined in the previous section using Stata 13. The interpretation of the results is omitted to avoid repetition.

Improved `marginsplot` Command in Stata 14

The improved `marginsplot` command in Stata 14 now can automatically plot the estimated probabilities of being in each category of an ordinal response variable in a single graph, or it can plot them separately with the `predict (outcome())` option. Previously in Stata 13, we needed to plot each estimated probability separately and then combine them into a single graph using the `graph combine` command.

The complicated command syntax in Stata 13 is now replaced by a single-word command `marginsplot` in Stata 14. Figure 4.3 shows the estimated probabilities when $Y = 1, 2, 3,$ and 4.

As we can see in Figure 4.3, the estimated probabilities of being in all four categories of health status are plotted in a single graph, and they look more straightforward than those in Figure 4.2.

Computing the Estimated Probabilities With the `mtable` Command in Stata 13 and Later

For Stata 13 and 14 users, an alternative way to compute estimated probabilities for each category of an ordinal response variable is to use `mtable`, which is part of the `SPost13` package (Long & Freese, 2014). Like the `margins` command in Stata 14, the `mtable` command simultaneously computes margins or estimated probabilities for all categories, which is more convenient than the `margins` command in Stata 13. The command `mtable, at(educ = (8 13 16)) atmeans ci` is used to replicate the results in the preceding section. In the syntax, the `ci` option requests the confidence interval. For more details on how to use this command, type `help mtable`.

```
. *mtable in Stata 13 and later
. quietly ologit healthre i.maritals edu age i.male

. mtable, at(educ = (8 13 16)) atmeans ci

Expression: Pr(healthre), predict(outcome())

          |     educ        1        2        3        4
----------+-----------------------------------------------
   Pr(y)  |        8    0.108    0.321    0.435    0.136
      ll  |        8    0.081    0.280    0.399    0.108
      ul  |        8    0.135    0.362    0.470    0.164
   Pr(y)  |       13    0.058    0.219    0.487    0.236
      ll  |       13    0.046    0.196    0.458    0.212
      ul  |       13    0.071    0.243    0.516    0.259
   Pr(y)  |       16    0.040    0.164    0.480    0.316
      ll  |       16    0.030    0.143    0.451    0.285
      ul  |       16    0.049    0.186    0.509    0.347

Specified values of covariates

          |        1.                   1.
          |  maritals      age         male
----------+-----------------------------------
 Current  |      .468     48.2         .444
```

Figure 4.3 Estimated Probabilities of Being in Categories 1, 2, 3, and 4 for educ
With Stata 14

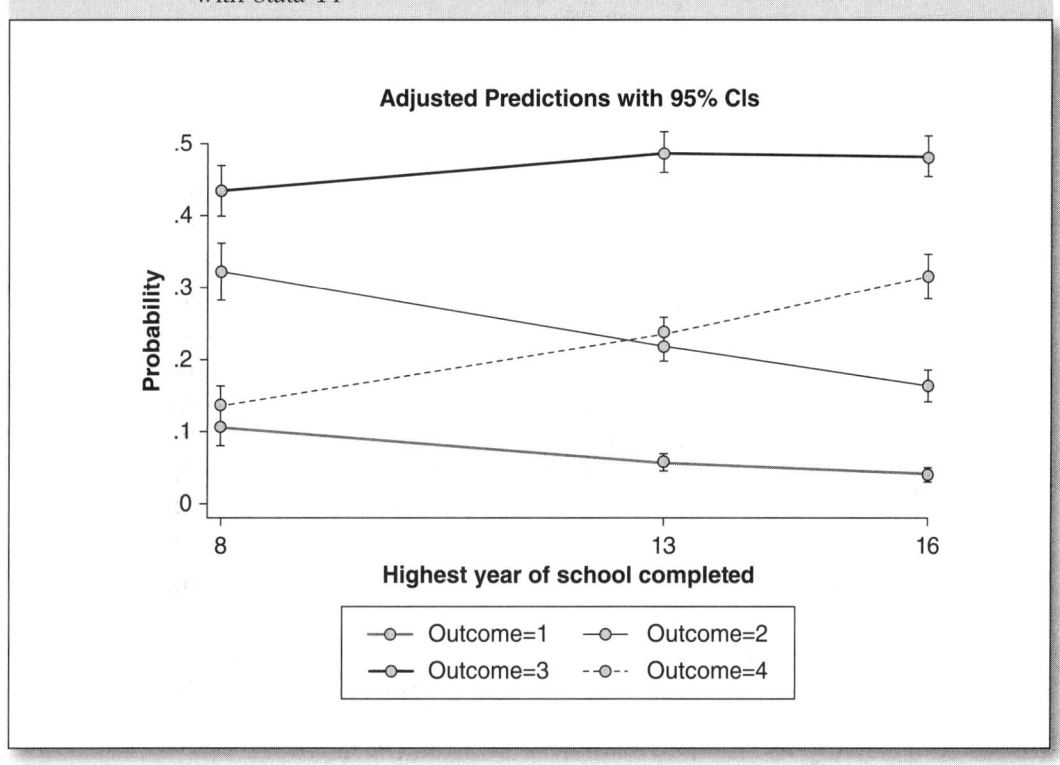

Note: CI = confidence interval.

The output contains two tables: One is for the estimated probabilities for each category numbered from 1 to 4 with the lower (ll) and upper limits (ul) of the confidence intervals, and the other is for the specified values, the means of the other three predictor variables. The estimated probabilities when $Y = 1, 2, 3$, and 4 for the predictor variable educ at the value of 8 are .108, .321, .435, and .136, respectively, when holding the other predictor variables at their means.

4.3.3 Model Comparisons Using the Log Likelihood Ratio Test and Other Fit Statistics

The log likelihood ratio test or the deviance difference test is used to compare the full model and the one-predictor model. We fit the single-predictor model and save the estimated log likelihood first, and then we fit the full model and save the estimated log likelihood. Finally, we compare the two models using the log likelihood ratio test. The following output is displayed.

```
. quietly xi: ologit healthre i.maritals

. estimates store model1

. quietly xi: ologit healthre i.maritals age educ i.male

. estimates store fullmodel

. lrtest model1 fullmodel

Likelihood-ratio test                              LR chi2(3)  =    107.44
(Assumption: model1 nested in fullmodel)           Prob > chi2 =     0.0000
```

The first command `quietly xi: logit healthre i.maritals` tells Stata to fit the single-predictor model quietly without showing the output. The command `estimates store model1` tells Stata to store the estimated log likelihood statistic for the previous model with a name of model1. Next, the command `quietly xi: logit healthre i.maritals age educ i.male` tells Stata to fit the full model quietly without providing the output. By using the command `estimates store fullmodel`, we store the estimated log likelihood statistics for the full model and name it "fullmodel". Finally, the command `lrtest model1 fullmodel` performs the log likelihood ratio test comparing the two models. If you do not specify `fullmodel` in the command, then the latest estimation results will be used for the comparison, so the command `lrtest model1` compares model1 with the latest model.

The log likelihood ratio chi-square test $\chi^2_{(3)}$ = 107.44, $p < .001$. The same result can be obtained if you compute it using the following equation:

$$G = D_{\text{Reduced}} - D_{\text{Full}} = -2 \times [-1566.197 - (-1512.479)] = 107.44, \text{df} = 4 - 1 = 3$$

Recall the `fitstat` command provides more fit statistics of the model. We can also compare them between the nested models. The procedure is to fit the reduced model first, save the model fit statistics using the `fitstat` command, and then fit the full model. Finally, compare the fit statistics between the two models.

The command `quietly xi: logit healths i.maritals` quietly fits the single-predictor model without showing the output. The command `quietly fitstat, saving(mod1)` requests the fit statistics and saves the results with a name "mod1". Keep in mind the name cannot be more than five characters, so give it a short name. The next command `quietly xi: logit healths i.maritals age educ i.male` fits the full model without showing the output. The fourth command `quietly fitstat, saving(full)` saves the results with a name of full. The command actually can be omitted, but it is good to have so you know the steps of analysis you need to follow. The final command `fitstat, using(mod1)` compares the final model (full) with the reduced model (mod1). The output is shown as follows.

```
. quietly xi: ologit healthre i.maritals

. quietly fitstat, saving(mod1)

. quietly xi: ologit healthre i.maritals age educ i.male

. quietly fitstat, saving(full)

. fitstat, using(mod1)
```

	Current	Saved	Difference
Log-likelihood			
Model	-1512.479	-1566.197	53.719
Intercept-only	-1579.407	-1579.407	0.000
Chi-square			
D (df=1293/1296/-3)	3024.957	3132.395	-107.437
LR (df=4/1/3)	133.856	26.419	107.437
p-value	0.000	0.000	0.000
R2			
McFadden	0.042	0.008	0.034
McFadden (adjusted)	0.038	0.006	0.032
McKelvey & Zavoina	0.105	0.021	0.084
Cox-Snell/ML	0.098	0.020	0.078
Cragg-Uhler/Nagelkerke	0.107	0.022	0.085
Count	0.464	0.457	0.007
Count (adjusted)	0.013	0.000	0.013
IC			
AIC	3038.957	3140.395	-101.437
AIC divided by N	2.338	2.416	-0.078
BIC (df=7/4/3)	3075.148	3161.075	-85.927
Variance of			
e	3.290	3.290	0.000
y-star	3.676	3.361	0.316

Note: Likelihood-ratio test assumes saved model nested in current model.

Difference of 85.927 in BIC provides very strong support for current model.

In the preceding box containing measures-of-fit statistics, the first column displays the names of various model fit statistics, the second column displays values of model fit statistics for the final model (labeled Current), the third column shows those for the one-predictor model (labeled Saved), and the last column displays the differences in the fit statistics between these two models (labeled Difference).

By examining the differences between the models, we find the deviance difference (labeled D) equals the difference between the log likelihood ratio test (labeled LR). All pseudo R^2 measures improve from the one-predictor model to the full model, which indicates the full model has a better fit than the one-predictor model. The AIC and BIC statistics decrease from the one-predictor model to the full model. Recall that smaller AIC and BIC statistics mean a better fit of the model.

4.4 Making Publication-Quality Tables

```
.quietly xi: ologit healthre i.maritals

.outreg2 using chap4out, e(ll df_m chi2) addstat(Pseudo R-squared, `e(r2_p)')
word replace

.outreg2 using chap4out, eform word append

.quietly xi: ologit healthre i.maritals age educ i.male

.outreg2 using chap4out, e(ll df_m chi2) addstat(Pseudo R-squared, `e(r2_p)') word
append

.outreg2 using chap4out, eform word append

.seeout
```

The first command `quietly xi: ologit healthre i.maritals` estimates the single-predictor model without providing the output.

The second command `outreg2 using chap4out, e(ll df_m chi2) addstat(Pseudo R-squared, `e(r2_p)')` word replace tells Stata to create a regression table for the estimate and save it to a file named chap4out. The `e(ll df_m chi2)` option asks Stata to include the estimated log likelihood value, degrees of freedom, and the chi-square test in the table. The `addstat` option asks Stata to include the pseudo $R^2$. The `word` option requests that the table be created in the Word document with the extension .rtf. The `replace` option asks Stata to replace any files with the same name as chap4out.

The third command `outreg2 using chap4out, eform word append` tells Stata to add the estimated odds ratios to the table as a separate column by using the `eform` option. The `append` option appends the results to the original table.

The fourth command `quietly xi: ologit healthre i.maritals age educ i.male` estimates the full model with four predictors without showing the output.

The fifth command `outreg2 using chap4out, e(ll df_m chi2) addstat(Pseudo R-squared, `e(r2_p)')` word append creates a regression table for the full model and appends it to the original table.

The sixth command `outreg2 using chap4out, eform word append` reports the odds ratios of the estimates for the full model and the results are appended to the original table by having an extra column.

The seventh command `seeout` opens the final table in the data browser. Click the blue link in the Stata results window, and you can see the created table.

These commands automatically produce Table 4.2, as shown here in its original format, presenting the results of both the single-predictor and the full logistic regression model.

With some edits to the original table, the final table is shown in Table 4.3. The edits include adding the model titles, labeling the logit coefficients and odds ratios, modifying variable names and names of fit statistics, and setting values up to three decimals.

Table 4.2   Results of the Proportional Odds Models: Single-Predictor Model and Full Model (Shown in Original Format Generated by Stata)

Variables	(1)	(2)	(3)	(4)
Healthre				
_Imaritals_1	0.533***	1.704***	0.514***	1.672***
	(0.104)	(0.177)	(0.106)	(0.177)
Age			−0.0193***	0.981***
			(0.00305)	(0.00299)
Educ			0.134***	1.144***
			(0.0173)	(0.0197)
_Imale_1			0.0531	1.054
			(0.105)	(0.111)
cut1				
Constant	−2.467***	0.0849***	−1.701***	0.182***
	(0.121)	(0.0102)	(0.298)	(0.0543)
cut2				
Constant	−0.736***	0.479***	0.125	1.133
	(0.0762)	(0.0365)	(0.286)	(0.324)

(Continued)

Table 4.2 (Continued)

Variables	(1)	(2)	(3)	(4)
cut3				
Constant	1.270***	3.563***	2.258***	9.563***
	(0.0824)	(0.294)	(0.294)	(2.809)
Observations	1,300	1,300	1,300	1,300
Pseudo R-squared	0.00836		0.0424	
Ll	−1566		−1512	
df_m	1		4	
chi2	26.42		133.9	

Note: Standard errors in parentheses.

***$p < 0.01$, **$p < 0.05$, *$p < 0.1$

Table 4.3   Results of the Proportional Odds Models: Single-Predictor Model and Full Model (Edited)

Variables	Single-Predictor Model		Full Model	
	b (SE(b))	OR(SE)	b (SE(b))	OR(SE)
Maritals	0.533***	1.704***	0.514***	1.672***
	(0.104)	(0.177)	(0.106)	(0.177)
Age			−0.019***	0.981***
			(0.003)	(0.003)
Educ			0.134***	1.144***
			(0.017)	(0.020)
Male			0.0531	1.054
			(0.105)	(0.111)
cut1				
α_1	−2.467***	0.085***	−1.701***	0.182***
	(0.121)	(0.010)	(0.298)	(0.054)

Variables	Single-Predictor Model		Full Model	
	b (SE(*b*))	OR(SE)	*b* (SE(*b*))	OR(SE)
cut2				
α_2	−0.736***	0.479***	0.125	1.133
	(0.076)	(0.037)	(0.286)	(0.324)
cut3				
α_3	1.270***	3.563***	2.258***	9.563***
	(0.082)	(0.294)	(0.294)	(2.809)
Observations	1,300	1,300	1,300	1,300
LR R^2	0.008		0.042	
Log likelihood	−1566		−1512	
df_m	1		4	
LR χ^2	26.42		133.90	

Note: Standard errors are shown in parentheses.

***$p < .01$.

4.5 Reporting the Results

Writing the results of ordinal logistic regression models is similar to that of binary logistic regression models.

First, describe the statistical method you used for data analysis, the dependent variable and the independent variables in the models, and your research hypothesis, or the purpose of your study. Second, report the model fit statistics, including but not limited to the log likelihood ratio statistic and the associated p value, and the pseudo R^2, followed by a concise statement of interpretation on whether the fitted model is better than the null model. If more fit statistics, such as various pseudo R^2 values, deviance statistics, and AIC and BIC statistics are computed, then include them in a table.

Third, report the parameter estimates for the predictor variables, their standard errors, the associated p values, and odds ratios either in a table or in the text. A table is preferable for models with multiple predictors. The odds ratios for each predictor should be interpreted.

You may have a table containing all the parameter estimates, p values, and odds ratios. If more than one model is fitted, then the results of all the competing models from the simple model to the full model should be presented in a table. The following is an example of summarizing results from the ordinal logistic regression model.

The ordinal logistic regression analysis was conducted to estimate the ordinal outcome variable, health status, from a set of predictor variables, such as marital status, years of education, age, and gender. A single-predictor model with marital status as the predictor was fitted first, and then the full model with all predictors was fitted. The log likelihood ratio test or the deviance difference test is used to compare the two models, $\chi^2_{(3)} = 107.44$, $p < .001$. The result indicated that the full model fitted data better than the single-predictor model. The results for both the single-predictor model and the full model are presented in Table 4.3.

For the full model, $LR\ \chi^2_{(4)} = 133.86$, $p < .001$, which indicated that the full model with four predictors provided a better fit than the null model with no independent variables in predicting the ordinal response variable. The likelihood ratio $R^2_L = .042$, which was larger than that of the one-predictor model but still small, suggesting that the relationship between the response variable, health status, and four predictors, was still small.

For the maritals predictor, OR = 1.672, which was greater than 1. It indicated that the odds of being above a particular category of health status (better health status) versus below that category for the married were 1.672 times as large as those for the unmarried when holding all the other predictors constant.

For the educ predictor, OR = 1.144, which was greater than 1. It indicated that the odds of being above a particular category of health status (better health status) increased by 1.144 for a one-unit increase in the predictor, education, when holding all the other predictors constant.

For the age predictor, OR = .981, which was less than 1. It indicated that the odds of being above a particular category of health status (better health status) decreased by a factor of .981 for a one-unit increase in the predictor, age, when all the other predictors remained constant. In other words, for a one-unit increase in age, the odds of being healthier decreased by 1.9%.

For the male predictor, $\beta = .053$, $p = .613$, which was not significantly different from 0; OR = 1.055, which almost equaled 1. It indicated that there is no relationship between being male and the cumulative odds of being in better health status. In other words, there was no significant difference in the odds of being in better health status between males and females.

4.6 Summary of Stata Commands in This Chapter

```
use chap4-gss2012, clear

*The one-predictor PO model
xi: ologit healthre i.maritals
xi: ologit healthre i.maritals, or
listcoef
fitstat
brant, detail

*The multiple-predictor PO model
xi: ologit healthre i.maritals educ age i.male
xi: ologit healthre i.maritals educ age i.male, or
listcoef
listcoef, reverse
fitstat
brant, detail

*margins and marginsplot in Stata 13
quietly ologit healthre i.maritals educ age i.male
margins, predict (outcome(1)) at(educ = (8 13 16)) atmeans vsquish
margins, predict (outcome(4)) at(educ = (8 13 16)) atmeans vsquish


*Combining plots using graph combine in Stata 13
quietly ologit healthre i.maritals educ age i.male
margins, predict (outcome(1)) at(educ = (8 13 16)) atmeans vsquish
marginsplot, name(plot1)
margins, predict (outcome(2)) at(educ = (8 13 16)) atmeans vsquish
marginsplot, name(plot2)
margins, predict (outcome(3)) at(educ = (8 13 16)) atmeans vsquish
marginsplot, name(plot3)
margins, predict (outcome(4)) at(educ = (8 13 16)) atmeans vsquish
marginsplot, name(plot4)
graph combine plot1 plot2 plot3 plot4, ycommon


*margins in Stata 14
quietly ologit healthre i.maritals educ age i.male
margins, at(educ = (8 13 16)) atmeans vsquish

*marginsplot in Stata 14
Marginsplot

*mtable in Stata 13 and later
quietly ologit healthre i.maritals educ age i.male
mtable, at(educ = (8 13 16)) atmeans ci

*Model comparison using the log likelihood ratio test
quietly xi: ologit healthre i.maritals age
estimates store model1
```

```
quietly xi: ologit healthre i.maritals age educ i.male
estimates store fullmodel
lrtest model1 fullmodel

*Model comparison using other fit statistics
quietly xi: ologit healthre i.maritals
quietly fitstat, saving(mod1)
quietly xi: ologit healthre i.maritals age educ i.male
quietly fitstat, saving(full)
fitstat, using(mod1)

*Making publication-quality tables
quietly xi: ologit healthre i.maritals
outreg2 using chap4out, e(ll df_m chi2) ///
addstat(Pseudo R-squared, `e(r2_p)')word replace
outreg2 using chap4out, eform word append
quietly xi: ologit healthre i.maritals age educ i.male
outreg2 using chap4out, e(ll df_m chi2) ///
addstat(Pseudo R-squared, `e(r2_p)') word append
outreg2 using chap4out, eform word append
seeout
```

4.7 EXERCISES

Use the GSS 2012 data for the following problems.

1. Conduct an analysis for a proportional odds model to estimate the ordinal response variable `fechld` from the five predictor variables `sex`, `educ`, `age`, `kidjob`, and `sibs`.

2. Identify the likelihood ratio test of the model, and interpret it.

3. Compute the deviance statistic for the model.

4. List two measures of pseudo R^2 and the AIC and BIC statistics computed by the `fitstat` command.

5. Identify the logit coefficient, the Wald z test, and the 95% confidence interval for the predictor variables `sex` and `educ`. Are they statistically significant?

6. Interpret the odds ratios for `sex`, `educ`, and `sibs`.

7. Use the `brant` command to test the proportional odds assumption, and interpret the results.

8. What are the important criteria you may use for model comparisons?

9. Make a publication-quality table containing the estimated logit coefficients and odds ratios.

10. Write a report to summarize the results from the output.

Partial Proportional Odds Models and Generalized Ordinal Logistic Regression Models

Objectives of This Chapter

This chapter presents partial proportional odds models and generalized ordinal logistic regression models when the proportional odds assumption is untenable. It begins with an introduction of the models, odds and odds ratio, model fit statistics, and interpretations of parameter estimates. After a description of the data, two examples on how to fit these two models using Stata are illustrated. Stata commands are explained, and output is interpreted in detail. This chapter focuses on fitting partial proportional odds models and generalized ordinal logistic regression models using Stata and on interpreting and presenting the results. After reading this chapter, you should be able to

- Identify when a partial proportional odds model or a generalized ordinal logistic regression model is used.
- Conduct analysis for both models using Stata.

(Continued)

(Continued)

- Interpret the output.
- Interpret the model in terms of odds ratios.
- Compute and plot the estimated probabilities using the `margins` and `marginsplot` commands, respectively.
- Compare models using the likelihood ratio test and other fit statistics.
- Present results in publication-quality tables using Stata.
- Write the results for publication.

5.1 Partial Proportional Odds Models and Generalized Ordinal Logistic Regression Models: An Introduction

In Chapter 4, we discussed the proportional odds (PO) model, which is used to estimate the cumulative probability of being at or below a particular level of the ordinal response variable or the complementary probability of being beyond that particular level. This model follows the assumption that the effect of each predictor is the same across the categories of the ordinal response variable. In other words, for each predictor, its effect on the ln odds of being at or below any category remains the same within the model. This restriction is referred to as the proportional odds assumption or the parallel lines assumption.

This assumption is strict and is often violated in real data analysis since the score test is strongly affected by sample size and the number of covariate patterns (Allison, 1999, 2012; Menard, 2010; O'Connell, 2006), for example, including continuous covariates as the predictors. What can we do if this assumption is violated? The first step is to identify the reason why the assumption does not hold by looking at the corresponding underlying binary models. We examine the coefficients for each predictor variable across the binary models and compare them with those from the proportional odds model (e.g., Allison, 1999, 2012; Bender & Grouven, 1998; Brant, 1990; Clogg & Shihadeh, 1994; Long, 1997; O'Connell, 2000, 2006). We also examine the pattern of the discrepancy and the number of predictor variables violating the assumption. Although not recommended, researchers may still continue to interpret the results with caution if the discrepancy is minor (Agresti, 2010). The next step is to look for any alternative models that are appropriate for the analysis when the PO assumption is untenable.

To deal with this issue, a better option is to fit the partial proportional odds (PPO) model or the generalized ordinal logit model (Fu, 1998; Liu & Koirala, 2012; Peterson & Harrell, 1990; Williams, 2006). In the PPO model, not all predictor variables violate the PO assumption, so the effects of those predictors violating the assumption are allowed to vary across categories. The generalized ordinal logit model can be considered an extreme case of the PPO model, and it allows the effect of each explanatory variable to vary. The original PPO model proposed by Peterson and Harrell (1990) specifies an interaction between a predictor variable that violates the PO assumption and different categories of the ordinal outcome variable and requires data restructuring. The programs developed by Fu (1998) and Williams (2006) make the analysis of generalized ordinal logit models and PPO models much easier. Their generalized ordinal logistic models relax the PO assumption by allowing the effect of each explanatory variable to vary across different cut points of the ordinal outcome variable without data restructuring. Williams's `gologit2` program (2006) for Stata was an extension of Fu's `gologit` (1998), and it is more powerful. It can estimate the generalized ordered logit model, PPO model, PO model, and logistic regression model within one program.

The generalized ordinal logistic regression model is an extension of the PO model by relaxing the PO assumption for all predictor variables. This model is expressed as follows:

$$\ln(Y_j') = \ln\left(\frac{\pi_j(\mathbf{x})}{1 - \pi_j(\mathbf{x})}\right) = \alpha_j + \left(\beta_{1j}X_1 + \beta_{2j}X_2 + \ldots + \beta_{pj}X_p\right) \tag{5.1}$$

Equation 5.1 also can be expressed as the one proposed by Fu (1998) and Williams (2006):

$$\text{logit} [\pi(Y > j \mid x_1, x_2, \ldots, x_p)] = \ln\left(\frac{\pi(Y > j \mid x_1, x_2, \ldots, x_p)}{\pi(Y \le j \mid x_1, x_2, \ldots, x_p)}\right) = \alpha_j + \left(\beta_{1j}X_1 + \beta_{2j}X_2 + \ldots + \beta_{pj}X_p\right) \tag{5.2}$$

where in both equations α_j are the intercepts or cut points and β_{1j}, β_{2j}, ..., β_{pj} are the logit coefficients. This model estimates the odds of being beyond a certain category relative to being at or below that category. A positive logit coefficient normally indicates that an individual is more likely to be in a higher category rather than in a lower category of the outcome variable. To estimate the odds of being at or below a particular category, however, the signs before both the intercepts and logit coefficients in Equation 5.2 need to be reversed.

In this expression, all of the effects of the explanatory variables are allowed to vary across each cut point. If some of these effects are found to be stable, they can be held constant like those in the PO model. So the PPO model is preferred to the generalized ordinal logistic regression if some predictor variables violate the assumption and their effects are estimated freely across different categories of the ordinal response variable.

5.1.1 Odds and Odds Ratios

The partial proportional odds model and the generalized ordinal logit model estimate the odds of being beyond a certain category versus being at or below that category. The cumulative odds in partial proportional odds models and generalized ordinal logit models are comparing $p(Y > j)$ and $p(Y \le j)$. In other words, the odds of being beyond a category are the probability of being above a category divided by the probability of being at or below that category:

$$\text{Odds}\ (Y > j) = \frac{p(Y > j)}{p(Y \le j)}$$

Odds $(Y > 1)$ equal the ratio of probability of being above category 1 to the probability of being at or below this category. The probability $p(Y > 1) = p(2) + p(3) + p(4)$ and the probability $p(Y \le 1) = p(1)$:

$$\text{Odds}\ (Y > 1) = \frac{p(Y > 1)}{1 - p(Y > 1)} = \frac{p(2) + p(3) + p(4)}{p(1)}$$

Odds $(Y > 2)$ equal the ratio of probability of being at or below category 2 to the probability of being above this category. Since $p(Y > 2) = p(3) + p(4)$ and $p(Y \le 2) = p(1) + p(2)$, the odds of being beyond category 2 can be expressed as follows:

$$\text{Odds}\ (Y > 2) = \frac{p(Y > 2)}{1 - p(Y > 2)} = \frac{p(3) + p(4)}{p(1) + p(2)}$$

Odds $(Y > 3)$ equal the ratio of probability of being above category 3 to the probability of being at or below this category. The equation is as follows:

$$\text{Odds}\ (Y > 3) = \frac{p(Y > 3)}{1 - p(Y > 3)} = \frac{p(4)}{p(1) + p(2) + p(3)}$$

Table 5.1 presents the logits, odds, and category comparisons for the PPO model/generalized ordinal logistic model for the health status with four levels.

Odds Ratios in PPO Models/Generalized
Ordinal Logistic Regression Models

Just like the odds ratios in the PO model, the odds ratios of being above a particular category in the PPO model and generalized ordinal logistic regression model are the exponentiated logit coefficients. An odds ratio is the change in the odds of being above a particular category for a one-unit increase from any value of x to the value of $(x + 1)$.

Table 5.1 Category Comparisons for the Partial Proportional Odds Model/ Generalized Ordinal Logistic Model With Four Levels of Health Status ($j = 1, 2, 3, 4$)

Category	Logit $P(Y > j)$	Odds	Probability Comparisons
Level 1	logit $P(Y > 1)$	$\dfrac{P(Y>1)}{P(Y\leq1)}$	Categories 2 through 4 vs. Category 1
Level 2	logit $P(Y > 2)$	$\dfrac{P(Y>2)}{P(Y\leq2)}$	Categories 3 and 4 vs. Categories 1 and 2
Level 3	logit $P(Y > 3)$	$\dfrac{P(Y>3)}{P(Y\leq3)}$	Category 4 vs. Categories 1, 2, and 3

5.1.2 Goodness of Fit

Since partial proportional odds models and generalized ordinal logistic regression models are extensions of proportional odds models, measures of fit statistics for the latter, such as the deviance, log likelihood ratio test, and pseudo R^2 measures, can be applied to the former models. The user-written programs `gologit` (Fu, 1998) and `gologit2` (Williams, 2006) do not work with the `listcoef` command (Long & Freese, 2014), so only limited fit statistics are available.

5.1.3 Interpretation of Model Parameter Estimates

The odds ratio in partial proportional odds models and generalized ordinal logistic regression models can be interpreted in the same way as that in the proportional odds regression. It can be interpreted as the change in the predicted logit or the log odds of being above a particular category relative to being at or below that category for a one-unit increase in the predictor variable.

Positive logit regression coefficients indicate positive relationships between the predictor variables and the logit function of the probability of being above a category. A positive coefficient corresponds to an OR greater than 1. So the odds of being beyond a particular category increase for a one-unit increase in the predictor variable.

When the logit coefficient itself is negative, the exponentiated coefficient, the OR, is less than 1. This means that the odds of being beyond a particular category decrease for a one-unit increase in the predictor variable.

When the logit coefficient equals 0, OR equals 1. It indicates that there is no relationship between the predictor and the odds.

To estimate the odds of being at or below a particular category, however, the signs before both the intercepts and logit coefficients in Equation 5.2 need to be reversed. Taking the inverse of the odds of being beyond a particular category gives us the odds of being at or below that category.

5.2 Research Example and Description of the Data and Sample

Research Problem and Questions: This chapter focuses on the same research problem as that in Chapter 4. We will still investigate the relationships between the ordinal response variable, health status, and four predictor variables, marital status, the highest education, age, and gender. Unlike in Chapter 4, however, here the research interest will focus on using the PPO model and the generalized ordinal logistic regression model when the PO assumption is violated. The research question is as follows: Do the four predictor variables significantly predict the ordinal response variable, health status? Specifically, do the four predictor variables significantly predict the cumulative odds and then the cumulative probabilities of being above a particular level of health status when the proportional odds assumption is violated?

Description of the Data and Sample: The data for the following analyses were the General Social Survey 2012 (GSS 2012). The same variables for the proportional odds model from Chapter 4 are used for data analysis in this chapter.

5.3 Partial Proportional Odds Models and Generalized Ordinal Logistic Models With Stata: Commands and Output

5.3.1 Stata Commands and Output

The Stata Commands `gologit` *and* `gologit2`

The user-written program `gologit` (Fu, 1998) can be used to fit the generalized ordinal logistic regression models. The other program `gologit2` (Williams, 2006) is an extension of `gologit` and can be used to fit both the partial proportional odds models and the generalized ordinal logistic models. Just like the commands for other regression models, the syntax for the generalized ordinal logistic regression is the command `gologit`, or `gologit2`, followed by the dependent variable and the independent variables. To fit a partial proportional odds model, the command `gologit2` is used. Since both are user-written programs, you need to install them first by typing `ssc install gologit` and then `ssc install gologit2` before running them. Please note that only the updated `gologit2` currently works with the factor variable coding and the `margins` and `margins-plot` commands. If you have a previous version installed and would like to update it, then type `ssc install gologit2, replace`. The latest `gologit2` command (updated in 2015) requires Stata 11.2 or higher and works with Stata 14. Users with older versions of Stata (e.g., Stata 9 or 10) should install the `gologit29` program.

The Partial PO Model: One-Predictor Model

Recall that the Brant tests for the PO assumption in Chapter 4 are violated for the overall model. After examining each predictor variable, we find that the assumption is upheld for `maritals, age,` and `male`, whereas it is violated for `educ`. For `educ`, the Brant test $\chi^2_{(1)} = 12.98$, $p = .002$, which is significant. Just like the one-predictor PO model, in the following one-predictor PPO model, we still choose `maritals` as the predictor variable. The varying effect of `educ`, which violates the PO assumption, will be explained once we fit the multiple-predictor PPO model.

The command `gologit2 healthre i.maritals, autofit lrforce store (mod1)` tells Stata to fit the partial proportional odds model for the ordinal outcome variable `healthre` using the predictor variable `maritals`. In the syntax, the factor variable notation for the categorical variable `maritals, i.maritals`, is used since `gologit2` has been updated by the author to work with factor variables. Users with older versions of Stata need to use the `xi` prefix command for categorical predictor variables.

The `autofit` option is used to select automatically the variables that meet the PO assumption and constrain those variables violating the assumption. Other options alternative to the `autofit` command include the `pl` command and the `npl` command. With the `pl` option, it assumes that all predictor variables meet the PO assumption, so the results are the same as those estimated from the `ologit` command, whereas the `npl` command assumes that all predictor variables violate the PO assumption, so the results are identical to those estimated from the `gologit` command. You can also specify predictor variables that meet the PO assumption with the `pl(varlist)` option or variables violating the assumption with the `npl(varlist)` option where the names of the variables are placed in the parenthesis. The `lrforce` command reports the likelihood ratio test statistics even when it is not available. For example, it reports the result when the sample size is different between nested models. The `store(mod1)` command stores the estimates for the following model comparison and names it `mod1`. The results of the one-predictor model are presented as follows.

```
. gologit2 healthre i.maritals, autofit lrforce store(mod1)

------------------------------------------------------------------------
Testing parallel lines assumption using the .05 level of significance...

Step  1:  Constraints for parallel lines imposed for 1.maritals (P Value = 0.7707)
Step  2:  All explanatory variables meet the pl assumption

Wald test of parallel lines assumption for the final model:

 ( 1)   [1]1.maritals - [2]1.maritals = 0
 ( 2)   [1]1.maritals - [3]1.maritals = 0

        chi2(  2) =      0.52
      Prob > chi2 =    0.7707

An insignificant test statistic indicates that the final model
does not violate the proportional odds/ parallel lines assumption

If you re-estimate this exact same model with gologit2, instead
of autofit you can save time by using the parameter

pl(0b.maritals 1.maritals)

------------------------------------------------------------------------

Generalized Ordered Logit Estimates        Number of obs   =      1,300
                                            LR chi2(1)      =      26.42
                                            Prob > chi2     =     0.0000
Log likelihood = -1566.1973                 Pseudo R2       =     0.0084

 ( 1)   [1]1.maritals - [2]1.maritals = 0
 ( 2)   [2]1.maritals - [3]1.maritals = 0
------------------------------------------------------------------------
   healthre |     Coef.   Std. Err.      z    P>|z|    [95% Conf. Interval]
------------+-----------------------------------------------------------
1           |
   maritals |
          1 |  .5327629   .1041722    5.11   0.000   .3285891   .7369367
            |
      _cons |  2.466528   .1205342   20.46   0.000   2.230285   2.702771
------------+-----------------------------------------------------------
2           |
   maritals |
          1 |  .5327629   .1041722    5.11   0.000   .3285891   .7369367
            |
      _cons |  .7358156   .0761853    9.66   0.000   .5864951   .8851361
------------+-----------------------------------------------------------
3           |
   maritals |
          1 |  .5327629   .1041722    5.11   0.000   .3285891   .7369367
            |
      _cons | -1.270491   .082395   -15.42   0.000  -1.431983     -1.109
------------------------------------------------------------------------
```

Interpreting the Output

The output includes two sections. The first section displays the result of the test for the PO assumption, and the second section presents the results of the generalized ordinal logistic regression model. At the beginning of the output, the proportional odds assumption or the parallel assumption is tested at the .05 significance level. It starts with the sentence "Testing parallel lines assumption using the .05 level of significance". This is just the same as the Brant test. At each step, only one predictor variable is tested. Since this is a one-predictor model, we only need to look at Step 1. The *p* value for the tested predictor maritals (labeled "P Value = 0.7707") is not significant. It indicates that the PO assumption is met for this variable so there will be a constraint imposed on this variable in the model. Therefore, the estimated logit coefficients for this variable will be the same across three binary logistic models for the dichotomized ordinal response variable. Step 2 states that "All explanatory variables meet the pl assumption" since the maritals variable is the only predictor in the model.

Next is the Wald test of the parallel lines or proportional odds assumption for the final model, which is the equivalent to the Brant test for the overall model.

The first equation [1]1.maritals - [2]1.maritals = 0 tests whether the coefficient from two binary models, numbered 1 and 2, are the same. The second equation [2]1.maritals - [3]1.maritals = 0 tests whether the coefficient from two binary models, numbered 2 and 3, are the same. $\chi^2_{(2)} = .52, p = .771$, indicates that the proportional odds assumptions for the model are upheld. The same result can be found from that of the Brant test in Chapter 4.

At the end of the Wald test of the PO assumption, the output lists the command pl(0b.maritals 1.maritals). It means that the pl command with the factor variable 1.maritals in the parenthesis produces the same result as that of the autofit command.

The second section of the output shows the results of the generalized ordinal logistic regression model. The top right of the table displays the number of observations in the dataset, the log likelihood ratio test statistic, the associated *p* value, and the pseudo R^2. The results are the same as those in the one-predictor PO model in Chapter 4.

Just like the PO model, the regression coefficient table displays the parameter estimates, standard errors, Wald *z* statistics, associated *p* values, and the 95% confident intervals of the parameter estimates. Unlike the regression tables for the PO models in Chapter 4, the table produced by the gologit2 command displays the parameter estimates for three underlying binary logistic models whose outcome variables are dichotomized from the ordinal response variable. The three models, numbered 1, 2, and 3, respectively, compare outcomes being above a particular category relative to being at or below that category. The first model compares categories

2 through 4 with category 1, the second model compares categories 3 and 4 with categories 1 and 2, and the third model compares category 4 with categories 1 through 3, respectively.

The logit regression coefficients for all three binary models are the same since the PO assumption for the predictor variable is not violated. $\beta = .533$, which indicates that for a one-unit increase in the predictor variable, the change in the logit or log odds of being beyond a category is .533.

The intercepts for three binary models, labeled "_cons", are 2.467, .736, and −1.270, respectively. Recall that the three cut points, _cut1, _cut2, and_cut3 in the PO model in Chapter 4, are −2.467, −.736, and 1.270, respectively. They are the same in the absolute values but have reversed signs because the PO model estimates the odds of being at or below a particular category compared with being above that category, whereas the generalized ordinal logistic regression model estimates the odds of being above a category versus at or below a category.

To request estimated odds ratios, we can add the option or to the original gologit2 command so the syntax is gologit2 healthre i.maritals, or. The command can be simplified by omitting the variable names. The output for the odds ratios is displayed as follows.

```
. gologit2, or

Generalized Ordered Logit Estimates            Number of obs    =      1,300
                                               LR chi2(1)       =      26.42
                                               Prob > chi2      =     0.0000
Log likelihood = -1566.1973                    Pseudo R2        =     0.0084

 ( 1)   [1]1.maritals - [2]1.maritals = 0
 ( 2)   [2]1.maritals - [3]1.maritals = 0
------------------------------------------------------------------------------
    healthre | Odds Ratio   Std. Err.      z    P>|z|     [95% Conf. Interval]
-------------+----------------------------------------------------------------
1            |
    maritals |
           1 |   1.703633    .1774712    5.11   0.000     1.389007    2.089525
             |
        _cons |   11.78147     1.42007   20.46   0.000     9.302521    14.92101
-------------+----------------------------------------------------------------
2            |
    maritals |
           1 |   1.703633    .1774712    5.11   0.000     1.389007    2.089525
             |
        _cons |   2.087184    .1590127    9.66   0.000     1.797677    2.423314
-------------+----------------------------------------------------------------
3            |
    maritals |
           1 |   1.703633    .1774712    5.11   0.000     1.389007    2.089525
             |
        _cons |   .2806936    .0231277  -15.42   0.000     .2388349    .3298886
------------------------------------------------------------------------------
```

The odds ratios for `maritals` are the same across all three binary models (labeled 1, 2 and 3, respectively). OR = 1.704, $p < .001$, indicates that the odds of being beyond a particular category of health status (better health status) versus being at or below that category (poorer health status) for the married are 1.704 times the odds for the unmarried.

The Partial PO Model: Multiple-Predictor Model

In the multiple-predictor model, all four predictor variables are included. The command `gologit2 healthre i.maritals educ age i.male, autofit lrforce store (mod2)` tells Stata to build the partial proportional odds model to estimate the ordinal outcome variable `healthre` using all four predictor variables `maritals`, `edu`, `age`, and `male`. The `store(mod2)` command in the option stores the estimates and names it mod2, which will be used for the likelihood ratio test to compare this model with the one-predictor model. The results of the multiple-predictor model are shown as follows.

```
. gologit2 healthre i.maritals educ age i.male, autofit lrforce store(mod2)

------------------------------------------------------------------------------
Testing parallel lines assumption using the .05 level of significance...

Step  1:  Constraints for parallel lines imposed for 1.maritals (P Value = 0.6899)
Step  2:  Constraints for parallel lines imposed for age (P Value = 0.1634)
Step  3:  Constraints for parallel lines imposed for 1.male (P Value = 0.1189)
Step  4:  Constraints for parallel lines are not imposed for
          educ (P Value = 0.00056)

Wald test of parallel lines assumption for the final model:

 ( 1)   [1]1.maritals - [2]1.maritals = 0
 ( 2)   [1]1.male - [2]1.male = 0
 ( 3)   [1]age - [2]age = 0
 ( 4)   [1]1.maritals - [3]1.maritals = 0
 ( 5)   [1]1.male - [3]1.male = 0
 ( 6)   [1]age - [3]age = 0

            chi2( 6) =     8.66
          Prob > chi2 =    0.1938

An insignificant test statistic indicates that the final model
does not violate the proportional odds/ parallel lines assumption

If you re-estimate this exact same model with gologit2, instead
of autofit you can save time by using the parameter

pl(0b.maritals 1.maritals 0b.male 1.male age)

------------------------------------------------------------------------------

Generalized Ordered Logit Estimates          Number of obs   =      1,300
                                              LR chi2(6)      =     147.50
                                              Prob > chi2     =     0.0000
Log likelihood = -1505.6567                   Pseudo R2       =     0.0467
```

```
( 1)   [1]1.maritals - [2]1.maritals = 0
( 2)   [1]1.male - [2]1.male = 0
( 3)   [1]age - [2]age = 0
( 4)   [2]1.maritals - [3]1.maritals = 0
( 5)   [2]1.male - [3]1.male = 0
( 6)   [2]age - [3]age = 0
```

healthre	Coef.	Std. Err.	z	P>|z|	[95% Conf. Interval]	
1						
maritals						
1	.5195306	.1060293	4.90	0.000	.3117169	.7273442
educ	.1737568	.034743	5.00	0.000	.1056619	.2418518
age	-.019021	.0030503	-6.24	0.000	-.0249994	-.0130426
male						
1	.0458268	.1049599	0.44	0.662	-.1598908	.2515443
_cons	1.212205	.4677279	2.59	0.010	.2954756	2.128935
2						
maritals						
1	.5195306	.1060293	4.90	0.000	.3117169	.7273442
educ	.1773769	.0214273	8.28	0.000	.1353802	.2193737
age	-.019021	.0030503	-6.24	0.000	-.0249994	-.0130426
male						
1	.0458268	.1049599	0.44	0.662	-.1598908	.2515443
_cons	-.6783392	.3257222	-2.08	0.037	-1.316743	-.0399354
3						
maritals						
1	.5195306	.1060293	4.90	0.000	.3117169	.7273442
educ	.0867662	.0209624	4.14	0.000	.0456806	.1278518
age	-.019021	.0030503	-6.24	0.000	-.0249994	-.0130426
male						
1	.0458268	.1049599	0.44	0.662	-.1598908	.2515443
_cons	-1.588762	.3346783	-4.75	0.000	-2.24472	-.9328049

Interpreting the Output

The first section tests the proportional odds assumption or the parallel assumption at the .05 significance level with each step testing one predictor variable. Step 1 tests the PO assumption for the maritals predictor. The p value for the tested predictor maritals (labeled "P Value = 0.6899") is not significant. It indicates that the PO assumption is met for this variable, so there will be a constraint imposed on this variable in the model. Steps 2 and 3 are for the predictors, age and male, respectively. Both p values are not significant (labeled "P Value = 0.1634" for age and "P Value = 0.1189" for male), which indicates that the PO assumption is met for the two variables. The final step, Step 4, tests the PO assumption for the predictor variable educ. The p value is statistically significant (labeled "P Value = 0.00056"), and it states that "Constraints for parallel lines are not imposed for educ", so this predictor violates the PO assumption and its effect will be allowed to vary across the three binary models.

The Wald test of the parallel lines or proportional odds assumption for the final model $\chi^2_{(6)} = 8.66$, $p = .1938$, indicates that the proportional odds assumption for the model is upheld. At the end of the Wald test of the PO assumption, the command pl(0b.maritals 1.maritals 0b.male 1.male age) tells us only three predictor variables, maritals, age, and male, meet the PO assumption. With the pl() option, the same result can be produced as that of the autofit command.

Just like with the single-predictor model, the second section of the output presents the generalized ordinal logit estimates. The top right of the table displays the summary fit statistics. The number of observations is still 1,300. The log likelihood ratio chi-square test $LR \chi^2_{(6)} = 147.50$, $p < .001$, indicates that the full model with four predictor variables provides a better fit than the null model with no independent variables in predicting the ordinal response variable. The likelihood ratio $R^2_L = .0467$ is still small. Compared with the one-predictor model $R^2_L = .0084$, the full model shows a better fit.

Just like with the single-predictor model, the coefficient table displays the parameter estimates of predictor variables for three underlying binary logistic models. The three models estimate the logit or log odds of being above a category versus at or below that category. Again, model 1 compares categories 2, 3, and 4 with category 1, model 2 compares categories 3 and 4 with categories 1 and 2, and model 3 compares category 4 with categories 1 through 3, respectively.

When reading the coefficients table, we need to look at the estimates for the predictor variables that meet the PO assumption and those violating the assumption separately since the effects of the former variables are constrained but the effects of those latter variables are free to vary. In the model, three predictor variables maritals, age, and male meet the PO assumption, so the equal-slope constraints are placed on those variables. Each of these three variables has the same logit regression coefficient across all three binary models. For maritals, $\beta = .520$; for age, $\beta = -.019$; and for male, $\beta = .046$.

The educ predictor is the only one that violates the PO assumption, so its effect is allowed to vary across the three binary models. The estimated logit coefficients are .174, .177, and .087 for each respective model. In model 1, the Wald z test = 5.00, $p < .001$; in model 2, Wald $z = 8.28$, $p < .001$; and in model 3, Wald $z = 4.14$, $p < .001$. The results of the Wald z tests indicate that logit coefficients for educ are significant across all three models.

To request estimated odds ratios, we can add the option or to the original gologit2 command. The command gologit2 healthre i.maritals edu age i.male, autofit lrforce or tells Stata to report the odds ratios and their standard errors with the Wald test of the PO assumption and the model fit statistics the same as those in the last model. The following output displays the odds ratio.

```
. gologit2 healthre i.maritals educ age i.male, autofit lrforce or

-----------------------------------------------------------------------------
Testing parallel lines assumption using the .05 level of significance...

Step  1:  Constraints for parallel lines imposed for 1.maritals (P Value = 0.6899)
Step  2:  Constraints for parallel lines imposed for age (P Value = 0.1634)
Step  3:  Constraints for parallel lines imposed for 1.male (P Value = 0.1189)
Step  4:  Constraints for parallel lines are not imposed for
          educ (P Value = 0.00056)

Wald test of parallel lines assumption for the final model:

 ( 1)   [1]1.maritals - [2]1.maritals = 0
 ( 2)   [1]1.male - [2]1.male = 0
 ( 3)   [1]age - [2]age = 0
 ( 4)   [1]1.maritals - [3]1.maritals = 0
 ( 5)   [1]1.male - [3]1.male = 0
 ( 6)   [1]age - [3]age = 0

          chi2(  6) =      8.66
        Prob > chi2 =     0.1938

An insignificant test statistic indicates that the final model
does not violate the proportional odds/ parallel lines assumption

If you re-estimate this exact same model with gologit2, instead
of autofit you can save time by using the parameter

pl(0b.maritals 1.maritals 0b.male 1.male age)

-----------------------------------------------------------------------------

Generalized Ordered Logit Estimates          Number of obs   =      1,300
                                              LR chi2(6)      =     147.50
                                              Prob > chi2     =     0.0000
Log likelihood = -1505.6567                   Pseudo R2       =     0.0467

 ( 1)   [1]1.maritals - [2]1.maritals = 0
 ( 2)   [1]1.male - [2]1.male = 0
 ( 3)   [1]age - [2]age = 0
 ( 4)   [2]1.maritals - [3]1.maritals = 0
 ( 5)   [2]1.male - [3]1.male = 0
 ( 6)   [2]age - [3]age = 0
```

healthre	Odds Ratio	Std. Err.	z	P>|z|	[95% Conf. Interval]	
1						
maritals						
1	1.681238	.1782606	4.90	0.000	1.365768	2.069577
educ	1.189766	.041336	5.00	0.000	1.111446	1.273605
age	.9811588	.0029928	-6.24	0.000	.9753105	.9870421
male						
1	1.046893	.1098817	0.44	0.662	.8522369	1.28601
_cons	3.360889	1.571981	2.59	0.010	1.343765	8.405912
2						
maritals						
1	1.681238	.1782606	4.90	0.000	1.365768	2.069577
educ	1.194081	.0255859	8.28	0.000	1.144972	1.245297
age	.9811588	.0029928	-6.24	0.000	.9753105	.9870421
male						
1	1.046893	.1098817	0.44	0.662	.8522369	1.28601
_cons	.5074591	.1652907	-2.08	0.037	.2680068	.9608515
3						
maritals						
1	1.681238	.1782606	4.90	0.000	1.365768	2.069577
educ	1.090642	.0228625	4.14	0.000	1.04674	1.136385
age	.9811588	.0029928	-6.24	0.000	.9753105	.9870421
male						
1	1.046893	.1098817	0.44	0.662	.8522369	1.28601
_cons	.2041782	.068334	-4.75	0.000	.1059572	.3934486

Interpreting the Output

The first section of the output presents the results of the parallel lines or PO assumption, which is the same as the one in the last output. We are interested in the odds ratios reported in the coefficients table in the second section.

First, let us take a look at the odds ratios for the predictor variables that meet the PO assumption. They are 1.681238, .9811588, and 1.046893 for marital status, age, and male, respectively. Among them, the odds ratios for marital status and age are significant, whereas the odds ratio for male is not significant. For marital status, Wald $z = 4.90$, $p < .001$; for age, Wald $z = -6.24$, $p < .001$; and for male, Wald $z = .44$, $p = .662$.

Second, we look at the predictor variables that violate the PO assumption. Only one predictor variable, educ, violates the assumption. The odds ratios for highest years of school completed are different across three binary models. They are 1.190, 1.194, and 1.091, respectively.

Interpreting the Odds Ratios of Being Beyond a Particular Category

The partial PO model and the generalized ordinal logistic model using the gologit2 command are parameterized to estimate the odds ratios of being beyond a particular category versus being at or below that category. The odds ratios in the partial PO model and the generalized ordinal logistic model can be interpreted in a similar way as that in the PO model. It can be interpreted as the change in the odds of being beyond a particular category for a one-unit increase in the predictor variable when holding all the other predictors constant.

A positive regression coefficient corresponds to an odds ratio of being beyond a particular category greater than 1. It means that the odds increase with the increase in the predictor variable. A negative regression coefficient corresponds to an odds ratio of being beyond a particular category less than 1. It indicates the odds of being beyond a category decrease with the increase in the predictor variable. A logit coefficient of zero corresponds to an odds ratio of one, which indicates there is no relationship between the predictor and the odds of being beyond a category, with other variables held fixed.

Unlike the PO model, the partial PO model allows the effects of some of the predictor variables to vary, so we need to interpret the odds ratios for the predictor variables that meet the PO assumption and those violating the assumption separately.

Three predictor variables, maritals, age, and male, meet the PO assumption in the model. Let us interpret their odds ratios first.

For the maritals predictor, $\beta = .520$, and its corresponding OR = 1.681. It indicates that the odds of being beyond a particular category of health status (better health status) versus being at or below that category (poorer health status) for the married are 1.681 times the odds for the unmarried when holding other predictors constant. If we interpret it in a percentage change, it means that the odds of being beyond a particular category of health status for the married are 68.1% higher than the odds for the unmarried when adjusting for other predictors.

For the age predictor, $\beta = -.019$, which is negative, so there is a negative relationship between age and the log odds of being above a category of health status; OR = .981, which is less than 1. It indicates that the odds of being beyond a particular category of health status (better health status) decrease by a factor of .981 for a one-unit increase in the predictor age when holding other variables constant. In other words, a one-unit increase in age is associated with a decrease by 1.9% in the odds of being healthier.

For the `male` predictor, $\beta = .046$, $p = .662$, which is not significantly different from 0; OR = 1.047, which almost equals 1. It indicates that there is no change in the odds for being male. In other words, there is no relationship between being male and the cumulative odds of being in better health status.

Next, let us interpret the predictor variables that violate the PO assumption. As identified from the output, only one predictor variable, `educ`, violates the assumption, so the odds ratios for `educ` are different across three binary models. The three odds ratios are 1.190, 1.194, and 1.091, respectively. Overall, the odds of being beyond a particular category of health status increase with the increase in `educ`. In other words, the number of years of school completed is associated with the odds of being in better health status. For each unit increase in `educ`, the odds increase by 19%, 19.4%, and 9.1% in each of the binary logistic models, respectively. The largest odds ratio is identified in the second binary model comparing categories 3 and 4 with categories 1 and 2, and the smallest odds ratio is found in the last binary model comparing category 4 with categories 1 through 3.

Interpreting the Odds Ratios of Being at or Below a Particular Category

In partial PO models and generalized ordinal logistic regression models, we need to reverse the signs before the intercepts and the logit coefficients. The equation is expressed as follows:

$$\ln\left(\frac{\pi(Y \le j \mid x_1, x_2, \ldots x_p)}{\pi(Y > j \mid x_1, x_2, \ldots x_p)}\right) = -\alpha_j - \left(\beta_{1j}X_1 + \beta_{2j}X_2 + \ldots \beta_{pj}X_p\right) \tag{5.3}$$

Exponentiating the logit coefficients with negative signs before them gives us the odds ratios of being at or below a particular category. It is obvious that the odds of being at or below a particular category $\exp(-\beta)$ are the inverse of the odds of being beyond that category $\exp(\beta)$ since $\exp(-\beta) = 1/\exp(\beta)$.

Although you can use the command `listcoef` with the `reverse` option to get the odds ratios of being at or below a particular category relative to beyond that category for the PO model after the `ologit` command, it currently does not work with the `gologit2` command. Computing the inverse of a number using Stata is a simple task with the `display` command.

Interpreting the Estimated Probabilities
With the `margins` Command in Stata 13

The updated `gologit2` command now works with the `margins` command in Stata 13 and 14. As explained in Chapter 4, the command `margins, predict (outcome(1)) at(educ = (8 13 16)) atmeans vsquish` tells Stata to compute the estimated probabilities for the category $Y = 1$ for the predictor variable educ at the values of 8, 13, and 16 when holding the other predictor variables at their means. Before running the margins command, the same partial PO model is fitted with the factor variable notation since the xi prefix command does not work with margins. You may notice that the original margins command for Stata 13 may provide slightly different results when it is executed in Stata 14. The results from the latest program are reported as follows.

```
. quietly gologit2 healthre i.maritals educ age male, autofit lrforce

. margins, predict (outcome(1)) at(educ = (8 13 16)) atmeans vsquish

Adjusted predictions                               Number of obs   =       1300
Model VCE    : OIM

Expression   : Pr(healthre==1), predict(outcome(1))
1._at        : 0.maritals        =     .5315385  (mean)
               1.maritals        =     .4684615  (mean)
               educ              =            8
               age               =         48.2  (mean)
               male              =     .4438462  (mean)
2._at        : 0.maritals        =     .5315385  (mean)
               1.maritals        =     .4684615  (mean)
               educ              =           13
               age               =         48.2  (mean)
               male              =     .4438462  (mean)
3._at        : 0.maritals        =     .5315385  (mean)
               1.maritals        =     .4684615  (mean)
               educ              =           16
               age               =         48.2  (mean)
               male              =     .4438462  (mean)

------------------------------------------------------------------------------
             |            Delta-method
             |    Margin   Std. Err.      z    P>|z|     [95% Conf. Interval]
-------------+----------------------------------------------------------------
        _at  |
          1  |   .124644   .0197317     6.32   0.000     .0859705    .1633175
          2  |   .0563617  .0064995     8.67   0.000     .043623     .0691004
          3  |   .0342498  .0060306     5.68   0.000     .0224301    .0460696
------------------------------------------------------------------------------
```

In the output, for the first combination (labeled 1._at), when educ = 8 and the other predictor variables are held at their means, the estimated margin or the probability when $Y = 1$ is .125.

For the second combination (labeled 2._at), when educ = 13 and the other three predictor variables are held at their means, the estimated probability when $Y = 1$ is .056.

For the third combination (labeled 3._at), when educ = 16 and the other predictor variables are held at their means, the estimated probability when $Y = 1$ is .034.

These three predicted probabilities are significantly different from 0 since $p < .001$.

The expected probabilities for other categories (i.e., 2, 3, and 4) of the ordinal response variable can be computed in a similar way with the predict (outcome()) option. To save space, the output for the expected probabilities for categories 2 and 3 is omitted, and the results for the expected probability for category 4 are only displayed in the following output.

```
. margins, predict (outcome(4)) at(educ = (8 13 16)) atmeans vsquish

Adjusted predictions                                    Number of obs    =        1300
Model VCE      : OIM

Expression    : Pr(healthre==4), predict(outcome(4))
1._at         : 0.maritals       =      .5315385  (mean)
                1.maritals       =      .4684615  (mean)
                educ             =             8
                age              =          48.2  (mean)
                male             =      .4438462  (mean)
2._at         : 0.maritals       =      .5315385  (mean)
                1.maritals       =      .4684615  (mean)
                educ             =            13
                age              =          48.2  (mean)
                male             =      .4438462  (mean)
3._at         : 0.maritals       =      .5315385  (mean)
                1.maritals       =      .4684615  (mean)
                educ             =            16
                age              =          48.2  (mean)
                male             =      .4438462  (mean)

------------------------------------------------------------------------------
             |            Delta-method
             |    Margin   Std. Err.      z    P>|z|     [95% Conf. Interval]
-------------+----------------------------------------------------------------
         _at |
          1  |   .1754153    .020218     8.68   0.000     .1357888    .2150417
          2  |   .2471459   .0125134    19.75   0.000     .2226202    .2716717
          3  |   .2986795   .0163069    18.32   0.000     .2667186    .3306404
------------------------------------------------------------------------------
```

In the output, when educ = 8, 13, and 16 and other predictor variables are held at their means, the margins or estimated probabilities for category 4 are .175, .247, and .299, respectively. They are all statistically different from zero. To visualize the results, the command marginsplot is used directly after the margins command. Figure 5.1 shows the estimated probabilities when $Y = 4$ (i.e., excellent health condition).

Figure 5.1 shows that with the increase of the highest year of school completed, the probabilities of being at category 4 increase.

Figure 5.1 Estimated Probabilities When $Y = 4$ for educ at 8, 13, and 16

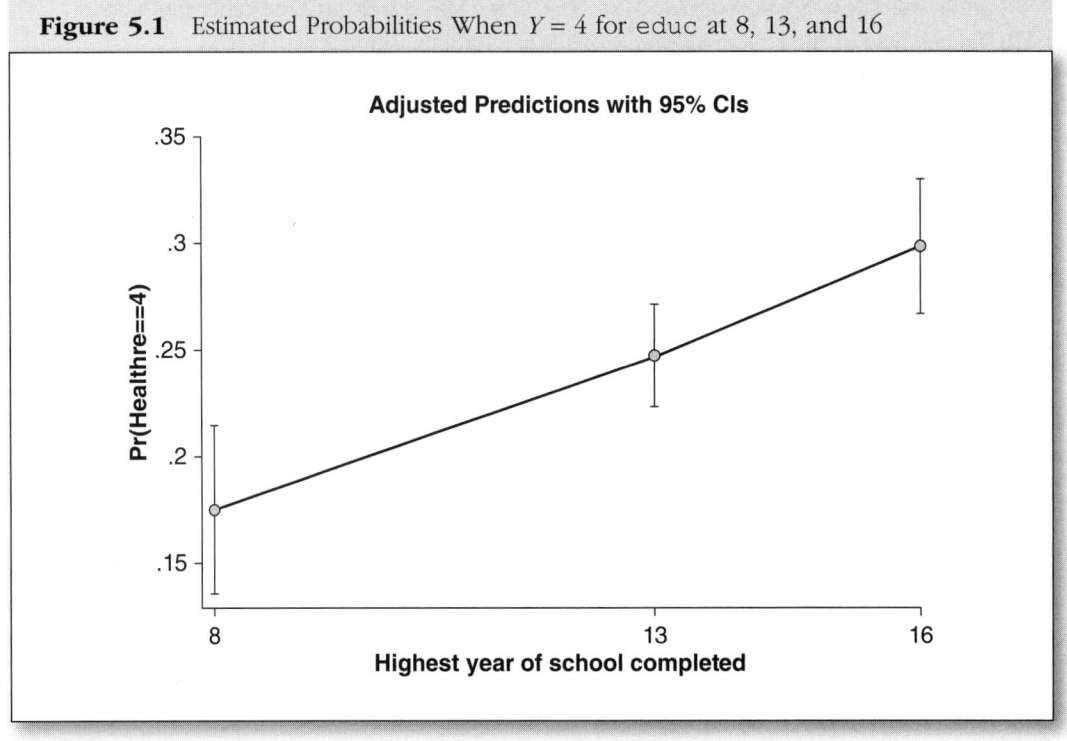

Note: CI = confidence interval.

As explained in Chapter 4, after each of the four plots is created using `margins-plot`, we can use the `graph combine` command to combine all four graphs for the estimated probabilities of being in each ordinal outcome (i.e., $Y = 1, 2, 3,$ and 4) into a single graph. Figure 5.2 is created with the following commands.

```
margins, predict (outcome(1)) at(educ = (8 13 16)) atmeans vsquish
marginsplot, name(mplot1)
margins, predict (outcome(2)) at(educ = (8 13 16)) atmeans vsquish
marginsplot, name(mplot2)
margins, predict (outcome(3)) at(educ = (8 13 16)) atmeans vsquish
marginsplot, name(mplot3)
margins, predict (outcome(4)) at(educ = (8 13 16)) atmeans vsquish
marginsplot, name(mplot4)
graph combine mplot1 mplot2 mplot3 mplot4, ycommon
```

Figure 5.2 Estimated Probabilities of Being in Categories 1, 2, 3, and 4 for educ

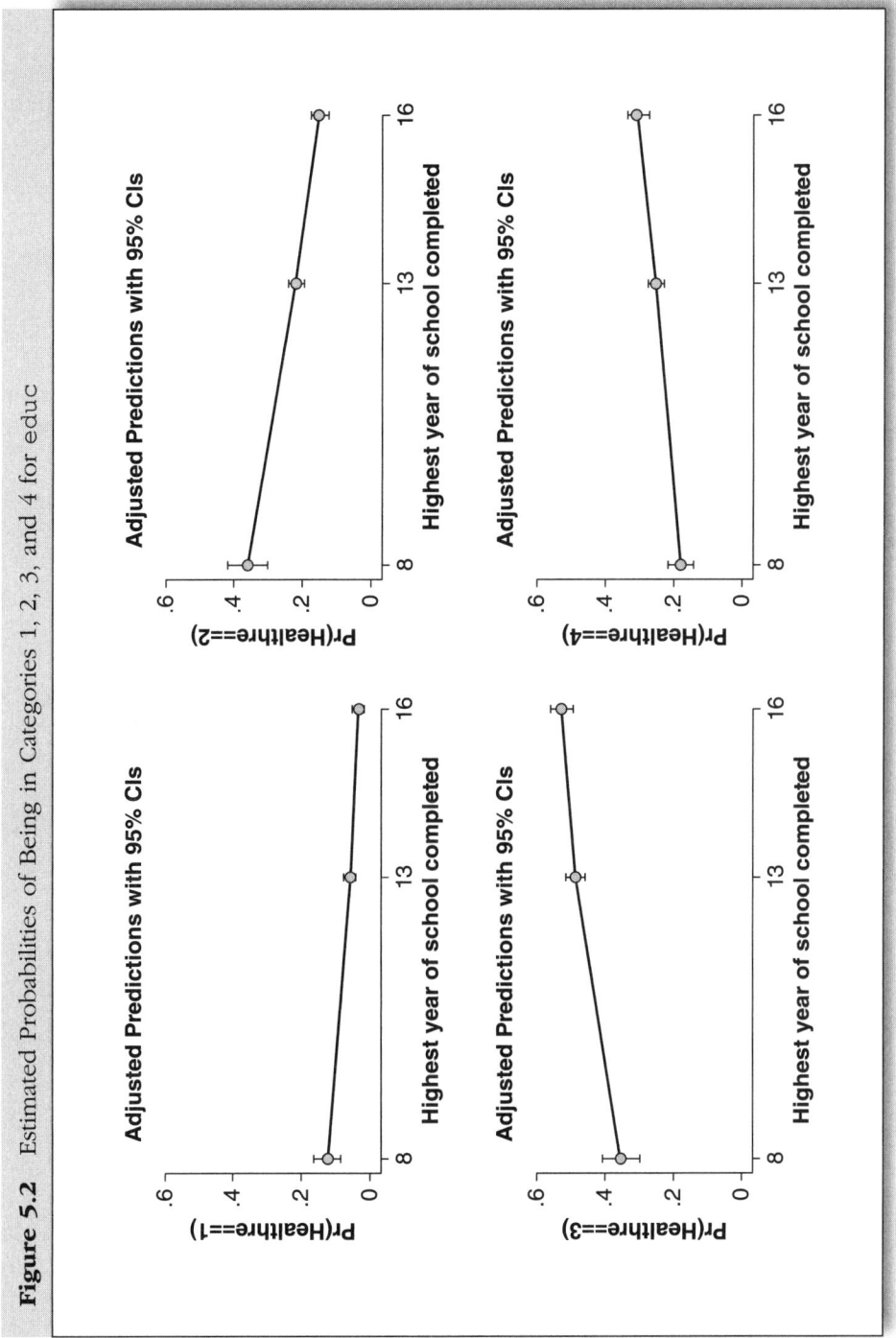

Note: CI = confidence interval.

The command `graph combine mplot1 mplot2 mplot3 mplot4, ycommon`, combines all four plots named mplot1 to mplot4, which are created by the `marginsplot` command.

As we can see, with the increase of the highest year of school completed, the probabilities of being in categories 1 and 2 decrease, whereas the probabilities of being in categories 3 and 4 increase.

Estimated Probabilities With the Improved *margins* and *marginsplot* Commands in Stata 14

In Stata 13, we need to compute the estimated probabilities for each category of an ordinal response variable separately. With the improved `margins` and `marginsplot` commands for ordinal response variables in Stata 14, we can simultaneously obtain and plot the results for all categories of the ordinal response variable without requesting the `predict (outcome())` option.

To replicate the results in the preceding section, the command `margins, at(educ = (8 13 16)) atmeans vsquish` tells Stata 14 to compute the estimated probabilities for all four categories of the ordinal response variable when $Y = 1$, 2, 3, and 4 for the predictor variable educ at the values of 8, 13, and 16 when holding the other predictor variables at their means. The output is shown as follows.

```
. *margins and marginsplot in Stata 14

. margins, at(educ = (8 13 16)) atmeans vsquish

Adjusted predictions                              Number of obs    =     1,30
Model VCE       : OIM

1._predict      : Pr(healthre==1), predict(pr outcome(1))
2._predict      : Pr(healthre==2), predict(pr outcome(2))
3._predict      : Pr(healthre==3), predict(pr outcome(3))
4._predict      : Pr(healthre==4), predict(pr outcome(4))
1._at           : 0.maritals      =    .5315385  (mean)
                  1.maritals      =    .4684615  (mean)
                  educ            =           8
                  age             =        48.2  (mean)
                  male            =    .4438462  (mean)
2._at           : 0.maritals      =    .5315385  (mean)
                  1.maritals      =    .4684615  (mean)
                  educ            =          13
                  age             =        48.2  (mean)
                  male            =    .4438462  (mean)
3._at           : 0.maritals      =    .5315385  (mean)
                  1.maritals      =    .4684615  (mean)
                  educ            =          16
                  age             =        48.2  (mean)
                  male            =    .4438462  (mean)
```

		Delta-method					
		Margin	Std. Err.	z	P>|z|	[95% Conf. Interval]	
_predict#_at							
1	1	.124644	.0197317	6.32	0.000	.0859705	.1633175
1	2	.0563617	.0064995	8.67	0.000	.043623	.0691004
1	3	.0342498	.0060306	5.68	0.000	.0224301	.0460696
2	1	.3534737	.0294035	12.02	0.000	.295844	.4111034
2	2	.2176288	.0120106	18.12	0.000	.1940884	.2411691
2	3	.1471933	.0127722	11.52	0.000	.1221602	.1722264
3	1	.346467	.0289403	11.97	0.000	.2897451	.403189
3	2	.4788636	.0147205	32.53	0.000	.450012	.5077153
3	3	.5198774	.0179745	28.92	0.000	.4846479	.5551068
4	1	.1754153	.020218	8.68	0.000	.1357888	.2150417
4	2	.2471459	.0125134	19.75	0.000	.2226202	.2716717
4	3	.2986795	.0163069	18.32	0.000	.2667186	.3306404

The first part of the output lists the notes for the predicted probabilities for each of the four ordinal categories (labeled from 1._predict to 4._predict) and three combinations of the predictor variables (labeled from 1._at to 3._at). The second part of the output is the table for the margins or estimated probabilities. The results are the same as those in the previous section using Stata 13 if the estimated probabilities for each of the four categories are combined.

To create a margins plot using Stata 13 in the preceding section, we need to plot each estimated probability separately using the marginsplot command and then combine them into a single graph using the graph combine command. With the improved marginsplot command in Stata 14, we now can automatically plot the estimated probabilities of being in each category of an ordinal response variable in a single graph. Figure 5.3 shows the estimated probabilities when $Y = 1, 2, 3,$ and 4.

The same estimated probabilities for all categories can be computed automatically using the mtable command (Long & Freese, 2014), which works in Stata 13 and 14. The results estimated by the margins command in Stata 14 can be replicated using the syntax mtable, at(educ = (8 13 16)) atmeans ci. The output is omitted here. See Chapter 4 for more details on this command.

Model Comparison Using the Log Likelihood Ratio Test

```
. lrtest mod1 mod2

Likelihood-ratio test                              LR chi2(5)   =    121.08
(Assumption: mod1 nested in mod2)                  Prob > chi2 =     0.0000
```

Figure 5.3   Estimated Probabilities of Being in Categories 1, 2, 3, and 4 for educ
With Stata 14

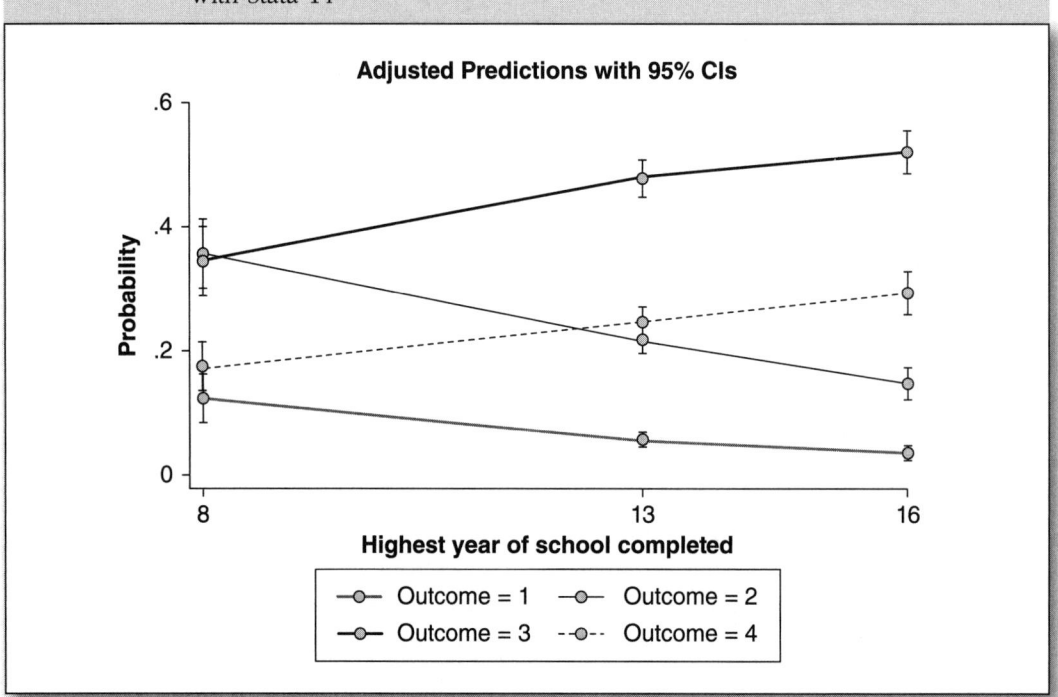

The command lrtest mod1 mod2 performs the log likelihood ratio test comparing the single-predictor model and the full model. The log likelihood ratio chi-square test $\chi^2_{(5)}$ = 121.08, p < .001. The same result can be obtained if you compute it using the following equation:

$$G = D_{\text{Reduced}} - D_{\text{Full}} = -2 \times [-1566.1973 - (-1505.6567)] = 121.08, \text{df} = 6 - 1 = 5$$

5.4. Generalized Ordinal Logistic Regression Models With Stata: An Example

5.4.1 Stata Commands and Output

The command xi: gologit healthre i.maritals educ age male tells Stata to estimate a generalized ordinal logistic regression model with the outcome variable healthre and four predictor variables using the gologit command. The xi prefix command is required since the gologit command does not work with the factor variable notation. This model relaxes the proportionality assumption by allowing the logits coefficients of all predictor variables to vary across the ordinal response variable. Similar to a series of underlying binary logistic regression models, where the data are dichotomized across different categories, the effects of the predictor variables estimated by the generalized ordinal logistic regression model could vary freely. In the command, no equal constraints are put on all four predictors, so each predictor will have three different coefficients. Please note that the current `gologit` command does not work with the `margins` command. The output is displayed as follows.

```
. xi: gologit healthre i.maritals educ age male
i.maritals        _Imaritals_0-1    (naturally coded; _Imaritals_0 omitted)
Iteration 0:  Log Likelihood = -1579.4068
Iteration 1:  Log Likelihood = -1520.3229
Iteration 2:  Log Likelihood =  -1501.484
Iteration 3:  Log Likelihood = -1501.3663
Iteration 4:  Log Likelihood = -1501.3663
Iteration 5:  Log Likelihood = -1501.3663


Generalized Ordered Logit Estimates                Number of obs   =     1300
                                                   Model chi2(12)  =   156.08
                                                   Prob > chi2     =   0.0000
Log Likelihood = -1501.3663303                     Pseudo R2       =   0.0494

------------------------------------------------------------------------------
    healthre |      Coef.   Std. Err.      z    P>|z|     [95% Conf. Interval]
-------------+----------------------------------------------------------------
mleq1        |
_Imaritals_1 |   .4348576   .2418395     1.80   0.072    -.0391391    .9088542
        educ |   .1720031   .0351942     4.89   0.000     .1030237    .2409824
         age |  -.0270331   .0064289    -4.20   0.000    -.0396335   -.0144327
        male |  -.2864255   .2332502    -1.23   0.219    -.7435875    .1707365
        _cons |   1.852722    .602811     3.07   0.002     .6712339     3.03421
-------------+----------------------------------------------------------------
mleq2        |
_Imaritals_1 |   .4557347   .1315739     3.46   0.001     .1978546    .7136147
        educ |   .1770585   .0213208     8.30   0.000     .1352705    .2188465
         age |  -.0215827   .0036757    -5.87   0.000     -.028787   -.0143784
        male |  -.0746521    .130169    -0.57   0.566    -.3297786    .1804745
        _cons |   -.462046   .3461825    -1.33   0.182    -1.140551    .2164593
-------------+----------------------------------------------------------------
mleq3        |
_Imaritals_1 |   .5816019   .1294529     4.49   0.000     .3278789     .835325
        educ |   .0867118   .0207867     4.17   0.000     .0459706    .1274529
         age |  -.0152159   .0038937    -3.91   0.000    -.0228475   -.0075844
        male |   .1934242   .1281262     1.51   0.131    -.0576986    .4445469
        _cons |   -1.86184   .3544246    -5.25   0.000    -2.556499    -1.16718
------------------------------------------------------------------------------
```

The same results can be obtained if you use the `gologit2` command with the npl option, for example, `gologit2 healthre i.maritals, npl lrforce`.

To fit a generalized ordinal logistic regression model, the `gologit2` command is recommended since it works with both the factor variable coding and the `margins` command.

Interpreting the Output

The top of the output displays the iterative process of log likelihood estimation, which is the same as other logistic regression models. The maximum log likelihood value is −1501.366.

The top right of the output displays the number of observations in the dataset, the log likelihood ratio test statistic and the associated p value, and the pseudo R^2.

The number of observations is 1,300. The log likelihood ratio chi-square test statistic `LR chi2 (12)` = 156.08. It is the difference in the −2 log likelihood (−2LL) between the current model that contains the predictor variables and the intercepts (LL at Iteration 5 = −1501.366) and the null model that contains only the intercepts (LL at Iteration 0 = −1579.407).

The associated p value with the log likelihood ratio chi-square test `Prob > chi2` = `0.0000` indicates that the null hypothesis is rejected. Therefore, the estimated model with four predictor variables provides a better fit than the null model with no independent variables in predicting cumulative odds of being above a category of health status.

The likelihood ratio R^2_L = .0494, which is the pseudo R^2, suggesting that the relationship between the response variable, health status, and the predictors is small.

The generalized ordinal logistic regression table displays the coefficients for the predictor variable and the intercepts, their standard errors, the Wald z statistics, the associated p values, and the 95% confident intervals. Three binary logistic regression equations, labeled `mleq1`, `mleq2`, and `mleq3`, respectively, estimate outcomes dichotomizing the ordinal response variable.

The logit coefficients of all four variables are different across three equations/models. For example, the regression coefficients for the first predictor variable, marital status, are .435, .456, and .582, respectively. A total of 12 logit coefficients are estimated so it will be tedious to interpret each coefficient.

Without retyping the variables, names, the command `xi: gologit, or` tells Stata to report the odds ratios. The output is displayed as follows.

```
. xi: gologit, or

Generalized Ordered Logit Estimates              Number of obs    =      1300
                                                 Model chi2(12)   =    156.08
                                                 Prob > chi2      =    0.0000
Log Likelihood =   -1501.3663303                 Pseudo R2        =    0.0494
```

```
------------------------------------------------------------------------
   healthre | Odds Ratio   Std. Err.      z    P>|z|    [95% Conf. Interval]
------------+-----------------------------------------------------------
mleq1       |
_Imaritals_1 |  1.544743    .3735798    1.80   0.072    .961617    2.481478
       educ |  1.187682    .0417995    4.89   0.000   1.108518    1.272499
        age |   .973329    .0062574   -4.20   0.000    .9611416    .9856709
       male |   .750943    .1751576   -1.23   0.219    .4754053   1.186178
------------+-----------------------------------------------------------
mleq2       |
_Imaritals_1 |  1.577332    .2075356    3.46   0.001   1.218785    2.041357
       educ |  1.193701    .0254506    8.30   0.000   1.144846    1.24464
        age |   .9786485   .0035972   -5.87   0.000    .9716234    .9857244
       male |   .9280663   .1208055   -0.57   0.566    .7190829   1.197786
------------+-----------------------------------------------------------
mleq3       |
_Imaritals_1 |  1.788902    .2315785    4.49   0.000   1.388021    2.305563
       educ |  1.090582    .0226696    4.17   0.000   1.047044    1.135931
        age |   .9848992   .0038349   -3.91   0.000    .9774115    .9924443
       male |  1.213397    .155468     1.51   0.131    .9439344   1.559783
------------------------------------------------------------------------
```

Just like the logit coefficients, the corresponding odds ratios of all four predictor variables vary across three binary models. The odds ratios can still be interpreted as the change in the odds comparing probabilities of being beyond a category versus at or below that category for a one-unit change in the predictor variable when holding all other predictors constant. For example, the odds ratios for age across three equations are .973, .979, and .985, respectively. They are similar since the PO assumption test is tenable for this predictor variable. They can be interpreted as the odds of being above a category decrease by 2.7%, 2.1%, and 1.5% across three comparisons, respectively, for a one-unit increase in age.

In the earlier partial PO model, we estimate 9 parameters, including 6 unique coefficients and 3 intercepts, whereas in the generalized ordinal logistic model, we estimate 15 parameters, including 12 unique coefficients and 3 intercepts. Compared with the generalized ordinal logistic model using gologit, the partial PO model using gologit2 is more parsimonious since it estimates six fewer parameters.

5.5 Making Publication-Quality Tables

```
1.  .quietly xi: gologit2 healthre i.maritals, autofit lrforce store(mod1)
2.  .quietly xi: gologit2 healthre i.maritals educ age i.male, autofit lrforce
    store(mod2)
```

```
3.  .outreg2 [mod1 mod2] using chap5out, e(ll df_m chi2) addstat(Pseudo R-squared,
    `e(r2_p)') long word replace
4.  .outreg2 [mod1 mod2] using chap5oddsout, eform e(ll df_m chi2) addstat(Pseudo
    R-squared, `e(r2_p)') long word replace
5.  .seeout
```

The first command `quietly xi: gologit2 healthre i.maritals, autofit lrforce store(mod1)` estimates the single-predictor model without providing the output and then saves the estimates with the name mod1.

The second command `quietly xi: gologit2 healthre i.maritals educ age i.male, autofit lrforce store(mod2)` tells Stata to estimates the full model without providing the output and then saves the estimates in a file named mod2.

The third command `outreg2[mod1 mod2] using chap5out, e(ll df_m chi2)addstat(Pseudo R-squared, `e(r2_p)') long word replace` tells Stata to create a regression table for the estimates from the two models and save it to a file named chap5out (Table 5.2). The option `e(ll df_m chi2)` asks Stata to include the estimated log likelihood value, degrees of freedom, and the chi-square statistic in the table. The `addstat` option requests the pseudo $R^2$. The `long` option tells Stata to save the estimates in one column for each underlying binary model. The `word` option requests that the table be created in the Microsoft Word document with the extension .rtf. The `replace` option asks Stata to replace any files with the same name as chap5out.

The fourth command `outreg2 [mod1 mod2] using chap5oddsout, eform e(ll df_m chi2) addstat (Pseudo R-squared, `e(r2_p)') long word replace` tells Stata to make a new table containing the estimated odds ratios by using the `eform` option (Table 5.4).

The fifth command `seeout` opens the final table in the data browser. By clicking the blue link in the Stata results window, you can see the created table, which is saved in the current working directory. The default is the tab delimited text file with the .txt extension.

The first three commands automatically produce Table 5.2, as shown here in its original format, presenting the logit coefficients for the PPO models, including both the single-predictor model and the full model.

With the edits to the model titles, labels for logit coefficients, variable names, model names, names of fit statistics, and decimals, the final table (Table 5.3) is displayed.

The fourth command using the `eform` option automatically produces Table 5.4, as shown here in its original format.

Table 5.5 is the final table with some edits. Please note that the pseudo R^2 for the single-predictor model is incorrectly displayed as that for the full model in the original table. This error was corrected in the edited table.

Table 5.2 Results (Coefficients) of the PPO Models: Single-Predictor Model and Full Model Using gologit2 ($Y >$ cat. j vs. $Y \leq$ cat. j) (Shown in Original Format Generated by Stata)

Variables	(1) mod1 healthre	(2) mod2 healthre
1		
_Imaritals_1	0.533***	0.520***
	(0.104)	(0.106)
Educ		0.174***
		(0.0347)
Age		−0.0190***
		(0.00305)
_Imale_1		0.0458
		(0.105)
2		
_Imaritals_1	0.533***	0.520***
	(0.104)	(0.106)
Educ		0.177***
		(0.0214)
Age		−0.0190***
		(0.00305)
_Imale_1		0.0458
		(0.105)
3		
_Imaritals_1	0.533***	0.520***
	(0.104)	(0.106)

(Continued)

Table 5.2 (Continued)

Variables	(1) mod1 healthre	(2) mod2 healthre
Educ		0.0868***
		(0.0210)
Age		−0.0190***
		(0.00305)
_Imale_1		0.0458
		(0.105)
1		
Constant	2.467***	1.212***
	(0.121)	(0.468)
2		
Constant	0.736***	−0.678**
	(0.0762)	(0.326)
3		
Constant	−1.270***	−1.589***
	(0.0824)	(0.335)
Observations	1,300	1,300
Pseudo R-squared	0.0467	0.0467
Ll	−1,566	−1,506
df_m	1	6
chi2	26.42	147.5

Note: Standard errors in parentheses.

***$p < 0.01$, **$p < 0.05$, *$p < 0.1$

Table 5.3   Results of the Partial Proportional Odds Model Using gologit2 ($Y >$ cat. j vs. $Y \leq$ cat. j): Coefficients (Edited)

Variables	(1) Single-predictor Model b (SE(b))	(2) Full Model b (SE(b))
Model 1 ($Y > 1$ vs. $Y \leq 1$)		
Maritals	0.533***	0.520***
	(0.104)	(0.106)
Educ		0.174***
		(0.0347)
Age		−0.019***
		(0.003)
Male		0.046
		(0.105)
Model 2 ($Y > 2$ vs. $Y \leq 2$)		
Maritals	0.533***	0.520***
	(0.104)	(0.106)
Educ		0.177***
		(0.021)
Age		−0.0190***
		(0.003)
Male		0.046
		(0.105)
Model 3 ($Y > 3$ vs. $Y \leq 3$)		
Maritals	0.533***	0.520***
	(0.104)	(0.106)
Educ		0.087***
		(0.021)

(Continued)

Table 5.3 (Continued)

Variables	(1) Single-predictor Model b (SE(b))	(2) Full Model b (SE(b))
Age		−0.019***
		(0.003)
Male		0.046
		(0.105)
α_1	2.467***	1.212***
	(0.121)	(0.468)
α_2	0.736***	−0.678**
	(0.0762)	(0.326)
α_3	−1.270***	−1.589***
	(0.082)	(0.335)
Observations	1,300	1,300
Pseudo R^2	0.008	0.047
Ll	−1,566	−1,506
df_m	1	6
chi2	26.42	147.5

Note: Standard errors in parentheses.

***$p < .01$.

Table 5.4 Results of the Partial Proportional Odds Model Using gologit2 ($Y >$ cat. j vs. $Y \leq$ cat. j): Odds Ratios (Shown in Original Format Generated by Stata)

Variables	(1) mod1 healthre	(2) mod2 healthre
1		
_Imaritals_1	1.704***	1.681***
	(0.177)	(0.178)

Variables	(1) mod1 healthre	(2) mod2 healthre
Educ		1.190***
		(0.0413)
Age		0.981***
		(0.00299)
_Imale_1		1.047
		(0.110)
2		
_Imaritals_1	1.704***	1.681***
	(0.177)	(0.178)
Educ		1.194***
		(0.0256)
Age		0.981***
		(0.00299)
_Imale_1		1.047
		(0.110)
3		
_Imaritals_1	1.704***	1.681***
	(0.177)	(0.178)
Educ		1.091***
Age		0.981***
		(0.00299)
_Imale_1		1.047
		(0.110)
1		
Constant	11.78***	3.361***

(Continued)

Table 5.4 (Continued)

Variables	(1) mod1 healthre	(2) mod2 healthre
	(1.420)	(1.572)
2		
Constant	2.087***	0.507**
	(0.159)	(0.165)
3		
Constant	0.281***	0.204***
	(0.0231)	(0.0683)
Observations	1,300	1,300
Pseudo R-squared	0.0467	0.0467
Ll	−1,566	−1,506
df_m	1	6
chi2	26.42	147.5

Note: Standard errors in parentheses.

***$p < 0.01$, **$p < 0.05$, *$p < 0.1$

Table 5.5 Results of the Partial Proportional Odds Model Using gologit2 ($Y >$ cat. j vs. $Y \leq$ cat. j): Odds Ratios (Edited)

Variables	(1) Single-Predictor Model OR (SE)	(2) Full Model OR(SE)
Model 1 ($Y > 1$ vs. $Y \leq 1$)		
Maritals	1.704***	1.681***
	(0.177)	(0.178)

Variables	(1) Single-Predictor Model OR (SE)	(2) Full Model OR(SE)
Educ		1.190***
		(0.041)
Age		0.981***
		(0.003)
Male		1.047
		(0.110)
Model 2 ($Y > 2$ vs. $Y \leq 2$)		
Maritals	1.704***	1.681***
	(0.177)	(0.178)
Educ		1.194***
		(0.026)
Age		0.981***
		(0.003)
Male		1.047
		(0.110)
Model 3 ($Y > 3$ vs. $Y \leq 3$)		
Maritals	1.704***	1.681***
	(0.177)	(0.178)
Educ		1.091***
		(0.023)
Age		0.981***
		(0.003)
Male		1.047
		(0.110)

(Continued)

Table 5.5   (Continued)

Variables	(1) Single-Predictor Model OR (SE)	(2) Full Model OR(SE)
α_1	11.78***	3.361***
	(1.420)	(1.572)
α_2	2.087***	0.507**
	(0.159)	(0.165)
α_3	0.281***	0.204***
	(0.023)	(0.068)
Observations	1,300	1,300
Pseudo R^2	0.008	0.047
Ll	−1,566	−1,506
df_m	1	6
chi2	26.42	147.5

Note: Standard errors of OR in parentheses.

***$p < .01$.

5.6 Reporting the Results

Reporting the results of the PPO model and the generalized ordinal logistic regression model is similar to that of the PO models. You need to test the PO assumption and justify why the PPO model or the generalized ordinal logistic regression model needs to be used to address your research question(s). Report the results of the Brant test for the PO assumption, and briefly explain the predictor variables violating the assumption.

Unlike the PO models, you need to report the logit coefficients and corresponding odds ratios for the predictor variables for the underlying binary logistic models. In the PPO model, the logit coefficients of some predictor variables may vary across the binary models, whereas the logit coefficients of all predictors vary freely in the generalized ordinal logistic regression model. For each binary model, label the category comparisons, such as above category *j* versus at or below category *j*.

Have a table containing all the parameter estimates, their standard errors, associated p values, and the odds ratios. The fit statistics, such as the log likelihood ratio chi-square test statistic and the likelihood ratio R^2, need be reported in the table. If more than one model is fitted, the results of all the competing models may also be presented in the table.

In the body of the text, briefly interpret the model fit statistics. In addition, interpret the odds ratios of the estimates across binary comparisons or splits. In the interpretation, explain the comparisons between categories, for example, whether the odds of being above a category versus at or below that category, or the inversed odds of being at or below a category. The following is an example of summarizing the results for the partial PO model.

The partial proportional odds model was fitted to estimate the ordinal outcome variable, health status, from a set of predictor variables, such as marital status, years of education, age, and gender. This model was used since it allows the effects of some predictor variables to vary when the proportional odds assumption (PO) does not hold. The omnibus Brant test of the PO assumption for the overall model (see Chapter 4), $\chi^2_{(8)} = 19.45$, $p = .013$, indicated that the PO assumption for the full model was violated. Examining each predictor variable, we found that three predictor variables `maritals`, `age`, and `male` met the assumption, while this assumption was untenable for the predictor `educ`. For `educ`, the Brant test $\chi^2_{(2)} = 12.98$, $p < .01$, which was significant. Therefore, fitting a partial PO model rather than a PO model might be a better option.

The log likelihood ratio chi-square test statistic $LR\ \chi^2_{(6)} = 147.50$, $p < .001$, indicated that the full model with four predictors provided a better fit than the null model with no independent variables in predicting the ordinal response variable. The likelihood ratio $R^2_L = .0467$ was still small.

Table 5.3 presents the coefficients and standard errors of the predictor variables. Three predictor variables `maritals`, `age`, and `male` meeting the PO assumption had the same logit regression coefficients across all three binary models. For `maritals`, $\beta = .520$, $p < .001$; for `age`, $\beta = -.019$, $p < .001$; and for `male`, $\beta = .046$, $p > .05$.

Table 5.5 presents the odds ratios and standard errors of the predictor variables. For the `maritals` predictor, OR = 1.681, which indicated that the odds of being beyond a particular category of health status for the married were 68.1% higher than the odds for the unmarried when adjusting for other predictors.

For the `age` predictor, OR = .981, which indicated that the odds of being beyond a particular category of health status (better health status) decreased by a factor of .981 for a one-unit increase in the predictor `age` when holding other variables constant.

(Continued)

(Continued)

For the `male` predictor, OR = 1.047, *p* = .662, which was not significantly different from 1. This indicated that there was no relationship between being male and the cumulative odds of being in better health status.

The `educ` predictor was the only one violating the PO assumption, so its effects varied across the three binary models. The estimated logit coefficients were .174, .177, and .087 for each respective model. In model 1, the Wald *z* test = 5.00, *p* < .001; in model 2, Wald *z* = 8.28, *p* < .001; and in model 3, Wald *z* = 4.14, *p* < .001. The results of the Wald *z* tests indicated that logit coefficients for `educ` were significant across all three models.

The odds ratios for `educ` were different across three binary models. The three odds ratios are 1.190, 1.194, and 1.091, respectively. Overall, the odds of being beyond a particular category of health status increased with the increase in `educ`. In other words, the number of years of school completed was associated with the odds of being in better health status. For each unit increase in `educ`, the odds increased by 19%, 19.4%, and 9.1% in each of the binary logistic models, respectively. The largest odds ratio was identified in the second binary model comparing categories 3 and 4 with categories 1 and 2, and the smallest odds ratio was found in the last binary model comparing category 4 with categories 1 through 3.

5.7 Summary of Stata Commands in This Chapter

```
use chap5-gss2012, clear

*The one-predictor partial PO model
gologit2 healthre i.maritals, autofit lrforce store(mod1)

*The multiple-predictor partial PO model
gologit2 healthre i.maritals educ age i.male, autofit lrforce store(mod2)
gologit2 healthre i.maritals educ age i.male, autofit lrforce or

*Model comparison using the log likelihood ratio test
lrtest mod1 mod2

*The generalized ordinal logistic regression model
xi: gologit healthre i.maritals educ age male
xi: gologit, or

*margins and marginsplot in Stata 13
quietly gologit2 healthre i.maritals educ age male, autofit lrforce
margins, predict (outcome(1)) at(educ = (8 13 16)) atmeans vsquish
marginsplot
margins, predict (outcome(4)) at(educ = (8 13 16)) atmeans vsquish
marginsplot
```

```
*margins plots combined in Stata 13
margins, predict (outcome(1)) at(educ = (8 13 16)) atmeans vsquish
marginsplot, name(mplot1)
margins, predict (outcome(2)) at(educ = (8 13 16)) atmeans vsquish
marginsplot, name(mplot2)
margins, predict (outcome(3)) at(educ = (8 13 16)) atmeans vsquish
marginsplot, name(mplot3)
margins, predict (outcome(4)) at(educ = (8 13 16)) atmeans vsquish
marginsplot, name(mplot4)
graph combine mplot1 mplot2 mplot3 mplot4, ycommon

*margins and marginsplot in Stata 14
quietly gologit2 healthre i.maritals educ age male, autofit lrforce
margins, at(educ = (8 13 16)) atmeans vsquish
marginsplot

*Making publication-quality tables using outreg2
quietly xi: gologit2 healthre i.maritals, autofit lrforce store(mod1)
quietly xi: gologit2 healthre i.maritals educ age i.male, autofit ///
lrforce store(mod2)
outreg2 [mod1 mod2] using chap5out, e(ll df_m chi2) ///
addstat(Pseudo R-squared, `e(r2_p)') long word replace
outreg2 [mod1 mod2] using chap5oddsout, eform e(ll df_m chi2) ///
addstat(Pseudo R-squared, `e(r2_p)') long word replace
seeout
```

5.8 EXERCISES

Use the GSS 2012 data for the following problems.

1. Conduct an analysis for a proportional odds model to estimate the ordinal response variable happy from the four predictor variables sex, educ, age, and satfin.

2. Use the brant command to test the proportional odds assumption for the model and each predictor. Is the assumption violated?

3. With the same variables, conduct an analysis for a partial proportional odds (PPO) model.

4. Identify predictor variable(s) violating the proportional odds or the parallel lines assumption.

5. Identify the likelihood ratio test of the model, and interpret it.

6. List the logit coefficients, the Wald z tests, and the 95% confidence intervals for the predictor variables violating the assumption.

7. Interpret the odds ratios for `educ` and `satfin`.

8. Make a publication-quality table containing the estimated logit coefficients and odds ratios.

9. Write a report to summarize the results from the output.

10. Fit a generalized ordinal logistic regression model using the same variables as those in the preceding PPO model in Exercise 3. Which model is more parsimonious?

CHAPTER 6

Continuation Ratio Models

Objectives of This Chapter

This chapter discusses continuation ratio models. After introducing the models, conditional probabilities, odds, and odds ratios are reviewed; model fit statistics and interpretations of parameter estimates also are discussed. This chapter shows how to conduct the analysis using Stata with step-by-step instructions. Stata commands are explained, and output is interpreted. This chapter also illustrates how the results are displayed in publication-quality tables using the Stata command and reported in text. It focuses on fitting continuation ratio models with Stata, as well as on interpreting and presenting the results. After reading this chapter, you should be able to

- Identify when a continuation ratio model is used.
- Fit the model using `ocratio` and `seqlogit` commands.
- Interpret the output.
- Interpret the model in terms of odds ratios.
- Compute and plot the estimated probabilities using the `margins` and `marginsplot` commands, respectively.
- Compare models using the likelihood ratio test and AIC.
- Present results in publication-quality tables using Stata.
- Write the results for publication.

6.1 Continuation Ratio Models: An Introduction

In the previous two chapters, we introduced the proportional odds (PO) model, the partial proportional odds (PPO) model, and the generalized ordinal logistic regression model. These models estimate either the cumulative odds of being at or below a particular category of the ordinal response variable or the odds of being beyond a particular level, which is the complementary direction. Correspondingly, these models compare the probabilities of being at or below a category and the probabilities of being above that category, or vice versa. In some situations, our interest of research is focused on a particular category rather than at or below that category; given that an individual must pass through lower categories before achieving a higher one, the continuation ratio (CR) model (Agresti, 2010; Allison, 2012; Ananth & Kleinbaum, 1997; Armstrong & Sloan, 1989; Fienberg, 1980; Fullerton, 2009; Greenland, 1994; Hardin & Hilbe, 2007, 2012; Liu, O'Connell, & Koirala, 2011; Long & Freese, 2006, 2014; O'Connell, 2006; Tutz, 1991, 2012) is a better choice than other models. The CR model is often called the sequential model or sequential logit model (Buis, 2007; Liao, 1994; Tutz, 1991, 2012), or a special case of the latter (Long & Freese, 2014). It is also called the model for nested dichotomies in Fox (2008). The CR model is particularly useful when analyzing the educational attainment data illustrated in Allison (1999) and Long and Freese (2014) when the transitions between different stages are obvious.

In the CR model, the ordinal categories represent successive stages or proficiency levels; for example, educational attainment from a high-school diploma, to a bachelor's degree, to a master's degree, to a doctorate degree. A person needs to be awarded the lower level degree before he or she receives the doctorate degree. This model estimates the odds of being in a certain category versus being beyond that category. In terms of probability, this model estimates the conditional probability of being in a category given that an individual has been in that category or beyond. In addition, it estimates the conditional probability of being beyond a category given that a person has attained that particular category since these two conditional probabilities are complementary.

When estimating the conditional probability of being beyond a category given that an individual has attained that particular category, that is, $\pi(Y > j \mid Y \geq j \mid)$, the CR model can be expressed in this form (Allison, 1999; O'Connell, 2006; Tutz, 2012):

$$\ln\left(\frac{\pi\left(Y > j \mid x_1, x_2, ..., x_p\right)}{\pi\left(Y = j \mid x_1, x_2, ..., x_p\right)}\right) = \alpha_j + \beta_1 X_1 + \beta_2 X_2 + ... + \beta_p X_p \qquad (6.1)$$

where $\pi(Y > j \mid x_1, x_2, ..., x_p)$ is the conditional probability of being beyond a category j, conditional on being in that category, given a set of predictors. $j = 1, 2, ..., J - 1$. α_j are the cut points, and $\beta_1, \beta_2, ..., \beta_p$ are the logit coefficients. Before the model is fitted, the dataset needs to be restructured following two steps (Allison, 1999; O'Connell, 2006). First, separate sub-datasets need to be constructed with the binary outcome variable being beyond a category coded as 1 and 0 otherwise. Individuals who have not advanced to a particular proficiency level are dropped at each stage. If the ordinal dependent variable has j categories, $J - 1$ sub-datasets should be created. Second, these datasets are combined into one dataset with a new binary outcome variable with 1 = beyond a particular category.

The CR model also estimates the odds of being in a particular category j relative to being beyond that category. In this situation, the CR model can be formulated in a different form (Ananth & Kleinbaum, 1997; Armstrong & Sloan, 1989; Fienberg, 1980; Liu, O'Connell, & Koirala, 2011; Long & Freese, 2006):

$$\ln\left(\frac{\pi\left(Y = j \mid x_1, x_2, \ldots, x_p\right)}{\pi\left(Y > j \mid x_1, x_2, \ldots, x_p\right)}\right) = \alpha_j + (-\beta_1 X_1 - \beta_2 X_2 - \ldots - \beta_p X_p) \quad (6.2)$$

where $\pi(Y = j \mid x_1, x_2, \ldots, x_p)$ is the conditional probability of being in category j, conditional on being in that category or beyond, given a set of predictors. $j = 1, 2, \ldots, J - 1$. α_j are the cut points, and $\beta_1, \beta_2 \ldots \beta_p$ are the logit coefficients. Just like the PO model, the CR model here also assumes that the logit coefficients are parallel across ordinal categories. This model is also called the constrained CR model (Cole & Ananth, 2001). The `ocratio` command in Stata (Wolfe, 1998) follows this form to fit the CR model, which is known as the forward CR model (Bender & Benner, 2000). Compared with other statistical software, Stata does not require data restructuring before model fitting, which makes data analysis of the CR model much easier.

The CR model may be parameterized differently to estimate the odds of being in a particular category j relative to being below that category (Hosmer, Lemeshow, & Sturdivant, 2013), which is not equivalent to the models for Equations 6.1 and 6.2. The Stata `ocratio` command cannot be used to fit this model directly, so the introduction of this model is beyond the scope of this chapter.

6.1.1 Conditional Probabilities, Odds, and Odds Ratios

The PO model estimates $J - 1$ cumulative odds and then the cumulative probabilities of being at or below a category of the ordinal response variable with j levels. When the ordinal response variable, health status, has four levels from 1 to 4, the proportional odds model estimates three cumulative probabilities, which include $P(Y \leq 1)$, $P(Y \leq 2)$, and $P(Y \leq 3)$. The cumulative probabilities of being beyond a category can also be estimated since they are the complementary probabilities of being at or below a particular category.

Unlike the cumulative probabilities in the PO model, the conditional probabilities are estimated in the CR logit model. The CR model estimates conditional probabilities of being in category j, conditional on being at or beyond that category, that is,

$P(Y = j \mid Y \geq j)$. This CR model can also estimate the conditional probability of being beyond a category given that the individual has achieved that particular category since $P(Y > j \mid Y \geq j)$ is the complementary form of $P(Y = j \mid Y \geq j)$.

Another difference between the CR model and the PO model is the change in the sample size. The number of observations increases from the PO model to the CR model due to different comparisons between categories. When the ordinal response variable has four categories, the comparisons include category 1 versus categories 2, 3, and 4; category 2 versus categories 3 and 4; and category 3 versus category 4 (Table 6.1 provides the comparisons among four health status levels).

The logit or log odds of being in a category relative to being beyond that category $\ln[P(Y = j)/P(Y > j)]$ is the conditional logit. The odds in the CR model are the probability of being in a category divided by the probability of being beyond that category. In other words, they compare the probability when $p(Y = j)$ and $p(Y > j)$.

Since the probability of being in a category in the CR model is a conditional probability, the odds are the conditional odds:

$$\text{Odds} = \frac{p(Y = j)}{p(Y > j)}$$

The odds ($Y = 1 \mid Y \geq 1$ vs. $Y > 1 \mid Y \geq 1$), abbreviated as the odds ($Y = 1 \mid Y \geq 1$), equal the ratio of the conditional probability of being in category 1 to the probability of being beyond this category. Since the probability $p(Y > 1) = p(2) + p(3) + p(4)$, the conditional odds of being in category 1, given an individual has reached this category, can be expressed in the following form:

$$\text{Odds } (Y = 1 \mid Y \geq 1) = \frac{p(Y = j)}{p(Y > j)} = \frac{p(1)}{p(2) + p(3) + p(4)}$$

Similarly, the odds ($Y = 2 \mid Y \geq 2$) equal the ratio of conditional probability of being in category 2 to the probability of being beyond this category. Since $p(Y > 2) = p(3) + p(4)$, the conditional odds of being in category 2, given an individual has reached this category, can be expressed as follows:

$$\text{Odds } (Y = 2 \mid Y \geq 2) = \frac{p(Y = 2)}{p(Y > 2)} = \frac{p(2)}{p(3) + p(4)}$$

The odds ($Y = 3 \mid Y \geq 3$) equal the ratio of conditional probability of being in category 3 to the probability of being beyond this category. Using the same method, we get the following equation:

$$\text{Odds } (Y = 3 \mid Y \geq 3) = \frac{p(Y = 3)}{p(Y > 3)} = \frac{p(3)}{p(4)}$$

Table 6.1 presents the logits, odds, and category comparisons for the continuation ratio model for the ordinal response variable with four levels.

Table 6.1   Category Comparisons for the Continuation Ratio Model With Four Levels of Health Status ($j = 1, 2, 3,$ and 4)

Category	Logit $P(Y = j \mid Y \geq j)$	Odds	Probability Comparisons
Level 1	logit $P(Y = 1 \mid Y \geq 1)$	$\dfrac{P(Y = 1)}{P(Y > 1)}$	Category 1 vs. Categories 2, 3, and 4
Level 2	logit $P(Y = 2 \mid Y \geq 2)$	$\dfrac{P(Y = 2)}{P(Y > 2)}$	Category 2 vs. Categories 3 and 4
Level 3	logit $P(Y = 3 \mid Y \geq 3)$	$\dfrac{P(Y = 3)}{P(Y > 3)}$	Category 3 vs. Category 4

Odds Ratios in the CR Model

Similar to the odds ratios in other logistic regression models, the odds ratio in the CR model is the change in the odds for a one-unit increase from any value of x to the value of $(x + 1)$. The odds ratios in the CR model can be either the odds of being in a category relative to being above that category, or their inverse, the odds of being above a category versus in that category. The former is the exponentiated negative logit coefficient $\exp(-\beta)$, and the latter is the exponentiated logit coefficient $\exp(\beta)$.

6.1.2 Goodness-of-Fit Statistics

Model fit statistics in the CR model, such as the likelihood ratio test and pseudo R^2, are reported. Other fit statistics, such as the Hosmer–Lemeshow goodness-of-fit test and Pulkstenis–Robinson (2004) modification, are currently unavailable in the CR model. Following the suggestion made by Hilbe (2009), in addition to the likelihood ratio test, the user-written `aic` command for the Akaike's Information Criterion (AIC) is also used for the comparison of model fit.

6.1.3 Interpretation of Model Parameter Estimates

As with other logistic regression models, the regression coefficient in the CR model is also the logit coefficient. Since the odds ratios in the CR model are different from other models, so is the interpretation of the logit coefficients. A logit coefficient can be interpreted as the change in the predicted logit or the log odds of being above a particular category versus being in a category for a one-unit increase in the predictor variable. Recall that the signs before logit coefficients in the equation of the CR model (Equation 6.2) are negative, so the logit coefficient with a preceding negative sign can be interpreted as the change in the log odds of being in a particular category versus being above that category for a one-unit increase in the predictor variable.

To get the odds ratio (OR) of being above a category versus being in that category, we can exponentiate the logit coefficient without the negative sign exp(β). Conversely, to get the OR of being in a particular category relative to being above that category, we exponentiate the logit coefficient with a preceding negative sign exp(−β). It is obvious that the conditional odds of being in a particular category relative to being above that category are the inverse of the odds of being above that category versus being in that category.

Positive logit coefficients indicate positive relationships between the predictor variables and the logit or log odds of being above a category versus being in that category. A positive coefficient corresponds to an OR greater than 1. So the odds of being beyond a particular category increase for a one-unit increase in the predictor variable.

A negative logit coefficient corresponds to the exponentiated coefficient, the OR, less than 1. This means that the odds of being beyond a particular category decrease for a one-unit increase in the predictor variable.

A logit coefficient of 0 corresponds to the OR equal to 1, which indicates that there is no relationship between the predictor and the log odds and, thus, that there is no change in the odds with an increase in the predictor variable.

To estimate the OR of being in a particular category versus being above that category, however, the signs before both the intercepts and logit coefficients in Equation 6.2 need to be reversed. Taking the inverse of the conditional odds of being beyond a particular category gives us the conditional odds of being in that category.

6.2 Research Example and Description of the Data and Sample

Research Problem and Questions: In this chapter, the purpose of the research example is still to investigate the relationships between the ordinal response variable, health status, and four predictor variables, marital status, the highest education, age, and gender. Unlike in the previous two chapters, however, here the research interest will focus on the odds of being in a particular level of health status versus being above that level using the CR model. The research question is as follows: Do the four predictor variables significantly predict the ordinal response variable, health status? Specifically, do the four predictor variables significantly predict the conditional probabilities of being in a particular level of health status given that higher levels of health status have been achieved?

Description of the Data and Sample: The data for the following analyses are from the General Social Survey 2012 (GSS 2012). For illustration purposes, the same variables from the previous two chapters are used for data analysis in this chapter.

The ordinal response variable is `healthre`, the recoded variable of health (health status) with four ordinal categories; the four predictor variables are `maritals`, `educ`, `age`, and `male`.

6.3 Continuation Ratio Models With Stata: Commands and Output

The Stata Command `ocratio`

The Stata user-written command `ocratio` is used for the analysis of continuation ratio models. Since this is a user-written program (Wolfe, 1998), you need to install it first by typing `ssc install ocratio` before fitting the model.

The syntax is the command `ocratio` followed by the dependent variable and the independent variable(s). For example, the command `ocratio y x` tells Stata to fit a continuation ratio model to estimate the ordinal dependent variable y with an independent variable x. The command has three link functions, logit, probit, and cloglog, with the logit as the default function. Since the `margins` command does not work with `ocratio`, its functions will not be introduced in this chapter.

The user-written command `aic` in Hardin and Hilbe (2012) is used to compute the AIC statistic. To install it, you also need to install other programs for their book since they are combined. Type `net from http://www.stata-press.com/data/hh3`, then type `net install glm3-ado`, and finally type `net install glm3-ado2`. An alternative option is to use the `display` command to compute AIC, which will be demonstrated later in the chapter.

6.3.1 The CR Model With the `logit` Link: One-Predictor Model

The command xi: `ocratio healthre i.maritals, link(logit)` tells Stata to fit the continuation ratio model for the ordinal response variable `healthre` with the predictor variable `maritals`. The prefix command xi is used for the categorical variable `maritals` since the factor variables introduced since Stata version 11 do not work with the `ocratio` command, and `i.maritals` indicates that the predictor variable `maritals` is categorical. The option `link(logit)` tells us the link function is the logit function, which is the default for the model. The command `aic` is used to compute the AIC statistic. The Stata output for the single-predictor continuation ratio model is displayed as follows.

```
. xi: ocratio healthre i.maritals, link(logit)
i.maritals          _Imaritals_0-1          (naturally coded; _Imaritals_0 omitted)

Continuation-ratio logit Estimates                      Number of obs =      3459
                                                        chi2(1)       =     26.93
                                                        Prob > chi2   =    0.0000
Log Likelihood = -1565.941                              Pseudo R2     =    0.0085


------------------------------------------------------------------------------
    healthre |      Coef.   Std. Err.      z    P>|z|     [95% Conf. Interval]
-------------+----------------------------------------------------------------
_Imaritals_1 |   .4641666   .0901228     5.15   0.000     .2875291    .6408041
------------------------------------------------------------------------------
       _cut1 |  -2.490959   .1184066               (Ancillary parameters)
       _cut2 |  -1.025142   .0782388
       _cut3 |   .7743144   .0831244
------------------------------------------------------------------------------
```

Interpreting the Output

Similar to the output in other logistic regression models, when the prefix command xi is used, the coding of the categorical variable maritals is displayed at the beginning of the output. It shows the variable is coded as _Imaritals_0-1 and the first dummy variable is omitted from the model (naturally coded; _Imaritals_0 omitted).

The top right of the logistic regression table provides the number of observations in the dataset, the log likelihood ratio test statistic and the associated p value, and the pseudo R^2. The dataset has 3,459 observations for the analysis. LR chi2 (1) = 26.93, which is the log likelihood ratio chi-square test statistic. The associated p value with the log likelihood ratio chi-square test Prob > chi2 = 0.0000 indicates that the one-predictor model provides a better fit than the null model with no independent variables in predicting the log odds and then the conditional probabilities for health status.

The pseudo R^2 = .0085, which is the likelihood ratio R^2_L, suggesting that the relationship between the response variable, health status, and the predictor, marital status, is small.

The continuation ratio logit estimates table displays the parameter estimates for the predictor variable, its standard error, the Wald z statistic, the associated p value, and the 95% confident interval. The cut points or intercepts and their standard errors are also displayed.

For the predictor variable maritals, Wald z = .4641666/.0901228 = 5.15. The associated p value, P>|z|=.000, so we reject the null hypothesis. Therefore, the predictor variable, marital status, is a significant predictor of the ordinal outcome variable, health status. The 95% confidence interval of the regression coefficient is [.2875291,

.6408041], which does not contain the value of 0, indicating that the predictor is significantly different from 0.

The logit regression coefficient $\beta = .464$ indicates for a one-unit increase in the predictor variable, the change in the logit or log odds of being beyond a category is .464. The corresponding odds ratio, which offers a better interpretation, will be explained later.

Without computing the odds ratios manually, we can request the odds ratios with the `eform` option. The syntax is as follows:

```
xi: ocratio healthre i.maritals, link(logit) eform
```

The following output displays the odds ratios.

```
. xi: ocratio healthre i.maritals, link(logit) eform
i.maritals          _Imaritals_0-1       (naturally coded; _Imaritals_0 omitted)

Continuation-ratio logit Estimates                      Number of obs =     3459
                                                        chi2(1)       =    26.93
                                                        Prob > chi2   =   0.0000
Log Likelihood = -1565.941                              Pseudo R2     =   0.0085

-----------------------------------------------------------------------------
   healthre | Odds ratio   Std. Err.      z    P>|z|      [95% Conf. Interval]
------------+----------------------------------------------------------------
_Imaritals_1 |   1.590688   .1433573    5.15   0.000      1.333129    1.898006
-----------------------------------------------------------------------------
   _cut1    |   -2.490959   .1184066              (Ancillary parameters)
   _cut2    |   -1.025142   .0782388
   _cut3    |    .7743144   .0831244
-----------------------------------------------------------------------------

. aic
AIC Statistic =    .9060079
```

Interpreting the Output

In the output, the summary of the model fit statistics is the same as that in the previous output. The continuation ratio logit estimates table reports the odds ratio, standard error, Wald z statistic, associated p value, and the 95% confidence interval for the predictor variable, marital status (labeled _Imaritals_1).

The odds ratio for the predictor variable, marital status (labeled _Imaritals_1), is 1.591, the Wald $z = 5.19$, P>|z| = .000, which is significant. The odds ratio is the exponentiated coefficient exp(.464). The 95% confidence interval of the odds ratio is [1.333, 1.898], which does not contain the odds ratio of 1, indicating it is significant.

The AIC statistic = .906. It can also be computed using the `display` command since AIC equals $-2(LL_m - k)/n$. The output is displayed as follows.

```
. *AIC = -2(LLm - k) / n
. display -2*(-1565.941-1)/3459
.90600809
```

Interpreting the Odds Ratio of Stopping in a Particular Category

The estimated logit regression coefficient $\beta = .464$, $z = 5.15$, $p < .001$, indicates that there is a positive relationship between the predictor variable, marital status, and the logit or log odds of being above a health status level. By substituting the value of the coefficient into Equation 6.2, $[\pi(Y = j \mid Y \geq j)] = \alpha_j + (-\beta_1 X_1)$, we get logit $[\pi(Y = j \mid Y \geq j)] = \alpha_j - .464$ (marital$s$). OR $= e^{(-.464)} = .629$, indicating that the odds of being in a category compared with being in higher categories for the married are .629 times the odds for the unmarried. In other words, the odds of being in a certain health status level rather than moving to a higher level are higher for the unmarried than the married.

Interpreting the Odds Ratio of Being Above a Particular Category

To estimate the odds and then the conditional probabilities of being beyond a category of health status, we need to change the signs before the cut points or intercepts and the estimated logit coefficients.

logit $[\pi(Y > j \mid Y \geq j)] = -\alpha_j + .464$ (`marital`s). By exponentiating the coefficient .464, we obtain the OR, which is 1.590, indicating that the odds of being in higher categories versus staying in a category for the married are 1.590 times the odds for the unmarried. In other words, the married have higher odds of being in better health condition.

The CR Model With the `cloglog` Link

In addition to the logit link, the CR model can also be fitted using the complementary log-log (clog-log) link with the cumulative option. The CR model with the complementary log-log link is actually the discrete-time proportional hazards model for the event history analysis or survival analysis (Allison, 1999; O'Connell, 2006). It estimates the hazard ratio (HR) rather than the odds ratio (OR) of being in a particular category relative to advancing to a higher category. In the event history analysis or survival analysis, the hazard or failure rate is defined as a function of experiencing an event for an individual in a time period. It is the conditional probability of an event occurrence or failure in a time interval, given that the individual has not experienced that event or has survived by that time (Mills, 2011; Singer & Willet, 2003).

In a binary logistic regression, the logit is the log odds of success. In other words, it is the log odds when the outcome variable $Y = 1$, whereas the complementary log-log function, clog-log, is $\log(-\log(1 - p))$, the log of the negative log of the complementary probability.

Similarly, the logit in the CR model is the log odds of being in a particular category where odds compare conditional probabilities of being in a particular category with those of being above that category, whereas the clog-log in the CR model is the log of the negative log of the complementary probability where the complementary probability equals 1 minus the conditional probability of being in a particular category conditional on being at or above that category (i.e., $1 - P(Y = j \mid Y \geq j)$).

In the command xi: ocratio healthre i.maritals, link(cloglog) cumulative, the option link(cloglog) requests the clog-log link function and the option cumulative requests the cut points or intercepts. The Stata output for the continuation ratio model with the clog-log link is presented as follows.

```
. xi: ocratio healthre i.maritals, link(cloglog) cumulative
i.maritals          _Imaritals_0-1       (naturally coded; _Imaritals_0 omitted)

Ordered cloglog Estimates                          Number of obs =     3459
                                                   chi2(1)       =    26.17
                                                   Prob > chi2   =   0.0000
Log Likelihood = -1566.323                         Pseudo R2     =   0.0083


------------------------------------------------------------------------------
   healthre |     Coef.   Std. Err.      z    P>|z|     [95% Conf. Interval]
------------+-----------------------------------------------------------------
_Imaritals_1 |   .3424795   .0673307    5.09   0.000     .2105137    .4744453
------------+-----------------------------------------------------------------
       _cut1 |  -2.571242   .1128291          (Ancillary parameters)
       _cut2 |   -.981268   .0795076
       _cut3 |   .4415468   .0625614
------------------------------------------------------------------------------

. aic
AIC Statistic =    .9062291
```

Interpreting the Output

The summary of fit statistics is the same as that in the CR model with the logit function. The log likelihood ratio chi-square test with 1 degree of freedom $LR\ \chi^2_{(1)} = 26.17$, $p < .001$, indicates that the full model with one predictor provides a better fit than the null model with no independent variables. The pseudo $R^2 = .0083$, which suggests that the relationship between the response variable, health status, and the predictor, marital status, is small. The AIC statistic is .906.

In the ordered cloglog estimates table, the estimated clog-log coefficient $\beta = .342$, Wald $z = 5.09$, $p < .001$, indicates that the logit coefficient of marital status is significant. Since clog-log $[\pi(Y = j \mid Y \geq j)] = \log(-\log(1 - \pi)) = \alpha_j + (-\beta_1 X_1)$, we calculated $\log(-\log(1 - \pi)) = \alpha_j - .342$ (maritals). By exponentiating $-.342$, $\exp(-.342)$, we get $HR = e^{(-.342)} = .710$, which indicates that the hazard of being in a particular health status level rather than beyond that level for the married is .710 times the hazard for the unmarried, so the hazard of stopping in a certain health status level rather than moving to a higher level is smaller for the married than the unmarried. In other words, the married are more likely to move to better health levels.

With the addition of the option eform, the following output displays the hazard ratios rather than odds ratios.

```
. xi: ocratio healthre i.maritals, link(cloglog) eform cumulative
i.maritals          _Imaritals_0-1      (naturally coded; _Imaritals_0 omitted)

Ordered cloglog Estimates                          Number of obs =      3459
                                                   chi2(1)       =     26.17
                                                   Prob > chi2   =    0.0000
Log Likelihood = -1566.323                         Pseudo R2     =    0.0083


-------------------------------------------------------------------------------
    healthre |  Haz. ratio    Std. Err.      z     P>|z|     [95% Conf. Interval]
-------------+-----------------------------------------------------------------
_Imaritals_1 |    1.408435    .094831      5.09   0.000     1.234312    1.607122
-------------+-----------------------------------------------------------------
       _cut1 |   -2.571242    .1128291            (Ancillary parameters)
       _cut2 |    -.981268    .0795076
       _cut3 |    .4415468    .0625614
-------------------------------------------------------------------------------
```

Interpreting the Output

The ordered cloglog estimates table reports the hazard ratio, its standard error, the Wald z statistic, the associated p value, and the 95% confidence interval for the predictor variable, marital status (labeled _Imaritals_1).

The hazard ratio for the predictor variable is 1.408, the Wald $z = 5.09$, $P > |z| = .000$, which is significant. The hazard ratio is the exponentiated coefficient $\exp(.342)$. The 95% confidence interval of the hazard ratio is [1.234, 1.607], which does not contain the hazard ratio of 1, indicating it is significant.

Interpretation of the hazard ratio is similar to that of the odds ratio. The hazard ratio is 1.541, indicating that the hazard of being in higher categories versus staying in a category for the married is 1.541 times the hazard for the unmarried. In other words, the married are more likely to be in better health conditions.

6.3.2 The CR Model With the `logit` Link: Multiple-Predictor Model

The command `xi: ocratio healthre i.maritals edu age male, link(logit)` fits the CR model with the logit link using four predictor variables maritals, edu, age, and `male`. The fitted model is referred to as the full model. The results of the full model are shown as follows.

```
. xi: ocratio healthre i.maritals edu age male, link(logit)
i.maritals            _Imaritals_0-1        (naturally coded; _Imaritals_0 omitted)

Continuation-ratio logit Estimates                    Number of obs =      3459
                                                      chi2(4)       =    122.11
                                                      Prob > chi2   =    0.0000
Log Likelihood =  -1518.35                            Pseudo R2     =    0.0387

------------------------------------------------------------------------------
   healthre |      Coef.   Std. Err.       z    P>|z|     [95% Conf. Interval]
------------+-----------------------------------------------------------------
_Imaritals_1|   .4724301    .092114     5.13   0.000     .2918899    .6529703
       educ |   .1075093   .0147409     7.29   0.000     .0786176     .136401
        age |  -.0157278   .0026189    -6.01   0.000    -.0208606   -.0105949
       male |   .0895459   .0914022     0.98   0.327    -.0895991    .2686909
------------+-----------------------------------------------------------------
      _cut1 |   -1.84848    .2630132            (Ancillary parameters)
      _cut2 |  -.3167183    .2501399
      _cut3 |   1.616745    .2588347
------------------------------------------------------------------------------

. aic
AIC Statistic =    .8802257
```

Interpreting the Output

The number of observations is still the same as the one for the single-predictor model, which means that there are no missing data. The log likelihood ratio chi-square test $LR\ \chi^2_{(4)} = 122.11$, $p < .001$, indicates that the full model with four predictor variables provides a better fit than the null model with no independent variables in predicting conditional probability for health status. The likelihood ratio $R^2_L = .0387$, which is larger than that of the single-predictor model, but it is still small.

The continuation ratio logit estimates table displays the parameter estimates for four predictor variables. Of them, the coefficients of two variables `maritals` and `educ` are positive, one coefficient is negative, and one is not significant.

For `maritals`, $\beta = .472$, Wald $z = 5.13$, $p < .001$, which is significant. The logit or log odds of being above a particular category versus being in that category for the married is .472 times the logit for the unmarried when holding other variables constant.

For educ, β = .108, Wald z = 7.29, p < .001, which indicates that the logit coefficients of both marital status and education are significant. The logit or log odds of being above a particular category versus being in that category increases by .108 for a one-unit increase in years of education (educ) when holding other variables constant.

The coefficient of age is negative, β = −.016, Wald z = −6.01, p < .001, which indicates the relationship between the predictor variable, age, and the logit of the conditional probability of being in better health status is significantly negative. The logit or log odds of being above a particular category versus being in that category decreases by −.016 for a one-unit increase in age when holding other variables constant.

One coefficient is not significant. For male, β = .090, Wald z = .98, p = .327, which is not significant, indicating there is no significant relationship between being male and the logit of the conditional probability.

The AIC statistic = .880. It can also be computed using the display command. The output is displayed as follows.

```
. *AIC = -2(LLm - k) / n
. display -2*(-1518.35-4)/3459
.8802255
```

The odds ratios are reported in the following with the eform option.

```
. xi: ocratio healthre i.maritals edu age male, link(logit) eform
i.maritals          _Imaritals_0-1       (naturally coded; _Imaritals_0 omitted)

Continuation-ratio logit Estimates                 Number of obs  =      3459
                                                   chi2(4)        =    122.11
                                                   Prob > chi2    =    0.0000
Log Likelihood =  -1518.35                         Pseudo R2      =    0.0387

------------------------------------------------------------------------------
   healthre |  Odds ratio    Std. Err.      z    P>|z|     [95% Conf. Interval]
------------+-----------------------------------------------------------------
_Imaritals_1 |   1.603887    .1477405     5.13   0.000     1.338956    1.921239
       educ |   1.113501    .016414      7.29   0.000     1.081791    1.146141
        age |   .9843953    .002578     -6.01   0.000     .9793554     .989461
       male |   1.093678    .0999645     0.98   0.327     .9142976    1.308251
------------+-----------------------------------------------------------------
      _cut1 |  -1.84848     .2630132              (Ancillary parameters)
      _cut2 |   -.3167183   .2501399
      _cut3 |   1.616745    .2588347
------------------------------------------------------------------------------
```

Interpreting Odds Ratios of Being Above a Particular Category

Among the four predictor variables, marital status (maritals) and highest years of education (educ) are positively associated with the odds of being beyond a particular health status level; the predictor age (age) is negatively associated with the odds; and the predictor variable being male (male) is not significantly associated with the odds.

In terms of odds ratios, the odds for the married of being in a better health status level relative to being in that level are 1.604 times the odds for the unmarried (OR = 1.604) after controlling for the effects of other predictors in the model. Similarly, the odds of being above a particular category versus being in that category increase by 1.114 or 11.4% for a one-unit increase in years of education (educ) when holding other variables constant.

On the other hand, for every one-unit increase in age, the odds of being in a better health status level decrease by a factor of .984 (OR = .984) when holding the effects of the other variables constant. In other words, people with an older age are less likely to advance to a better health status level or they are more likely to stop in a particular health status level.

The odds ratio for the predictor male is .984, $p = .327$, which is not significant, indicating being male does not influence the odds of being in a higher category.

Model Comparison Using the Likelihood Ratio Test

The command lrtest does not work with the ocratio command so we need to compute the log likelihood ratio test statistic comparing the single-predictor model and the full model with the following equation:

$$G = D_{Reduced} - D_{Full} = -2 \times [-1565.941 - (-1518.35)] = 95.182$$

$$df = 4 - 1 = 3$$

The log likelihood ratio chi-square test $\chi^2_{(3)} = 95.182$, $p < .001$. The same result can be obtained if using the display command in Stata, which can be used as a calculator. The output is as follows.

```
. display -2*(-1565.941-(-1518.35))

  95.182
```

Model Comparison Using the AIC Statistic

In addition, the AIC statistics can be used for model comparisons using the aic command (Hilbe, 2009; Hardin & Hilbe, 2012). Compared with the single-predictor model (.906), the AIC statistic indicates that the full model fits the data slightly better (.880).

6.3.3 Fitting Continuation Ratio
Models Using the `seqlogit` Command

In addition to the `ocratio` command, the `seqlogit` command, which is another user-written program (Buis, 2007), can be used to fit continuation ratio models. You need to install it first by typing `ssc install seqlogit`. It works in Stata 11 or later versions. Stata 9 and 10 users need to install the old `seqlogit10` program. For more details on how to use this command, type `help seqlogit`.

The syntax of the `seqlogit` command includes more options than that of the `ocratio` command since the former command can fit more complex sequential logit models. To fit the continuation ratio model introduced in this chapter, we need to specify the `tree` option with the transition points or category comparisons in the `seqlogit` command syntax. As introduced in 6.1.1, the example of the continuation ratio model compares category 1 with categories 2, 3, and 4; category 2 with categories 3 and 4; and category 3 with category 4. It can be specified as `tree (1: 2 3 4, 2: 3 4, 3: 4)`. The comparisons between categories are separated by a colon, and each comparison or transition is separated by a comma.

The command `seqlogit healthre i.maritals edu age i.male, tree (1: 2 3 4, 2: 3 4, 3: 4)` tells Stata to fit the sequential logit model or the CR model using four predictor variables `maritals`, `educ`, `age`, and `male`. The factor variable notation rather than the `xi` prefix command is used here since the `seqlogit` command requires Stata 11 or later versions, and it works with the factor variable notation. The output is shown as follows.

```
. seqlogit healthre i.maritals educ age i.male, tree (1: 2 3 4, 2: 3 4, 3:4)

Transition tree:

Transition 1: 1 : 2 3 4
Transition 2: 2 : 3 4
Transition 3: 3 : 4

Computing starting values for:

Transition 1
Transition 2
Transition 3

Iteration 0:    log likelihood = -1504.8399
Iteration 1:    log likelihood = -1504.8399    (backed up)
```

```
                                        Number of obs    =       1,300
                                        LR chi2(12)      =      149.13
   Log likelihood = -1504.8399          Prob > chi2      =      0.0000


   -------------------------------------------------------------------------
      healthre |    Coef.    Std. Err.      z    P>|z|    [95% Conf. Interval]
   ------------+------------------------------------------------------------
   _2_3_4v1    |
     1.maritals |   .4719058   .2413072    1.96   0.051   -.0010476    .9448592
          educ |     .16038   .0338494    4.74   0.000    .0940363    .2267237
           age |  -.0256337   .0063526   -4.04   0.000   -.0380845   -.0131829
        1.male |  -.2242893   .2338906   -0.96   0.338   -.6827065    .2341279
         _cons |   1.882017   .6012492    3.13   0.002    .7035901    3.060444
   ------------+------------------------------------------------------------
   _3_4v2      |
     1.maritals |   .4551559   .1447193    3.15   0.002    .1715112    .7388006
          educ |   .1537196   .0230543    6.67   0.000    .1085339    .1989052
           age |  -.0174213   .0039551   -4.40   0.000   -.0251731   -.0096696
        1.male |   .0344913   .1436101    0.24   0.810   -.2469793    .315962
         _cons |  -.1680425   .3774757   -0.45   0.656   -.9078813    .5717962
   ------------+------------------------------------------------------------
   _4v3        |
     1.maritals |   .4910952   .1380011    3.56   0.000    .220618     .7615725
          educ |   .0303252   .0226669    1.34   0.181   -.0141011    .0747514
           age |  -.0084522   .0041549   -2.03   0.042   -.0165956   -.0003087
        1.male |   .2289075   .137111     1.67   0.095   -.0398251    .4976401
         _cons |   -.927911   .3840353   -2.42   0.016   -1.680606   -.1752157
   -------------------------------------------------------------------------
```

Interpreting the Output

The transition tree is displayed at the beginning of the output. The three transitional points include transitions from category 1 to categories 2 through 4, from category 2 to categories 3 and 4, and from categories 3 to 4. The log likelihood value, the number of observations, the log likelihood ratio test statistic, and the associated p value are reported next.

At the bottom of the output, the regression coefficient table for the fitted model displays the parameter estimates, standard errors, Wald z statistics, associated p values, and the 95% confident intervals of the parameter estimates. This table includes three subsections. Each section displays the parameter estimates for an underlying binary logistic model, which reflects the category comparison specified in the transition tree. The first binary model estimates the logit or log odds of being in categories 2, 3, and 4 versus 1; the second model estimates the logit of being in categories 3 and 4 versus 2; and the third model estimates the logit of being in category 4 versus 3. In other words, these three binary models estimate the logit or log odds of being above a particular category versus being in that category.

The logit coefficient table here looks similar to that of the generalized ordinal logistic regression model introduced in Chapter 5 since the logit coefficients of all predictor variables are allowed to vary freely across the underlying binary models. For example, the estimated logit coefficients for `maritals` are .472, .455, and .491 for each binary model. Therefore, the estimated sequential logit model in this example is called the unconstrained continuation ratio model. If an ordinal response variable has J levels in an unconstrained continuation ratio model, we need to estimate $J - 1$ logit coefficients for each predictor variable. A more parsimonious model is the constrained continuation ratio model where the logit coefficients for each predictor variable are constrained to be equal across the underlying binary comparisons. The CR model introduced in the preceding sections is the constrained model, which is the focus in this chapter. Similar to the PO model, the logit coefficients for each predictor variable are assumed to be equal in the constrained CR model.

How can we fit the same CR model introduced earlier in this chapter using the `seqlogit` command? In other words, how can we fit the constrained CR model? To replicate the results previously estimated by the `ocratio` command, we need to constrain the logit coefficients for each predictor to be equal across the three binary logistic regression models with the `constraint` option in the `seqlogit` command syntax. Before executing the `seqlogit` command, we also need to use the `constraint define` command or `constraint` to define eight constraints, with two for each predictor variable. The total number of constraints is determined by both the number of predictor variables and the number of transitions or comparisons in the transition tree. See Chapter 7 for more details on how to use the `constraint` command.

The command for the first constraint `[_2_3_4v1]1.maritals = [_3_4v2]1.maritals` tells Stata to constrain the logit coefficient in the model labeled `[_2_3_4v1]`, comparing categories 2, 3, and 4 with category 1, to be the same as the coefficient for the model labeled `[_3_4v2]`, comparing categories 3 and 4 with category 2. Similarly, the second constraint `[_3_4v2]1.maritals = [_4v3]1.maritals` is defined so that the logit coefficient in the model labeled `[_3_4v2]` is the same as that for the model labeled `[_4v3]`. The remaining six constraints are for the other three predictor variables and can be defined in a similar way.

The command `seqlogit healthre i.maritals educ age i.male, tree (1: 2 3 4, 2: 3 4, 3: 4) constraints (1/8)` tells Stata to fit the CR model with eight constraints defined earlier. The output is displayed as follows.

```
. constraint define 1 [_2_3_4v1]1.maritals=[_3_4v2]1.maritals

. constraint define 2 [_3_4v2]1.maritals=[_4v3]1.maritals

. constraint define 3 [_2_3_4v1]educ=[_3_4v2]educ

. constraint define 4 [_3_4v2]educ=[_4v3]educ
```

```
. constraint define 5 [_2_3_4v1]age=[_3_4v2]age

. constraint define 6 [_3_4v2]age=[_4v3]age

. constraint define 7 [_2_3_4v1]1.male=[_3_4v2]1.male

. constraint define 8 [_3_4v2]1.male=[_4v3]1.male

. seqlogit healthre i.maritals educ age i.male, tree (1: 2 3 4, 2: 3 4, 3:4)
constraints (1/8)

Transition tree:

Transition 1: 1 : 2 3 4
Transition 2: 2 : 3 4
Transition 3: 3 : 4

Computing starting values for:

Transition 1
Transition 2
Transition 3

Iteration 0:   log likelihood =  -1597.629
Iteration 1:   log likelihood = -1518.7136
Iteration 2:   log likelihood = -1518.3505
Iteration 3:   log likelihood = -1518.3504
```

```
                                        Number of obs      =      1,300
                                        Wald chi2(0)       =         .
Log likelihood = -1518.3504             Prob > chi2        =         .

 ( 1)   [_2_3_4v1]1.maritals - [_3_4v2]1.maritals = 0
 ( 2)   [_3_4v2]1.maritals - [_4v3]1.maritals = 0
 ( 3)   [_2_3_4v1]educ - [_3_4v2]educ = 0
 ( 4)   [_3_4v2]educ - [_4v3]educ = 0
 ( 5)   [_2_3_4v1]age - [_3_4v2]age = 0
 ( 6)   [_3_4v2]age - [_4v3]age = 0
 ( 7)   [_2_3_4v1]1.male - [_3_4v2]1.male = 0
 ( 8)   [_3_4v2]1.male - [_4v3]1.male = 0
```

healthre	Coef.	Std. Err.	z	P>|z|	[95% Conf. Interval]	
_2_3_4v1						
1.maritals	.4724299	.0921141	5.13	0.000	.2918895	.6529703
educ	.1075093	.0147409	7.29	0.000	.0786175	.136401
age	-.0157278	.0026189	-6.01	0.000	-.0208606	-.0105949
1.male	.0895459	.0914023	0.98	0.327	-.0895993	.268691
_cons	1.84848	.2630136	7.03	0.000	1.332983	2.363977
_3_4v2						
1.maritals	.4724299	.0921141	5.13	0.000	.2918895	.6529703
educ	.1075093	.0147409	7.29	0.000	.0786175	.136401
age	-.0157278	.0026189	-6.01	0.000	-.0208606	-.0105949

```
  1.male |  .0895459   .0914023    0.98   0.327    -.0895993    .268691
    _cons |  .3167184   .2501401    1.27   0.205    -.1735471   .8069839
-----------+----------------------------------------------------------------
_4v3       |
1.maritals |  .4724299   .0921141    5.13   0.000     .2918895   .6529703
     educ |  .1075093   .0147409    7.29   0.000     .0786175    .136401
      age | -.0157278   .0026189   -6.01   0.000    -.0208606  -.0105949
   1.male |  .0895459   .0914023    0.98   0.327    -.0895993    .268691
    _cons | -1.616745   .2588349   -6.25   0.000    -2.124052  -1.109438
-----------+----------------------------------------------------------------
```

The beginning of the output displays the eight constraints defined, followed by the results of the seqlogit command with the tree and constraints option. The logit coefficients table displays the logit coefficients for four predictor variable for the three binary models. The logit coefficients for each predictor variable are the same across the binary models. Compared with the output of the multiple-predictor CR model produced by the ocratio command in the preceding section, the estimated logit coefficients are the same, whereas the intercepts produced by the seqlogit command have opposite signs.

To request the odds ratios, the or option is added to the command syntax. The output is shown as follows. To save space, only the odds ratios table is displayed.

```
. seqlogit healthre i.maritals educ age i.male, tree (1: 2 3 4, 2: 3 4, 3:4)
constraints (1/8) or
(first portion of the output omitted)
------------------------------------------------------------------------------
  healthre | Odds Ratio  Std. Err.      z    P>|z|    [95% Conf. Interval]
-----------+------------------------------------------------------------------
_2_3_4v1   |
1.maritals |  1.603887   .1477406    5.13   0.000    1.338955   1.921239
     educ |  1.113501   .0164141    7.29   0.000    1.081791   1.146141
      age |  .9843953    .002578   -6.01   0.000    .9793554    .989461
   1.male |  1.093677   .0999646    0.98   0.327    .9142975   1.308251
    _cons |  6.350161   1.670179    7.03   0.000    3.792339   10.63316
-----------+------------------------------------------------------------------
_3_4v2     |
1.maritals |  1.603887   .1477406    5.13   0.000    1.338955   1.921239
     educ |  1.113501   .0164141    7.29   0.000    1.081791   1.146141
      age |  .9843953    .002578   -6.01   0.000    .9793554    .989461
   1.male |  1.093677   .0999646    0.98   0.327    .9142975   1.308251
    _cons |  1.372616   .3433463    1.27   0.205    .8406775   2.241138
-----------+------------------------------------------------------------------
_4v3       |
1.maritals |  1.603887   .1477406    5.13   0.000    1.338955   1.921239
     educ |  1.113501   .0164141    7.29   0.000    1.081791   1.146141
      age |  .9843953    .002578   -6.01   0.000    .9793554    .989461
   1.male |  1.093677   .0999646    0.98   0.327    .9142975   1.308251
    _cons |  .1985439   .0513901   -6.25   0.000    .1195462    .3297443
------------------------------------------------------------------------------
```

In the odds ratios table, the odds ratios for each predictor variable are the same across the three underlying binary logistic regression models labeled _2_3_4v1, _3_4v2, and _4v3. The odds ratios for maritals, educ, age, and male are 1.604, 1.114, .984, and 1.094, respectively, which are the same as those estimated by the ocratio command in the preceding section.

Computing the Estimated Probabilities With the margins *Command*

Although the seqlogit command works with margins, the syntax for the margins command is not as simple as in previous chapters. Recall that in a CR model we estimate the conditional probability of being in a category given that an individual has been in that category or beyond. This conditional probability of being in a category is the probability of not passing a transition in the CR model with the seqlogit command, whereas the conditional probability of being above that category is the probability of passing a transition.

In the predict command, which is one of the seqlogit postestimation commands, we need to specify the trpr, transition, and choice options. For example, transition(1) specifies the first transition or category comparison, which is shown in the transition tree in the original seqlogit syntax. In this example, transition 1 (1: 2 3 4) compares category 1 with categories 2, 3, and 4, or vice versa. The choice option specifies the choice with 0 as the reference group and 1 as the outcome group. So if we specify choice(0) after transition(1), we are interested in the estimated probability of being in category 1, the reference group coded as 0, rather than in the probability of being in categories 2, 3, and 4, the outcome group coded as 1. If we specify choice(1), we can calculate the probability of being in categories 2, 3, and 4. For more details on how to specify these options in predict, type help seqlogit postestimation.

The command margins, at(educ = (8 13 16)) atmeans predict(trpr transition(1) choice(0)) tells Stata to calculate the estimated probabilities of being in category 1 in the first transition tree, comparing categories 2, 3, and 4 with category 1, for the predictor variable educ at the values of 8, 13, and 16 when holding the other predictor variables at their means. The output is displayed as follows.

```
. margins, at(educ = (8 13 16)) atmeans predict(trpr transition(1) choice(0))

Adjusted predictions                              Number of obs   =        1300
Model VCE    : OIM

Expression   : Probability of choosing 0 in transition transition(1),
               predict(trpr transition(1) choice(0))

1._at        : 0.maritals     =     .5315385  (mean)
               1.maritals     =     .4684615  (mean)
               educ           =            8
```

```
               age            =        48.2  (mean)
               0.male         =     .5561538  (mean)
               1.male         =     .4438462  (mean)

2._at        : 0.maritals     =     .5315385  (mean)
               1.maritals     =     .4684615  (mean)
               educ           =          13
               age            =        48.2  (mean)
               0.male         =     .5561538  (mean)
               1.male         =     .4438462  (mean)

3._at        : 0.maritals     =     .5315385  (mean)
               1.maritals     =     .4684615  (mean)
               educ           =          16
               age            =        48.2  (mean)
               0.male         =     .5561538  (mean)
               1.male         =     .4438462  (mean)
```

```
------------------------------------------------------------------------------
             |             Delta-method
             |     Margin   Std. Err.      z    P>|z|     [95% Conf. Interval]
-------------+----------------------------------------------------------------
         _at |
          1  |   .0987194    .0118632    8.32   0.000     .0754679    .1219709
          2  |   .0601385    .0065546    9.18   0.000     .0472918    .0729853
          3  |   .0442936    .0053655    8.26   0.000     .0337775    .0548098
------------------------------------------------------------------------------
```

The results in the output look similar to those estimated in Chapters 4 and 5. The estimated transition probabilities when $Y = 1$ in the first transition tree for the predictor variable educ at the values of 8, 13, and 16 are .099, .060, and .044, respectively, when holding the other predictor variables at their means.

The transition probabilities in the other two transitions (i.e., 2 and 3) can be computed in a similar way. For example, the margins command with the predict(trpr transition(2) choice(0)) option calculates the estimated probabilities of being in category 2 in the second transition tree, comparing categories 3 and 4 with category 2. To save space, the output is omitted.

Immediately after the margins command, the marginsplot command is used to plot the estimated transition probabilities for $Y = 1$ in the first transition tree, which is shown in Figure 6.1.

Figure 6.1 shows that with the increase of the highest year of school completed, the conditional probabilities of being in category 1 in the first transition decrease. As illustrated in previous chapters, the graph combine command can be used to combine graphs for the estimated conditional probabilities for all categories. The resulting graph is omitted here. Please note that the marginsplot command in Stata 14 currently can only plot one graph each time from the results estimated by the margins program with the predict(trpr transition() choice()) option.

Figure 6.1 Estimated Conditional Probabilities When $Y = 1$ in Transition 1 for educ at 8, 13, and 16

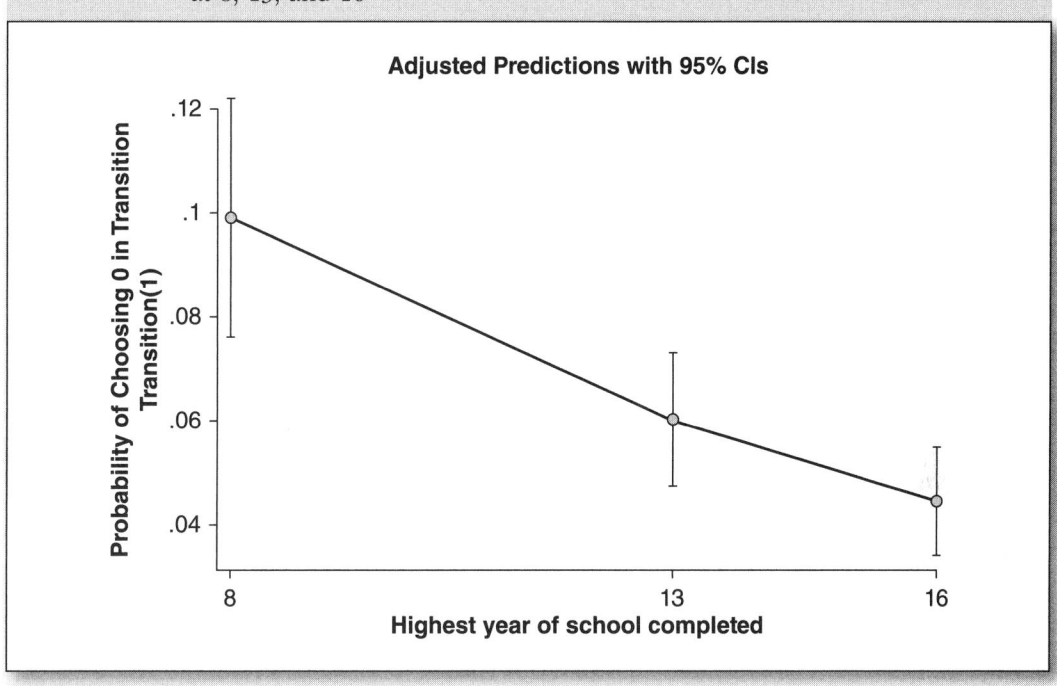

Note: CI = confidence interval.

6.4 Making Publication-Quality Tables

```
1.  .quietly xi: ocratio healthre i.maritals, link(logit)
2.  .outreg2 using chap6out, word replace
3.  .outreg2 using chap6out, eform word append
4.  .quietly xi: ocratio healthre i.maritals educ age i.male, link(logit)
5.  .outreg2 using chap6out, word append
6.  .outreg2 using chap6out, eform word append
7.  .seeout
```

The first command `quietly xi: ocratio healthre i.maritals, link(logit)` estimates the single-predictor model without providing the output.

The second command `outreg2 using chap6out, word replace` tells Stata to create a regression table for the estimates and save it to a file named chap6out. The `word` option requests that the table be created in the Microsoft Word document with the extension .rtf. The `replace` option asks Stata to replace any files with the same name as chap6out.

The third command `outreg2 using chap6out, eform word append` tells Stata to request the estimated odds ratios using the `eform` option, and it appends the results to the original table with the `append` option.

The fourth command `quietly xi: ocratio healthre i.maritals educ age i.male, link(logit)` estimates the full model with four predictors without showing the output.

The fifth command `outreg2 using chap6out, word append` creates a regression table for the full model and appends it to the original table.

The sixth command `outreg2 using chap6out, eform word append` reports the odds ratios of the estimates for the full model and appends the results to the original table.

The seventh command `seeout` opens the final table in the data browser.

The previous commands automatically produce Table 6.2, as shown here in its original format, presenting the results for both the single-predictor CR model and the full model.

Table 6.2   Results of the Continuation Ratio Models: Single-Predictor and Full Models ($Y >$ cat. j vs. $Y =$ cat. j) (Shown in Original Format Generated by Stata)

Variables	(1)	(2)	(3)	(4)
	.	.	.	.
_cut1	−2.491***	0.0828***	−1.848***	0.157***
	(0.118)	(0.00981)	(0.263)	(0.0414)
_cut2	−1.025***	0.359***	−0.317	0.729
	(0.0782)	(0.0281)	(0.250)	(0.182)
_cut3	0.774***	2.169***	1.617***	5.037***
	(0.0831)	(0.180)	(0.259)	(1.304)
_Imaritals_1	−0.464***	0.629***	−0.472***	0.623***
	(0.0901)	(0.0567)	(0.0921)	(0.0574)
Educ			−0.108***	0.898***
			(0.0147)	(0.0132)
Age			0.0157***	1.016***
			(0.00262)	(0.00266)
_Imale_1			−0.0895	0.914
			(0.0914)	(0.0836)

Note: Standard errors in parentheses.

***$p < 0.01$, **$p < 0.05$, *$p < 0.1$

After editing the model titles, labels for logit coefficients, odds ratios, variable names, and the decimals, and adding model fit statistics, the final table (Table 6.3) is presented.

Table 6.3 Results of the Continuation Ratio Models: Single Predictor and Full Models (Y > cat. *j* vs. Y = cat. *j*) (Edited)

Variables	Single-Predictor Model		Full Model	
	b (SE(b))	OR	*b* (SE(b))	OR
_cut1 (α_1)	−2.491***	0.083***	−1.848***	0.157***
	(0.118)	(0.010)	(0.263)	(0.041)
_cut2 (α_2)	−1.025***	0.359***	-0.317	0.729
	(0.078)	(0.028)	(0.250)	(0.182)
_cut3 (α_3)	0.774***	2.169***	1.617***	5.037***
	(0.083)	(0.180)	(0.259)	(1.304)
Maritals	−0.464***	0.629***	−0.472***	0.623***
	(0.090)	(0.057)	(0.092)	(0.057)
Educ			−0.108***	0.898***
			(0.015)	(0.013)
Age			0.0157***	1.016***
			(0.003)	(0.003)
Male			−0.090	0.914
			(0.091)	(0.084)
Observations	3,459		3,459	
LR R²	.009		.039	
Log likelihood	−1,565.941		−1,518.35	
LR χ²	$\chi^2_1 = 29.93$		$\chi^2_4 = 122.11$	

Note: Standard errors in parentheses.

***$p < .01$.

6.5 Reporting the Results

Writing the results for the continuation ratio models is similar to that for the proportional odds models. First, describe the statistical method you use for data analysis and justify why it is needed for your research. Also describe the dependent variable and the independent variables in the models, and your research hypothesis or the purpose of your study. Furthermore, report and interpret the available model fit statistics, including the log likelihood ratio chi-square test statistic and the pseudo R^2.

Second, have a summary table containing all the parameter estimates, their standard errors, p values or whether the estimates are significant, and odds ratios.

Third, in the body of the text, concisely interpret the parameter estimates for the predictor variables. When interpreting the odds ratios of the estimates, pay attention to the directions of the comparisons. Explain clearly the odds comparing the probabilities of being in a higher category versus being in a particular category, or vice versa.

Fourth, when more than one model is fitted, results of all the competing models may be presented in the table. The following is an example of summarizing results for the continuation ratio model with all four predictors (full model).

The continuation ratio model was fitted to estimate the ordinal outcome variable, health status, from a set of predictor variables, including marital status, years of education, age, and gender. This model was used to address the category comparisons between outcomes being in a particular category and those being above that category.

The log likelihood ratio chi-square test statistic $LR\,\chi^2_{(4)} = 122.11$, $p < .001$, indicated that the full model with four predictor variables provided a better fit than the null model with no independent variables in predicting conditional probability for health status. The likelihood ratio $R^2_L = .0387$, which was small.

Table 6.3 displays the parameter estimates for four predictor variables. Of them, the coefficients of two variables maritals ($\beta = .472$, Wald $z = 5.13$, $p < .001$) and educ ($\beta = .108$, Wald $z = 7.29$, $p < .001$) were positively significant, the coefficient of age ($\beta = -.016$, Wald $z = -6.01$, $p < .001$) was negatively significant, and male ($\beta = .090$, Wald $z = .98$, $p = .327$) was not significant.

In terms of odds ratios, the odds of being in a better health status level relative to being in that level for the married were 1.604 times the odds for the unmarried (OR = 1.604) after controlling for the effects of other predictors in the model. Similarly, the odds of being above a particular category versus being in that category increased by 1.114 or 11.4% for a one-unit increase in years of education (educ) when holding other variables constant. On the other hand, for every one-unit increase in age, the odds of being in a better health status level decreased by a factor of .984 (OR = .984) when holding the effects of the other variables constant. In addition, the odds ratio for the predictor male was .984, $p = .327$, which was not significant, indicating being male did not influence the odds of being in a higher category.

6.6 Summary of Stata Commands in This Chapter

```
use chap6-gss2012, clear

*The one-predictor CR model with logit link
xi: ocratio healthre i.maritals, link(logit)
xi: ocratio healthre i.maritals, link(logit) eform
aic

*AIC = -2(LLm - k) / n
display -2*(-1565.941-1)/3459

*The one-predictor CR model with clog-log link
xi: ocratio healthre i.maritals, link(cloglog) cumulative
aic
xi: ocratio healthre i.maritals, link(cloglog) eform cumulative

*The multiple-predictor CR model with logit link
xi: ocratio healthre i.maritals educ age male, link(logit)
xi: ocratio healthre i.maritals educ age male, link(logit) eform

*AIC = -2(LLm - k) / n
display -2*(-1518.35-4)/3459

*Likelihood ratio test
display -2*(-1565.941-(-1518.35))

*Fitting the CR model using seqlogit
seqlogit healthre i.maritals educ age i.male, tree (1: 2 3 4, 2: 3 4, 3:4)

*Fitting the constrained CR model
*Step 1: Define constraints
constraint define 1 [_2_3_4v1]1.maritals=[_3_4v2]1.maritals
constraint define 2 [_3_4v2]1.maritals=[_4v3]1.maritals
constraint define 3 [_2_3_4v1]educ=[_3_4v2]educ
constraint define 4 [_3_4v2]educ=[_4v3]educ
constraint define 5 [_2_3_4v1]age=[_3_4v2]age
constraint define 6 [_3_4v2]age=[_4v3]age
constraint define 7 [_2_3_4v1]1.male=[_3_4v2]1.male
constraint define 8 [_3_4v2]1.male=[_4v3]1.male

*Step2: Fitting the same CR model with the constraints option
seqlogit healthre i.maritals educ age i.male, tree (1: 2 3 4, 2: 3 4, 3:4) ///
constraints (1/8)
seqlogit healthre i.maritals educ age i.male, tree (1: 2 3 4, 2: 3 4, 3:4) ///
constraints (1/8) or

*margins and marginsplot in Stata 13 and later
margins, at(educ = (8 13 16)) atmeans predict(trpr transition(1) choice(0))
marginsplot

*Making publication-quality tables using outreg2
```

```
quietly xi: ocratio healthre i.maritals, link(logit)
outreg2 using chap6out, word replace
outreg2 using chap6out, eform word append
quietly xi: ocratio healthre i.maritals educ age i.male, link(logit)
outreg2 using chap6out, word append
outreg2 using chap6out, eform word append
```

6.7 EXERCISES

Use the GSS 2012 data for the following problems.

1. Conduct an analysis for a continuation ratio model using the `ocratio` command to examine the relationship between the outcome variable `degree` and three predictor variables `sex`, `class`, and `sibs`.

2. Identify the likelihood ratio test of the model, and interpret it.

3. Identify the logit coefficients, the Wald z tests, and the 95% confidence intervals for `sex` and `sibs`. Are they statistically significant?

4. Interpret the odds ratios for `class` and `sibs`.

5. Make a publication-quality table containing the estimated logit coefficients and odds ratios.

6. Write a report to summarize the results from the output.

7. Fit the same CR model as that in Exercise 1 using the `seqlogit` command.

CHAPTER 7

Adjacent Categories Logistic Regression Models

Objectives of This Chapter

This chapter presents adjacent categories logistic regression models. After introducing the models, the odds, odds ratios, and model fit statistics are reviewed, and how to interpret parameter estimates is discussed. Next, the multinomial logistic regression model is briefly introduced, and the transformation from the multinomial logistic model to the adjacent categories model is discussed. After a description of the data, this chapter illustrates how to conduct the analysis using Stata with step-by-step instructions. Stata `mlogit` and `constraint` commands are explained, and output is interpreted. This chapter also illustrates how the results are displayed in publication-quality tables using the `outreg2` command and reported in text. It focuses on fitting adjacent categories logistic regression models with Stata, as well as on interpreting and presenting the results. After reading this chapter, you should be able to

- Determine when an adjacent categories model is used.
- Conduct analysis using Stata.
- Interpret the output.
- Interpret the model in terms of odds ratios.
- Compute and plot the estimated probabilities using the `margins` and `marginsplot` commands, respectively.
- Compute the estimated probabilities using the `mtable` command.
- Compare models using the likelihood ratio test and other fit statistics.
- Present results in publication-quality tables using Stata.
- Write the results for publication.

7.1 Adjacent Categories Models: An Introduction

The name of the adjacent categories model explains itself. It focuses on the pairs of adjacent categories for ordinal response variables. Specifically, the adjacent categories model (Agresti, 2010; Clogg & Shihadeh, 1994; Goodman, 1983; Hosmer & Lemeshow, 2000; O'Connell, 2006) estimates the odds of being in a category versus being in an adjacent category of an ordinal response variable. This is different from the proportional odds model, which compares lower categories relative to higher categories, and the continuation ratio model, which compares a particular category with all the other higher categories. The adjacent categories model compares a category $j + 1$ with the next lower category j, or a category j with the next higher category $j + 1$, for two adjacent categories since they are just in opposite directions. If an ordinal response variable has J levels, then there are $J - 1$ pairs of adjacent categories, which is the number of comparisons between these pairs of adjacent categories. For example, an ordinal response variable, such as health status, has four levels. There are three pairs of adjacent categories, comparing category 2 with category 1, category 3 with category 2, and category 4 with category 3.

The adjacent categories model can be written as:

$$\text{logit}[\pi(Y = j+1 \mid x_1, x_2, \ldots, x_p)] = \ln\left(\frac{\pi\left(Y = j+1 \mid x_1, x_2, \ldots, x_p\right)}{\pi\left(Y = j \mid x_1, x_2, \ldots, x_p\right)}\right)$$

$$= \alpha_j + \beta_1 X_1 + \beta_2 X_2 + \ldots + \beta_p X_p \qquad (7.1)$$

where j is the number of the categories of the ordinal response variable from 1 to $J - 1$. α_j are the intercepts, and $\beta_1, \beta_2, \ldots, \beta_j$ are the logit coefficients for each predictor variable. The adjacent categories model is a special case of multinomial logistic regression with constrained regression coefficients. Although the logit coefficients in the multinomial logistic regression model are allowed to vary when comparing pairs of adjacent categories, they are assumed to be equal across the ordinal categories in the adjacent categories (AC) model.

7.1.1 Odds and Odds Ratios in AC Models

In previous chapters, we introduced the proportional odds (PO) model and the continuation ratio (CR) model. Let us now look at the difference in the odds among the PO model, the CR model, and the AC model.

In PO models, we estimate the odds of being at or below a particular category ($Y \leq j$) versus being above that category ($Y > j$). The odds of being at or below a category in ordinal logistic regression are the probability of being at or below a category divided by the probability of being above that category:

$$\text{Odds } (Y \le j) = \frac{p(Y \le j)}{p(Y > j)} \tag{7.2}$$

The CR model estimates the conditional probabilities of being in category j, conditional on being at or beyond that category; that is, $P(Y = j \mid Y \ge j)$. The odds in the CR model are the ratio of the probability of being in a category to the probability of being beyond that category:

$$\text{Odds } (Y = j \mid Y \ge j) = \frac{p(Y = j)}{p(Y > j)} \tag{7.3}$$

In comparison, in adjacent categories models, we estimate the odds of adjacent categories, which are the ratio of the probabilities of one category to the next lower category. The odds can be expressed as:

$$\text{Odds } (Y = j + 1 \text{ vs. } Y = j) = \frac{p(Y = j+1)}{p(Y = j)} \tag{7.4}$$

where j can be any categories from 0 to J categories.

Or:

$$\text{Odds } (Y = j \text{ vs. } Y = j - 1) = \frac{p(Y = j)}{p(Y = j - 1)} \tag{7.5}$$

where j is larger than 0.

For example, an outcome variable, health status, has four ordinal levels from 1 to 4, with 1 = poor, 2 = fair, 3 = good, and 4 = excellent. We estimate the following three odds: the odds of being in category 2 versus 1, the odds of being in category 3 versus 2, and the odds of being in category 4 versus 3.

Odds $(Y = 2 \text{ vs. } Y = 1)$ equal the ratio of probability of being in category 2 to the probability of being in category 1:

$$\text{Odds } (Y = 2 \text{ vs. } 1) = \frac{p(Y = 2)}{p(Y = 1)} = \frac{p(2)}{p(1)}$$

Odds $(Y = 3 \text{ vs. } Y = 2)$ equal the ratio of probability of being in category 3 to the probability of being in category 2. It can be expressed as follows:

$$\text{Odds } (Y = 3 \text{ vs. } 2) = \frac{p(Y = 3)}{p(Y = 2)} = \frac{p(3)}{p(2)}$$

Odds ($Y = 4$ vs. $Y = 3$) equal the probability of being in category 4 divided by the probability of being in category 3. It can be expressed as follows:

$$\text{Odds } (Y = 4 \text{ vs. } 3) = \frac{p(Y = 4)}{p(Y = 3)} = \frac{p(4)}{p(3)}$$

In sum, the odds of being in category 2 compared with category 1 is the probability comparison between category 2 and category 1; the odds of being in category 3 versus category 2 is the probability comparison between categories 3 and 2; and the odds of being in category 4 versus category 3 is the probability comparison between categories 4 and 3. Generally, the odds in the CR model compare two adjacent probabilities; that is, $p(Y = j + 1)$ and $p(Y = j)$. Table 7.1 presents the logits, odds, and category comparisons for the adjacent categories model for the ordinal response variable with four levels.

Table 7.1   Category Comparisons for the Adjacent Categories Model With Four Levels of Health Status ($j = 1, 2, 3, 4$)

Category	Logit $P(Y = j + 1 \mid j)$	Odds	Probability Comparisons
Level 2	logit $P(Y = 2)$	$\dfrac{P(Y = 2)}{P(Y = 1)}$	Category 2 vs. category 1
Level 3	logit $P(Y = 3)$	$\dfrac{P(Y = 3)}{P(Y = 2)}$	Category 3 vs. category 2
Level 4	logit $P(Y = 4)$	$\dfrac{P(Y = 4)}{P(Y = 3)}$	Category 4 vs. category 3

Odds Ratios in Adjacent Categories Logistic Regression

Just like the odds ratios in the PO model and the CR model, the odds ratios in the adjacent categories model are also the exponentiated logistic coefficients $\exp(\beta)$. Since the odds in the AC model are the probability comparisons between the higher category and the lower category for two adjacent categories, the odds ratio is the ratio of the two odds when the value of a predictor is ($x + 1$) relative to the odds when the predictor has a value of x. It is the change in the odds of being in a higher category relative to the next lower category for a one-unit increase in a predictor variable when holding others constant.

7.1.2 Goodness-of-Fit Statistics

Since the AC model is a special case of the multinomial logistic regression model, called the constrained multinomial model, the model fit statistics for the latter model such as the log likelihood statistic, deviance statistic, several types of pseudo R^2 statistics, log likelihood ratio test, and Akaike information criterion (AIC) and Bayesian information criterion (BIC) statistics can also be applied to the AC model.

As with other logistic regression models, the log likelihood ratio test can be used to compare nested models. As explained in the previous chapters, the log likelihood ratio statistic is the difference between the deviance or –2LL for the reduced model and that for the full model. The likelihood ratio test statistic follows a chi-square distribution with the degrees of freedom of the distribution equal to the difference in the number of parameters between these two models. A significant likelihood ratio chi-square test indicates that the full model rather than the reduced model is preferred.

In addition, the AIC and BIC statistics can be used to compare either nested or non-nested models. Different versions of the AIC and BIC statistics are reported using the `fitstat` command (Long & Freese, 2006, 2014). For example, AIC, AIC × n, AIC used by Stata, BIC, BIC′, and BIC used by Stata are reported in `SPost9` (Long & Freese, 2006), whereas AIC, AIC divided by N, and BIC are reported in the latest `SPost13` (Long & Freese, 2014). See Chapter 4 for a detailed discussion on the AIC and BIC statistics.

7.1.3 Interpretation of Model Parameter Estimates

The estimated regression coefficients in the AC model are the logit coefficients, which are the log odds of being in a higher category relative to the next lower category for any pairs of adjacent categories. It is the change in the predicted logit or the log odds for a one-unit increase in a predictor variable when holding others constant. Exponentiating the logit coefficients gives us the odds ratios. The odds ratio in the AC model can be interpreted in a similar way as other logistic regression models. However, they are different when the odds in these models are comparing different categories. In the AC model, an odds ratio is interpreted as the change in the predicted odds of being in a higher category for a pair of adjacent categories for a one-unit increase in the predictor variable.

When the OR is larger than 1, the odds of being in a higher category versus the next lower category for a pair of adjacent categories increase for a one-unit increase in the predictor variable.

When the OR is less than 1, the odds of being in a higher category versus the next lower category decrease for a one-unit increase in the predictor variable.

When the OR equals 1, there is no relationship between the predictor and the odds, so there is no change in the odds when the values of the predictor variable change.

What about the odds of being in a lower category for any pair of the adjacent categories? They are just the reciprocal of the odds of being in a higher category for a pair of the adjacent categories. Whereas the latter compares a higher category $j + 1$ with the next lower category j, the odds of being in a lower category compare the category j with the next higher category $j + 1$.

To get the OR of being in a lower category, we need to exponentiate the logit coefficient with a negative sign before that, that is, $\exp(-\beta)$, or take the reciprocal of the odds of being in a higher category for a pair of the adjacent categories.

No official program has been developed to estimate the AC model in Stata. Since the AC model is treated as the constrained multinomial logistic regression, it can be estimated via the multinomial logistic regression model with imposed constraints. The multinomial model will be introduced first, and then how to set up the constraints to formulate the AC model will be explained.

7.1.4 From the Multinomial Logistic Model to the AC Model

The Multinomial Logistic Model

The multinomial logistic regression model is used to estimate the nominal response variables with more than two categories. The odds in this model compare any category with the baseline category or the reference category. It can be treated as a combination of a series of binary logistic regression models with a particular category = 1 and the base category = 0. When there are J categories, it estimates $J - 1$ binary logistic regression models simultaneously. This model can be expressed as follows:

$$\ln\left(\frac{\pi\left(Y = j \mid x_1, x_2, ..., x_p\right)}{\pi\left(Y = J \mid x_1, x_2, ..., x_p\right)}\right) = \alpha_j + \beta_{j1}X_1 + \beta_{j2}X_2 + ... + \beta_{jp}X_p \qquad (7.6)$$

where $j = 1, 2, ..., J - 1$; J is the base category, which can be any category but is generally the highest one; α_j are the intercepts, and $\beta_{j1}, \beta_{j2}, ..., \beta_{jp}$ are the logit coefficients for each predictor variable at each comparison. Since the model includes $J - 1$ comparisons, it estimates $J - 1$ logit coefficients for each predictor. A more detailed introduction of this model can be seen in Chapter 12.

Transformation Between the AC Model and the Multinomial Logistic Regression Model

The AC model is a constrained multinomial logistic regression model. The AC model focuses on the comparison between adjacent categories, whereas the multinomial logistic regression model compares a category with the baseline category. When the logit coefficients of the predictor variables in the AC model are constrained to be equal, we can easily transform the equations of this model into those for a multinomial model, and vice versa. The transformation between these two models is shown as follows.

Let us see a simple AC model with one predictor variable:

$$\ln\left(\frac{\pi(Y = j+1)}{\pi(Y = j)}\right) = \alpha_j + \beta_1 X_1 \tag{7.7}$$

When the ordinal response variable has four categories from 1 to 4, we estimate the following three binary logistic regression models simultaneously for the AC model, comparing three pairs of adjacent categories. They are expressed as follows:

$$\ln\left(\frac{\pi(Y = 2)}{\pi(Y = 1)}\right) = \alpha_1 + \beta_1 X_1$$

$$\ln\left(\frac{\pi(Y = 3)}{\pi(Y = 2)}\right) = \alpha_2 + \beta_1 X_1$$

$$\ln\left(\frac{\pi(Y = 4)}{\pi(Y = 3)}\right) = \alpha_3 + \beta_1 X_1$$

The intercept for each model is different, but the logit coefficient is the same, so these models estimate three different intercepts (i.e., α_1, α_2, and α_3) and one regression coefficient β_1.

In a multinomial logistic regression model, if we choose the lowest level, category 1, as the baseline category, then we estimate three logit coefficients for a predictor variable, comparing categories 2 and 1, categories 3 and 1, and categories 4 and 1. We can formulate three equations for the multinomial model based on these for the AC model shown earlier.

The first equation for the multinomial model is the same as the first one for the AC model since both equations compare categories 2 and 1.

The second equation for the multinomial model compares categories 3 and 1. It can be easily obtained by adding the first two equations for the AC model together:

$$\ln\left(\frac{\pi(Y = 2)}{\pi(Y = 1)}\right) + \ln\left(\frac{\pi(Y = 3)}{\pi(Y = 2)}\right) = (\alpha_1 + \beta_1 X_1) + (\alpha_2 + \beta_1 X_1)$$

The left side of the equation:

$$\ln\left(\frac{\pi(Y = 2)}{\pi(Y = 1)}\right) + \ln\left(\frac{\pi(Y = 3)}{\pi(Y = 2)}\right) = \ln\left[\left(\frac{\pi(Y = 2)}{\pi(Y = 1)}\right) \times \left(\frac{\pi(Y = 3)}{\pi(Y = 2)}\right)\right] = \ln\left(\frac{\pi(Y = 3)}{\pi(Y = 1)}\right)$$

The right side of the equation:

$$(\alpha_1 + \beta_1 X_1) + (\alpha_2 + \beta_1 X_1) = (\alpha_1 + \alpha_2) + 2\beta_1 X_1$$

Putting the two sides together, we obtain the following equation:

$$\ln\left(\frac{\pi(Y = 3)}{\pi(Y = 1)}\right) = (\alpha_1 + \alpha_2) + 2\beta_1 X_1$$

This is the second equation for the multinomial model. The intercept in this model equals the sum of intercepts from the first two adjacent categories models comparing categories 2 and 1 and categories 3 and 2. The regression coefficient in this model equals twice the regression coefficient in the adjacent category model.

If we add all three binary models for the adjacent category model using the same method, then we get the third equation for the multinomial model comparing categories 4 versus 1:

$$\ln\left(\frac{\pi(Y = 4)}{\pi(Y = 1)}\right) = (\alpha_1 + \alpha_2 + \alpha_3) + 3\beta_1 X_1$$

The intercept in this model equals the sum of intercepts of all three adjacent models comparing categories 2 and 1, 3 and 2, and 4 and 3. The coefficient in this model equals three times the regression coefficient in the adjacent category model.

Therefore, to estimate the AC model for an ordinal response variable with four categories, we just need to estimate the multinomial model by constraining the logit coefficients in the equations for those binary logistic regression models. Specifically, the logit coefficient in the model comparing categories 3 and 1 equals twice the logit coefficient in the model comparing categories 2 and 1, and the logit coefficient comparing categories 4 and 1 equals three times the logit coefficient comparing categories 2 and 1. Table 7.2 presents the model transformation between the AC model and the multinomial logistic model.

To estimate the AC model for an ordinal response variable, we follow three steps.

First, we estimate the constrained multinomial logistic regression by specifying the constraints on the logit coefficients, that is, $(J - 1) \times \beta_1$. If the ordinal response has J categories $(1, 2, ..., J)$, $J - 2$ constraints need to be specified on the logit coefficients for each predictor since no constraint is needed for the first equation comparing categories 2 and 1.

Second, we determine the intercepts and the logit coefficients for the AC model from the estimated multinomial logistic model. The intercept and the logit coefficient for the first response function (Equation 1) for the AC model and the multinomial logistic model are the same. The intercepts for the successive response equations of

Table 7.2   Model Transformation Between the AC Model and the Multinomial Logistic Model

Equations/ Submodels	Adjacent Categories Logistic Model Category Comparisons		Multinomial Logistic Model Category Comparisons	
Equation 1	Categories 2 vs. 1	$\ln\left(\frac{\pi(Y=2)}{\pi(Y=1)}\right) = \alpha_1 + \beta_1 X_1$	Categories 2 vs. 1	$\ln\left(\frac{\pi(Y=2)}{\pi(Y=1)}\right) = \alpha_1 + \beta_1 X_1$
Equation 2	Categories 3 vs. 2	$\ln\left(\frac{\pi(Y=3)}{\pi(Y=2)}\right) = \alpha_2 + \beta_1 X_1$	Categories 3 vs. 1	$\ln\left(\frac{\pi(Y=3)}{\pi(Y=1)}\right) = (\alpha_1 + \alpha_2) + 2\beta_1 X_1$
Equation 3	Categories 4 vs. 3	$\ln\left(\frac{\pi(Y=4)}{\pi(Y=3)}\right) = \alpha_3 + \beta_1 X_1$	Categories 4 vs. 1	$\ln\left(\frac{\pi(Y=4)}{\pi(Y=1)}\right) = (\alpha_1 + \alpha_2 + \alpha_3) + 3\beta_1 X_1$

the AC model are computed by subtracting the intercepts from previous response equations of the multinomial logistic model.

Third, we formulate the response equations comparing adjacent categories by using the intercepts and coefficients obtained in the second step.

If we constrain the logit coefficients for any pairs of adjacent categories to be the same, following the simple transformation displayed in Table 7.2, then a general equation for the constrained multinomial logistic regression model can be written as follows:

$$\ln\left(\frac{\pi(Y=j|x_1, x_2,...,x_p)}{\pi(Y=1|x_1, x_2,...,x_p)}\right) = (\alpha_1 + \alpha_2 +...+ \alpha_{j-1}) + (J-1)\beta_1 X_1 + (J-1)\beta_2 X_2$$
$$+ ... + (J-1)\beta_p X_p \tag{7.8}$$

where $j = 2, ..., J-1$. The coefficients in this model equal $(J-1)$ times the coefficients in the AC model. So if we know a logit coefficient for a predictor variable in a constrained multinomial logistic regression model, then the corresponding coefficient for that predictor in the AC model can be obtained by dividing $(J-1)$ from the coefficient in the former model.

Sometimes the baseline category is 0 when an ordinal response variable includes level 0 and has $j + 1$ categories. This equation can be rewritten as:

$$\ln\left(\frac{\pi(Y=j|x_1, x_2,...,x_p)}{\pi(Y=0|x_1, x_2,...,x_p)}\right) = (\alpha_1 + \alpha_2 +...+ \alpha_j) + J \times \beta_1 X_1 + J \times$$
$$\beta_2 X_2 + ... + J \times \beta_p X_p \tag{7.9}$$

where $j = 1, 2, ..., J$. The coefficients in this model equal J times the coefficients in the AC model.

Understanding whether the ordinal response variable includes level 0 is important since we need to know how many constraints are needed, and how to specify the constraints in the equation when estimating an AC model from a constrained multinomial model.

7.2 Research Example and Description of the Data and Sample

Just like the research examples in Chapters 4 through 6, this example will investigate the relationships between the ordinal response variable, health status, and four predictor variables. Unlike in the previous chapters, however, the research interest in this chapter focuses on the odds comparing adjacent categories by using the AC model. The research question is as follows: Do the four predictor variables significantly predict the odds comparing adjacent levels of health status?

The data for the following analyses were taken from the General Social Survey 2012 (GSS 2012). The same variables for Chapters 4 through 6 are used for data analysis in this chapter.

- `healthre`: the recoded variable of health (health status) with four ordinal categories (1 = poor health, 2 = fair health, 3 = good health, and 4 = excellent health)
- `maritals`: the recoded variable of `marital` (marital status) with 1 = currently married and 0 = not currently married
- `educ`: the highest education
- `age`: respondent's age
- `male`: recoded variable of sex with 1 = male and 0 = female

7.3 Adjacent Categories Models With Stata: Commands and Output

The Stata Command `mlogit`

Currently, no official program in Stata has been developed for the AC model, but we can estimate it indirectly from the multinomial logistic model by using the command `mlogit` with constraints. The command `mlogit` is used for multinomial logistic regression, and the syntax for this model is the command `mlogit`, followed by the dependent variable and the independent variable(s).

7.3.1 Multinomial Logistic Regression Using Stata

First, let us estimate a multinomial logistic regression without any constraints. Then, a constrained model will be fitted for the AC model.

The command xi: mlogit healthre i.maritals, baseoutcome(1) tells Stata to conduct the multinomial logistic regression to estimate the outcome variable healthre from the predictor variable maritals. Here, the ordinal response variable, health status, is treated as nominal. The option baseoutcome(1) specifies the lowest category 1 as the reference group. The default is the category with the highest value, but you can specify any category as the referent. For Stata 13 and 14 users, the prefix xi command can be omitted. The output is displayed as follows.

```
. xi: mlogit healthre i.maritals, baseoutcome(1)
i.maritals            _Imaritals_0-1      (naturally coded; _Imaritals_0 omitted)

Iteration 0:   log likelihood = -1579.4068
Iteration 1:   log likelihood = -1566.0026
Iteration 2:   log likelihood = -1565.9373
Iteration 3:   log likelihood = -1565.9373

Multinomial logistic regression                   Number of obs   =       1300
                                                   LR chi2(3)      =      26.94
                                                   Prob > chi2     =     0.0000
Log likelihood = -1565.9373                        Pseudo R2       =     0.0085

------------------------------------------------------------------------------
   healthre |     Coef.   Std. Err.     z     P>|z|    [95% Conf. Interval]
------------+-----------------------------------------------------------------
1           | (base outcome)
------------+-----------------------------------------------------------------
2           |
_Imaritals_1|  .1179594   .2598065    0.45   0.650   -.3912521    .6271708
       _cons|  1.153672   .1575446    7.32   0.000    .8448903    1.462454
------------+-----------------------------------------------------------------
3           |
_Imaritals_1|   .400345   .2428648    1.65   0.099   -.0756612    .8763513
       _cons|  1.80426    .1482347   12.17   0.000    1.513725    2.094794
------------+-----------------------------------------------------------------
4           |
_Imaritals_1|  .8701996   .2528992    3.44   0.001    .3745263    1.365873
       _cons|  1.02692    .1600771    6.42   0.000    .713175     1.340666
------------------------------------------------------------------------------
```

Interpreting the Output

The top of the output is the same as it is for any logistic regression model. Since the maximum likelihood estimation is used for this model, it shows the iterations for the estimation and the maximum log likelihood value for the model.

On the top right of the multinomial logistic regression table, it displays the number of observations in the dataset, the log likelihood ratio test statistic and the associated p value, and the pseudo R^2. The dataset has 1,300 cases for the analysis. LR chi2 (3) = 26.94.

The associated p value with the log likelihood ratio chi-square test `Prob > chi2 = 0.0000` indicates that the null hypothesis, that the one-predictor model is not better than the null model with no independent variables, is rejected. Therefore, the one-predictor model provides a better fit than the null model with no independent variables in predicting health status.

The likelihood ratio R^2_L = .0085 suggests that the relationship between the response variable, health status, and the predictor, marital status, is small.

Similar to other logistic regression models, the multinomial logistic regression table displays the parameter estimates for the predictor variable and the intercepts, their standard errors, the Wald z statistics, the associated p values, and the 95% confident intervals for the parameter estimates. Different from other models, the table displays the parameter estimates for three binary logistic models comparing each category with the base category. The three models are numbered 2, 3, and 4. Since the base category is 1, these three equations compare categories 2 and 1, 3 and 1, and 4 and 1, respectively.

Based on the parameter estimates in the output, the three equations can be expressed as:

$$\ln\left(\frac{\pi(Y=2)}{\pi(Y=1)}\right) = 1.154 + .118 \text{ maritals}$$

$$\ln\left(\frac{\pi(Y=3)}{\pi(Y=1)}\right) = 1.804 + .400 \text{ maritals}$$

$$\ln\left(\frac{\pi(Y=4)}{\pi(Y=1)}\right) = .870 + 1.027 \text{ maritals}$$

The null hypothesis for the Wald test is that the coefficient of the predictor variable is zero, and the alternative hypothesis is that the coefficient of the predictor variable is significantly different from zero.

The first equation, labeled "2" in the output, compares category 2 and category 1. For the `maritals` predictor variable, Wald z = .45. The associated p value `P>|z|=.650`, so we fail to reject the null hypothesis. Therefore, the predictor variable, marital status, is not a significant predictor in predicting the log odds of being in category 2 versus 1. The 95% confidence interval of the regression coefficient is [−.3912521, .6271708].

The second equation, labeled "3" in the output, compares category 3 and category 1. For the same predictor variable `maritals`, Wald z = 1.65. The associated p value `P>|z|=.099`, so we fail to reject the null hypothesis. Therefore, the predictor variable, marital status, is not a significant predictor in predicting the log odds of being in categories 3 versus 1. The 95% confidence interval of the regression coefficient is [−.0756612, .8763513].

The third equation, labeled "4" in the output, compares category 4 and category 1. For `maritals`, Wald $z = 3.44$. The associated p value, P>|z|=.001, so we reject the null hypothesis. Therefore, the predictor variable, marital status, is a significant predictor in predicting the log odds of being in categories 4 versus 1. The 95% confidence interval of the regression coefficient is [.3745263, 1.365873].

The `listcoef` command, which is part of the SPost13 package (Long & Freese, 2014), gives us the odds comparing any two of the four categories. The output is displayed as follows.

```
. listcoef

mlogit (N=1300): Factor change in the odds of healthre

Variable: _Imaritals_1 (sd=0.499)
-----------------------------------------------------------------------------
                            |       b        z     P>|z|      e^b    e^bStdX
----------------------------+------------------------------------------------
1            vs 2           |  -0.1180   -0.454    0.650    0.889      0.943
1            vs 3           |  -0.4003   -1.648    0.099    0.670      0.819
1            vs 4           |  -0.8702   -3.441    0.001    0.419      0.648
2            vs 1           |   0.1180    0.454    0.650    1.125      1.061
2            vs 3           |  -0.2824   -1.900    0.057    0.754      0.869
2            vs 4           |  -0.7522   -4.573    0.000    0.471      0.687
3            vs 1           |   0.4003    1.648    0.099    1.492      1.221
3            vs 2           |   0.2824    1.900    0.057    1.326      1.151
3            vs 4           |  -0.4699   -3.451    0.001    0.625      0.791
4            vs 1           |   0.8702    3.441    0.001    2.387      1.544
4            vs 2           |   0.7522    4.573    0.000    2.122      1.456
4            vs 3           |   0.4699    3.451    0.001    1.600      1.264
-----------------------------------------------------------------------------
```

The first column shows one category versus the other for all possible comparisons. The columns labeled "b", "z", and "P>|z|" display the logit coefficients, z statistics, and associated p values for each comparison. The column labeled e^b lists the odds ratios.

With the `adjacent` option, the `listcoef` command provides the odds comparing pairs of adjacent categories as follows.

```
. listcoef, adjacent

mlogit (N=1300): Factor change in the odds of healthre

Variable: _Imaritals_1 (sd=0.499)
```

		b	z	P>|z|	e^b	e^bStdX
1	vs 2	-0.1180	-0.454	0.650	0.889	0.943
2	vs 1	0.1180	0.454	0.650	1.125	1.061
2	vs 3	-0.2824	-1.900	0.057	0.754	0.869
3	vs 2	0.2824	1.900	0.057	1.326	1.151
3	vs 4	-0.4699	-3.451	0.001	0.625	0.791
4	vs 3	0.4699	3.451	0.001	1.600	1.264

The logit coefficients comparing adjacent categories 2 versus 1, 3 versus 2, and 4 versus 3 are .118, .282, and .470, respectively. The corresponding odds ratios for the comparisons are 1.125, 1.326, and 1.600, respectively.

The fitstat command, which is also part of the SPost13 package, produces various model fit statistics, such as the log likelihood statistic, deviance statistic, log likelihood ratio test, six types of pseudo R^2 statistics, and AIC and BIC statistics.

```
. fitstat

                          |     mlogit
--------------------------+------------
Log-likelihood            |
                   Model  |   -1565.937
          Intercept-only  |   -1579.407
--------------------------+------------
Chi-square                |
       Deviance (df=1294) |    3131.875
             LR (df=3)    |      26.939
                 p-value  |       0.000
--------------------------+------------
R2                        |
                McFadden  |       0.009
     McFadden (adjusted)  |       0.005
            Cox-Snell/ML  |       0.021
 Cragg-Uhler/Nagelkerke   |       0.022
                   Count  |       0.457
         Count (adjusted) |       0.000
--------------------------+------------
IC                        |
                     AIC  |    3143.875
        AIC divided by N  |       2.418
             BIC (df=6)   |    3174.895
```

7.3.2 Single-Predictor AC Model Using Stata

Next, we estimate the AC model using the mlogit command. Since the AC model is a constrained multinomial logistic model, we need to specify the constraints first.

Recall that the coefficients in the constrained multinomial model are $(J - 1)$ times as large as the coefficients in the AC model. In this example, the ordinal response has four categories, so two constraints $(J - 2 = 2)$ need to be specified for each predictor. The first equation comparing categories 2 and 1 does not need any constraints.

The command for the first constraint, `constraint 1 [3]_Imaritals_1 = 2*[2]_Imaritals_1`, tells Stata to constrain the logit coefficient in the model labeled [3], comparing categories 3 and 1, to be twice as large as the coefficient for the model labeled [2], comparing categories 2 and 1.

The second constraint specifies `constraint 2 [4]_Imaritals_1 = 3*[2]_Imaritals_1` so that the logit coefficient in the model labeled [4], which compares categories 4 and 1, is three times the coefficient for the model labeled [2], comparing categories 2 and 1.

There are several tips for writing the commands for imposing constraints. First, number all the constraints since you need to list the numbers in the option of constraints in the following command syntax when fitting the model: for example, constraint 1. Second, the names of the predictor variables should be exactly the same ones in the model. If there are any indicator or dummy variables, type the names shown in the model that start with _I. Third, the number of the submodels or equations can be found from the multinomial logistic model without constraints and each number should be placed in brackets. For example, [2] and [3] represent models 2 and 3. It is recommended to fit a regular multinomial logistic model without any constraints first since you can get the right variable names and model numbers, which will be used for defining the constraints.

The two constraints are specified in the option for the original `mlogit` command as `constraints (1 2)`. The command for the constrained multinomial logistic model is `xi: mlogit healthre i.maritals, baseoutcome(1) constraints (1 2)`. The output for the constrained multinomial logistic model is displayed as follows.

```
. constraint 1 [3]_Imaritals_1=2*[2]_Imaritals_1

. constraint 2 [4]_Imaritals_1=3*[2]_Imaritals_1

. xi: mlogit healthre i.maritals, baseoutcome(1) constraints (1 2)
i.maritals          _Imaritals_0-1     (naturally coded; _Imaritals_0 omitted)

Iteration 0:   log likelihood = -1579.4068
Iteration 1:   log likelihood = -1566.9518
Iteration 2:   log likelihood = -1566.8533
Iteration 3:   log likelihood = -1566.8533

Multinomial logistic regression              Number of obs   =       1300
                                              Wald chi2(1)    =      24.41
Log likelihood = -1566.8533                   Prob > chi2     =     0.0000
```

```
( 1)   - 2*[2]_Imaritals_1 + [3]_Imaritals_1 = 0
( 2)   - 3*[2]_Imaritals_1 + [4]_Imaritals_1 = 0
```

healthre	Coef.	Std. Err.	z	P>|z|	[95% Conf. Interval]
1	(base outcome)					
2						
_Imaritals_1	.3311804	.0670296	4.94	0.000	.1998048	.4625561
_cons	1.080686	.126381	8.55	0.000	.8329834	1.328388
3						
_Imaritals_1	.6623609	.1340593	4.94	0.000	.3996095	.9251122
_cons	1.707501	.1251289	13.65	0.000	1.462253	1.952749
4						
_Imaritals_1	.9935413	.2010889	4.94	0.000	.5994143	1.387668
_cons	1.002311	.1469283	6.82	0.000	.7143368	1.290285

Interpreting the Constrained Multinomial Logistic Model

The first three lines show the echoed commands for the constraints and the model. The top of the output looks similar to that of the other logistic regression models. The steps of the iterations for the maximum likelihood estimation are shown at the top left. The upper-right corner lists a summary of model fit statistics, including the number of observations, Wald chi-square test, and the associated p value.

The multinomial logistic regression table is displayed at the bottom of the output. Just above the table, two constraints numbered (1) and (2) are expressed as a linear combination of regression coefficients from different equations below. The multinomial logistic regression table is different from that of the binary logistic regression since it displays the parameter estimates for three different model comparisons. The output for model 1 is omitted since category 1 is the reference group. The three models numbered 2, 3, and 4 compare categories 2 and 1, 3 and 1, and 4 and 1, respectively.

Based on the parameter estimates in the output, the three equations for the constrained multinomial logistic regression can be expressed as:

$$\ln\left(\frac{\pi(Y=2)}{\pi(Y=1)}\right) = 1.081 + .331 \ \_\text{Imaritals}\_1$$

$$\ln\left(\frac{\pi(Y=3)}{\pi(Y=1)}\right) = 1.708 + .662 \ \_\text{Imaritals}\_1$$

$$\ln\left(\frac{\pi(Y=4)}{\pi(Y=1)}\right) = 1.002 + .994 \ \_\text{Imaritals}\_1$$

Recall that the logit coefficient in the second equation (model 3) is constrained to be twice as large as that in the first equation (model 2): $.331 \times 2 = .662$; and the logit coefficient in the third equation (model 4) is constrained to be three times that in the first equation (model 2): $.331 \times 3 = .993$.

We can easily transform the three equations for the constrained multinomial logistic regression model into the corresponding equations for the AC model. Let us determine the intercepts and the logit coefficients for the AC model from the estimated multinomial logistic model.

Model 2 is identical to that in the constrained multinomial logistic model:

$$\ln\left(\frac{\pi(Y=2)}{\pi(Y=1)}\right) = 1.081 + .331 \ \_\text{Imaritals}\_1$$

For model 3, the intercept is computed by subtracting it from that in model 2: $1.708 - 1.081 = .627$:

$$\ln\left(\frac{\pi(Y=3)}{\pi(Y=2)}\right) = (1.708 - 1.081) + .331 \ \_\text{Imaritals}\_1 = .627 + .331 \ \_\text{Imaritals}\_1$$

The intercept for model 4 is computed by subtracting it from those in the previous two models: $1.002 - 1.708 = -.706$:

$$\ln\left(\frac{\pi(Y=4)}{\pi(Y=3)}\right) = (1.002 - 1.708) + .331 \ \_\text{Imaritals}\_1$$
$$= -.706 + .331 \ \_\text{Imaritals}\_1$$

To get the odds ratios for any pairs of adjacent categories, we need to exponentiate the logit coefficient $\exp(.331) = 1.393$.

We can also use the `listcoef` command with the option `adjacent`. The results table is displayed as follows.

```
. listcoef, adjacent

mlogit (N=1300): Factor change in the odds of healthre

Variable: _Imaritals_1 (sd=0.499)
----------------------------------------------------------------------------
                        |       b        z      P>|z|       e^b    e^bStdX
------------------------+---------------------------------------------------
1          vs 2         |  -0.3312   -4.941    0.000      0.718     0.848
2          vs 1         |   0.3312    4.941    0.000      1.393     1.180
2          vs 3         |  -0.3312   -4.941    0.000      0.718     0.848
3          vs 2         |   0.3312    4.941    0.000      1.393     1.180
3          vs 4         |  -0.3312   -4.941    0.000      0.718     0.848
4          vs 3         |   0.3312    4.941    0.000      1.393     1.180
----------------------------------------------------------------------------
```

The command `fitstat` produces a more detailed summary of the model fit statistics. The results are presented as follows, which are similar to those for the previous unconstrained multinomial logistic model.

```
. fitstat

                          |       mlogit
--------------------------+------------
Log-likelihood            |
                   Model  |    -1566.853
--------------------------+------------
Chi-square                |
      Deviance (df=1296)  |     3133.707
             Wald (df=1)  |       24.412
                 p-value  |        0.000
--------------------------+------------
R2                        |
                   Count  |        0.457
         Count (adjusted) |        0.000
--------------------------+------------
IC                        |
                     AIC  |     3141.707
         AIC divided by N |        2.417
             BIC (df=4)   |     3162.387
```

The deviance statistic is 3,133.707, and the AIC and BIC are 3,141.707 and 3,162.387, respectively. They will be compared with those of the multiple-predictor model in the next section.

7.3.3 Making Publication-Quality Tables for the Single-Predictor AC Model

```
. outreg2 using chap7out, e(ll df_m chi2)word long replace
chap7out.rtf
dir : seeout

. outreg2 using chap7out, eform word long append
chap7out.rtf
dir : seeout
```

The command `outreg2` is used to produce the publication-quality table for the estimated model. The command `outreg2 using chap7out, e(ll df_m chi2) word long replace` tells Stata to create a regression table for the estimates and save it to a Word file named chap7out. The command `e(ll df_m chi2)` in the option asks Stata to include the estimated log likelihood value, degrees of freedom, and the chi-square test in the table. The `word` option requests that the table be created in a Word document with the extension .rtf. The `long` option needs to be used since three binary models are estimated. The intercepts and coefficients will be properly listed in one column by using this option. The `replace` option asks Stata to replace any files with the same name as chap7out.

The second command `outreg2 using chap7out, eform word long append` requests Stata to add the estimated odds ratios to the table as a separate column by using the `eform` option. In the Stata output window, if you click the blue link of the file, chap7out.rtf, then you open the Word file containing the table. Clicking dir: `seeout` opens the table in the data browser, which is the tab delimited text file with the .txt extension. Both files are saved in the current working directory.

These commands automatically produce Table 7.3, as shown here in its original format, presenting the results for the single-predictor constrained multinomial logistic regression model.

Table 7.3 Results of the Constrained Multinomial Logistic Regression Model: Single Predictor (Shown in Original Format Generated by Stata)

Variables	(1)	(2)
2		
_Imaritals_1	0.331***	1.393***
	(0.0670)	(0.0933)
Constant	1.081***	2.947***
	(0.126)	(0.372)
3		
_Imaritals_1	0.662***	1.939***
	(0.134)	(0.260)
Constant	1.708***	5.515***
	(0.125)	(0.690)
4		
_Imaritals_1	0.994***	2.701***
	(0.201)	(0.543)
Constant	1.002***	2.725***
	(0.147)	(0.400)
Observations	1,300	1,300
Ll	−1,567	
df_m	1	
chi2	24.41	

Note: Standard errors in parentheses.

***$p < 0.01$, **$p < 0.05$, *$p < 0.1$

After editing the variable names, the decimals, the labels for the logit coefficients and odds ratios, and the model fit statistics, the final table (Table 7.4) is presented.

We assume that the logit coefficient for each predictor in the AC model is equal across the ordinal response categories. Based on the transformation from the

Table 7.4 Results of the Constrained Multinomial Logistic Regression Model: Single Predictor (Edited)

Variables	b (SE(b))	OR
Model 1(Y = 2 vs. Y = 1)		
_Imaritals_1	0.331***	1.393***
	(0.0670)	(0.093)
α_1	1.081***	
	(0.126)	
Model 2 (Y = 3 vs. Y = 1)		
_Imaritals_1	0.662***	1.939***
	(0.134)	(0.260)
α_2	1.708***	
	(0.125)	
Model 3 (Y = 4 vs. Y = 1)		
_Imaritals_1	0.994***	2.701***
	(0.201)	(0.543)
α_3	1.002***	
	(0.147)	
Observations	1,300	1,300
LR R^2	.008	
Log likelihood	−1,567	
χ^2_1	24.41	

Notes. Y = 1 is the base category. Standard errors are shown in parentheses.

***p < .01.

Table 7.5 Results of the Adjacent Categories Logistic Regression: Single-Predictor Model

Variables	b (SE(b))	OR
α_1	1.081***	
α_2	.627***	
α_3	−.706***	
_Imaritals_1	0.331***	1.393***
	(0.067)	(0.093)
Observations	1,300	1,300
LR R^2	.008	
Log likelihood	−1,567	
χ^2_1	24.41	

Note. Standard errors are shown in parentheses.

***$p < .01$.

constrained multinomial logistic regression model into the AC model (see Table 7.2), the results of the AC model are displayed in Table 7.5.

7.3.4 Adjacent Categories Models With Stata: Multiple-Predictor Model

The Stata Command constraint

The full model includes four predictors, so we need to impose eight constraints with two constraints for each predictor. The total number of constraints is determined by both the number of predictor variables and the number of categories of the ordinal response variable. It equals $p \times (J - 2)$, where p is the number of predictor variables and J is the number of categories of the ordinal response variable. The number of categories of the ordinal response variables needs to be subtracted by two because among the $J - 1$ binary models, no constraints are needed for the first model comparing categories 2 and 1 (category 1 is assumed to be the base category).

The first two constraints are the same as those in the one-predictor model. The constraints 3 and 4 are for the predictor educ. [3]educ=2*[2]educ, which means that the coefficient for educ in model 3 is twice as large as that in model 2, and [4] educ=3*[2]educ means that the coefficient for educ in model 4 is three times as large as that in model 2. The constraints 5 and 6 and 7 and 8 are for the predictor variables age and male, respectively. The variable male is the indictor variable with two categories, so it shows as _Imale_1 in the model.

The Stata Command mlogit

The command mlogit is followed by the ordinal response variable healthre and four predictor variables. The xi prefix is used since there are two categorical predictors, maritals and male. This prefix command can be omitted when categorical predictor variables are coded as dummy variables or when the coding for factor variables is used for Stata 13 and 14 users. The option baseoutcome(1) specifies category 1 as the reference group, and constraints (1/8) specifies constraints from 1 to 8.

```
. constraint 1 [3]_Imaritals_1=2*[2]_Imaritals_1

. constraint 2 [4]_Imaritals_1=3*[2]_Imaritals_1

. constraint 3 [3]educ=2*[2]educ

. constraint 4 [4]educ=3*[2]educ

. constraint 5 [3]age=2*[2]age

. constraint 6 [4]age=3*[2]age

. constraint 7 [3]_Imale_1=2*[2]_Imale_1

. constraint 8 [4]_Imale_1=3*[2]_Imale_1

.
. *Fitting a multiple-predictor model
. xi: mlogit healthre i.maritals educ age i.male, baseoutcome(1) constraint (1/8)
i.maritals        _Imaritals_0-1    (naturally coded; _Imaritals_0 omitted)
i.male            _Imale_0-1        (naturally coded; _Imale_0 omitted)

Iteration 0:   log likelihood = -1579.4068
Iteration 1:   log likelihood = -1514.4886
Iteration 2:   log likelihood = -1511.7514
Iteration 3:   log likelihood = -1511.7382
Iteration 4:   log likelihood = -1511.7382

Multinomial logistic regression              Number of obs   =       1300
                                              Wald chi2(4)    =     117.18
Log likelihood = -1511.7382                   Prob > chi2     =     0.0000
```

```
( 1)   - 2*[2]_Imaritals_1 + [3]_Imaritals_1 = 0
( 2)   - 3*[2]_Imaritals_1 + [4]_Imaritals_1 = 0
( 3)   - 2*[2]educ + [3]educ = 0
( 4)   - 3*[2]educ + [4]educ = 0
( 5)   - 2*[2]age + [3]age = 0
( 6)   - 3*[2]age + [4]age = 0
( 7)   - 2*[2]_Imale_1 + [3]_Imale_1 = 0
( 8)   - 3*[2]_Imale_1 + [4]_Imale_1 = 0
```

healthre	Coef.	Std. Err.	z	P>|z|	[95% Conf. Interval]
1	(base outcome)					
2						
_Imaritals_1	.3380758	.0703209	4.81	0.000	.2002494	.4759022
educ	.0870895	.01138	7.65	0.000	.0647851	.109394
age	-.0124675	.0019899	-6.27	0.000	-.0163676	-.0085674
_Imale_1	.0348856	.0691053	0.50	0.614	-.1005583	.1703295
_cons	.6690359	.2164337	3.09	0.002	.2448337	1.093238
3						
_Imaritals_1	.6761516	.1406418	4.81	0.000	.4004988	.9518044
educ	.1741791	.0227601	7.65	0.000	.1295702	.218788
age	-.024935	.0039798	-6.27	0.000	-.0327352	-.0171349
_Imale_1	.0697713	.1382106	0.50	0.614	-.2011165	.340659
_cons	.7473622	.3764106	1.99	0.047	.009611	1.485113
4						
_Imaritals_1	1.014227	.2109626	4.81	0.000	.6007482	1.427707
educ	.2612686	.0341401	7.65	0.000	.1943553	.328182
age	-.0374025	.0059696	-6.27	0.000	-.0491028	-.0257023
_Imale_1	.1046569	.2073159	0.50	0.614	-.3016748	.5109885
_cons	-.6230892	.5604743	-1.11	0.266	-1.721599	.4754203

Interpreting Stata Output of the Fitted Model

The top right of the output displays the summary of fit statistics. The number of observations is still 1,300. The Wald chi-square test `Wald chi2 (4) = 117.18`. The associated p value `Prob > chi2 = 0.0000`, which indicates that the null hypothesis is rejected. Therefore, at least one predictor variable is significant in predicting health status.

Eight constraints of the logit coefficients are displayed following the log likelihood value of -1511.7382. Next, the table of the coefficients for the multinomial logistic regression model is displayed.

Based on the estimated coefficients, the three equations for the model can be expressed as follows:

$$\ln\left(\frac{\pi(Y=2)}{\pi(Y=1)}\right) = .669 + .338 \text{ _Imaritals_1} + .087 \text{ educ} -.012 \text{ age} + .035 \text{ _Imale_1}$$

$$\ln\left(\frac{\pi(Y=3)}{\pi(Y=1)}\right) = .747 + .676 \text{ \_Imaritals\_1} + .174 \text{ educ} -.025 \text{ age} + .070 \text{ \_Imale\_1}$$

$$\ln\left(\frac{\pi(Y=4)}{\pi(Y=1)}\right) = -.623 + 1.014 \text{ \_Imaritals\_1} + .261 \text{ educ} -.037 \text{ age} +$$
$$.105 \text{ \_Imale\_1}$$

Computing Intercepts and the Logit
Coefficients for 3 (J – 1 = 3) AC Equations

The intercepts and the logit coefficients for three AC equations can be determined from the equations for the multinomial logistic regression model presented previously. The logit coefficients are constrained to be equal across the three equations for the AC model. They are the same as the ones estimated by the first equation for the multinomial model. The intercept for each successive model is computed by subtracting it from that from the previous model. For example, the intercept for the second equation (α_2) for the AC model equals the intercept from the second equation for the multinomial model (.747) – the intercept for the first equation for the multinomial model (.669) = .078. Table 7.6 presents the results for the intercepts and logit coefficients for three equations of the AC model.

From the output for the multiple-predictor model, across all three models labeled 2, 3, and 4, all four predictor variables except the variable male are significantly different from 0 in predicting the odds of being in that category versus the base category.

In all three models, for the male predictor variable, Wald $z = .50$. The associated p value is P>|z|=.614, so we fail to reject the null hypothesis. Therefore, the predictor variable of being a male is not a significant predictor in predicting the log odds of

Table 7.6 Intercepts and the Logit Coefficients for Three (*J* – 1 = 3) AC Equations

Variables for the AC Model	Transformation From the Multinomial Logistic Model	Intercepts and Coefficients for the AC Model
α_1	.669	.669
α_2	.747 – .669	.078
α_3	−.623 – .747	−1.37
_Imaritals_1	.338	.338
Educ	.087	.087
Age	−.012	−.012
_Imale_1	.035	.035

being in categories 2 versus 1, the log odds of being in categories 3 versus 1, and the log odds of being in categories 4 versus 1.

The `listcoef` command with the adjacent option computes the odds of two adjacent categories for all four predictor variables. The following results are displayed.

```
. listcoef, adjacent

mlogit (N=1300): Factor change in the odds of healthre

Variable: _Imaritals_1 (sd=0.499)
------------------------------------------------------------------------
              |         b         z      P>|z|        e^b     e^bStdX
--------------+---------------------------------------------------------
1      vs  2  |   -0.3381    -4.808     0.000       0.713       0.845
2      vs  1  |    0.3381     4.808     0.000       1.402       1.184
2      vs  3  |   -0.3381    -4.808     0.000       0.713       0.845
3      vs  2  |    0.3381     4.808     0.000       1.402       1.184
3      vs  4  |   -0.3381    -4.808     0.000       0.713       0.845
4      vs  3  |    0.3381     4.808     0.000       1.402       1.184
------------------------------------------------------------------------

Variable: educ (sd=3.151)
------------------------------------------------------------------------
              |         b         z      P>|z|        e^b     e^bStdX
--------------+---------------------------------------------------------
1      vs  2  |   -0.0871    -7.653     0.000       0.917       0.760
2      vs  1  |    0.0871     7.653     0.000       1.091       1.316
2      vs  3  |   -0.0871    -7.653     0.000       0.917       0.760
3      vs  2  |    0.0871     7.653     0.000       1.091       1.316
3      vs  4  |   -0.0871    -7.653     0.000       0.917       0.760
4      vs  3  |    0.0871     7.653     0.000       1.091       1.316
------------------------------------------------------------------------

Variable: age (sd=17.432)
------------------------------------------------------------------------
              |         b         z      P>|z|        e^b     e^bStdX
--------------+---------------------------------------------------------
1      vs  2  |    0.0125     6.265     0.000       1.013       1.243
2      vs  1  |   -0.0125    -6.265     0.000       0.988       0.805
2      vs  3  |    0.0125     6.265     0.000       1.013       1.243
3      vs  2  |   -0.0125    -6.265     0.000       0.988       0.805
3      vs  4  |    0.0125     6.265     0.000       1.013       1.243
4      vs  3  |   -0.0125    -6.265     0.000       0.988       0.805
------------------------------------------------------------------------

Variable: _Imale_1 (sd=0.497)
------------------------------------------------------------------------
              |         b         z      P>|z|        e^b     e^bStdX
--------------+---------------------------------------------------------
1      vs  2  |   -0.0349    -0.505     0.614       0.966       0.983
2      vs  1  |    0.0349     0.505     0.614       1.036       1.017
2      vs  3  |   -0.0349    -0.505     0.614       0.966       0.983
3      vs  2  |    0.0349     0.505     0.614       1.036       1.017
3      vs  4  |   -0.0349    -0.505     0.614       0.966       0.983
4      vs  3  |    0.0349     0.505     0.614       1.036       1.017
------------------------------------------------------------------------
```

The first column shows all possible comparisons between adjacent categories. The column labeled e^b lists the odds ratios. Table 7.7 presents the odds ratios comparing a higher category with the next lower category and their inversed odds ratios.

Interpreting the Odds Ratios of Being in a Higher Category (j + 1) Versus the Next Lower Category j

Both the mlogit command with the rrr option and the command listcoef, adjacent provide the odds ratios of being in a higher category versus the next lower category for a pair of adjacent categories. The odds ratios can be interpreted just like the other logistic regression models except that the comparisons between categories are different.

A positive logit regression coefficient corresponds to an odds ratio being a higher category versus the next lower category greater than 1, whereas a negative logit coefficient corresponds to an odds ratio less than 1. The odds of being in a higher category increase or decrease depending on whether an odds ratio is greater or less than 1 for a one-unit increase in the predictor variable when holding all the other predictors constant. If a logit coefficient is zero, then its corresponding odds ratio equals 1. This means that when all other variables are held constant, there is no relationship between the predictor and the odds of being in a higher category.

For the predictor, marital status, β = .338, which is positive; OR = 1.402, which is greater than 1. This indicates that the odds of being in a higher category of health status versus the next lower category for the married are 1.402 times as great as the odds for the unmarried when holding all the other predictors constant.

For the educ predictor, β = .087, which is positive; OR = 1.091, which is greater than 1. This indicates that the odds of being in a higher category of health status versus the next lower category increase by 1.091 for a one-unit increase in the predictor education when holding all the other predictors constant.

Table 7.7   Odds Ratios in the AC Model Comparing Adjacent Categories

Variables	OR Comparing a Higher Category Versus the Next Lower Category ($j + 1$ vs. j)	OR Comparing a Lower Category Versus the Adjacent Higher Category (j vs. $j + 1$)
_Imaritals_1	1.402***	.713***
Educ	1.091***	.917***
Age	.988***	1.013***
_Imale_1	1.036	.966

***p < .01.

For the age predictor, β = −.012, which is negative; OR = .988, which is less than 1. This indicates that the odds of being in a higher category of health status versus the next lower category decrease by a factor of .988 for a one-unit increase in the predictor age with all the other predictors remaining constant. In other words, for a one-unit increase in age, the odds of being healthier decrease by 1.2%.

For the male predictor, β = .035, p = .614, which is not significantly different from 0; OR = 1.036, which almost equals 1. This indicates that there is no relationship between being a male and the odds of being in a higher category of health status versus the next lower category. In other words, there is no significant difference between the male and the female in the odds of being in better health status versus the next lower category (i.e., categories 2 vs. 1, categories 3 vs. 2, and categories 4 vs. 3).

Interpreting the Odds Ratios of Being in a Lower Category for the AC Model

Just as the odds in other ordinal regression models have directions, they also have directions in the AC model. In other words, we can reverse the order of categories that the odds compare. In AC models, we can either estimate the odds of being in a higher category versus the next lower category or the odds of being in a lower category versus the next higher category since they are reciprocals of each other. To compute the odds of being in a lower category versus the next higher category, we exponentiate the negative coefficient exp(−β). The same results are displayed if we use the listcoef command with the adjacent option since the odds for both directions are presented.

For the maritals predictor, OR = .713, which is less than 1. This indicates that the odds of being in a lower category out of the adjacent categories of health status for the married are .713 times the odds for the unmarried when holding all the other predictors constant.

For the educ predictor, OR = .917, which is less than 1. This indicates that the odds of being in a lower category of health status (poorer health status) decrease by .917 for a one-unit increase in the predictor education when holding all the other predictors constant. In other words, the odds of being in a lower category decrease by 8.3% for a one-unit increase in education when the effects of other predictors are held constant in the model.

For the age predictor, OR = 1.013, which is greater than 1. This indicates that the odds of being in a lower category of health status increase by a factor of 1.013 for a one-unit increase in the predictor age when all the other predictors remain constant. In other words, for a one-unit increase in age, the odds of having poorer health increase by 1.3%.

For the male predictor, OR = .966 (p =.614), which is close to 1. This indicates that there is no relationship between being a male and the odds of being in a lower category of health status.

The command `fitstat` computes various model fit statistics. The following results are displayed.

```
Fitstat

                            |       mlogit
----------------------------+--------------
Log-likelihood              |
                   Model    |    -1511.738
----------------------------+--------------
Chi-square                  |
      Deviance (df=1293)    |     3023.476
            Wald (df=4)     |      117.185
                p-value     |        0.000
----------------------------+--------------
R2                          |
                   Count    |        0.464
        Count (adjusted)    |        0.013
----------------------------+--------------
IC                          |
                     AIC    |     3037.476
         AIC divided by N   |        2.337
              BIC (df=7)    |     3073.667
```

The deviance statistic for the multiple-predictor model is 3,023.476. The Wald chi-square test $\chi^2_{(4)} = 117.185$, $p < .001$, so we conclude that at least one logit coefficient in the multiple-predictor model is significantly different from zero. The AIC and BIC are 3,037.476 and 3,073.667, respectively. In addition, the command `fitstat` with the `using()` option can be used to compare various fit statistics between two models (see Chapter 3 for a more detailed explanation).

Interpreting the Estimated Probabilities
With the `margins` Command in Stata 13

The `margins` command can also be used to compute the estimated probabilities of being in a particular category of the ordinal response variable at specified values of predictor variables. The same `margins` syntax for the proportional odds model is used for the following analysis. The command `margins, predict (outcome(1))` `at(educ = (8 13 16)) atmeans vsquish` tells Stata to compute the estimated probabilities when the ordinal response variable $Y = 1$ for the predictor variable `educ` at the values of 8, 13, and 16 when holding the other predictor variables at their means.

In the output, when `educ` equals 8, 13, and 16, and other predictor variables are held at their means, the margins or estimated probabilities for being in poor

```
. margins, predict (outcome(1)) at(educ = (8 13 16)) atmeans vsquish

Adjusted predictions                              Number of obs   =        1300
Model VCE      : OIM

Expression     : Pr(healthre==1), predict(outcome(1))
1._at          : _Imaritals_1     =      .4684615  (mean)
                 educ             =             8
                 age              =          48.2  (mean)
                 _Imale_1         =      .4438462  (mean)
2._at          : _Imaritals_1     =      .4684615  (mean)
                 educ             =            13
                 age              =          48.2  (mean)
                 _Imale_1         =      .4438462  (mean)
3._at          : _Imaritals_1     =      .4684615  (mean)
                 educ             =            16
                 age              =          48.2  (mean)
                 _Imale_1         =      .4438462  (mean)

------------------------------------------------------------------------------
             |            Delta-method
             |    Margin   Std. Err.      z    P>|z|     [95% Conf. Interval]
-------------+----------------------------------------------------------------
         _at |
          1  |  .1193112   .0154363     7.73   0.000     .0890565    .1495658
          2  |  .0554071   .0063378     8.74   0.000     .0429852    .0678289
          3  |  .0328351   .0047591     6.90   0.000     .0235075    .0421627
------------------------------------------------------------------------------
```

health (i.e., category 1) are .119, .055, and .033, respectively. They are all statistically significant ($p < .001$).

We can also compute the margins or expected probabilities for the other three categories (i.e., 2, 3, and 4) of the ordinal response variable by specifying the option predict() option. With the predict (outcome(4)) option, we can compute the expected probabilities for category 4. To save space, only the output for the expected probabilities for category 4 is displayed as follows.

```
. margins, predict (outcome(4)) at(educ = (8 13 16)) atmeans vsquish

Adjusted predictions                              Number of obs   =        1300
Model VCE      : OIM

Expression     : Pr(healthre==4), predict(outcome(4))
1._at          : _Imaritals_1     =      .4684615  (mean)
```

```
            educ              =            8
            age               =         48.2  (mean)
            _Imale_1          =     .4438462  (mean)
2._at     : _Imaritals_1      =     .4684615  (mean)
            educ              =           13
            age               =         48.2  (mean)
            _Imale_1          =     .4438462  (mean)
3._at     : _Imaritals_1      =     .4684615  (mean)
            educ              =           16
            age               =         48.2  (mean)
            _Imale_1          =     .4438462  (mean)
```

```
---------------------------------------------------------------------
           |            Delta-method
           |    Margin   Std. Err.        z    P>|z|   [95% Conf. Interval]
-----------+---------------------------------------------------------
       _at |
         1 |  .1436775     .015311     9.38    0.000    .1136685    .1736865
         2 |  .2463828    .0122982    20.03    0.000    .2222788    .2704868
         3 |  .3197319    .0156434    20.44    0.000    .2890715    .3503924
---------------------------------------------------------------------
```

In the output, when educ equals 8, 13, and 16, and other predictor variables are held at their means, the margins or estimated probabilities for being in excellent health (i.e., category 4) are .144, .246, and .320, respectively. They are all statistically significant ($p < .001$). The marginsplot command is used to visualize the results presented. Figure 7.1 shows the estimated probabilities of being in excellent health condition (i.e., $Y = 4$).

After each separate graph is created using the marginsplot command, we can also plot the estimated probabilities of being in each category (i.e., $Y = 1, 2, 3$, and 4) in a single graph (Figure 7.2) by using the graph combine command. The following shows the complete commands.

```
margins, predict (outcome(1)) at(educ = (8 13 16)) atmeans vsquish
marginsplot, name(mplot1)
margins, predict (outcome(2)) at(educ = (8 13 16)) atmeans vsquish
marginsplot, name(mplot2)
margins, predict (outcome(3)) at(educ = (8 13 16)) atmeans vsquish
marginsplot, name(mplot3)
margins, predict (outcome(4)) at(educ = (8 13 16)) atmeans vsquish
marginsplot, name(mplot4)
graph combine mplot1 mplot2 mplot3 mplot4, ycommon
```

Figure 7.1 Estimated Probabilities of Being in Category 4 for educ at 8, 13, and 16

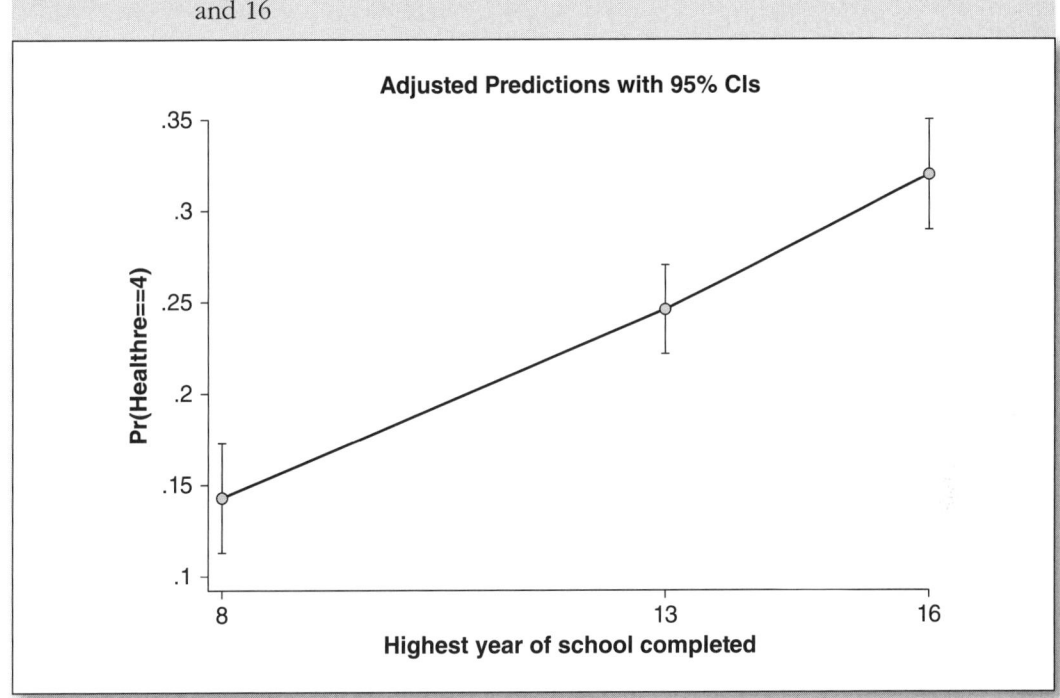

The command `graph combine mplot1 mplot2 mplot3 mplot4, ycom-`
mon combines all four plots named mplot1 to mplot4, which are created by the `mar-`
`ginsplot` command.

Estimated Probabilities With the Improved
`margins` *and* `marginsplot` *Commands in Stata 14*

As explained in Chapter 4, the updated `margins` and `marginsplot` commands
in Stata 14 are more powerful than the previous ones in Stata 13. Rather than computing estimated probabilities for each category of an ordinal response variable separately
by running the `margins` command multiple times, we can run the command in Stata
14 once, and it simultaneously provides the results of the estimated probabilities for
all categories of the ordinal response variable.

To replicate the results in the preceding section, the command `margins,`
`at(educ = (8 13 16))` `atmeans vsquish` tells Stata 14 to compute the estimated probabilities for $Y = 1, 2, 3,$ and 4 for the predictor variable educ at the values

Figure 7.2 Estimated Probabilities of Being in Categories 1, 2, 3, and 4 for educ

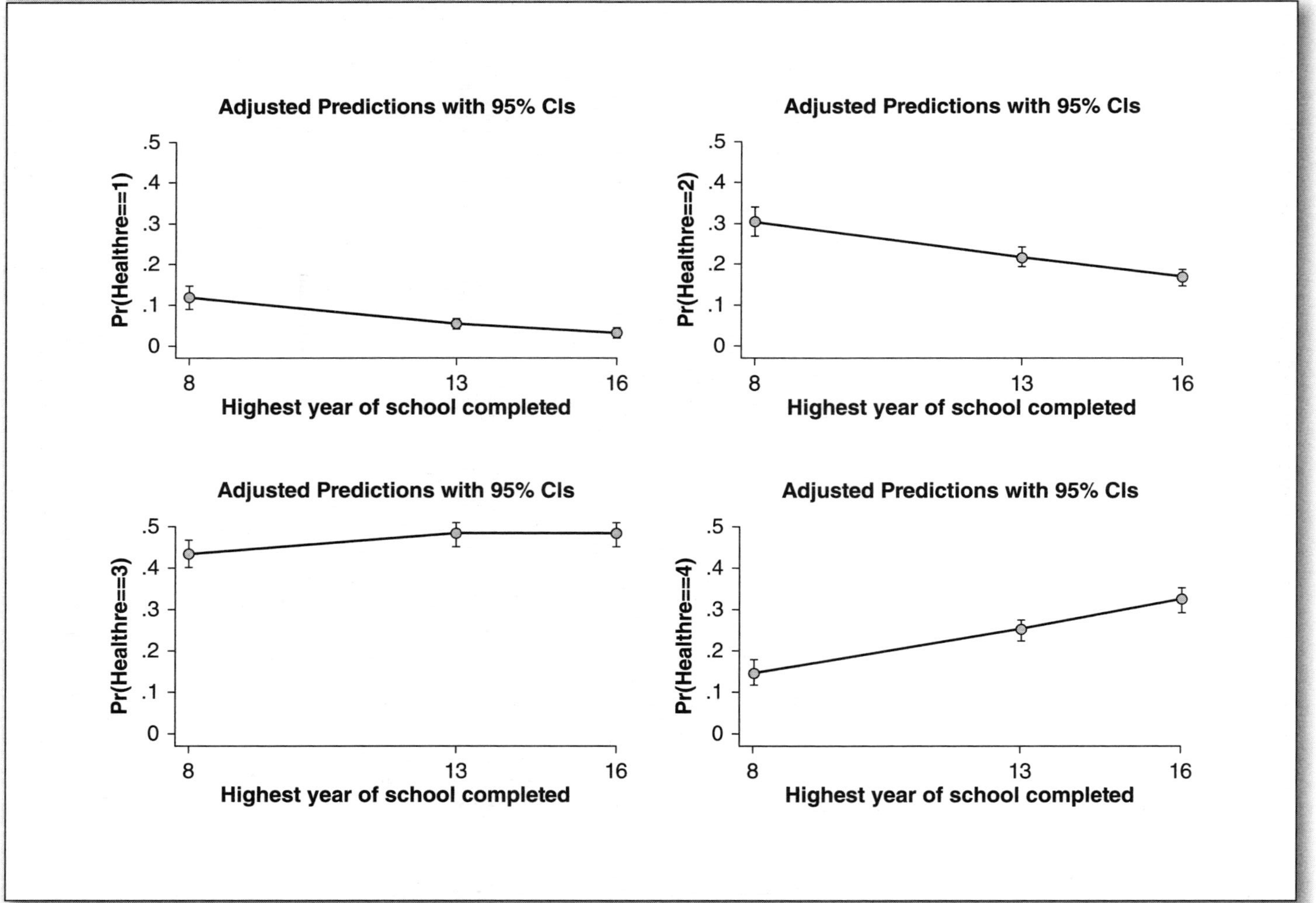

of 8, 13, and 16 when holding the other predictor variables at their means. The output is shown as follows.

```
. *margins and marginsplot in Stata 14
. margins, at(educ = (8 13 16)) atmeans vsquish

Adjusted predictions                              Number of obs    =     1,300
Model VCE     : OIM

1._predict    : Pr(healthre==1), predict(pr outcome(1))
2._predict    : Pr(healthre==2), predict(pr outcome(2))
3._predict    : Pr(healthre==3), predict(pr outcome(3))
4._predict    : Pr(healthre==4), predict(pr outcome(4))
1._at         : _Imaritals_1    =     .4684615  (mean)
                educ            =            8
                age             =         48.2  (mean)
                _Imale_1        =     .4438462  (mean)
2._at         : _Imaritals_1    =     .4684615  (mean)
                educ            =           13
                age             =         48.2  (mean)
                _Imale_1        =     .4438462  (mean)
3._at         : _Imaritals_1    =     .4684615  (mean)
                educ            =           16
                age             =         48.2  (mean)
                _Imale_1        =     .4438462  (mean)

------------------------------------------------------------------------------
             |            Delta-method
             |   Margin   Std. Err.      z    P>|z|    [95% Conf. Interval]
-------------+----------------------------------------------------------------
_predict#_at |
         1 1 |  .1193112   .0154363     7.73   0.000    .0890565    .1495658
         1 2 |  .0554071   .0063378     8.74   0.000    .0429852    .0678289
         1 3 |  .0328351   .0047591     6.90   0.000    .0235075    .0421627
         2 1 |  .3050308   .0188587    16.17   0.000    .2680684    .3419931
         2 2 |  .2189477   .0118616    18.46   0.000    .1956994    .2421959
         2 3 |  .1684926   .0111275    15.14   0.000    .146683     .1903022
         3 1 |  .4319806   .0167857    25.73   0.000    .3990811    .46488
         3 2 |  .4792625   .0144571    33.15   0.000    .4509271    .5075978
         3 3 |  .4789404   .0144496    33.15   0.000    .4506196    .5072611
         4 1 |  .1436775   .015311      9.38   0.000    .1136685    .1736865
         4 2 |  .2463828   .0122982    20.03   0.000    .2222788    .2704868
         4 3 |  .3197319   .0156434    20.44   0.000    .2890715    .3503924
------------------------------------------------------------------------------
```

The number of the predicted probabilities for the four ordinal categories (labeled from 1._predict to 4._predict) and three combinations of the predictor variables (labeled from 1._at to 3._at) are listed at the top of the output. The table for the margins or estimated probabilities is displayed at the bottom. As we can see, the results produced by Stata 14 are the same as those combined in the previous section using Stata 13. The interpretation of the results is omitted to save space.

Improved `marginsplot` Command in Stata 14

The improved `marginsplot` command in Stata 14 now can automatically plot the estimated probabilities of being in each category of an ordinal response variable in a single graph, which makes the previously complex steps for the same graphing in Stata 13 simpler. With just a single-word `marginsplot` command in Stata 14, the estimated probabilities when $Y = 1, 2, 3$, and 4 are displayed in Figure 7.3. As we can see, Figure 7.3 is an improved margins plot for Figure 7.2.

Computing the Estimated Probabilities With the `mtable` Command in Stata 13 and Later

The same estimated probabilities for each category of the ordinal response variable in the preceding section can be computed automatically using the `mtable` command (Long & Freese, 2014). See Chapter 4 for more details on this command.

```
. mtable, at(educ = (8 13 16)) atmeans ci

Expression: Pr(healthre), predict(outcome())

         |    educ        1        2        3        4
---------+-------------------------------------------------
 Pr(y) |       8    0.119    0.305    0.432    0.144
    ll |       8    0.089    0.268    0.399    0.114
    ul |       8    0.150    0.342    0.465    0.174
 Pr(y) |      13    0.055    0.219    0.479    0.246
    ll |      13    0.043    0.196    0.451    0.222
    ul |      13    0.068    0.242    0.508    0.270
 Pr(y) |      16    0.033    0.168    0.479    0.320
    ll |      16    0.024    0.147    0.451    0.289
    ul |      16    0.042    0.190    0.507    0.350

Specified values of covariates

         | _Imaritals_1      age   _Imale_1
---------+---------------------------------
 Current |        .468      48.2       .444
```

In the output, the estimated probabilities when $Y = 1, 2, 3$, and 4 for the predictor variable educ at the value of 8 are .119, .305, .432, and .144, respectively, when holding the other predictor variables at their means. Accordingly, the estimated

Figure 7.3 Estimated Probabilities of Being in Categories 1, 2, 3, and 4 for educ
With Stata 14

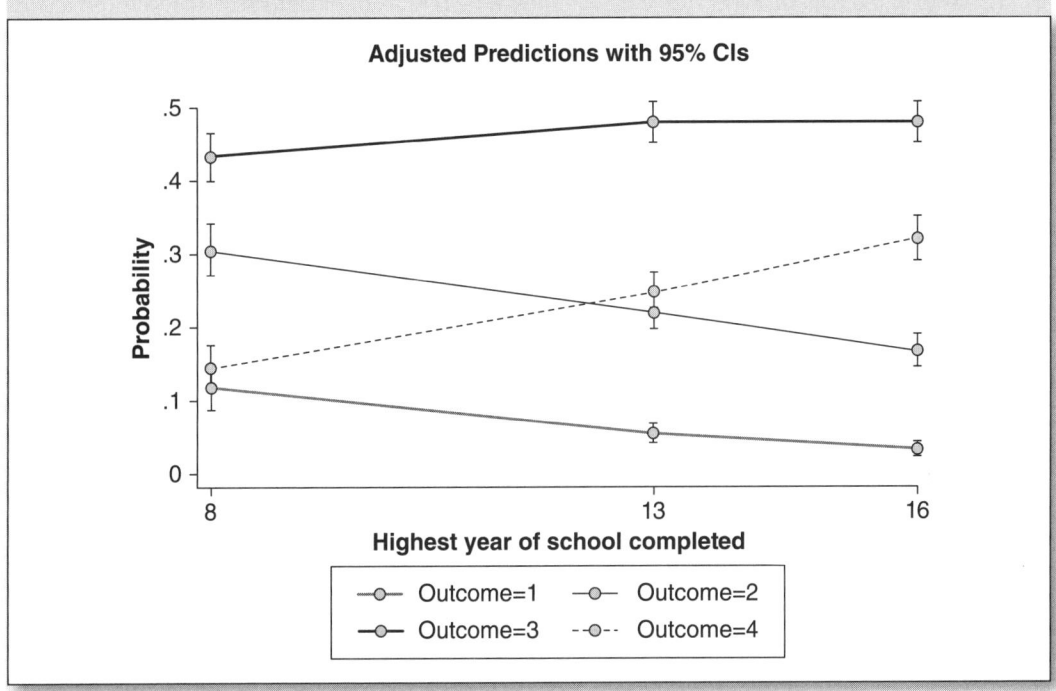

probabilities for all four categories for educ at the values of 13 and 16 can be interpreted in a similar way.

7.3.5 Making Publication-Quality
Tables for the Multiple-Predictor Model

```
1.  . outreg2 using chap7out_full, e(ll df_m chi2) word long replace
2.  . outreg2 using chap7out_full, eform word long append
chap7out_full.rtf
dir : seeout
```

The command is used immediately after the adjacent categories model is fitted. The
command outreg2 using chap7out_full, e(ll df_m chi2) word long
replace tells Stata to create a regression table for the full model and save it to a Word file
named chap7out_full. With the option eform, the odds ratios can be requested by using
the following command: outreg2 using chap7out_full, eform word long

append. Clicking the blue link of the file chap7out_full.rtf opens the Word file, which is saved in the current working directory. Clicking dir: seeout opens the tab delimited format table in the data browser. Table 7.8, as shown here in its original format, displays the results for the full model for the constrained multinomial logistic regression.

Table 7.8   Results of the Constrained Multinomial Logistic Regression Model: Full Model (Shown in Original Format Generated by Stata)

Variables	(1)	(2)
2		
_Imaritals_1	0.338***	1.402***
	(0.0703)	(0.0986)
Educ	0.0871***	1.091***
	(0.0114)	(0.0124)
Age	−0.0125***	0.988***
	(0.00199)	(0.00197)
_Imale_1	0.0349	1.036
	(0.0691)	(0.0716)
Constant	0.669***	1.952***
	(0.216)	(0.423)
3		
_Imaritals_1	0.676***	1.966***
	(0.141)	(0.277)
Educ	0.174***	1.190***
	(0.0228)	(0.0271)
Age	−0.0249***	0.975***
	(0.00398)	(0.00388)
_Imale_1	0.0698	1.072
	(0.138)	(0.148)
Constant	0.747**	2.111**
	(0.376)	(0.795)
4		
_Imaritals_1	1.014***	2.757***
	(0.211)	(0.582)
Educ	0.261***	1.299***
	(0.0341)	(0.0443)

Variables	(1)	(2)
Age	−0.0374***	0.963***
	(0.00597)	(0.00575)
_Imale_1	0.105	1.110
	(0.207)	(0.230)
Constant	−0.623	0.536
	(0.560)	(0.301)
Observations	1,300	1,300
Ll	−1,512	
df_m	4	
chi2	117.2	

Note: Standard errors in parentheses.

***$p < 0.01$, **$p < 0.05$, *$p < 0.1$

By applying the same edits as for Table 7.4, the final table (Table 7.9) is then presented.

Table 7.9 Results of the Constrained Multinomial Logistic Regression Model: Full Model (Edited)

Variables	b (SE(b))	OR
Model 1 ($Y = 2$ vs. $Y = 1$)		
_Imaritals_1	0.338***	1.402***
	(0.070)	(0.100)
Educ	0.087***	1.091***
	(0.011)	(0.012)
Age	−0.013***	0.988***
	(0.002)	(0.002)
_Imale_1	0.035	1.036
	(0.069)	(0.072)
α_1	0.669***	
	(0.216)	

(Continued)

Table 7.9 (Continued)

Variables	b (SE(b))	OR
Model 2 (Y = 3 vs. Y = 1)		
_Imaritals_1	0.676***	1.966***
	(0.141)	(0.277)
Educ	0.174***	1.190***
	(0.023)	(0.027)
Age	−0.025***	0.975***
	(0.004)	(0.004)
_Imale_1	0.070	1.072
	(0.138)	(0.148)
α_2	0.747**	
	(0.376)	
Model 3 (Y = 4 vs. Y = 1)		
_Imaritals_1	1.014***	2.757***
	(0.211)	(0.582)
Educ	0.261***	1.299***
	(0.034)	(0.044)
Age	−0.037***	0.963***
	(0.006)	(0.006)
_Imale_1	0.105	1.110
	(0.207)	(0.230)
α_3	−0.623	
	(0.560)	
Observations	1,300	
LR R^2	.043	
Log likelihood	−1512	
Wald χ^2_4	117.2	

Note: Standard errors are shown in parentheses.

**$p < .05$, ***$p < .01$.

Based on the transformation from the constrained multinomial logistic regression model into the AC model (see Table 7.2), the results of the multiple-predictor AC model (full model) are displayed in Table 7.10.

Table 7.10   Results of the Adjacent Categories Logistic Regression: Full Model

Variables	b (SE(b))	OR
α_1	.669***	
α_2	.078	
α_3	−1.37 ***	
_Imaritals_1	0.338***	1.402***
	(0.070)	(0.099)
educ	0.087***	1.091***
	(0.011)	(0.012)
age	−0.013***	0.988***
	(0.002)	(0.002)
_Imale_1	0.035	1.036
	(0.069)	(0.072)
Observations	1,300	1,300
LR R^2	.043	
Log likelihood	−1,512	
χ^2_4	117.2	

Note: Standard errors are shown in parentheses.

***$p < .01$.

7.4 Reporting the Results

Just like with the other logistic regression models for ordinal response variables, you may include the following information when reporting the results of the AC model for a research report. Please note that these guidelines are generic, so presentation of the results may vary across disciplines and journals.

First, describe the statistical method you use for data analysis, the dependent variable and the independent variables, and the purpose of your study. Since the AC model is estimated as a constrained multinomial logistic regression model, both methods should be explained. The constraints for the AC model also need to be reported. Equations for the AC model are optional.

Second, report the model fit statistics when available, such as the Wald chi-square test or the log likelihood ratio statistic, the associated p value, and the pseudo R^2 followed by a concise statement of interpretation.

Third, report the parameter estimates for the predictor variables, the Wald z statistics, the degrees of freedom, the associated p values, and the odds ratios in a table. In the text, interpret the logit coefficients and the associated odds ratios. The results table also needs to include a couple of the model fit statistics listed previously. More model fit statistics, such as extra pseudo R^2 values, and AIC and BIC statistics may be reported if the user-written command `fitstat` is used.

Fourth, if more than one model is fitted, then the results of all the competing models from the simple model to the full model can be presented in a table. The following is an example summarizing the results from the AC logistic regression model.

The adjacent categories logistic regression analysis was conducted to estimate the ordinal outcome variable, health status, from a set of predictor variables, such as marital status, years of education, age, and gender. The adjacent categories model focused on the odds comparing pairs of adjacent categories. It was estimated as a constrained multinomial logistic regression model by imposing constraints on the logit coefficients. The Wald chi-square test statistic *Wald* $\chi^2_{(4)}$ = 117.2, $p < .001$, which indicated that the overall model with four predictors was significant. Across the three equations for the AC model, all four predictor variables, except the `male` variable, were significantly different from 0 in predicting the odds of being in that category versus the base category.

The results of the full model for the adjacent categories logistic regression are presented in Table 7.10. For the predictor marital status, OR = 1.402, which indicated that the odds of being in a higher category of health status versus the next lower category for the married were 1.402 times as large as those for the unmarried when holding all the other predictors constant.

For the predictor education, OR = 1.091, which indicated that the odds of being in a higher category of health status versus the next lower category increased by 1.091 for a one-unit increase in the predictor education when holding all the other predictors constant.

For the predictor age, OR = .988, which was less than 1. This indicated that the odds of being in a higher category of health status versus the next lower category decreased by a factor of .988 for a one-unit increase in the predictor age when holding all the other predictors constant. In other words, for a one-unit increase in age, the odds of being healthier decreased by 1.2%.

For the predictor male, $\beta = .035$, $p = .614$, which was not significantly different from 0; OR = 1.036, which almost equals 1. This indicated that there was no relationship between being a male and the odds of being in a higher category of health status versus the next lower category.

7.5 Summary of Stata Commands in This Chapter

```
use chap7-gss2012, clear

*Single-predictor multinomial model
xi: mlogit healthre i.maritals, baseoutcome(1)
listcoef
listcoef, adjacent
fitstat

*Single-predictor AC Model
constraint 1 [3]_Imaritals_1=2*[2]_Imaritals_1
constraint 2 [4]_Imaritals_1=3*[2]_Imaritals_1
xi: mlogit healthre i.maritals, baseoutcome(1) constraints (1 2)
listcoef, adjacent
fitstat

*Making publication-quality tables using outreg2
outreg2 using chap7out, e(ll df_m chi2) word long replace
outreg2 using chap7out, eform word long append

*Multiple-predictor AC Model
*Specifying constraints
constraint 1 [3]_Imaritals_1=2*[2]_Imaritals_1
constraint 2 [4]_Imaritals_1=3*[2]_Imaritals_1
constraint 3 [3]educ=2*[2]educ
constraint 4 [4]educ=3*[2]educ
constraint 5 [3]age=2*[2]age
constraint 6 [4]age=3*[2]age
constraint 7 [3]_Imale_1=2*[2]_Imale_1
constraint 8 [4]_Imale_1=3*[2]_Imale_1

*Fitting a multiple-predictor model
xi: mlogit healthre i.maritals educ age i.male, baseoutcome(1) constraint (1/8)
listcoef, adjacent
fitstat

*margins and marginsplot in Stata 13
margins, predict (outcome(1)) at(educ = (8 13 16)) atmeans vsquish
marginsplot, name(mplot1)
margins, predict (outcome(2)) at(educ = (8 13 16)) atmeans vsquish
marginsplot, name(mplot2)
margins, predict (outcome(3)) at(educ = (8 13 16)) atmeans vsquish
marginsplot, name(mplot3)
```

```
margins, predict (outcome(4)) at(educ = (8 13 16)) atmeans vsquish
marginsplot, name(mplot4)

*margins plots combined in Stata 13
graph combine mplot1 mplot2 mplot3 mplot4, ycommon

*margins and marginsplot in Stata 14 after refitting the AC model
margins, at(educ = (8 13 16)) atmeans vsquish
marginsplot

*mtable in Stata 13 and later
mtable, at(educ = (8 13 16)) atmeans ci

*Making publication-quality tables using outreg2
outreg2 using chap7out_full, e(ll df_m chi2) word long replace
outreg2 using chap7out_full, eform word long append
```

7.6 EXERCISES

Use the GSS 2012 data for the following problems.

1. Conduct an analysis for an adjacent categories logistic regression model to examine the relationship between the ordinal response variable fechld and the predictor variables educ, kidjob, and sibs.

2. Recode the values of fechld by reversing the order and defining strongly disagree to be 1 and strongly agree to be 4. Specify the recoded category 1 (strongly disagree) as the base category when fitting the model.

3. Based on the output, write the equations for the constrained multinomial logistic regression.

4. From the equations for the multinomial logistic regression model, compute the intercepts and the logit coefficients for the equations of the AC model.

5. Use the listcoef command with the adjacent option to compute the odds of two adjacent categories.

6. Interpret the odds ratios.

7. Produce a publication-quality table with the outreg2 command displaying the results for the constrained multinomial logistic regression.

8. Make a publication-quality table displaying results for the AC model.

9. Write a report to summarize the results for the AC model.

CHAPTER 8

Stereotype Logistic Regression Models

Objectives of This Chapter

This chapter discusses the stereotype logistic regression model. It first introduces the advantages of this model, ordinality constraints, odds and odds ratio, measures of goodness of fit, and interpretation of parameter estimates. After a description of the data, it illustrates two examples of stereotype logistic regression models using Stata with detailed instructions and compares these models with the proportional odds model. It also illustrates how the results are displayed in publication-quality tables using the `outreg2` command and reported in text. This chapter focuses on fitting stereotype logistic regression models with Stata and interpreting and presenting the results. After reading this chapter, you should be able to

- Determine when a stereotype logistic regression model is used.
- Fit stereotype logistic regression models using Stata.
- Interpret the output.
- Interpret the model in terms of odds ratios.
- Compute and plot the estimated probabilities using the `margins` and `marginsplot` commands, respectively.
- Compute the estimated probabilities using the `mtable` command.
- Compare models using the likelihood ratio test and other fit statistics.
- Present results in publication-quality tables using Stata.
- Write the results for publication.

8.1 Stereotype Logistic Regression Models: An Introduction

The stereotype logistic (SL) regression model is an alternative model for ordinal response variables when the parallel lines or the proportional odds (PO) assumption does not hold. It was first developed by Anderson (1984) and later introduced by Greenland (1994), Hilbe (2009), Kuss (2006), Liu (2014), Long and Freese (2006, 2014), and Menard (2010). As discussed in Chapter 4, a proportional odds model assumes that the underlying binary models, which dichotomize the ordinal response variable, have the same coefficients. In other words, the logit coefficients for each predictor are the same, and thus, graphically the lines for the coefficients are parallel across the ordinal categories. When the PO assumption is violated, one option is to fit the partial proportional odds (PPO) model or the generalized ordinal logit model (Fu, 1998; Liu & Koirala, 2012; Peterson & Harrell, 1990; Williams, 2006), which has been discussed in Chapter 5. Another alternative is the SL model, which is the focus of this chapter.

The SL model can be treated as an extension of both the multinomial logistic regression model and the PO model. First, the SL model is like the multinomial logistic model since they both estimate the odds of being at a particular category compared with the baseline category. Although either of the two models can be used when the proportional odds assumption is violated, compared with the multinomial logistic regression model, the SL model estimates fewer parameters and, thus, is more parsimonious. Second, similar to the PO model, the SL model estimates the ordinal response variable rather than the nominal outcome variable when given a set of predictors. However, the SL model does not assume the PO assumption and allows the effect of each predictor to vary across the ordinal categories.

The SL model has three advantages: First, it relaxes the PO assumption and imposes constraints or scale parameters on the logit coefficients to ensure the ordinal nature of the outcome variable. Second, when researchers are unsure whether an assumed ordinal response variable is truly ordinal due to the subjective judgment of the categories, the SL model can be used to test whether these underlying categories can be collapsed or if the order can be adjusted or reversed. Finally, the SL model can be applied to analyze the ordinal data when the ordinal response variable has more than one dimension.

The Stereotype Logistic Model

Anderson's SL model (1984) can be written in the following form:

$$\text{logit}[\pi(j, J)] = \ln\left(\frac{\pi\left(Y = j | x_1, x_2, ..., x_p\right)}{\pi\left(Y = J | x_1, x_2, ..., x_p\right)}\right) = \alpha_j - \phi_j\left(\beta_1 X_1 + \beta_2 X_2 + ... + \beta_p X_p\right) \qquad (8.1)$$

where $j = 1, 2, ..., J - 1$; J is the baseline or reference category, which is the last category here, but it can be the first category or any of the other categories decided by the researcher; Y is the ordinal response variable with categories from j to J; α_j are the intercepts; $\beta_1, \beta_2, ..., \beta_p$ are the logit coefficients for the predictors, $X_1, X_2, ..., X_p$, respectively; and ϕ_j are the constraints or scale parameters that are used to ensure the outcome variable is ordinal if the following condition is satisfied:

$$1 = \phi_1 > \phi_2 > \phi_3 > ... \phi_{J-1} > \phi_J = 0 \qquad (8.2)$$

These constraints are ordered from 1 to 0 with each constraint getting smaller. The first constraint ϕ_1 is set to be 1, and the last constraint ϕ_J is equal to 0 so that the estimated SL model can be identified. For example, when the ordinal response variable has four categories, only two constraints (i.e., ϕ_2 and ϕ_3) need to be estimated and they meet the following condition: $1 = \phi_1 > \phi_2 > \phi_3 > \phi_4 = 0$. If any two pairs of the constraints are the same, then these two categories are indistinguishable and are thus collapsed into one. For example, if $\phi_2 = \phi_3$, these two categories (categories 2 and 3) can be grouped together. The ordinality of the constraints can be tested in the model so that researchers can decide whether any categories need to be merged or reordered.

What happens when the constraints (ϕ_j) are multiplied by a logit coefficient (β) of a predictor variable? By using the same example when the ordinal response variable has four categories, we obtain $\phi_1\beta$, $\phi_2\beta$, $\phi_3\beta$, and $\phi_4\beta$ and $\phi_1\beta > \phi_2\beta > \phi_3\beta > \phi_4\beta$. With a negative sign before the logit coefficients, $\phi_j\beta$ can be ordered in an opposite direction so that $-\phi_1\beta < -\phi_2\beta < -\phi_3\beta < -\phi_4\beta$. Since the constraints are ordered with the first constraint set to 1 and the last one set to 0, we can easily see that the function of the constraints or scale parameters here is to ensure the ordinality of the model.

To calculate the odds of being in a category j versus a category m, we just need to take the exponential of $[(\alpha_j - \alpha_m) - (\phi_j - \phi_m)\beta]$. When the category m becomes the baseline category J, we just need to substitute it into the equation. Since $\phi_j = 0$, we get $[(\alpha_j - 0) - (\phi_j - 0)\beta] = \alpha_j - \phi_j\beta$. By exponentiating $(-\phi_j\beta)$, we get the odds of being in a category j versus the baseline category J for a unit change in a predictor.

Equation 8.1 is the form for Anderson's one-dimensional SL model, which was generally referred to as the SL model in literature. Anderson (1984) also argued that an ordinal response variable could be more than one dimension and therefore proposed the multidimensional SL model. If the ordinal outcome variable has J categories, the maximum dimensions would be $J - 1$. The multidimensional SL model with $J - 1$ dimensions is actually equal to the multinomial logistic regression model. Researchers may feel confused with an SL model with more than one dimension. Long and Freese (2006, 2014) provided detailed explanation on one-dimensional versus higher dimensional SL models. The one-dimensional SL can be used when the categories of an ordinal response variable are ordered on one dimension. However, the categories can be ordered on more than one dimension, which may need a higher dimensional SL model. For example, a five-point Likert scale from strongly disagree to strongly agree can be ordered by three dimensions: either strongly agree or disagree, either agree or disagree, and neutral. The one-dimensional SL model is the only focus in this chapter for the simplicity of model building and interpretation.

Connection With the Multinomial Logistic Model

The multinomial logistic model can be expressed as follows:

$$\ln\left(\frac{\pi\left(Y = j | x_1, x_2, ..., x_p\right)}{\pi\left(Y = J | x_1, x_2, ..., x_p\right)}\right) = \alpha_j + \beta_{j1}X_1 + \beta_{j2}X_2 + ... + \beta_{jp}X_p \tag{8.3}$$

where $j = 1, 2, ..., J - 1$; J is the base category, which can be any category but is generally the highest one; α_j are the intercepts; and $\beta_{j1}, \beta_{j2}, ..., \beta_{jp}$ are the logit coefficients for each comparison. The model estimates $j - 1$ logit coefficients for each predictor.

Lunt (2001) considered the SL model to be the constrained multinomial logistic model and developed the Stata `soreg` program before the official Stata `slogit` program was implemented. Compared with the multinomial logistic regression model in Equation 8.3, the left side of the logit link function for the SL model in Equation 8.1 looks the same since both the SL model and the multinomial model estimate the odds of being in a particular category versus the baseline category. Examining the systematic component (linear predictors) in both models, it is obvious that the logit coefficients β_j in the multinomial logistic model correspond to $(-\phi_j(\beta))$ in the SL model. As explained, the constraints (ϕ_j) in the SL model are used to ensure that the model is ordinal.

Which model is more parsimonious? In a multinomial logistic regression model, when the nominal outcome variable has j categories and p predictors, we need to estimate $(J - 1)$ intercepts and $(J - 1) \times p$ coefficients. The total number of parameters = $(J - 1) + (J - 1) \times p = (J - 1) \times (1 + p)$. In the SL model, when the ordinal outcome variable has j categories and p predictors, we estimate $(J - 1)$ intercepts, $(J - 2)$ scale parameters, and p coefficients. The number of scale parameters (ϕ_j) = $J - 2$ since ϕ_1 and ϕ_J are constrained to be 1 and 0 and do not need to be estimated. The total number of parameters = $[(J - 1) + (J - 2) + p] = (2J - 3 + p)$. Fewer parameters are estimated in the SL model so this model, is more parsimonious than the multinomial logistic model.

8.1.1 Odds and Odds Ratios in Stereotype Logistic Regression Models

Unlike the proportional odds model, which estimates the odds of being at or below a particular category versus being above that category, but similar to the multinomial logistic model, the SL model estimates the logit or log odds of being in a particular category relative to the baseline category. The odds in the SL model can be defined as the ratio of the probability of being in a particular category to the probability of being in the base category. This is expressed as:

$$\text{Odds } (Y = j \text{ vs. } Y = J) = \frac{p(Y = j)}{p(Y = J)} \tag{8.4}$$

where j can be any category from 1 to $J - 1$ categories.

In the following example, an ordinal response variable, health status, has four ordinal levels from 1 to 4, with 1 = poor, 2 = fair, 3 = good, and 4 = excellent. We estimate three odds with category 4 as the base or reference category: the odds of being in category 1 versus category 4, the odds of being in category 2 versus category 4, and the odds of being in category 3 versus category 4.

Odds $(Y = 1$ vs. $Y = 4)$ equal the ratio of probability of being in category 1 to the probability of being in category 4:

$$\text{Odds } (Y = 1 \text{ vs. } 4) = \frac{p(Y = 1)}{p(Y = 4)} = \frac{p(1)}{p(4)}$$

Odds $(Y = 2$ vs. $Y = 4)$ and odds $(Y = 3$ vs. $Y = 4)$ can be expressed in a similar way:

$$\text{Odds } (Y = 2 \text{ vs. } 4) = \frac{p(Y = 2)}{p(Y = 4)} = \frac{p(2)}{p(4)}$$

$$\text{Odds } (Y = 3 \text{ vs. } 4) = \frac{p(Y = 3)}{p(Y = 4)} = \frac{p(3)}{p(4)}$$

In sum, the odds in the SL model compare the probability of being in a particular category $p(Y = j)$ with the probability of being in the base category $p(Y = J)$. Please note that the logit or log odds of being in a particular category versus the base category in the SL model need to be multiplied by negative scale parameters $-\phi_j$. Table 8.1 presents the scale parameters, logits, odds, and category comparisons for the SL model for the ordinal response variable with four levels.

Table 8.1 Category Comparisons for the Stereotype Logistic Regression Model With Four Levels of Health Status $(j = 1, 2, 3, 4)$

Equation	Scale Parameters (ϕ_j)	Logit $P(Y = j$ vs. $J)$	Odds	Probability Comparisons
1	$\phi_1 = 1$	logit $P(Y = 1$ vs. 4)	$\dfrac{P(Y = 1)}{P(Y = 4)}$	Category 1 vs. category 4
2	ϕ_2	logit $P(Y = 2$ vs. 4)	$\dfrac{P(Y = 2)}{P(Y = 4)}$	Category 2 vs. category 4
3	ϕ_3	logit $P(Y = 3$ vs. 4)	$\dfrac{P(Y = 3)}{P(Y = 4)}$	Category 3 vs. category 4

Odds Ratios in Stereotype Logistic Regression

Unlike the multinomial logistic model, the logit coefficients estimated by the SL model have scale parameters ϕ_j; thus, the odds ratio of being in a category j versus the baseline category J is obtained by taking the exponential of $[(\alpha_j - \alpha_J) - (\phi_j - \phi_J)\beta]$. Since the estimated α_J and ϕ_J equal 0, the equation could be simplified to be $(\alpha_j - \phi_j\beta)$. Exponentiating $(-\phi_j\beta)$ gives us the odds ratio of being in a category j versus the baseline J for a unit change in a predictor variable when holding other predictor variables constant.

8.1.2 Model Fit Statistics

Model fit statistics for the SL model are limited. Currently, Stata reports the log likelihood statistic, Wald chi-square test, and the log likelihood ratio test statistic for model comparisons. Using the `fitstat` command by Long and Freese (2014), other model fit statistics, such as the deviance statistic and the AIC and BIC statistics, can also be provided. However, no pseudo R^2 statistics are available for the model.

Just as with the models in the previous chapters, the log likelihood ratio test can be applied to compare nested models. It is defined as the difference between the deviance or –2LL for the reduced model and that for the full model. The log likelihood ratio test statistic follows a chi-square distribution with the degrees of freedom of the distribution equal to the difference in the number of parameters between these two models. In addition, the AIC and BIC statistics reported by the `estat ic` command or the `fitstat` command (Long & Freese, 2014) can be used for model comparisons. Smaller values of AIC and BIC statistics indicate a better fit of the model.

8.1.3 Interpretation of Model Parameter Estimates

A logit coefficient in the SL model is the log odds of being in a particular category relative to the base category. The logit coefficients need to be multiplied by the constraints or the scale parameters $-\phi_j$. After multiplication, they can be interpreted as the change in the predicted logit or the log odds for a one-unit increase in a predictor variable when holding other predictor variables constant. Exponentiating the product of the logit coefficients and the negative scale parameters gives us the odds ratios of being in a category j versus the baseline J. The interpretation of odds ratios in the SL model is similar to that of other logistic regression models. The odds ratios are the change in the predicted odds of being in a particular category compared with the base category for a one-unit increase in the predictor variable when holding other predictor variables constant.

When an OR is larger than 1, the odds of being in a particular category versus the base category increase for a one-unit increase in the predictor variable.

When an OR is less than 1, the odds of being in a particular category versus the base category decrease for a one-unit increase in the predictor variable.

An OR of 1 indicates that there is no relationship between the predictor variable and the estimated odds.

The odds of being in the base category compared with a particular category can also be estimated since they are just the reciprocal of the odds of being in a particular category versus the base category. These two odds are different in the order when comparing categories. The odds of being in a particular category versus the base category compare category j and the base category J, whereas the odds of being in the base category compared with a particular category compare categories in the reversed order, that is, the base category J versus a particular category j.

To get the odds ratio of being in the base category relative to a particular category, we can either exponentiate ($\phi_j \beta$) or take the reciprocal of the odds of being in a particular category versus the base category.

8.2 Research Example and Description of the Data and Sample

In this chapter, the purpose of the research example is still to investigate the relationships between the ordinal response variable, health status, and four predictor variables. Different from previous chapters, however, the SL model is used for data analysis since it relaxes the proportional odds assumption.

The data for the following analyses is from the General Social Survey 2012 (GSS 2012). The same variables from previous chapters are used for data analysis in this chapter.

- `healthre`: the recoded variable of health (health status) with four ordinal categories (1 = poor health, 2 = fair health, 3 = good health, and 4 = excellent health)
- `maritals`: the recoded variable of marital (marital status) with 1 = currently married and 0 = not currently married
- `educ`: the highest level of education
- `age`: respondent's age
- `male`: recoded variable of sex with 1 = male and 0 = female

8.3 Stereotype Logistic Regression Models With Stata: Commands and Output

The Stata Command `slogit`

The Stata command `slogit` is used for the analysis of stereotype logistic regression models. In the syntax, the command `slogit` is followed by the dependent variable and the independent variable(s). For example, the command `slogit y x` tells Stata to fit a stereotype logistic model to estimate the ordinal dependent variable y with an independent variable x. For more details on how to use this command, type the `help slogit` command.

8.3.1 The SL Model: One-Predictor Model

The command `slogit healthre educ, dim(1)` tells Stata to fit the stereotype logistic regression model for the ordinal response variable `healthre` with the predictor variable `educ`. The option `dim(1)` tells us this is the one-dimensional model. The Stata output for the single-predictor stereotype logistic regression model is displayed as follows.

```
. slogit healthre educ, dim(1)

Iteration 0:   log likelihood = -1983.9528   (not concave)
Iteration 1:   log likelihood = -1551.0726   (not concave)
Iteration 2:   log likelihood = -1540.6004
Iteration 3:   log likelihood = -1537.8896
Iteration 4:   log likelihood = -1537.4133
Iteration 5:   log likelihood = -1537.3428
Iteration 6:   log likelihood = -1537.3413
Iteration 7:   log likelihood = -1537.3413

Stereotype logistic regression                Number of obs   =       1300
                                               Wald chi2(1)    =      43.09
Log likelihood = -1537.3413                    Prob > chi2     =     0.0000

 (1)  [phi1_1]_cons = 1
-------------------------------------------------------------------------------
    healthre |     Coef.   Std. Err.      z    P>|z|     [95% Conf. Interval]
-------------+-----------------------------------------------------------------
        educ |   .252143   .0384131    6.56   0.000     .1768548    .3274313
-------------+-----------------------------------------------------------------
     /phi1_1 |         1          .       .      .            .           .
     /phi1_2 |  .7496808   .1191255    6.29   0.000     .516199    .9831625
     /phi1_3 |  .1387562   .0855631    1.62   0.105    -.0289444    .3064569
     /phi1_4 |         0   (base outcome)
-------------+-----------------------------------------------------------------
     /theta1 |  1.849693   .4938876    3.75   0.000     .881691    2.817695
     /theta2 |  2.285555   .3784856    6.04   0.000    1.543737    3.027373
     /theta3 |  1.025498   .3325926    3.08   0.002     .3736289   1.677368
     /theta4 |         0   (base outcome)
-------------------------------------------------------------------------------
(healthre=4 is the base outcome)
```

Interpreting the Output

Just as with the other logistic regression models, the top of the output displays a series of iterations for the log likelihood estimation. The maximum log likelihood value is –1537.34 (`Iteration 7: log likelihood = -1537.3413`).

The top right of the logistic regression table provides the number of observations in the dataset, the Wald chi-square test statistic and the associated p value, and the pseudo R^2. The sample for the analysis has 1,300 cases. `Wald chi2 (1) = 43.09`, which is the Wald chi-square test statistic.

The null hypothesis of the Wald chi-square test for the overall model is that the predictor variable does not contribute to the model, and the alternative hypothesis is that the one-predictor model is significant.

The associated p value for the Wald chi-square test is significant: `Prob > chi2 = 0.0000`, which indicates that the null hypothesis is rejected. Therefore, the predictor variable in the one-predictor model significantly predicts the logit or log odds of being in a particular category relative to the base category.

The stereotype logistic regression table displays the parameter estimates for the predictor variable, the constraints, and the intercepts, their standard errors, the Wald z statistics, the associated p values, and the 95% confidence intervals of the parameter estimates.

The null hypothesis for the Wald z test is that the coefficient of the predictor variable is zero, and the alternative hypothesis is that the coefficient of the predictor variable is significantly different from zero.

For the single-predictor variable `educ`, Wald $z = 6.56$. The associated p value, `P>|z|=.000`, so we reject the null hypothesis. Therefore, the predictor variable `educ` is a significant predictor of the ordinal outcome variable, health status. The 95% confidence interval of the regression coefficient is [`.1768548, .3274313`]. It does not contain 0, which indicates the coefficient is significantly different from 0.

The logit regression coefficient (labeled `Coef.`) $\beta = .252$. Before interpreting it, we need to multiply it by the constraints or scale parameters (labeled as `/phi1_1-/phi1_4`), which will be explained next.

These constraints or scale parameters (ϕ_j) are reported in the middle of the table. The first constraint (labeled as `/phi1_1`) is set to 1, and the fourth constraint (labeled as `/phi1_4`) is set to 0. The second and third constraints are $.7496808$ and $.1387562$, respectively. Recall that ϕ_j are a list of ordinality constraints and they satisfy the condition $1 = \phi_1 > \phi_2 > \phi_3 > \phi_4 = 0$.

Multiplying the regression coefficient β and the constraints $-\phi_j$ gives us the logit coefficients or log odds of being in a particular category versus the base category. Therefore, $(-\phi_1\beta)$, $(-\phi_2\beta)$, and $(-\phi_3\beta)$ in the model are the logit coefficients comparing categories 1 versus 4, 2 versus 4, and 3 versus 4, respectively.

$$(-\phi_1\beta) = -1 \times .252 = -.252$$

$$(-\phi_2\beta) = -.750 \times .252 = -.189$$

$$(-\phi_3\beta) = -.139 \times .252 = -.035$$

On the other hand, $(\phi_j\beta)$ are the logit coefficients or log odds of being in the base category versus a particular category after removing the negative sign. Therefore, $(\phi_1\beta)$, $(\phi_2\beta)$, and $(\phi_3\beta)$ are the logit coefficients comparing categories 4 with 1, 4 with 2, and 4 with 3, respectively.

$(\phi_1\beta) = 1 \times .252 = .252$, which indicates that for a one-unit increase in the predictor variable `educ`, the logit or log odds of being in category 4 versus category 1 increases by a factor of .252. Similarly, the logits comparing categories 4 with 2 and categories 4 with 3 increase by .189 and .035, respectively. The corresponding odds ratios, which offer a better interpretation, can be computed either manually or by using the `listcoef` command, which will be illustrated later in this chapter.

The regression table also reports the intercepts (labeled as /theta1-/theta4). They are the intercepts for each equation comparing a particular category with the base category. The base category by default is the highest outcome (e.g., `healthre` = 4 in this example), but you can specify any category as the base category.

Interpreting the Odds Ratios of Being in a Particular Category Versus the Base Category for the SL Model

The SL model estimates the logit odds of being in a category relative to the base-line category. By substituting the value of the coefficient into Equation 8.1:

$$\ln\left(\frac{\pi(Y = j \mid x_1)}{\pi(Y = J \mid x_1)}\right) = \alpha_j - \phi_j(\beta_1 X_1)$$

we get logit $[\pi(j, J \mid \text{educ})] = \alpha_j - \phi_j \times .252$ (educ).

Recall that the exponentiated $(-\phi_j\beta)$ is the odds ratio of being in a category j versus the baseline J for a one-unit change in a predictor variable. In this model, we define the odds ratio of being in category 1 compared with the base category 4 as OR (1, 4). OR(1, 4) = $e^{(-1 \times .252)}$ = $e^{(-.252)}$ = .777, which indicates that for a one-unit increase in years of education, the odds of being in category 1 of health condition versus the base category 4 decrease by a factor of .777. In other words, people with higher education levels are more likely to be in the highest health condition (category 4) rather than being in category 1.

The odds ratio of being in category 2 versus category 4, OR(2, 4) = $e^{(-.750 \times .252)}$ = $e^{(-.189)}$ = .828, which indicates that for a one-unit increase in years of education, the odds of being in category 2 of health condition versus the base category 4 decrease by a factor of .828. Similarly, the odds ratio of being in category 3 versus category 4, OR(3, 4) = $e^{(-.139 \times .252)}$ = $e^{(-.035)}$ = .966, which indicates that for a one-unit increase in years of education, the odds of being in category 3 of health condition versus the base category 4 decrease by a factor of .966.

Interpreting the Odds Ratios of Being in the Base Category Versus a Particular Category for the SL Model

The odds ratios can also be interpreted in the opposite direction when we reverse the order of the category comparisons. The odds of being in the base category J relative to a particular category j is the inverse of the odds of being in that category versus the baseline category. To estimate the odds of being in the base category relative to a particular category, we just need to change the signs before the intercepts and the estimated logits in Equation 8.1. The modified logit equation logit $[\pi(J, j \mid educ)] = -\alpha_j + \phi_j \times .252$ (educ). By exponentiating $(\phi_j\beta)$, we get the odds ratio of being in the base category J versus any other category for a one-unit change in a predictor variable.

OR $(4, 1) = e^{(1 \times .252)} = 1.287$, which indicates that the odds of being in base category 4 compared with category 1 increase by 1.287 with a one-unit increase in years of education. The odds ratio of being in the base category 4 relative to being in level 2, OR$(4, 2) = e^{(.750 \times .252)} = e^{(.189)} = 1.208$, which indicates that for a one-unit increase in years of education, the odds of being in the base category 4 of health condition versus category 2 increase 1.208. Finally, OR$(4, 3) = e^{(.139 \times .252)} = e^{(.035)} = 1.036$, which indicates that for a one-unit increase in years of education, the odds of being in the base category 4 of health condition versus category 3 increase 1.036. In summary, the increase in years of education is associated with the odds of being in excellent health condition (base category 4) relative to any other lower health condition.

Different from the `ologit` command, the `or` option for displaying odds ratios does not work with `slogit`. Without computing the odds ratios manually, we can obtain the same results by using the `listcoef` command with the `expand` option, which is part of the `SPost13` package (Long & Freese, 2014). The output is displayed as follows.

```
. listcoef, expand

slogit (N=1300): Factor change in odds

  Odds of: 4 vs 1

----------------------------------------------------------------------------
              |        b        z     P>|z|      e^b    e^bStdX      SDofX
--------------+-------------------------------------------------------------
         educ |   0.2521    6.564    0.000     1.287     2.213      3.151
--------------+-------------------------------------------------------------
phi           |
       phi1_1 |   1.0000       .        .         .         .          .
       phi1_2 |   0.7497    6.293    0.000         .         .          .
       phi1_3 |   0.1388    1.622    0.105         .         .          .
--------------+-------------------------------------------------------------
theta         |
       theta1 |   1.8497    3.745    0.000         .         .          .
       theta2 |   2.2856    6.039    0.000         .         .          .
       theta3 |   1.0255    3.083    0.002         .         .          .
----------------------------------------------------------------------------

slogit (N=1300): Factor change in the odds of healthre
```

```
Variable: educ (sd=3.151)
----------------------------------------------------------------------------
                          |        b      z     P>|z|      e^b    e^bStdX
--------------------------+-------------------------------------------------
1          vs  2          |   -0.0631  -1.740    0.082     0.939     0.820
1          vs  3          |   -0.2172  -6.024    0.000     0.805     0.504
1          vs  4          |   -0.2521  -6.564    0.000     0.777     0.452
2          vs  1          |    0.0631   1.740    0.082     1.065     1.220
2          vs  3          |   -0.1540  -6.276    0.000     0.857     0.615
2          vs  4          |   -0.1890  -6.810    0.000     0.828     0.551
3          vs  1          |    0.2172   6.024    0.000     1.243     1.982
3          vs  2          |    0.1540   6.276    0.000     1.167     1.625
3          vs  4          |   -0.0350  -1.511    0.131     0.966     0.896
4          vs  1          |    0.2521   6.564    0.000     1.287     2.213
4          vs  2          |    0.1890   6.810    0.000     1.208     1.814
4          vs  3          |    0.0350   1.511    0.131     1.036     1.117
----------------------------------------------------------------------------
```

In the output, the first table displays the odds ratio for the predictor variable `educ` comparing the base category 4 with category 1 (labeled as `Odds of: 4 vs 1` on the top). OR (labeled as `e^b`) = 1.287.

In the second table, the first column shows all possible comparisons between two categories. The column labeled `b` lists the logit coefficients, which have taken the constraints or scale parameters into account. The column labeled `e^b` lists the odds ratios. For example, the odds ratios comparing category 1 with the base category 4, category 2 with category 4, and category 3 with category 4 are `.777`, `.828`, and `.966`, respectively.

```
. fitstat

                              |    slogit
------------------------------+-------------
Log-likelihood                |
                   Model      |  -1537.341
           Intercept-only     |  -1579.407
------------------------------+-------------
Chi-square                    |
       Deviance (df=1294)     |   3074.683
            Wald (df=1)       |     43.086
                p-value       |      0.000
------------------------------+-------------
R2                            |
                McFadden      |      0.027
     McFadden (adjusted)      |      0.023
           Cox-Snell/ML       |      0.063
Cragg-Uhler/Nagelkerke        |      0.069
------------------------------+-------------
```

```
IC                    |
             AIC |      3086.683
    AIC divided by N |         2.374
         BIC (df=6)  |      3117.703
```

Model Fit Statistics

The `fitstat` command reports model fit statistics. The deviance statistic is 3,074.683. The AIC statistic, which is also reported as the AIC used by Stata in the previous `SPost9`, is 3,086.683. The AIC divided by N is 2.374, and the corresponding BIC (also reported as the BIC used by Stata in `SPost9`) is 3,117.703. They will be interpreted when we compare them with those for the full model.

The AIC and BIC statistics can also be computed from the postestimation `estat ic` command. The output is shown as follows.

```
. estat ic

Akaike's information criterion and Bayesian information criterion

------------------------------------------------------------------------------
      Model |    Obs    ll(null)   ll(model)     df          AIC          BIC
------------+-----------------------------------------------------------------
    _fsest_0 |   1300   -1579.407   -1537.341      6     3086.683     3117.703
------------------------------------------------------------------------------
               Note:  N=Obs used in calculating BIC; see [R] BIC note
```

8.3.2 The SL Model: Multiple-Predictor Model

The command `slogit healthre educ maritals age male, dim(1)` tells Stata to predict the ordinal response variable `healthre` from four predictor variables `educ`, `maritals`, `age`, and `male` using stereotype logistic regression. Please note that `maritals` and `male` are dummy variables so they can be directly included in the model. The output is shown as follows.

```
. slogit healthre educ maritals age male, dim(1)

Iteration 0:   log likelihood = -1725.2127  (not concave)
Iteration 1:   log likelihood = -1536.0552  (not concave)
Iteration 2:   log likelihood = -1514.7837
Iteration 3:   log likelihood = -1509.2965
Iteration 4:   log likelihood = -1508.9675
Iteration 5:   log likelihood =  -1508.956
Iteration 6:   log likelihood =  -1508.956
```

```
Stereotype logistic regression                    Number of obs   =      1300
                                                  Wald chi2(4)    =     76.92
Log likelihood =  -1508.956                       Prob > chi2     =    0.0000

 (1)  [phi1_1]_cons = 1
-----------------------------------------------------------------------------
    healthre |     Coef.   Std. Err.      z    P>|z|     [95% Conf. Interval]
-------------+---------------------------------------------------------------
        educ |   .2542666   .0352995     7.20   0.000     .1850809    .3234523
    maritals |    .882704   .2099854     4.20   0.000     .4711402    1.294268
         age |  -.0339957    .006026    -5.64   0.000    -.0458065   -.0221849
        male |    .052946   .1941454     0.27   0.785     -.327572    .433464
-------------+---------------------------------------------------------------
     /phi1_1 |          1  (constrained)
     /phi1_2 |   .7305334   .0870447     8.39   0.000     .5599289    .901138
     /phi1_3 |   .2345265   .0624922     3.75   0.000     .1120441    .357009
     /phi1_4 |          0  (base outcome)
-------------+---------------------------------------------------------------
     /theta1 |   .5996329   .5221377     1.15   0.251    -.4237382    1.623004
     /theta2 |   1.371411   .3982693     3.44   0.001     .5908179    2.152005
     /theta3 |   1.116292   .2093595     5.33   0.000      .705955    1.526629
     /theta4 |          0  (base outcome)
-----------------------------------------------------------------------------
(healthre=4 is the base outcome)
```

The Wald chi-square test $\chi^2_{(4)} = 76.92$, $p < .001$, indicates that at least one of the four predictors in the full model is significant in predicting the ordinal response variable.

The stereotype logistic regression table displays the parameter estimates for the four predictor variables, the constraints, and the intercepts, their standard errors, the Wald z statistics, the associated p values, and the 95% confidence intervals of the parameter estimates. Among four predictor variables, the logit coefficients for educ, maritals, and age are statistically significant (P>|z| = .000), whereas the coefficient for male is not significant ($p = .785$).

The four constraints are 1, .731, .235, and 0, respectively. They are useful when computing the logit coefficients comparing a particular category versus the base category with the formula $(-\phi_j\beta)$.

With the command listcoef, expand, we get detailed results of logit coefficients and odds ratios (exponentiated coefficients) for predictor variables when taking the constraints (ϕ_j) into account. Without the expand option, we only get the odds ratios comparing the base category 4 with category 1. The expand option provides odds ratios comparing any pairs of the four categories. The following output is displayed.

```
. listcoef, expand

slogit (N=1300): Factor change in odds

  Odds of: 4 vs 1

--------------------------------------------------------------------------
             |        b        z    P>|z|       e^b   e^bStdX     SDofX
-------------+------------------------------------------------------------
        educ |   0.2543    7.203    0.000     1.290     2.228     3.151
    maritals |   0.8827    4.204    0.000     2.417     1.554     0.499
         age |  -0.0340   -5.641    0.000     0.967     0.553    17.432
        male |   0.0529    0.273    0.785     1.054     1.027     0.497
-------------+------------------------------------------------------------
phi          |
      phi1_1 |   1.0000        .        .         .         .         .
      phi1_2 |   0.7305    8.393    0.000         .         .         .
      phi1_3 |   0.2345    3.753    0.000         .         .         .
-------------+------------------------------------------------------------
theta        |
      theta1 |   0.5996    1.148    0.251         .         .         .
      theta2 |   1.3714    3.443    0.001         .         .         .
      theta3 |   1.1163    5.332    0.000         .         .         .
--------------------------------------------------------------------------

slogit (N=1300): Factor change in the odds of healthre

Variable: educ (sd=3.151)
-----------------------------------------------------------------------
                        |        b        z    P>|z|       e^b   e^bStdX
------------------------+----------------------------------------------
1           vs 2        |  -0.0685   -2.482    0.013     0.934     0.806
1           vs 3        |  -0.1946   -6.000    0.000     0.823     0.542
1           vs 4        |  -0.2543   -7.203    0.000     0.775     0.449
2           vs 1        |   0.0685    2.482    0.013     1.071     1.241
2           vs 3        |  -0.1261   -5.721    0.000     0.882     0.672
2           vs 4        |  -0.1858   -7.217    0.000     0.830     0.557
3           vs 1        |   0.1946    6.000    0.000     1.215     1.846
3           vs 2        |   0.1261    5.721    0.000     1.134     1.488
3           vs 4        |  -0.0596   -3.445    0.001     0.942     0.829
4           vs 1        |   0.2543    7.203    0.000     1.290     2.228
4           vs 2        |   0.1858    7.217    0.000     1.204     1.795
4           vs 3        |   0.0596    3.445    0.001     1.061     1.207
-----------------------------------------------------------------------

Variable: maritals (sd=0.499)
-----------------------------------------------------------------------
                        |        b        z    P>|z|       e^b   e^bStdX
------------------------+----------------------------------------------
1           vs 2        |  -0.2379   -2.224    0.026     0.788     0.888
1           vs 3        |  -0.6757   -4.160    0.000     0.509     0.714
1           vs 4        |  -0.8827   -4.204    0.000     0.414     0.644
2           vs 1        |   0.2379    2.224    0.026     1.269     1.126
```

2	vs 3		-0.4378	-4.257	0.000	0.645	0.804
2	vs 4		-0.6448	-4.242	0.000	0.525	0.725
3	vs 1		0.6757	4.160	0.000	1.965	1.401
3	vs 2		0.4378	4.257	0.000	1.549	1.244
3	vs 4		-0.2070	-2.622	0.009	0.813	0.902
4	vs 1		0.8827	4.204	0.000	2.417	1.554
4	vs 2		0.6448	4.242	0.000	1.906	1.380
4	vs 3		0.2070	2.622	0.009	1.230	1.109

Variable: age (sd=17.432)

			b	z	P>|z|	e^b	e^bStdX
1	vs 2		0.0092	2.357	0.018	1.009	1.173
1	vs 3		0.0260	5.217	0.000	1.026	1.574
1	vs 4		0.0340	5.641	0.000	1.035	1.809
2	vs 1		-0.0092	-2.357	0.018	0.991	0.852
2	vs 3		0.0169	5.364	0.000	1.017	1.342
2	vs 4		0.0248	5.843	0.000	1.025	1.542
3	vs 1		-0.0260	-5.217	0.000	0.974	0.635
3	vs 2		-0.0169	-5.364	0.000	0.983	0.745
3	vs 4		0.0080	3.060	0.002	1.008	1.149
4	vs 1		-0.0340	-5.641	0.000	0.967	0.553
4	vs 2		-0.0248	-5.843	0.000	0.975	0.649
4	vs 3		-0.0080	-3.060	0.002	0.992	0.870

Variable: male (sd=0.497)

			b	z	P>|z|	e^b	e^bStdX
1	vs 2		-0.0143	-0.274	0.784	0.986	0.993
1	vs 3		-0.0405	-0.274	0.784	0.960	0.980
1	vs 4		-0.0529	-0.273	0.785	0.948	0.974
2	vs 1		0.0143	0.274	0.784	1.014	1.007
2	vs 3		-0.0263	-0.273	0.785	0.974	0.987
2	vs 4		-0.0387	-0.272	0.786	0.962	0.981
3	vs 1		0.0405	0.274	0.784	1.041	1.020
3	vs 2		0.0263	0.273	0.785	1.027	1.013
3	vs 4		-0.0124	-0.269	0.788	0.988	0.994
4	vs 1		0.0529	0.273	0.785	1.054	1.027
4	vs 2		0.0387	0.272	0.786	1.039	1.019
4	vs 3		0.0124	0.269	0.788	1.012	1.006

The first table in the output displays the odds ratios for four predictor variables comparing the base category 4 with category 1 (labeled as `Odds of: 4 vs 1` on the top). ORs (labeled as `e^b`) for `educ`, `maritals`, `age`, and `male` are 2.228, 1.554, .553, and 1.027, respectively.

The next four tables display the odds ratios comparing any of the two categories for the four predictor variables with one table for each of them. We are interested in the odds ratios comparing a particular category (1, 2, or 3) with the base category 4, or the odds ratios comparing the base category 4 with any other category.

Just as with the single-predictor SL model, the odds ratios comparing a particular category with the base category are the exponentiated $(-\phi_j\beta)$, whereas the odds ratios comparing the base category 4 with any other category are the exponentiated $(\phi_j\beta)$. For each of the four predictor variables, we obtain the odds of being in the baseline category J versus any other category. For example, for the `educ` predictor, the odds ratio of being in base category 4 compared with category 1, OR (4, 1) = $e^{(1 \times .254)}$ = 1.290; the odds ratio of being in base category 4 compared with category 2, OR(4, 2) = $e^{(.731 \times .254)}$ = $e^{(.186)}$ = 1.204; and the odds ratio of being in base category 4 compared with category 3, OR(4, 3) = $e^{(.235 \times .254)}$ = $e^{(.060)}$ = 1.061. The odds ratios for the other two predictors can be calculated in a similar way. Summarizing the results directly from the output obtained with the `listcoef, expand` command, Table 8.2 displays the odds comparing the base category and the other categories for all four predictor variables.

Table 8.2 Odds Ratios for All Four Predictor Variables Across Three Comparisons $(Y = J$ vs. $Y = j)$

Category Comparisons	$Y = 4$ vs. $Y = 1$	$Y = 4$ vs. $Y = 2$	$Y = 4$ vs. $Y = 3$
Variables	OR	OR	OR
Educ	1.290**	1.204**	1.061**
Maritals	2.417**	1.906**	1.230**
Age	.967**	.976**	.992**
Male	1.054	1.040	1.013

$**p < .01.$

Interpreting the Odds Ratios of Being in the Base Category Versus a Particular Category

The odds ratios in the SL model can be interpreted in a similar way as other logistic regression models except that the SL model compares different categories. Just as with binary logistic regression models, a positive logit regression coefficient in the SL model corresponds to an odds ratio greater than 1, a negative coefficient is associated with an odds ratio less than 1, and a coefficient of 0 corresponds to an odds ratio of 1.

The odds ratio can be interpreted as the change in the odds of being in the baseline category versus any other category for a one-unit increase in the predictor variable when holding other predictors constant. For each predictor variable, there are $J - 1$

comparisons, so there are $J - 1$ odds ratios. Among four predictor variables, three of them (all but male) are statistically significant across three comparisons.

In terms of odds ratio, the odds of being in the base category 4 versus category 1 increase by a factor of 1.290 with a one-unit increase in educ. For the predictor maritals, OR = 2.417, which indicates that the odds of being in the base category 4 versus category 1 for the married are 2.417 times the odds for the unmarried when holding all the other predictors constant. However, for age, OR = .967, which is less than 1, indicating that the odds decrease by a factor of .967 for a one-unit increase in age when holding the effects of the other variables constant. In addition, for male, OR = 1.054, p = .785, which is not significant, indicating that there is no relationship between the predictor and the odds.

The odds ratio for each predictor also needs to be interpreted across three comparisons. For example, regarding educ, the odds ratios of being in category 4 versus category 1, category 4 versus category 2, and category 4 versus category 3 are 1.290, 1.204, and 1.061, respectively. Therefore, for each one-unit increase in educ, the increases in the odds across three comparisons are 1.290, 1.204, and 1.061, respectively. The odds ratios for the other three variables can be interpreted in a similar way.

Interpreting the Odds Ratios of Being in a Particular Category Versus the Base Category

The odds ratios comparing a particular category versus the base category can be obtained by exponentiating $(-\phi_j\beta)$, which are the inverse of the odds ratios above comparing the base category 4 with any other category. The output of the listcoef command with the expand option also provides odds ratios comparing categories in the opposite direction, that is, a particular category versus the base category. Table 8.3 summarizes the odds ratios for all predictor variables across three comparisons.

Table 8.3   Odds Ratios for All Four Predictor Variables Across Three Comparisons ($Y = j$ vs. $Y = J$)

Category Comparisons	$Y = 1$ vs. $Y = 4$	$Y = 2$ vs. $Y = 4$	$Y = 3$ vs. $Y = 4$
Variables	OR	OR	OR
Educ	0.776**	0.831**	0.942**
Maritals	0.414**	0.525**	0.813**
Age	1.035**	1.025**	1.008**
Male	0.948	0.962	0.987

**p < .01.

For educ, the odds ratios of being in category 1 versus category 4, category 2 versus category 4, and category 3 versus category 4 are .776, .831, and .942, respectively. Therefore, for each one-unit increase in educ, the odds across the three comparisons decrease by .776, .831, and .942, respectively. Similarly, for each one-unit increase in maritals, the odds across three comparisons decrease by .414, .525, and .813, respectively. On the other hand, for each one-unit increase in age, the odds across three comparisons increase by 1.035, 1.025, and 1.008, respectively. In addition, the odds ratios for males are still close to 1 and are not statistically significant.

The postestimation command estat summarize provides the descriptive statistics for predictor variables in the model. The results are displayed as follows.

```
. estat summarize

  Estimation sample slogit              Number of obs =    1300

  -----------------------------------------------------------------
      Variable |        Mean     Std. Dev.       Min         Max
  -------------+---------------------------------------------------
      healthre |    2.928462      .854435          1           4
         educ  |    13.51308     3.150666          0          20
      maritals |    .4684615     .4991964          0           1
          age  |        48.2      17.4318         18          89
         male  |    .4438462     .4970279          0           1
  -----------------------------------------------------------------
```

Another postestimation command estat ic computes the AIC and BIC statistics, which are displayed in the following output.

```
. estat ic

  Akaike's information criterion and Bayesian information criterion

  -----------------------------------------------------------------------------
      Model |     Obs    ll(null)   ll(model)      df          AIC          BIC
  ----------+------------------------------------------------------------------
    _fsest_0 |    1300   -1579.407   -1508.956       9     3035.912     3082.443
  -----------------------------------------------------------------------------
               Note:  N=Obs used in calculating BIC; see [R] BIC note
```

The AIC and BIC statistics are 3,035.912 and 3,082.443, respectively, which are equal to those reported as AIC and BIC used by Stata if you use the following fitstat command. The fit statistics for the full model are displayed as follows.

```
. fitstat

                              |      slogit
------------------------------+-------------
Log-likelihood                |
                    Model |     -1508.956
            Intercept-only |     -1579.407
------------------------------+-------------
Chi-square                    |
        Deviance (df=1291) |      3017.912
             Wald (df=4) |        76.919
                 p-value |         0.000
------------------------------+-------------
R2                            |
                  McFadden |         0.045
       McFadden (adjusted) |         0.039
            Cox-Snell/ML |         0.103
   Cragg-Uhler/Nagelkerke |         0.113
------------------------------+-------------
IC                            |
                     AIC |      3035.912
          AIC divided by N |         2.335
             BIC (df=9) |      3082.443
```

Compared with the single-variable SL model, both AIC and AIC divided by N indicate that the full model fits the data better. The AIC and AIC divided by N are 3,035.912 and 2.335, respectively, for the full model, and those for the single model are 3,086.683 and 2.374. This result is also supported by the model comparison using the BIC statistic.

Interpreting the Estimated Probabilities
With the margins Command in Stata 13

The margins command works with stereotype logistic regression. By using this command, we can estimate the adjusted margins or the predicted probabilities for each category of the ordinal response variable at the specified values of the predictor variables. The command margins, predict (outcome(1)) at(educ = (8 13 16)) atmeans vsquish tells Stata to compute the estimated probabilities when the ordinal response variable $Y = 1$ for the predictor variable educ at the values of 8, 13, and 16 when holding the other predictor variables at their means. The predict (outcome(1)) option estimates probability for the first category of the ordinal response variable $Y = 1$. The at(educ = (8 13 16)) option specifies the values for the predictor variable educ. The vsquish option makes the output compact.

```
. margins, predict (outcome(1)) at(educ = (8 13 16)) atmeans vsquish

Adjusted predictions                                    Number of obs    =      1300
Model VCE      : OIM

Expression    : Pr(healthre==1), predict(outcome(1))
1._at         : educ              =            8
                maritals          =    .4684615  (mean)
                age               =        48.2  (mean)
                male              =    .4438462  (mean)
2._at         : educ              =           13
                maritals          =    .4684615  (mean)
                age               =        48.2  (mean)
                male              =    .4438462  (mean)
3._at         : educ              =           16
                maritals          =    .4684615  (mean)
                age               =        48.2  (mean)
                male              =    .4438462  (mean)

------------------------------------------------------------------------------
              |            Delta-method
              |    Margin   Std. Err.      z    P>|z|    [95% Conf. Interval]
--------------+---------------------------------------------------------------
         _at  |
           1  |  .1228388    .016475     7.46   0.000    .0905484    .1551292
           2  |  .0564793   .0070214     8.04   0.000    .0427177    .0702409
           3  |  .0329162   .0058851     5.59   0.000    .0213816    .0444508
------------------------------------------------------------------------------
```

When educ = 8, and the other predictor variables are held at their means (maritals = .468, age = 48.2, and male = .444), the estimated margin or the probability (labeled as Margin) for $Y = 1$ is .123.

When educ = 13, and the other three predictor variables are held at their means, the estimated probability for $Y = 1$ is .056.

When educ = 16, and the other predictor variables are held at their means, the estimated probability for $Y = 1$ is .033.

All three predicted margins or probabilities are significantly different from 0 since $p < .001$. They are plotted using the marginsplot, name(splot1) command. Figure 8.1 shows the estimated probabilities when $Y = 1$ (i.e., poor health condition).

The graph shows that with the increase of the years of education, the probability of being in poor health condition (category 1) decreases. In other words, people with higher levels of education are less likely associated with poor health conditions.

The expected probabilities for other categories (i.e., 2, 3, and 4) of the ordinal response variable can be computed in a similar way. For example, the margins

Figure 8.1   Estimated Probabilities When $Y = 1$ for educ at 8, 13, and 16

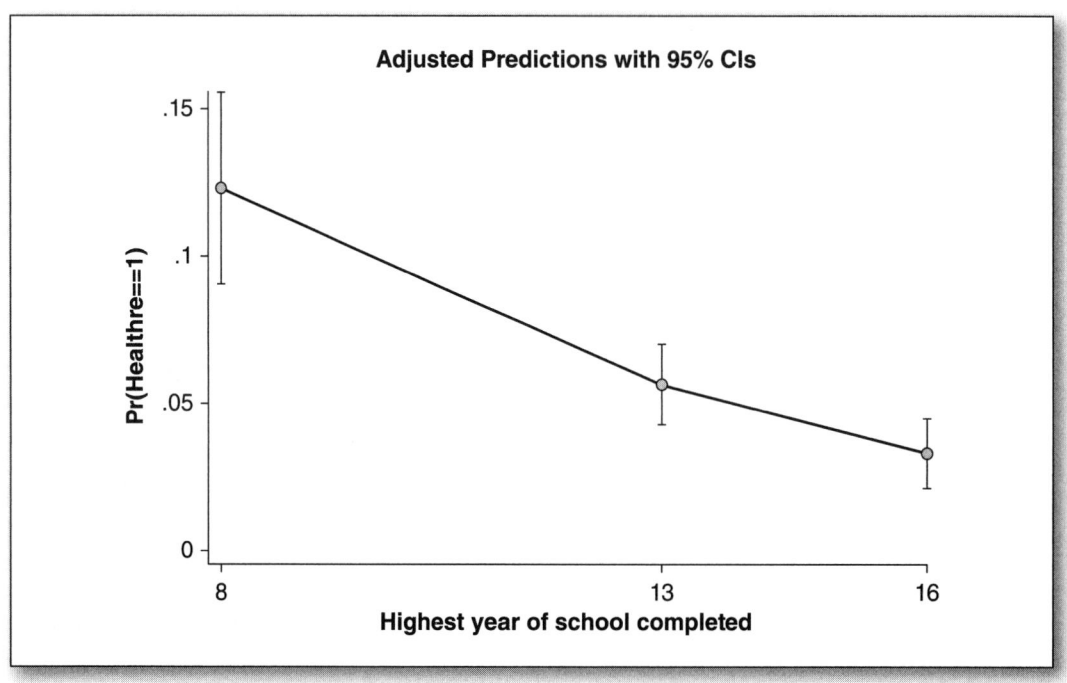

Note: CI = confidence interval.

command with the option `predict (outcome(4))` computes the estimated probability for the fourth category of the ordinal response variable. To save space, the output for the expected probabilities for categories 2 and 3 is omitted. The following output shows the results for the expected probability for category 4.

```
. margins, predict (outcome(4)) at(educ = (8 13 16)) atmeans vsquish

Adjusted predictions                                    Number of obs   =        1300
Model VCE    : OIM

Expression   : Pr(healthre==4), predict(outcome(4))
1._at        : educ            =           8
               maritals        =    .4684615  (mean)
               age             =        48.2  (mean)
               male            =    .4438462  (mean)
2._at        : educ            =          13
               maritals        =    .4684615  (mean)
               age             =        48.2  (mean)
               male            =    .4438462  (mean)
```

```
3._at            : educ              =           16
                   maritals          =    .4684615 (mean)
                   age               =        48.2 (mean)
                   male              =    .4438462 (mean)

--------------------------------------------------------------------------------
                 |             Delta-method
                 |     Margin   Std. Err.        z    P>|z|    [95% Conf. Interval]
-----------------+--------------------------------------------------------------
             _at |
              1  |   .1550568   .0175055      8.86    0.000     .1207466    .189367
              2  |    .254201   .0130575     19.47    0.000     .2286087   .2797933
              3  |   .3176714   .0153544     20.69    0.000     .2875774   .3477654
```

When educ = 8, 13, and 16, and other predictor variables are held at their means, the margins or estimated probabilities for category 4 are .155, .254, and .318, respectively. They are all statistically different from zero. To visualize the estimated probabilities of being in category 4, the command marginsplot, name(splot4) is used next. Figure 8.2 shows the estimated probabilities when $Y = 4$ (i.e., excellent health condition).

Figure 8.2   Estimated Probabilities When $Y = 4$ for educ at 8, 13, and 16

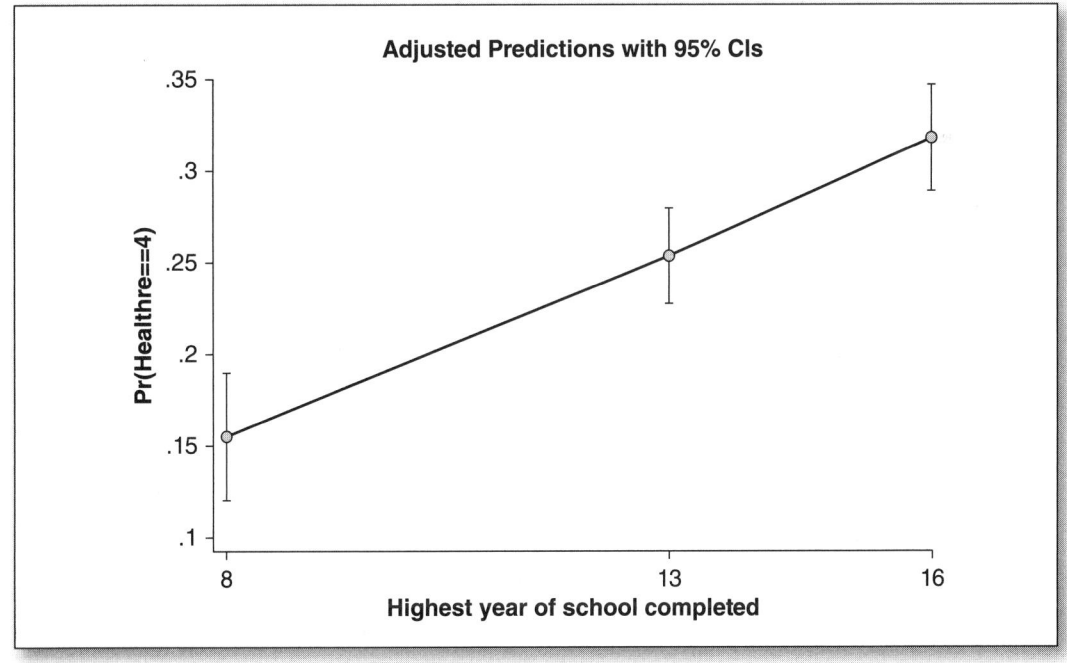

The graph shows that with the increase of the years of education, the probabilities of being in excellent health condition (category 4) increase. In other words, people with a higher level of education are more likely to be in excellent health condition.

By using the graph combine command after each separate graph is drawn, we can plot the estimated probabilities of being in each category (i.e., $Y = 1, 2, 3,$ and 4) in a single graph (Figure 8.3). The command graph combine splot1 splot2 splot3 splot4, ycommon combines all four plots, named splot1 to splot4, which are created by the marginsplot command. The detailed commands are provided as follows.

```
margins, predict (outcome(1)) at(educ = (8 13 16)) atmeans vsquish
marginsplot, name(splot1)
margins, predict (outcome(2)) at(educ = (8 13 16)) atmeans vsquish
marginsplot, name(splot2)
margins, predict (outcome(3)) at(educ = (8 13 16)) atmeans vsquish
marginsplot, name(splot3)
margins, predict (outcome(4)) at(educ = (8 13 16)) atmeans vsquish
marginsplot, name(splot4)
graph combine splot1 splot2 splot3 splot4, ycommon
```

Computing the Estimated Probabilities
With the Improved margins Command in Stata 14

As explained in Chapter 4, the updated margins command in Stata 14 simultaneously provides the results of the estimated probabilities for each category of an ordinal response variable instead of running the program multiple times with the separate results in Stata 13.

To replicate the results in the preceding section, the command margins, at(educ = (8 13 16)) atmeans vsquish tells Stata 14 to compute the estimated probabilities for $Y = 1, 2, 3,$ and 4 for the predictor variable educ at the values of 8, 13, and 16 when holding the other predictor variables at their means. The output is shown as follows.

```
. *Margins and marginsplot in Stata 14
. quietly slogit healthre educ i.maritals age i.male, dim(1)
. margins, at(educ = (8 13 16)) atmeans vsquish

Adjusted predictions                              Number of obs    =    1,300
Model VCE     : OIM

1._predict   : Pr(healthre==1), predict(pr outcome(#1))
2._predict   : Pr(healthre==2), predict(pr outcome(#2))
3._predict   : Pr(healthre==3), predict(pr outcome(#3))
4._predict   : Pr(healthre==4), predict(pr outcome(#4))
```

```
1._at          : educ          =              8
                 0.maritals    =      .5315385  (mean)
                 1.maritals    =      .4684615  (mean)
                 age           =          48.2  (mean)
                 0.male        =      .5561538  (mean)
                 1.male        =      .4438462  (mean)
2._at          : educ          =             13
                 0.maritals    =      .5315385  (mean)
                 1.maritals    =      .4684615  (mean)
                 age           =          48.2  (mean)
                 0.male        =      .5561538  (mean)
                 1.male        =      .4438462  (mean)
3._at          : educ          =             16
                 0.maritals    =      .5315385  (mean)
                 1.maritals    =      .4684615  (mean)
                 age           =          48.2  (mean)
                 0.male        =      .5561538  (mean)
                 1.male        =      .4438462  (mean)
```

		Delta-method				
	Margin	Std. Err.	z	P>|z|	[95% Conf. Interval]	
_predict#_at						
1 1	.1228388	.016475	7.46	0.000	.0905484	.1551292
1 2	.0564793	.0070214	8.04	0.000	.0427177	.0702409
1 3	.0329162	.0058851	5.59	0.000	.0213816	.0444508
2 1	.3326184	.0249069	13.35	0.000	.2838018	.381435
2 2	.2154176	.0120607	17.86	0.000	.191779	.2390563
2 3	.1541953	.0129068	11.95	0.000	.1288984	.1794923
3 1	.3894861	.0261481	14.90	0.000	.3382367	.4407354
3 2	.4739021	.0147497	32.13	0.000	.4449932	.5028109
3 3	.495217	.0162006	30.57	0.000	.4634645	.5269696
4 1	.1550568	.0175055	8.86	0.000	.1207466	.189367
4 2	.254201	.0130575	19.47	0.000	.2286087	.2797933
4 3	.3176714	.0153544	20.69	0.000	.2875774	.3477654

The `margins` command is executed immediately after the multiple-predictor SL model is fitted. In the output, the notes for the predicted probabilities for each of the four ordinal categories (labeled from 1._predict to 4._predict) and three combinations of the predictor variables (labeled from 1._at to 3._at) are listed first. The table for the margins or estimated probabilities is displayed at the bottom. As we can see, the results produced by Stata 14 are the same as those combined in the previous section using Stata 13.

Improved `marginsplot` Command in Stata 14

The improved `marginsplot` command in Stata 14 now can automatically plot the estimated probabilities of being in each category of an ordinal response variable

Figure 8.3 Estimated Probabilities of Being in Categories 1, 2, 3, and 4 for educ

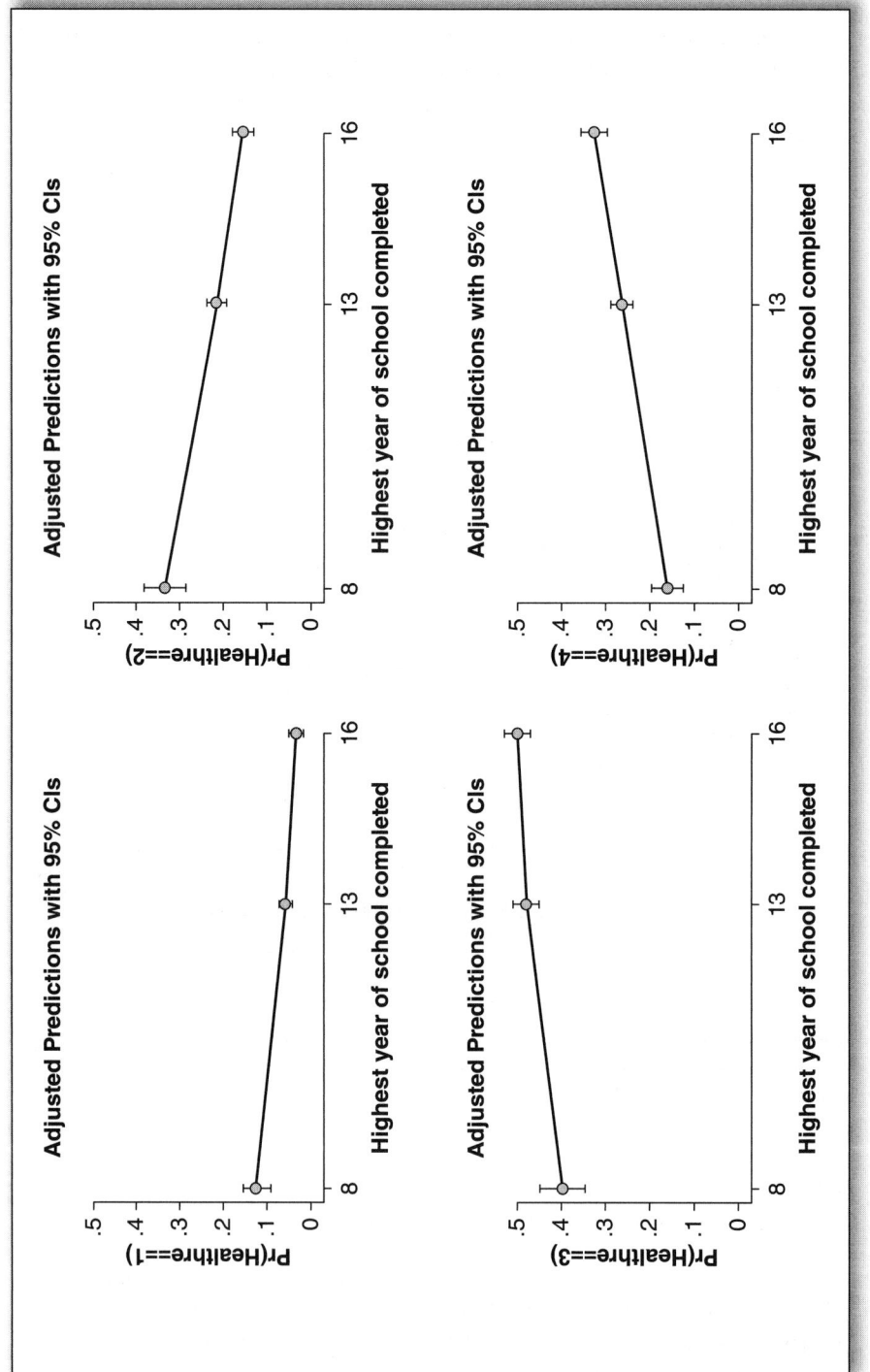

Note: CI = confidence interval.

in a single graph. Figure 8.4, which is produced with the `marginsplot` command in Stata 14, displays the estimated probabilities when $Y = 1$, 2, 3, and 4.

Figure 8.4 is an improved graph of Figure 8.3. Both figures tell us that with the increase of the highest year of school completed, the probabilities of being in categories 1 and 2 decrease, whereas the probabilities of being in categories 3 and 4 increase.

Figure 8.4 Estimated Probabilities of Being in Categories 1, 2, 3, and 4 for educ With Stata 14

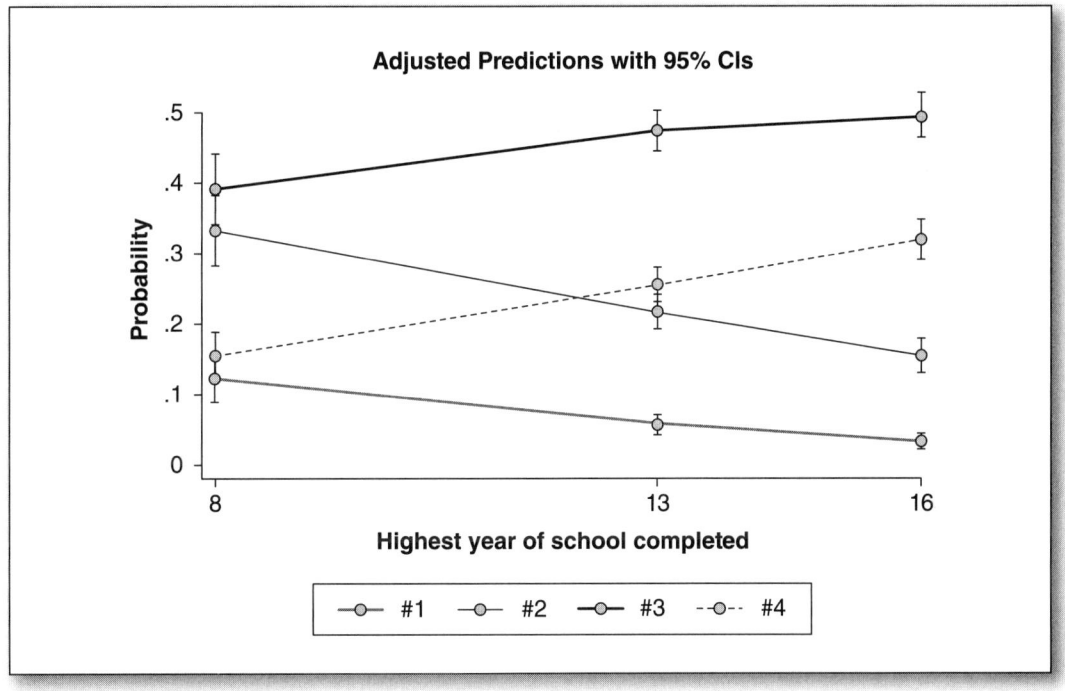

Note: CI = confidence interval.

Computing the Estimated Probabilities
With the `mtable` Command in Stata 13 and Later

To replicate the results in the preceding section, we can use `mtable` (Long & Freese, 2014) to compute simultaneously the estimated probabilities for all categories. This command works in both Stata 13 and 14. The output displayed as follows is more concise and readable than that by the `margins` command in Stata 14.

```
. mtable, at(educ = (8 13 16)) atmeans ci

Expression: Pr(healthre), predict(outcome())

        |     educ         1         2         3         4
--------+--------------------------------------------------
 Pr(y)  |        8     0.123     0.333     0.389     0.155
    ll  |        8     0.091     0.284     0.338     0.121
    ul  |        8     0.155     0.381     0.441     0.189
 Pr(y)  |       13     0.056     0.215     0.474     0.254
    ll  |       13     0.043     0.192     0.445     0.229
    ul  |       13     0.070     0.239     0.503     0.280
 Pr(y)  |       16     0.033     0.154     0.495     0.318
    ll  |       16     0.021     0.129     0.463     0.288
    ul  |       16     0.044     0.179     0.527     0.348

Specified values of covariates

          |        1.                  1.
          |   maritals       age      male
----------+----------------------------------
  Current |       .468      48.2      .444
```

8.3.3 Model Comparisons Using the Log Likelihood Ratio Test

The log likelihood ratio test, or the deviance difference test, is used to compare the full model and the one-predictor model. We fit the single-predictor model and save the estimated log likelihood first, and then we fit the full model and save the estimated log likelihood. Finally, we compare the two models using the log likelihood ratio test. The following output is displayed.

```
. quietly slogit healthre educ, dim(1)
. estimates store mod1

. quietly slogit healthre educ maritals age male, dim(1)

. estimates store mod2

. lrtest mod1 mod2

Likelihood-ratio test                            LR chi2(3)  =      56.77
(Assumption: mod1 nested in mod2)                Prob > chi2 =     0.0000
```

The log likelihood ratio chi-square test, $\chi^2_{(3)} = 56.77$, $p < .001$, which indicates that the full model with four predictor variables fits the data better than the single-predictor model.

8.4 Making Publication-Quality Tables

```
1.  . quietly slogit healthre educ, dim(1)
2.  . outreg2 using chpt8out, e(ll chi2) dec(3) long word replace
3.  . outreg2 using chpt8out, eform dec(3) long word append
4.  . quietly slogit healthre educ maritals age male, dim(1)
5.  . outreg2 using chpt8out, e(ll chi2) dec(3) long word append
6.  . outreg2 using chpt8out, eform dec(3) long word append
    chpt8out.rtf
    dir : seeout
```

The first command `quietly slogit healthre edu, dim(1)` estimates the single-predictor SL model without providing the output. The second command `outreg2 using chpt8out, e(ll chi2) dec(3) long word replace` tells Stata to create a regression table for the estimated coefficients and save it to a file named chpt8out. The option `e(ll chi2)` requests the estimated log likelihood value and the chi-square test statistic. The `dec(3)` option requests three decimal places for all statistics. The `word` option requests that the table be created in a Word document with the extension .rtf. The `long` option lists equation names and variable names in one column. The `replace` option asks Stata to replace any files with the same name as chpt8out.

The remaining three commands repeat the same process for the full model with four predictor variables, so the final table contains the estimated coefficients and odds ratios for both the single-predictor and the full model with all four predictor variables.

Clicking the blue link named `chpt8out.rtf` in the Stata results window opens the document containing the created table, which is saved in the current working directory. Clicking the other blue link named `dir: seeout` opens the final table in the data browser.

These six commands automatically produce Table 8.4, as shown here in its original format, presenting the results of both the single-predictor and the full stereotype logistic regression model.

Table 8.4 Results of the Stereotype Logistic Regression Models: Single-Predictor Model and Full Model (Shown in Original Format Generated by Stata)

Variables	(1)	(2)	(3)	(4)
dim1				
Educ	0.252***	1.287***	0.254***	1.290***
	(0.038)	(0.049)	(0.035)	(0.046)
Maritals			0.883***	2.417***
			(0.210)	(0.508)
Age			−0.034***	0.967***
			(0.006)	(0.006)
Male			0.053	1.054
			(0.194)	(0.205)
phi1_1				
Constant	1.000	2.718	1.000	2.718
	(0.000)	(0.000)	(0.000)	(0.000)
phi1_2				
Constant	0.750***	2.116***	0.731***	2.076***
	(0.119)	(0.252)	(0.087)	(0.181)
phi1_3				
Constant	0.139	1.149	0.235***	1.264***
	(0.086)	(0.098)	(0.062)	(0.079)
theta1				
Constant	1.850***	6.358***	0.600	1.821
	(0.494)	(3.140)	(0.522)	(0.951)
theta2				
Constant	2.286***	9.831***	1.371***	3.941***
	(0.378)	(3.721)	(0.398)	(1.570)
theta3				

Variables	(1)	(2)	(3)	(4)
Constant	1.025***	2.788***	1.116***	3.054***
	(0.333)	(0.927)	(0.209)	(0.639)
Observations	1,300	1,300	1,300	1,300
Ll	−1,537		−1,509	
chi2	43.09		76.92	

Note: Standard errors in parentheses.

***$p < 0.01$, **$p < 0.05$, *$p < 0.1$

After adding the model titles, the labels for logit coefficients and odds ratios, the names for the constraints and the intercepts on the original table, and the names of fit statistics, the final table is displayed as in Table 8.5.

Table 8.5 Results of the Stereotype Logistic Regression Models: Single-Predictor Model and Full Model (Edited)

Variables	Single-Predictor Model		Full Model	
	b (SE(b))	OR	b (SE(b))	OR
dim1				
Educ	0.252***	1.287***	0.254***	1.290***
	(0.038)	(0.049)	(0.035)	(0.046)
Maritals			0.883***	2.417***
			(0.210)	(0.508)
Age			−0.034***	0.967***
			(0.006)	(0.006)
Male			0.053	1.054
			(0.194)	(0.205)
phi1_1				
(ϕ_1)	1.000		1.000	

(Continued)

Table 8.5   (Continued)

Variables	Single-Predictor Model		Full Model	
	b (SE(b))	OR	b (SE(b))	OR
	(0.000)		(0.000)	
phi1_2				
(ϕ_2)	0.750***		0.731***	
phi1_3				
(ϕ_3)	0.139		0.235***	
	(0.086)		(0.062)	
theta1				
(α_1)	1.850***		0.600	
	(0.494)		(0.522)	
theta2				
(α_2)	2.286***		1.371***	
	(0.378)		(0.398)	
theta3				
(α_3)	1.025***		1.116***	
	(0.333)		(0.209)	
Observations	1,300	1,300	1,300	1,300
Log likelihood	−1,537		−1,509	
Wald χ^2	43.09		76.92	

Note: Standard errors in parentheses.

***$p < .01$.

8.5 Reporting the Results

Writing the results of stereotype logistic regression models is similar to other logistic regression models.

First, describe the statistical method you used for data analysis, the dependent variable and the independent variables in the models, and your research hypothesis or the purpose of your study. Explain why stereotype logistic regression is appropriate for the analysis. Second, report model fit statistics, such as the Wald chi-square statistic and the associated p value, followed by your interpretation of whether the fitted model is better than the null model. If more than one model is developed, compare model fit statistics, such as likelihood ratio test statistics, deviance statistics, and AIC and BIC statistics.

Third, report the parameter estimates for the predictor variables, their standard errors, the associated p values, and odds ratios either in a table or in the text. A table is preferable for models with multiple predictors. In addition, report the constraints or scale parameters and explain how they are used for computing the odds ratios comparing a particular category versus the base category. Furthermore, the odds ratios for each predictor should be interpreted. If more than one model is fitted, the results of all the competing models from the simple model to the full model should be presented in a table. The following is an example of summarizing the results for the stereotype logistic regression model.

The stereotype logistic regression analysis was conducted to estimate the ordinal outcome variable, health status, from a set of predictor variables, such as marital status, years of education, age, and gender. This model was used since it relaxed the proportional odds assumption. A single-predictor model with years of education as the predictor was fitted first, and then the full model with all predictors was fitted. The log likelihood ratio test or the deviance difference test was used to compare the two models, $\chi^2_{(3)} = 56.77$, $p < .001$. The result indicated that the full model fitted data better than the single-predictor model.

The odds ratios comparing the base category with a particular category are the exponentiated $(\phi_j\beta)$. Table 8.2 displays the odds comparing the base category and the other categories for all four predictor variables for the full model. The results for both the single-predictor model and the full model are presented in Table 8.5.

For the full model, Wald $\chi^2_{(4)} = 76.92$, $p < .001$, which indicated that at least one predictor variable in the full model was significant in predicting the ordinal response variable.

For the predictor, years of education, the odds ratios of being in category 4 versus category 1, category 4 versus category 2, and category 4 versus category 3 were 1.290, 1.204, and 1.061, respectively. Therefore, for each one-unit increase in the predictor variable, the increases in the odds across three comparisons were 1.290, 1.204, and 1.061, respectively, when holding other predictor variables constant.

Similarly, the odds of being in the base category versus a particular category across three comparisons for the married were 2.417, 1.906, and 1.230 times the odds for the unmarried. On the other hand, for each one-unit increase in age, the odds across three comparisons decreased by .967, .976, and .992, respectively. In addition, the odds ratios for males were still close to 1 and were not statistically significant ($p = .785$), which indicated that there was no relationship between the predictor and the odds.

8.6 Summary of Stata Commands in This Chapter

```
use chap8-gss2012, clear

*The one-predictor SL model
slogit healthre educ, dim(1)
listcoef, expand
fitstat
estat ic

*The multiple-predictor SL model
slogit healthre educ maritals age male, dim(1)
listcoef, expand
estat summarize
estat ic
fitstat

*Margins and marginsplot in Stata 13
margins, predict (outcome(1)) at(educ = (8 13 16)) atmeans vsquish
marginsplot, name(splot1)
margins, predict (outcome(2)) at(educ = (8 13 16)) atmeans vsquish
marginsplot, name(splot2)
margins, predict (outcome(3)) at(educ = (8 13 16)) atmeans vsquish
marginsplot, name(splot3)
margins, predict (outcome(4)) at(educ = (8 13 16)) atmeans vsquish
marginsplot, name(splot4)

*margins plots combined in Stata 13
graph combine splot1 splot2 splot3 splot4, ycommon

*Margins and marginsplot in Stata 14
quietly slogit healthre educ i.maritals age i.male, dim(1)
margins, at(educ = (8 13 16)) atmeans vsquish
marginsplot

*mtable in Stata 13 and later
mtable, at(educ = (8 13 16)) atmeans ci

*Model comparison using the log likelihood ratio test
quietly slogit healthre educ, dim(1)
estimates store mod1
quietly slogit healthre educ maritals age male, dim(1)
estimates store mod2
lrtest mod1 mod2

*Making publication-quality tables using outreg2
quietly slogit healthre educ, dim(1)
outreg2 using chpt8out, e(ll chi2) dec(3) long word replace
outreg2 using chpt8out, eform dec(3) long word append

quietly slogit healthre educ maritals age male, dim(1)
outreg2 using chpt8out, e(ll chi2) dec(3) long word append
outreg2 using chpt8out, eform dec(3) long word append
seeout
```

8.7 EXERCISES

Use the GSS 2012 data for the following problems.

1. Conduct an analysis for a stereotype logistic regression model, and estimate the ordinal response variable happy from the four predictor variables sex, educ, age, and satfin.

2. Identify the Wald chi-square test of the model, and interpret it.

3. List constraints or scale parameters.

4. In the regression table, identify the logit coefficient, the Wald z test, and the 95% confidence interval for the predictor variables age and satfin. Are they statistically significant? What categories are they comparing?

5. Use the listcoef, expand command to compute the odds ratios. List the odds ratios for satfin comparing the base category versus each of the other two categories, and interpret them.

6. Make a publication-quality table containing the estimated logit coefficients and odds ratios.

7. Write a report to summarize the results from the output.

CHAPTER 9

Ordinal Logistic Regression With Complex Survey Sampling Designs

Objectives of This Chapter

This chapter presents ordinal logistic regression models with complex survey sampling designs. It starts with an introduction to the features of complex sampling designs followed by discussions of variance estimation in complex survey sampling. After a description of the research example and the data, various types of models are built, including the proportional odds model without weights, the model with weights, and the model with complex survey sampling designs. In addition, methods are discussed to deal with the issue of single sampling units with strata. This chapter also illustrates how the results are displayed in publication-quality tables using the Stata command and reported in text. It focuses on fitting proportional odds models with complex survey sampling designs using Stata, as well as on interpreting and presenting the results. After reading this chapter, you should be able to

- Understand features of complex survey sampling and weights.
- Conduct an analysis for data with weights using Stata.
- Conduct an analysis for data with complex sampling designs using Stata.
- Use appropriate methods to deal with singleton strata.
- Interpret the output.
- Present results in publication-quality tables using Stata.
- Write the results for publication.

9.1 Proportional Odds Models With Complex Survey Sampling Designs: An Introduction

The conventional proportional odds (PO) model assumes that the data are collected by using simple random sampling by which each sampling unit has an equal probability of being selected from the population. However, in some situations, researchers have used more complex sampling designs rather than simple random sampling, such as stratified sampling and multistage cluster sampling. Here are several examples of international or national datasets with complex sampling survey designs: the Programme for International Student Assessment (PISA) surveys collected data of 15-year-old students in internationally participating countries; the National Opinion Research Center (NORC) administered the General Social Survey (GSS) to collect demographic and attitude data of Americans; the National Center for Education Statistics (NCES) sponsored and carried out a series of studies, such as the Early Childhood Longitudinal Study–Kindergarten (ECLS-K), the National Educational Longitudinal Study of 1988 (NELS:88), the Educational Longitudinal Study of 2002 (ELS:2002), and the High School Longitudinal Study of 2009 (HSLS:2009). These designs had the common features of using strata, clusters, and unequal probability of selection in data collection.

Since complex survey sampling designs involve the use of different strata (e.g., geographic areas), clustered sampling techniques, and unequal selection probabilities, ignoring these features in data analysis may lead to biased estimates of parameters and incorrect variance estimates. In other words, the parameters and their variances may be either overestimated or underestimated.

Although several strategies were commonly used for design-based analysis of survey data, using specialized software to accommodate complex sampling designs was the most desirable choice, and the examples of conducting various statistical analyses, such as multiple regression, ordinal logistic regression, and structural equation modeling, have been illustrated (Hahs-Vaughn, 2005, 2006; Hahs-Vaughn & Lomax, 2006; Liu & Koirala, 2013; Muthén & Satorra, 1995; Stapleton, 2002, 2006, 2008; Thomas & Heck, 2001).

9.1.1 Features of Complex Surveys

Sampling Weights

Weights in survey sampling reflect different probabilities of selection. If we know the probability that the number of observations in a selected sample represents those in the population, by taking the inverse of it, we can obtain the sampling weight. In other words, a sampling weight is just the inverse of the probability of sample selection (Lohr, 1999). For example, if the probability of a sample selection is 1/5, the sampling weight is 5. With multistage sampling, which will be introduced later in this chapter, the probability of the overall selection is the product of the probability of selection at each stage. For example, in a two-stage cluster sampling design, the probability of selection for schools is 1/10 and the one for classes is 1/5. The overall probability of the selection = 1/10 × 1/5 = 1/50. Taking the inverse of 1/50 gives us 50, which is the sampling weight. In addition, weights can be used for nonresponse adjustment so that the nonresponses are weighted with those respondents with similar characteristics. They also can be used for the poststratification adjustment so that the sample would be more representative of the population. In Stata, the pweight command, which is specified within a bracket, is used for sampling weights. For example, [pweight = BYSTUWT] specifies the weight variable BYSTUWT as the sampling weight.

Strata

In stratified sampling, researchers first identify strata or subgroups with different characteristics in the population, and then they conduct simple random sampling within each stratum. For example, strata could be different geographic locations, age-groups, group sizes, types of occupation, or levels of medical conditions. Stratification can be done either during research design or after the survey data are collected. The latter is known as the poststratification, which classifies the sample using strata or subgroups and adjusts weights for each group. In Stata, the Strata() option needs to be used when the complex survey designs are specified.

Clusters

Cluster sampling is used when researchers do not have access to a sampling frame of all units in the population. In cluster sampling, we first identify clusters or groups rather than individual subjects and then randomly select these clusters. For example, if we want to select a sample of high-school students in public schools in a state, we can randomly select school districts and then get a sample of high-school students from these school districts. Cluster sampling may involve more than two stages, which is known as multistage sampling. For example, counties are randomly selected first, and then school districts and classes are randomly selected sequentially. In Stata, the primary sampling units (PSUs) are referred to as the clusters.

Before conducting data analysis for complex survey data with the svy command, we need to use the svyset command to specify the complex survey features discussed. For example, in a dataset, the variable for clusters is psu, the strata variable is strata, and the variable for the probability weight is weight. We can specify these features by using the svyset command as follows:

```
svyset psu [pweight=weight], strata(strata)
```

In this syntax, the svyset command is followed by psu, the variable name for the primary sampling units or clusters in the data; [pweight=weight] specifies the probability weight weight; and the option strata(strata) is separated from the command by a comma. An example on how to use the command is provided in 9.3.3.

9.1.2 Variance Estimation in Complex Survey Sampling

Two techniques are widely used for unbiased variance estimation in complex sampling survey designs, including linearization and replicated sampling methods (Lee & Forthofer, 2006; Levy & Lemeshow, 2008; Lohr, 1999). The linearization method is the Taylor series approximation, also known as the delta method (Kalton, 1983), whereas the replicated methods estimate the variance of a parameter by generating replicated subsamples and examining the variability of the subsample estimates. The replicated methods, also referred to as the resampling methods, include the balanced repeated replication (BRR), the jackknife repeated replication (JRR), and the bootstrap method (Lee & Forthofer, 2006; Levy & Lemeshow, 2008). This chapter focuses on the Taylor series approximation method only since it is implemented as the default in Stata.

In statistics, the Taylor series linearization is used to get a linear approximation to the nonlinear function or statistic, and then the variance of the function or statistic can be derived from the Taylor series approximation. Specifically, the variance estimation in complex survey sampling using the Taylor series expansion follows two steps. First, use the first-order Taylor series to get a linear approximation of the function. Second, estimate the variance of the parameter including complex survey features, such as strata, cluster, and weight variables. Therefore, the variance estimate is based on the weighted variance across primary sampling units within a stratum (Lee & Forthofer, 2006). In most general-purpose statistical software packages, researchers need to specify strata, cluster, and weights before fitting a statistical model.

To estimate the sampling variance of the parameter estimate in ordinal logistic regression, Binder (1983) developed a general formula for linear regression and generalized linear models with the complex survey data using the Taylor series, which was widely used and known as the sandwich variance estimator. The technical details of variance estimation are beyond the scope of this chapter. Refer to Binder (1983) and Heeringa, West, and Berglund (2010) for more information on this topic.

In Stata, when a stratum only contains a single sampling unit, standard errors of the parameters are estimated to be missing. To deal with this issue, any of the three singleunit() options (i.e., certainty, scaled, and centered) can be specified separately in the svyset command syntax.

9.2 Research Example and Description of the Data and Sample

In the following research example, we will still investigate the relationships between the ordinal response variable, health status, and four predictor variables. However, the research interest focuses on three different proportional odds models: the model without considering weights, the model with weights, and the model with complex survey sampling designs.

The data for the following analyses were taken from the General Social Survey 2012 (GSS 2012). The same variables from previous chapters are used for data analysis in this chapter:

- healthre: the recoded variable of health (health status) with four ordinal categories (1 = poor health, 2 = fair health, 3 = good health, and 4 = excellent health)
- maritals: the recoded variable of marital (marital status) with 1 = currently married and 0 = not currently married
- educ: the highest education
- age: respondent's age
- male: recoded variable of sex with 1 = male and 0 = female
- wtss: weight variable
- vpsu: primary sampling unit
- vstrat: strata variable

9.3 Data Analysis With Stata: Commands and Output

9.3.1 Proportional Odds Model
With Four Explanatory Variables Without Weights

With the command xi: ologit healthre i.maritals age educ i.male, we predict the ordinal response variable healthre from four predictor variables by using a proportional odds model. Neither weights nor complex survey sampling designs are specified in the command. This model is referred to as the proportional odds model without weights. As noted, the prefix xi command can be omitted for Stata 13 and 14 users. The output is displayed as follows.

```
. xi: ologit healthre i.maritals age educ i.male
i.maritals         _Imaritals_0-1      (naturally coded; _Imaritals_0 omitted)
i.male             _Imale_0-1          (naturally coded; _Imale_0 omitted)

Iteration 0:    log likelihood = -1579.4068
Iteration 1:    log likelihood = -1513.1847
Iteration 2:    log likelihood = -1512.4793
Iteration 3:    log likelihood = -1512.4786
Iteration 4:    log likelihood = -1512.4786

Ordered logistic regression                    Number of obs   =       1300
                                                LR chi2(4)      =     133.86
                                                Prob > chi2     =     0.0000
Log likelihood = -1512.4786                     Pseudo R2       =     0.0424

------------------------------------------------------------------------------
    healthre |      Coef.   Std. Err.      z    P>|z|     [95% Conf. Interval]
-------------+----------------------------------------------------------------
 _Imaritals_1 |   .5140058   .1058117     4.86   0.000     .3066187    .721393
         age |  -.0193069   .0030489    -6.33   0.000    -.0252826   -.0133313
        educ |   .1344542   .0172594     7.79   0.000     .1006265    .168282
   _Imale_1 |   .0530645    .104808     0.51   0.613    -.1523553   .2584843
-------------+----------------------------------------------------------------
       /cut1 |  -1.701386   .2976461                     -2.284762   -1.11801
       /cut2 |   .1246361   .2863089                     -.4365191   .6857913
       /cut3 |   2.257928   .2936829                      1.68232   2.833536
------------------------------------------------------------------------------
```

To request the odds ratios, we use the command $xi:$ $ologit$ with the or option. The output is shown as follows.

```
. xi: ologit, or

Ordered logistic regression                    Number of obs   =       1300
                                                LR chi2(4)      =     133.86
                                                Prob > chi2     =     0.0000
Log likelihood = -1512.4786                     Pseudo R2       =     0.0424

------------------------------------------------------------------------------
    healthre | Odds Ratio  Std. Err.      z    P>|z|     [95% Conf. Interval]
-------------+----------------------------------------------------------------
 _Imaritals_1 |   1.671975   .1769146     4.86   0.000     1.358823   2.057297
         age |   .9808782   .0029906    -6.33   0.000     .9750343   .9867572
        educ |   1.143912   .0197432     7.79   0.000     1.105863    1.18327
   _Imale_1 |   1.054498   .1105197     0.51   0.613     .8586831   1.294966
-------------+----------------------------------------------------------------
       /cut1 |  -1.701386   .2976461                     -2.284762   -1.11801
       /cut2 |   .1246361   .2863089                     -.4365191   .6857913
       /cut3 |   2.257928   .2936829                      1.68232   2.833536
------------------------------------------------------------------------------
```

Interpreting the Output

The log likelihood ratio chi-square test is *LR* $\chi^2_{(4)}$ = 133.86, *p* < .001, which indicates that the proportional odds model with four predictor variables has a better fit than the null model with no independent variables.

The odds ratios reported in the output compare the odds of being beyond a particular category versus being at or below that category. They are interpreted as the change (increase or decrease) in the odds associated with a one-unit increase in the predictor variable when holding other predictors constant. The interpretation for each predictor variable is omitted here since detailed interpretations are presented in Chapter 4.

9.3.2 Proportional Odds Model With Weights

Next, the same PO model with weights is fitted. To fit a PO model with weights only, the svyset command should not be used since we do not specify complex survey sampling designs. In this example, we only need to specify the probability weight wtss, which is the weight variable for GSS 2012 data, in the syntax as [pweight = wtss]. The result for the PO model with the estimation of weights is displayed as follows.

```
. xi:ologit healthre i.maritals age educ i.male [pw=wtss]
i.maritals          _Imaritals_0-1      (naturally coded; _Imaritals_0 omitted)
i.male              _Imale_0-1          (naturally coded; _Imale_0 omitted)

Iteration 0:    log pseudolikelihood = -1554.4855
Iteration 1:    log pseudolikelihood = -1494.0283
Iteration 2:    log pseudolikelihood = -1493.3735
Iteration 3:    log pseudolikelihood = -1493.3729
Iteration 4:    log pseudolikelihood = -1493.3729

Ordered logistic regression                     Number of obs   =       1300
                                                Wald chi2(4)    =      86.64
                                                Prob > chi2     =     0.0000
Log pseudolikelihood = -1493.3729               Pseudo R2       =     0.0393

-----------------------------------------------------------------------------
             |               Robust
    healthre |    Coef.    Std. Err.      z    P>|z|     [95% Conf. Interval]
-------------+---------------------------------------------------------------
 _Imaritals_1|   .6091842   .1240911     4.91   0.000     .3659702    .8523983
         age |  -.0224671   .0036419    -6.17   0.000    -.029605   -.0153292
        educ |   .1119547   .0227436     4.92   0.000     .067378    .1565314
    _Imale_1 |   .1230133   .1284353     0.96   0.338    -.1287152    .3747418
-------------+---------------------------------------------------------------
```

```
        /cut1 |   -2.103854    .372957                          -2.834836   -1.372872
        /cut2 |   -.3162062    .3617395                         -1.025203    .3927902
        /cut3 |    1.834523    .3732973                          1.102873    2.566172
----------------------------------------------------------------------------------

. xi:ologit, or

Ordered logistic regression                     Number of obs   =        1300
                                                Wald chi2(4)    =       86.64
                                                Prob > chi2     =      0.0000
Log pseudolikelihood = -1493.3729               Pseudo R2       =      0.0393


---------------------------------------------------------------------------------
             |               Robust
    healthre |  Odds Ratio   Std. Err.      z    P>|z|     [95% Conf. Interval]
-------------+-------------------------------------------------------------------
 _Imaritals_1 |   1.838931    .2281949    4.91   0.000     1.441912    2.345265
         age |    .9777834    .0035609   -6.17   0.000     .9708289    .9847877
        educ |   1.118462     .0254379    4.92   0.000       1.0697    1.169447
    _Imale_1 |   1.130899     .1452474    0.96   0.338     .8792243    1.454616
-------------+-------------------------------------------------------------------
        /cut1 |   -2.103854    .372957                          -2.834836   -1.372872
        /cut2 |   -.3162062    .3617395                         -1.025203    .3927902
        /cut3 |    1.834523    .3732973                          1.102873    2.566172
---------------------------------------------------------------------------------
```

Interpreting the Output

The PO model with sampling weights uses the pseudo-likelihood instead of the true likelihood in the maximum likelihood estimation. The Wald chi-square test is $\chi^2_{(4)} = 84.64$, $p < .001$, which indicates that the model with the four predictor variables is significant in predicting the ordinal response variable. The pseudo $R^2 = .039$.

The regression table reports the coefficients, the robust standard errors, the Wald z statistics, the associated p values, and the 95% confidence intervals. Among the four predictor variables, the coefficients of three variables are significantly different from zero. The coefficients for maritals and educ are .609 and .112, respectively, which are positive and indicate that the log odds of being above a particular category of health status versus being at or below that category increases with the increase of the value in each of these two predictor variables. The coefficient of age is −.022, which indicates that with the increase in age, the log odds of being above a particular category of health status decreases. The coefficient for male is .123 and is not significant ($z = .96$, $p = .338$).

The odds ratios are more interpretable than the logit coefficients. They are reported with the or option immediately after the original xi:ologit command.

For the maritals predictor, OR = 1.839, which is greater than 1. This indicates that the odds of being above a particular category of health status (i.e., being in better health status) versus being at or below that category for the married are

1.839 times as large as the odds for the unmarried when holding all the other predictors constant.

For the `age` predictor, OR = .978, which is less than 1. This indicates that the odds of being above a particular category of health status decrease by a factor of .978 for a one-unit increase in the predictor `age` when holding all the other predictors constant. In other words, for a one-unit increase in age, the odds of being healthier decrease by 2.2%.

For the `educ` predictor, OR = 1.118, which is greater than 1. This indicates that the odds of being above a particular category of health status increase by 11.8% for a one-unit increase in the predictor `educ` when holding other predictors constant.

For the `male` predictor, $\beta = .123$, $z = .96$, $p = .338$, which is not significantly different from 0; and OR = 1.131, which is not significantly different from 1. This indicates that there is no relationship between being a male and the cumulative of being in better health status.

Table 9.1 presents a comparison of PO model results with and without weighted estimation. Compared with the unweighted PO model, all the cut points/intercepts decrease when sampling weights are specified in the PO model. Regarding the logit

Table 9.1 Comparison of the PO Models With and Without Weighted Estimation

Variables	PO Model—Unweighted			PO Model With Weights		
	b (SE(b))	OR	P	b (SE(b))	OR	P
α_1	−1.701			−2.104		
α_2	.125			−.316		
α_3	2.258			1.835		
maritals	.514** (.106)	1.672	<.001	.609** (.124)	1.839	<.001
age	−.019** (.003)	.981	<.001	−.022** (.004)	.978	<.001
educ	.134** (.017)	1.144	<.001	.112** (.023)	1.118	<.001
male	.053 (.104)	1.054	.613	.123 (.128)	1.131	.338
LR R^2	.042			.039		
Model Fit	$\chi^2_{(4)} = 133.86$**			$\chi^2_{(4)} = 86.64$**		

Note: Standard errors in parentheses.

**$p < .05$.

coefficients, two coefficients (educ and age) decrease and the other two (maritals and male) increase. In addition, the standard errors of all four predictors increase. In other words, the standard errors are underestimated when weights are not applied.

Specifically, when the weights are applied to the PO model, the estimated logit regression coefficient for educ decreases by 16.4%, and its standard error increases by 35%, compared with those in the unweighted PO model; the logit coefficient for age decreases by 15.8%, and its standard error increases by 33%; the logit coefficient for maritals increases by 18.5%, and its standard error increases by 17%; in addition, the logit coefficient for male increases by 132% with an increase of 23% in its standard error.

9.3.3 Proportional Odds Model for Complex Survey Data Using the Stata svy Command

Stata's svy: prefix command for survey data is used to fit the PO model when taking all the elements of survey design features, such as strata, cluster, and the weight variable, into account. Before fitting the model, the svyset command needs to be employed by specifying the complex sampling design variables and weights. For more details on how to use this command, type the help svyset command. In this example, we specify these design features as follows:

```
svyset vpsu [pw=wtss], strata(vstrat)
```

In this syntax, the svyset command is followed by vpsu, the variable name for the primary sampling units or clusters in the data; [pw=wtss] specifies the probability weight wtss, which is the weight variable for the GSS 2012 data; and the option strata(vstrat) is separated from the command by a comma. Here the strata variable is vstrat. The result of the specified sampling design information and weights is presented as follows.

```
. svyset vpsu [pw=wtss], strata(vstrat)

      pweight: wtss
          VCE: linearized
     Strata 1: vstrat
        SU 1: vpsu
       FPC 1: <zero>
```

Following the svyset command, the command xi: svy: ologit heal-thre i.maritals age educ i.male is then used to fit the PO model for complex sampling survey data. In this syntax, the xi: prefix is used to create dummy variables for `maritals` and `male`; `svy: ologit` is the command for the PO model for the survey data; `healthre` is the ordinal response variable followed by four predictor variables. The results produced by `svy: ologit` are displayed as follows.

```
. xi: svy: ologit healthre i.maritals age educ i.male
i.maritals          _Imaritals_0-1      (naturally coded; _Imaritals_0 omitted)
i.male              _Imale_0-1          (naturally coded; _Imale_0 omitted)
(running ologit on estimation sample)

Survey: Ordered logistic regression

Number of strata    =       66              Number of obs     =        1300
Number of PSUs      =      132              Population size   = 1299.8077
                                            Design df         =          66
                                            F(   4,     63)   =       20.48
                                            Prob > F          =      0.0000

------------------------------------------------------------------------------
             |             Linearized
    healthre |     Coef.   Std. Err.      t    P>|t|     [95% Conf. Interval]
-------------+----------------------------------------------------------------
 _Imaritals_1|   .6091842   .1203283     5.06   0.000     .368941    .8494274
         age |  -.0224671   .0034727    -6.47   0.000    -.0294006  -.0155336
        educ |   .1119547   .0227363     4.92   0.000     .0665602   .1573492
    _Imale_1 |   .1230133   .1412158     0.87   0.387    -.158933    .4049597
-------------+----------------------------------------------------------------
       /cut1 |  -2.103854   .3422224    -6.15   0.000    -2.787123  -1.420585
       /cut2 |  -.3162062   .3434001    -0.92   0.361    -1.001827   .3694141
       /cut3 |   1.834523   .3512405     5.22   0.000     1.133248   2.535797
------------------------------------------------------------------------------
```

Interpreting the Output

As with other logistic regression models, the beginning of the output shows the coding note for two categorical variables `maritals` and `male`. The results of survey designs are shown below the title "`Survey: Ordered logistic regression`". The number of total observations for the analysis is 1,300. The number of strata is 66, and the number of PSUs is 132. The design degrees of freedom are 66, which equal the difference between the number of PSUs and the number of strata (132 − 66 = 66). Different from the log likelihood ratio chi-square test reported for the conventional PO model, here the adjusted Wald test for all parameters is reported, $F(4, 63) = 20.48$, $p < .001$, which indicates that the PO model with four predictor variables is significant in predicting the cumulative odds of being at or below a particular level of health status.

The survey ordinal logistic regression table displays the parameter estimates for the predictor variables and the cut points, their linearized standard errors, t statistics, associated p values, and 95% confident intervals. We first examine the logit coefficients (labeled Coef.) for the four predictor variables. Just like those in the PO model with weights, the coefficients of three variables are significantly different from zero. The column labeled Linearized Std. Err. lists the linearized standard errors, which indicates that the Taylor series approximation method is used to estimate the standard errors.

The logit coefficients for maritals, age, and educ are significant. For the maritals predictor, $\beta = .609$, $t = 5.06$, $p < .001$; for the age predictor, $\beta = -.022$, $t = -6.47$, $p < .001$; and for the educ predictor, $\beta = .112$, $t = 4.92$, $p < .001$. However, the effect of male is not significantly different from zero, $\beta = .123$, $t = .87$, $p = .387$.

After substituting the values of the estimated logit coefficients into the equation:

$$\ln\left(\frac{\pi\left(Y \le j \mid x_1, x_2, ..., x_p\right)}{\pi\left(Y > j \mid x_1, x_2, ..., x_p\right)}\right) = \alpha_j + (-\beta_1 X_1 - \beta_2 X_2 - ... - \beta_p X_p)$$

we get:

$$\text{logit}[\pi(Y \le j)] = \alpha_j + (-.609X_1 + .022X_2 - .112X_3 - .123X_4)$$

By exponentiating the negative logit coefficients (i.e., $e^{(-\beta)}$) we get the odds of being at or below a particular level of health status. Therefore, the odds of being at or below a particular category of health status (i.e., worse health status) versus being above that category (i.e., better health status) for the four predictors maritals, age, educ, and male are .544, 1.023, .894, and .884, respectively.

When estimating the odds of being at or below a health category, the three cut points are used to differentiate the adjacent categories of health status. $\alpha_1 = -2.104$, which is the cut point for the cumulative logit model for $Y \le 1$ (i.e., level 1 versus levels 2–4); α_2 is the cut point for the cumulative logit model for $Y \le 2$ (i.e., levels 1 and 2 versus levels 3 and 4); and the final α_3 is used as the cut point for the logit model when $Y \le 3$ (i.e., levels 1–3 versus level 4).

To estimate the odds of being above a particular category compared with being at or below that category, we need to reverse the signs before the cut points and the logit coefficients in the previous equation. The new equation can be expressed as follows:

$$\ln\left(\frac{\pi\left(Y > j \mid x_1, x_2, ..., x_p\right)}{\pi\left(Y \le j \mid x_1, x_2, ..., x_p\right)}\right) = -\alpha_j + (\beta_1 X_1 + \beta_2 X_2 + ... + \beta_p X_p)$$

By substituting the values of the estimated logit coefficients into the equation, we obtain:

$$\text{logit }[\pi(Y > j)] = -\alpha_j + .609X_1 - .022X_2 + .112X_3 + .123X_4$$

Odds ratios can be calculated by exponentiating the coefficients. An easier way is to use the Stata command. The command `xi: svy: ologit` with the `or` option (the ordinal response variable and the four predictor variables can be omitted from the syntax) reports the odds ratios for the four predictor variables. The output is displayed as follows.

```
. xi: svy: ologit, or

Survey: Ordered logistic regression

Number of strata    =         66        Number of obs     =         1300
Number of PSUs      =        132        Population size    =    1299.8077
                                        Design df          =           66
                                        F(    4,      63)  =        20.48
                                        Prob > F           =       0.0000


               |               Linearized
      healthre | Odds Ratio   Std. Err.      t    P>|t|     [95% Conf. Interval]
---------------+----------------------------------------------------------------
  _Imaritals_1 |   1.838931    .2212754    5.06   0.000     1.446202    2.338308
           age |   .9777834    .0033956   -6.47   0.000     .9710274    .9845864
          educ |   1.118462    .0254297    4.92   0.000     1.068825    1.170404
     _Imale_1  |   1.130899    .1597008    0.87   0.387     .8530535    1.499242
---------------+----------------------------------------------------------------
         /cut1 |  -2.103854    .3422224   -6.15   0.000    -2.787123   -1.420585
         /cut2 |  -.3162062    .3434001   -0.92   0.361    -1.001827    .3694141
         /cut3 |   1.834523    .3512405    5.22   0.000     1.133248    2.535797
```

In terms of odds ratios, an increase in `maritals` (OR = 1.839) and `educ` (OR = 1.113) is associated with an increase in the odds of being above a particular category of health status, rather than of being at or below that category. On the other hand, an increase in `age` (OR = .978) is associated with a decrease in the odds of being above a particular category of health status. In addition, being male (`male`) does not have an impact on the odds (OR = 1.331, z = .87, p = .387).

Comparison of Parameter and Standard Error Estimates From the PO Model for Complex Survey Data and the Conventional PO Model (Unweighted)

Table 9.2 provides the parameter and standard error estimates obtained from the PO model for complex sampling data using Stata `svy: ologit` and those from the unweighted PO model with Stata `ologit`. After sampling design variables and probability weights are applied to the PO model, the estimated logit coefficients and their standard errors are different from those in the conventional PO model. The logit

Table 9.2   Comparison Between the PO Model for Complex Survey Data Using Stata `svy: ologit` and the Conventional PO Model

Variables	PO Model—Unweighted			PO Model for Complex Survey Sampling		
	b (SE(*b*))	OR	*P*	*b* (SE(*b*))	OR	*P*
α_1	−1.701			−2.104		
α_2	.125			−.316		
α_3	2.258			1.835		
maritals	.514** (.106)	1.672	<.001	.609** (.120)	1.839	<.001
age	−.019** (.003)	.981	<.001	−.022** (.003)	.978	<.001
educ	.134** (.017)	1.144	<.001	.112** (.023)	1.118	<.001
male	.053 (.104)	1.054	.613	.123 (.141)	1.131	.387
LR R²	.042			—		
Model Fit	$\chi^2_{(4)} = 133.86$**			$F(4, 63) = 20.48$**		

Note: Standard errors in parentheses.

**$p < .05$.

coefficients for educ and age decrease, whereas the coefficients for maritals and male increase. In addition, the standard errors of the three predictor variables increase.

9.3.4 How to Deal With Singleton Strata

A singleton stratum means that a stratum contains only a single PSU. When Stata identifies that strata contain only one sampling unit for variance estimation, it reports the missing standard errors by default. To deal with this singleton PSU issue, the svyset command provides the other three options, including certainty, scaled, and centered. The first option singleunit(certainty) recognizes the single sampling unit in a stratum as a certainty unit (sampling unit chosen with 100% certainty), which contributes nothing to variance estimation across sampling units. The second option singleunit(scaled) is a scaled version of the first one, which uses the average variance of strata with multiple sampling units for the stratum with a single unit. The third option `singleunit(centered)` uses the grand mean across sampling units for variance estimation.

Each of these three options for the single unit is used separately in the `svyset` command since single sampling units resulted in missing standard errors in the model. And then Stata `svy: ologit` is used to conduct each ordinal logistic regression analysis. The results of the estimated standard errors using all `singleunit()` options are shown in the subsequent discussion. Since `singleunit(missing)` is the default option, the missing values for standard error estimations are also provided.

Research Example and Description of Data and Sample

In the following research example, we are interested in whether an ordinal outcome variable, mathematics proficiency of high-school students, can be predicted by four predictor variables: gender, whether a student's native language is English, students' socioeconomic status, and math self-efficacy. The ELS:2002 base-year data is used to fit a PO model with complex sampling designs. The following variables are used for data analysis:

- `Profnew`: mathematics proficiency level with six ordinal categories from 0 to 5
- `BYSEX`: gender
- `BYSTLANG`: whether a student's native language is English
- `BYSES2`: socioeconomic status
- `selfefficacy`: math self-efficacy
- `BYSTUWT`: weight variable for the students in the base-year data
- `PSU`: primary sampling unit
- `STRAT_ID`: strata variable

Model Fitting With Stata: Commands and Output

Before model fitting, the `svyset` command is used to set up the design features. The syntax is listed as follows:

```
svyset PSU [pweight = BYSTUWT] , strata (STRAT_ID)
```

In the `svyset` syntax, the variable name for the primary sampling units or clusters in the data is `PSU`; the probability weight `pweight` is the student weight for the based year data (`BYSTUWT`), and the strata is `START_ID`.

Next, the command `xi: svy: ologit Profnew i.BYSEX BYSTLANG BYSES2 selfefficacy` is used to fit the PO model to estimate the ordinal response variable `Profnew` from four predictor variables. The output is as follows.

```
. svyset PSU [pweight = BYSTUWT] , strata (STRAT_ID)

      pweight: BYSTUWT
          VCE: linearized
  Single unit: missing
     Strata 1: STRAT_ID
         SU 1: PSU
        FPC 1: <zero>

. xi: svy: ologit  Profnew i.BYSEX BYSTLANG BYSES2 selfefficacy
i.BYSEX            _IBYSEX_1-2          (naturally coded; _IBYSEX_1 omitted)
(running ologit on estimation sample)

Survey: Ordered logistic regression

Number of strata    =        361          Number of obs     =      12653
Number of PSUs      =        750          Population size   =  2641379.8
                                          Design df         =        389
                                          F(  0,    389)    =          .
                                          Prob > F          =          .


----------------------------------------------------------------------------
             |              Linearized
     Profnew |      Coef.   Std. Err.      t    P>|t|     [95% Conf. Interval]
-------------+--------------------------------------------------------------
   _IBYSEX_2 |  -.0062428          .        .       .             .          .
    BYSTLANG |   .5447152          .        .       .             .          .
      BYSES2 |   1.017793          .        .       .             .          .
selfefficacy |   .7196142          .        .       .             .          .
-------------+--------------------------------------------------------------
       /cut1 |  -1.025683          .        .       .             .          .
       /cut2 |   1.195922          .        .       .             .          .
       /cut3 |   2.317533          .        .       .             .          .
       /cut4 |   3.853195          .        .       .             .          .
----------------------------------------------------------------------------
Note: missing standard errors because of stratum with single sampling unit.
```

Since there is a singleton stratum in the data, the output only reports the coefficients and the standard errors are reported as missing. Next, the option singleunit(certainty) is specified in the svyset syntax, and we fit the PO model with the svy: prefix command. The output is displayed as follows.

```
. svyset PSU [pweight = BYSTUWT] , strata (STRAT_ID) singleunit(certainty)

      pweight: BYSTUWT
          VCE: linearized
  Single unit: certainty
     Strata 1: STRAT_ID
         SU 1: PSU
        FPC 1: <zero>
```

```
. xi: svy: ologit  Profnew i.BYSEX BYSTLANG BYSES2 selfefficacy
i.BYSEX          _IBYSEX_1-2          (naturally coded; _IBYSEX_1 omitted)
(running ologit on estimation sample)

Survey: Ordered logistic regression

Number of strata   =      361          Number of obs    =      12653
Number of PSUs     =      750          Population size  =  2641379.8
                                       Design df        =        389
                                       F(   4,    386)  =     421.00
                                       Prob > F         =     0.0000


------------------------------------------------------------------------------
             |              Linearized
     Profnew |      Coef.   Std. Err.      t    P>|t|     [95% Conf. Interval]
-------------+----------------------------------------------------------------
   _IBYSEX_2 |  -.0062428   .0424177    -0.15   0.883    -.0896395    .077154
    BYSTLANG |   .5447152   .0745531     7.31   0.000     .3981378   .6912925
      BYSES2 |   1.017793   .0361666    28.14   0.000     .9466865   1.088899
selfefficacy |   .7196142   .0278655    25.82   0.000     .6648283   .7744001
-------------+----------------------------------------------------------------
       /cut1 |  -1.025683   .1190051    -8.62   0.000    -1.259657  -.7917097
       /cut2 |   1.195922   .1107456    10.80   0.000     .9781874   1.413657
       /cut3 |   2.317533   .1129772    20.51   0.000     2.09541   2.539655
       /cut4 |   3.853195   .1185603    32.50   0.000     3.620096   4.086295
------------------------------------------------------------------------------
Note: strata with single sampling unit treated as certainty units.
```

In addition, the second option `singleunit(scaled)` is specified in the `svyset` syntax and the same PO model is fitted. The following output is produced.

```
. svyset PSU [pweight = BYSTUWT] , strata (STRAT_ID) singleunit(scaled)

      pweight: BYSTUWT
          VCE: linearized
  Single unit: scaled
     Strata 1: STRAT_ID
         SU 1: PSU
        FPC 1: <zero>

. xi: svy: ologit  Profnew i.BYSEX BYSTLANG BYSES2 selfefficacy
i.BYSEX          _IBYSEX_1-2          (naturally coded; _IBYSEX_1 omitted)
(running ologit on estimation sample)

Survey: Ordered logistic regression

Number of strata   =      361          Number of obs    =      12653
Number of PSUs     =      750          Population size  =  2641379.8
                                       Design df        =        389
                                       F(   4,    386)  =     419.84
                                       Prob > F         =     0.0000
```

```
----------------------------------------------------------------------------
              |              Linearized
      Profnew |     Coef.   Std. Err.       t    P>|t|    [95% Conf. Interval]
--------------+-------------------------------------------------------------
    _IBYSEX_2 |  -.0062428   .0424766    -0.15   0.883    -.0897552    .0772697
     BYSTLANG |   .5447152   .0746565     7.30   0.000     .3979343     .691496
       BYSES2 |   1.017793   .0362168    28.10   0.000     .9465878    1.088998
  selfefficacy |   .7196142   .0279042    25.79   0.000     .6647522    .7744762
--------------+-------------------------------------------------------------
        /cut1 |  -1.025683   .1191703    -8.61   0.000    -1.259982    -.791385
        /cut2 |   1.195922   .1108994    10.78   0.000     .9778852    1.413959
        /cut3 |   2.317533    .113134    20.48   0.000     2.095102    2.539963
        /cut4 |   3.853195   .1187248    32.45   0.000     3.619773    4.086618
----------------------------------------------------------------------------
Note: variance scaled to handle strata with a single sampling unit.
```

Finally, the option `singleunit(centered)` is specified in the `svyset` syntax. The output is displayed as follows.

```
. svyset PSU [pweight = BYSTUWT] , strata (STRAT_ID) singleunit(centered)

     pweight: BYSTUWT
         VCE: linearized
 Single unit: centered
    Strata 1: STRAT_ID
       SU 1: PSU
      FPC 1: <zero>

. xi: svy: ologit  Profnew i.BYSEX BYSTLANG BYSES2 selfefficacy
i.BYSEX           _IBYSEX_1-2           (naturally coded; _IBYSEX_1 omitted)
(running ologit on estimation sample)

Survey: Ordered logistic regression

Number of strata    =        361        Number of obs     =      12653
Number of PSUs      =        750        Population size   = 2641379.8
                                        Design df         =        389
                                        F(  4,    386)    =     420.88
                                        Prob > F          =     0.0000


----------------------------------------------------------------------------
              |              Linearized
      Profnew |     Coef.   Std. Err.       t    P>|t|    [95% Conf. Interval]
--------------+-------------------------------------------------------------
    _IBYSEX_2 |  -.0062428   .0424184    -0.15   0.883    -.0896407    .0771552
     BYSTLANG |   .5447152   .0745532     7.31   0.000     .3981374    .6912929
       BYSES2 |   1.017793   .0361677    28.14   0.000     .9466842    1.088902
  selfefficacy |   .7196142   .0278691    25.82   0.000     .6648213    .7744071
--------------+-------------------------------------------------------------
        /cut1 |  -1.025683   .1190074    -8.62   0.000    -1.259662    -.7917052
        /cut2 |   1.195922   .1107492    10.80   0.000     .9781804    1.413664
        /cut3 |   2.317533   .1129799    20.51   0.000     2.095405    2.539661
        /cut4 |   3.853195    .118566    32.50   0.000     3.620085    4.086306
----------------------------------------------------------------------------
Note: strata with single sampling unit centered at overall mean.
```

As we can see from this output, the linearized standard errors estimated with three options including certainty, scaled, and centered are almost identical in this example.

9.4 Making Publication-Quality Tables

```
. quietly xi: ologit Profnew i.BYSEX BYSTLANG BYSES2 selfefficacy
. outreg2 using complexout, e(ll chi2) dec(3) word replace
complexout.rtf
dir : seeout
. quietly xi: ologit Profnew i.BYSEX BYSTLANG BYSES2 selfefficacy [pweight = BYSTUWT]
. outreg2 using complexout, dec(3) word append
complexout.rtf
dir : seeout
. svyset PSU [pweight = BYSTUWT], strata (STRAT_ID) singleunit(certainty)

        pweight: BYSTUWT
            VCE: linearized
    Single unit: certainty
        Strata 1: STRAT_ID
            SU 1: PSU
           FPC 1: <zero>

. quietly xi: svy: ologit Profnew i.BYSEX BYSTLANG BYSES2 selfefficacy
. outreg2 using complexout, dec(3) word append
complexout.rtf
dir : seeout
```

The first command `quietly xi: ologit Profnew i.BYSEX BYSTLANG BYSES2 selfefficacy` estimates the conventional PO model without providing the output. The second command `outreg2 using complexout, e(ll chi2) dec(3) word replace` tells Stata to create a regression table and save it to a file named complexout.

The next two commands estimate the PO model with weights, and the estimates are appended to the created regression table in a separate column. Please note that since the pseudo-likelihood estimation is used for the PO model with sampling weights, the option `e(ll chi2)` is not requested.

The fifth command `svyset PSU [pweight = BYSTUWT], strata (STRAT_ID) singleunit(certainty)` sets up the design features. In the syntax, `PSU` is the primary sampling unit or cluster, `BYSTUWT` is the probability weight `pweight`, and `STRAT_ID` is the strata variable. The `singleunit(certainty)` option is used to deal with the singleton stratum.

The final two commands estimate the PO model for complex sampling designs with the `svy:` prefix command, and the estimates are added to the regression table.

The previous seven commands automatically produce Table 9.3, as shown here in its original format, which presents the results of all three fitted models.

Table 9.3　Results of the Conventional PO Model, PO Model With Weights, and PO Model for Complex Survey Sampling (Shown in Original Format Generated by Stata)

	(1)	(2)	(3)
Variables	profnew	profnew	profnew
_IBYSEX_2	−0.043	−0.006	−0.006
	(0.034)	(0.042)	(0.042)
BYSTLANG	0.162***	0.545***	0.545***
	(0.048)	(0.067)	(0.075)
BYSES2	1.003***	1.018***	1.018***
	(0.025)	(0.031)	(0.036)
selfefficacy	0.720***	0.720***	0.720***
	(0.021)	(0.026)	(0.028)
Constant cut1	−1.487***	−1.026***	−1.026***
	(0.081)	(0.104)	(0.119)
Constant cut2	0.765***	1.196***	1.196***
	(0.073)	(0.095)	(0.111)
Constant cut3	1.878***	2.318***	2.318***
	(0.075)	(0.096)	(0.113)
Constant cut4	3.441***	3.853***	3.853***
	(0.079)	(0.101)	(0.119)
Observations	11,726	11,726	12,653
Ll	−15,939		
chi2	3,293		

Note: Standard errors in parentheses.

***$p < 0.01$, **$p < 0.05$, *$p < 0.1$

After editing the variable names, adding the model names, adding the labels for the logit coefficients and standard errors to the column headings, and formatting the table, the final table is presented as in Table 9.4.

Table 9.4 Results of the Conventional PO Model, PO Model With Weights, and PO Model for Complex Survey Sampling (Edited)

Variables	Conventional PO Model b (SE(b))	PO Model With Weights b (SE(b))	PO Model With Sampling Designs b (SE(b))
BYSEX	−0.043	−0.006	−0.006
	(0.034)	(0.042)	(0.042)
BYSTLANG	0.162***	0.545***	0.545***
	(0.048)	(0.067)	(0.075)
BYSES2	1.003***	1.018***	1.018***
	(0.025)	(0.031)	(0.036)
Selfefficacy	0.720***	0.720***	0.720***
	(0.021)	(0.026)	(0.028)
α_1	−1.487***	−1.026***	−1.026***
	(0.081)	(0.104)	(0.119)
α_2	0.765***	1.196***	1.196***
	(0.073)	(0.095)	(0.111)
α_3	1.878***	2.318***	2.318***
	(0.075)	(0.096)	(0.113)
α_4	3.441***	3.853***	3.853***
	(0.079)	(0.101)	(0.119)
Observations	11,726	11,726	12,653
Model Fit	$\chi^2_{(4)} = 3{,}293$		

Note: Standard errors in parentheses.

***$p < .01$.

9.5 Reporting the Results

Writing results for proportional odds models with complex survey sampling is similar to that for the conventional proportional odds models except that the features of complex survey sampling need to be explained in text. The following are the generic guidelines for reporting the results. You may need to adjust your writing since your discipline or journals may have different requirements.

First, as with the reports for previous ordinal logistic regression models, describe the statistical method you used for data analysis, the dependent variable and the independent variables in the models, and your research hypothesis or the purpose of your study. Justify why the analysis with complex survey sampling is needed.

Second, explain the elements of the complex survey sampling, such as strata, clusters, and weight variables. In addition, explain how these elements are specified before fitting the model and what procedure is employed; for example, the `svyset` command needs to be used.

Third, report the F statistic with the associated degrees of freedom and the p value for the fitted model. Additionally, report the available number of strata and clusters (PSUs) from the output. If the model is fitted with weights only, report the Wald chi-square statistic and the associated p value.

Fourth, report the parameter estimates for the predictor variables, their standard errors, the associated p values, and the odds ratios in a table. When the conventional PO model is fitted for comparison purposes, you may report the results for all models in the table and interpret the change in the parameter estimates and standard errors. Briefly explain what method is used to estimate the standard errors for the parameters, for example, the Taylor series approximation method. In addition, interpret the odds ratios for each predictor.

Finally, if there is a single stratum in the data, report how to deal with this case since the estimated standard errors for the logit coefficients are missing. The following is an example of how to summarize the results for the PO model with complex survey sampling.

The proportional odds model for complex sampling designs was conducted to predict the ordinal outcome variable, health status, from a set of predictor variables, such as marital status, years of education, age, and gender. This model was used since the complex survey sampling designs involve the use of different strata, clustered sampling techniques, and unequal selection probabilities. Before fitting the model, the `svyset` command was employed by specifying the complex sampling design variables and weights. The Taylor series approximation method was used for variance estimation.

The number of strata was 66, and the number of primary sampling units (PSUs) was 132. The design degrees of freedom were 66, which equaled the difference between the number of PSUs and the number of strata (132 − 66 = 66). Table 9.2 provides the parameter and standard error estimates from the PO model for the complex sampling data. The adjusted Wald test for all parameters rather than the log likelihood ratio chi-square test for the conventional PO model was reported: $F(4, 63) = 20.48$, $p < .001$. This indicated that the PO model with four predictor variables was significant in predicting the cumulative odds of being at or below a particular level of health status.

The logit coefficients for `maritals,` `age,` and `educ` were significant. For the `maritals` predictor, $\beta = .609$, $t = 5.06$, $p < .001$; for the `age` predictor, $\beta = -.022$, $t = -6.47$, $p < .001$; and for the `educ` predictor, $\beta = .112$, $t = 4.92$, $p < .001$. However, the effect of `male` was not significantly different from zero, $\beta = .123$, $t = .87$, $p = .387$.

In terms of odds ratios, an increase in `maritals` (OR = 1.839) or `educ` (OR = 1.113) was associated with an increase in the odds of being above a particular category of health status rather than of being at or below that category. On the other hand, an increase in `age` (OR = .978) was associated with a decrease in the odds of being above a particular category of health status. In addition, being male (`male`) did not have an impact on the odds of being in a particular category of health status (OR = 1.331, $z = .87$, $p = .387$).

9.6 Summary of Stata Commands in This Chapter

```
use chap9-gss2012, clear

*Conventional PO model without weights
xi: ologit healthre i.maritals age educ i.male
xi: ologit, or

*PO model with weights
xi:ologit healthre i.maritals age educ i.male [pw=wtss]
xi: ologit, or

*PO model for complex survey sampling
svyset vpsu [pw=wtss], strata(vstrat)
xi: svy: ologit healthre i.maritals age educ i.male
xi: svy: ologit, or

*Deal with Singleton Stratum: Three Approaches

*use a different dataset
use chap9-els2002, clear

*certainty
svyset PSU [pweight = BYSTUWT] , strata (STRAT_ID) singleunit(certainty)
xi: svy: ologit  Profnew i.BYSEX BYSTLANG BYSES2 selfefficacy

*scaled
svyset PSU [pweight = BYSTUWT] , strata (STRAT_ID) singleunit(scaled)
xi: svy: ologit  Profnew i.BYSEX BYSTLANG BYSES2 selfefficacy

*centered
svyset PSU [pweight = BYSTUWT] , strata (STRAT_ID) singleunit(centered)
xi: svy: ologit  Profnew i.BYSEX BYSTLANG BYSES2 selfefficacy

*Making publication-quality tables using outreg2
quietly xi: ologit Profnew i.BYSEX BYSTLANG BYSES2 selfefficacy
outreg2 using complexout, e(ll chi2) dec(3) word replace
quietly xi: ologit Profnew i.BYSEX BYSTLANG BYSES2 selfefficacy [pweight = BYSTUWT]
```

```
outreg2 using complexout, dec(3) word append
svyset PSU [pweight = BYSTUWT], strata (STRAT_ID) singleunit(certainty)
quietly xi: svy: ologit Profnew i.BYSEX BYSTLANG BYSES2 selfefficacy
outreg2 using complexout, dec(3) word append
seeout
```

9.7 EXERCISES

Use the GSS 2012 data for the following problems.

1. Conduct an analysis for a proportional odds model with complex sampling survey designs to estimate the ordinal response variable fechld from the five predictor variables sex, educ, age, kidjob, and sibs. The probability weight for the data is wtss, and the strata variable is vstrat.

2. Specify the complex sampling design variables and weights using the svyset command.

3. What is the number of strata, and what is the number of primary sampling units (PSUs)?

4. What are the F test for the model and the p value? What do they tell us?

5. Interpret the odds ratios for sex, educ, and sibs.

6. Fit a conventional PO model without weights and the elements of survey designs using the same predictor variables as in Exercise 1. Compare the logit coefficient and its standard error of sex between the conventional PO model and the PO model with complex sampling survey designs.

7. Make a publication-quality table containing the estimated logit coefficients and odds ratios.

8. Write a report to summarize the results from the output.

CHAPTER **10**

Multilevel Modeling for Continuous and Binary Response Variables

Objectives of This Chapter

This chapter introduces multilevel modeling. To facilitate the understanding of multilevel models for ordinal response variables in the next chapter, this chapter presents multilevel modeling for continuous outcome variables and binary response variables. It starts with an introduction to multilevel modeling followed by an illustration of these two models with two research examples using Stata. In each example, when the multilevel model can be employed and how to specify a two-level model is discussed followed by the description of the research questions and data. Then several models, from the unconditional (null) model to the random-intercept model and random-coefficient model to the contextual models, are illustrated using Stata with step-by-step instructions. Stata commands are explained, and the output is interpreted for each model in detail. This chapter also illustrates how the results are displayed in publication-quality tables using the Stata command and reported in text. It focuses on model fitting with Stata, as well as on interpreting and presenting the results. After reading this chapter, you should be able to

- Determine when multilevel modeling for continuous and binary response variables is used.
- Formulate multilevel models.

(Continued)

(Continued)

- Conduct the multilevel modeling analysis for continuous and binary response variables using Stata.
- Interpret the output.
- Compute and interpret the intraclass correlation coefficient (ICC).
- Be familiar with model fitting strategies.
- Compare models using the log likelihood ratio test.
- Present results in publication-quality tables using Stata.
- Write the results for publication.

10.1 Multilevel Modeling: An Introduction

In previous chapters, we have focused on single-level analytic techniques for ordinal response variables. Multilevel modeling has been widely used in education, social, and behavioral sciences in recent years, and researchers are increasingly interested in applying this technique to analyze multilevel data in their research. This chapter will present multilevel modeling when the data structure has more than one level. It will start with an introduction to multilevel modeling for continuous outcome variables.

10.1.1 Multilevel Data Structure

Multilevel data, nested data, or hierarchical structured data have a data format in which observations at lower levels are nested within a higher level. For example, in businesses, employees are nested within companies; in educational research, student-level data are nested in school-level data; in medical science, patients are nested within hospitals; in political sciences, voters are nested within districts; and in sociology, families are nested within communities. Observations in the same group could be more homogeneous than those across different groups, and thus, the assumption of independence is violated. Another type of multilevel data structure occurs in longitudinal studies in which there are repeated measures for each subject. In this case, measures for multiple time points are nested within a subject. This type of analysis is known as the multilevel analysis for change (Singer & Willet, 2003). The focus of this text is the cross-sectional data structure.

What can multilevel modeling do? There are several advantages to using multilevel modeling. First, in multilevel modeling, variables at higher levels can be included in the model to estimate their relationships with the outcome variable. Second, we can examine whether an effect or slope of a variable at a lower level is allowed to vary among higher level variables. Third, we can also examine whether higher level variables moderate the relationships between lower level variables and the outcome variable.

10.1.2 Intraclass Correlation

With a multilevel data structure, the observations within a group or cluster may violate the assumption of independency. In other words, the observations within the same group or cluster may be more homogeneous than those in other groups or clusters. To justify why multilevel modeling is warranted, we also need to examine how much variance of the outcome variable is accounted for by groups or clusters. The intraclass correlation coefficient (ICC) is used as an index to measure the proportion of variance in the outcome variable explained by groups or clusters (Hox, 2010; Raudenbush & Bryk, 2002; Snijders & Bosker, 2012). It is the ratio of the between-group variance to the total variance. Its range is from 0 to 1. When it is close to 0, it means that using multilevel modeling might not be a good strategy for data analysis. A larger ICC provides strong evidence that this technique is needed.

10.1.3 Overview of a Basic Two-Level Model

Let us look at a basic two-level model with one predictor in each level. In the following example, researchers are interested in estimating the math achievement scores from a student-level variable, math self-efficacy, and a school-level variable (whether a school is public). The level 1 predictor variable is math self-efficacy (gceffic). Both the intercept and the slope of gceffic are allowed to vary randomly across schools. The level 2 predictor is whether a school is public or private (public). Following the convention of model specification by Raudenbush and Bryk (2002), a two-level model can be expressed as:

$$\text{Level 1: } Y_{ij} = \beta_{0j} + \beta_{1j}\texttt{gceffic}_{ij} + r_{ij} \qquad\qquad 10.1$$

$$\text{Level 2: } \beta_{0j} = \gamma_{00} + \gamma_{01}\texttt{public}_j + u_{0j} \qquad\qquad 10.2$$

$$\beta_{1j} = \gamma_{10} + \gamma_{11}\texttt{public}_j + u_{1j}$$

where Y_{ij} represents the math achievement score for the ith student in the jth school, β_{0j} is the level 1 intercept, the average math achievement score in the jth school, β_{1j} is the level 1 slope for gceffic in the jth school, and gceffic$_{ij}$ represents the value of math self-efficacy of the ith student in the jth school. r_{ij} is the random error, which is the deviation of the individual's math score from the average math score in the school.

The γ_{00} is the overall intercept of the outcome variable across schools. It is the predicted mean math achievement score controlling for the effect of the level 2 predictor (i.e., when the level 2 predictor variables are held constant at 0). The γ_{01} represents the effect of the level 2 variable `public` on the intercept. The γ_{10} represents the mean of the level 1 slope when the level 2 predictors are held constant at 0, and γ_{11} represents the effect of the level 2 predictor `public`. The γ_{11} is the cross-level interaction between `gceffic` and `public`, which moderates the effect of math self-efficacy on math achievement scores. The u_{0j} and u_{1j} are the random effects associated with the level 1 intercept and the slope of `gceffic` across schools, respectively. In other words, the level 1 intercept and the slope of `gceffic` are allowed to vary randomly across schools so their respective variances (i.e., between-group variance and slope variance) need to be estimated.

Fixed Effects Versus Random Effects

In multilevel modeling, fixed effects are the regression coefficients that estimate the relationships between the predictor variables and the outcome variable from the entire population (West, Welch, & Gałecki, 2014), whereas random effects are the randomly varying parameters across higher level units. For example, the random intercept (u_{0j}) in the previous example is a random deviation from the overall intercept, and the random coefficient (u_{1j}) is a random deviation from the overall fixed effect, the slope of `gceffic`.

What if an ordinary least-squares (OLS) regression instead of multilevel modeling is used? In other words, what will happen if random effects are not estimated? When a single-level regression analysis is conducted to analyze multilevel data, the precision of parameter estimates is compromised (Heck & Thomas, 2009). Heck and Thomas (2009) pointed out that multilevel modeling has four advantages over the OLS regression method. These advantages include incorporating regression equations at different levels into a single statistical model, more accurate estimates of standard errors, flexibility in specifying various models, and the capability of estimating different types of response variables.

By using multilevel modeling, we can estimate the influence of both the student-level and school-level predictors on the outcome variable. We can also investigate whether there are cross-level interactions between variables at different levels. In addition, we can estimate random effects by allowing the intercept and slopes of lower level predictors to vary randomly at higher levels. The variance and covariance components of the random effects can also be determined. For example, the estimated error variance for r_{ij} is the within-group variance, the estimated variance for u_{0j} is the intercept variance, which is the between-group variance, and the estimated variance for u_{1j} is the slope variance.

10.1.4 Model-Building Strategies

Although researchers may have their own strategies to build multilevel models, a common practice illustrated by Raudenbush and Bryk (2002), Snijders and Bosker (2012), and other publications (Garson, 2013; Heck & Thomas, 2009; Heck, Thomas, & Tabata, 2010; Kreft & de Leeuw, 1998; Luke, 2004; West, Welch, & Gałecki, 2014) is to start from a basic model and work up to more complex models. Specifically, this strategy starts with the unconditional means model with no level 1 and level 2 predictors (null model). This model is equivalent to the one-way random-effects analysis of variance (ANOVA) model. This model serves as the baseline model for future model comparisons. The unconditional means model estimates the overall average of the outcome variable across all subjects and the between-group and within-group variances. The variance between groups or clusters estimated from this model can be used to calculate the ICC so that we can decide whether multilevel modeling is needed. The between-group and within-group variances can also be used to compute the proportion of variance explained after the level 1 and level 2 predictors are added to the model. Next, we can add level 1 predictors and build a random-intercept model and a random-coefficient model. In the random-intercept model, only intercepts are allowed to vary freely in higher level clusters, and the level 1 slopes are fixed. In the random-coefficient model, both intercepts and coefficients of the level 1 predictors are allowed to vary across higher level clusters. Finally, we add level 2 predictors to the level 2 model so the random-coefficient model includes both level 1 and level 2 predictors. This model is referred to as the contextual model. If the model has more than two levels, then higher level predictors can be added.

Although the earlier simple-to-complex model-building strategy is commonly followed by researchers, you can decide whether all the steps need to be followed for your own research.

10.1.5 Model Fit Statistics

As with logistic regression models, several measures of goodness-of-fit statistics, such as the log-likelihood ratio test and the Akaike information criterion (AIC) and Bayesian information criterion (BIC) statistics, can be applied to multilevel modeling for continuous outcome variables. The following discussion is a brief review of these tests (see Chapter 3 for a more detailed description).

Log Likelihood Ratio Test

The log likelihood ratio test can be used to compare nested models. Models are nested when one model, the reduced model, is a special case of the other one, the full model. For example, more constraints can be put on parameters in one model than the other. A simple case is that one model (model 1) contains predictor variables X_1 and X_2, and the second model (model 2) contains an extra variable X_3. We conclude that model 1 is nested within model 2 since predictors in the former are the subset of the latter. In multilevel modeling, an unconditional model is nested within a random-intercept model, which is then nested within a random-coefficient model and finally a contextual model with both level 1 and level 2 predictor variables.

The log likelihood ratio test statistic is expressed as the difference in –2LL between nested models, where LL stands for the log likelihood value for the fitted model with either the full maximum likelihood (ML) estimation or the restricted maximum likelihood (REML) estimation. Since deviance equals –2LL, the log likelihood ratio test is also referred to as the difference in deviance, which follows a chi-square distribution, with the degrees of freedom of the distribution equaling the difference in the number of parameters between two nested models. The difference in deviance is often expressed as a generic form: G = Deviance for the reduced model – Deviance for the full model or $D_{Reduced} - D_{Full}$, where the reduced model has fewer variables and is nested within the full model. As with ordinal logistic regression models in previous chapters, we use the log likelihood ratio test to compare nested models from a simple model with one predictor to more complex models with multiple predictors. In multilevel modeling, we can also use the same test to compare a series of nested models from the unconditional means model to the random-intercept model to more complex models, such as the contextual models with level 1 and level 2 predictor variables. A significant log likelihood ratio chi-square test statistic indicates that a more complex model fits the data better than a simpler, nested model.

Information Criteria Indices: AIC and BIC

The AIC and the BIC statistics can be used to compare non-nested models. Both AIC and BIC statistics can be applied to multilevel modeling. The AIC statistic adjusts the deviance by the number of parameters. It is expressed as –2LL + 2k or Deviance + 2k, where k is the number of parameters. The BIC statistic is defined as BIC = –2LL + ln(n) × k = D_m + ln(n) × k, where n is the sample size and k is the number of parameters. Smaller AIC and BIC statistics indicate a better fit of the model.

10.1.6 Centering

The purpose of centering is to make the results more interpretative. It is often used when a predictor variable does not have a meaningful value of zero. By subtracting the mean of a predictor variable from each value, we obtain a meaningful zero for the predictor variable. Predictors at both levels of the model can be centered. Two types of centering are often used in multilevel modeling. One is grand-mean centering, and the other is group-mean centering. For the grand-mean centering, we subtract the grand mean of the predictor variable from each value of the sample. For example, when we use grand-mean centering of math efficacy (`efficacy`), we compute the overall mean of this variable and then subtract it from each score of efficacy. For the group-mean centering, we subtract the group mean, which is the mean of each group or cluster where individuals are nested from each score. For example, to group-mean center the predictor variable `efficacy`, we first compute the group mean for each school where a student belongs and then subtract the mean for each school (i.e., group mean) from each score of efficacy.

The choice of grand-mean centering and group-mean centering is complicated, and this topic has been widely discussed in the literature (Enders & Tofighi, 2007; Hofmann & Gavin, 1988; Hox, 2010; Kreft, de Leeuw, & Aiken, 1995; Luke, 2004; Ma, Ma, & Bradley, 2008; McCoach, 2010; Paccagnella, 2006). The advantage of grand-mean centering is that the subsequent multilevel models with this centering are mathematically equivalent to the models using raw scores without centering. It also makes the computation faster and reduces convergence problems (Hox, 2010). On the other hand, group-mean centering produces a model that is mathematically different from the raw score model. Hox (2010) suggested using group-mean centering with caution for novice users. Enders and Tofighi (2007) suggested that researchers use group-mean centering when level 1 variables and the interactions among them are the research interests, whereas grand-mean centering is a good choice if level 2 variables are the focus after controlling for level 1 variables. Therefore, the decision of using centering methods should be based on research questions or theories.

10.1.7 Data Structure for Model Fitting

In Stata, the data structure for multilevel modeling is a single dataset containing variables at different levels. For example, a two-level model needs a single dataset with both student-level and school-level variables and the former variables are nested with the latter. If the original student-level and school-level variables are saved in two separate datasets, then they need to be merged into one dataset in a format where students are nested within schools. This stacked data format requires that each school have multiple records, one for each student. For example, when 50 students are selected from a school, in the dataset, 50 students with different IDs (with each one having a row) are nested within the same school ID. Such a dataset needs to be created before model fitting.

10.2 Multilevel Modeling for Continuous Outcome Variables

10.2.1 Research Example and Research Questions

In the following example, researchers are interested in examining the relationships between high-school students' mathematics achievement and mathematics self-efficacy, school type, and school climate using the Educational Longitudinal Study of 2002 (ELS:2002) data. The student-level predictor variable is students' mathematics self-efficacy, and the two school-level predictor variables are school type and school climate. The following research questions will be addressed:

1. Can high-school students' mathematics scores be significantly predicted by students' mathematics self-efficacy?

2. Do school characteristics, such as school type and school climate, significantly impact math achievement?

3. Do mathematics scores vary across schools?

4. Does the relationship between mathematics self-efficacy and mathematics achievement vary across schools?

5. Are there any interaction effects between the two school-level variables (i.e., school type and school climate) and math self-efficacy? In other words, does school type or school climate moderate or influence the relationship between mathematics self-efficacy and mathematics achievement? Put in another way: Does the effect of mathematics self-efficacy on mathematics achievement vary across school type and school climate?

10.2.2 Description of the Data and Sample

The ELS:2002 base-year data are used for the following analyses. The variables are listed as follows:

- `mathach`: mathematics item response theory (IRT) estimated scores of high-school students
- `gceffic`: math self-efficacy (grand-mean centered)
- `public`: school type (1 = public, 0 = private and others)
- `csclimat`: school climate (grand-mean centered)

10.2.3 Multilevel Modeling for Continuous Outcome Variables With Stata: Commands and Output

The Stata Command `xtmixed`

The Stata command `xtmixed` is used for multilevel models with continuous outcome variables. The syntax includes three components. The first component is the command `xtmixed` followed by the dependent variable with or without the predictor variable(s). This is the fixed effects part of the model. The predictor variables from different levels are specified here, but the syntax itself does not tell the specific levels within which the variables belong. Next, two vertical lines (||) specify the random effects of the model. An identifier variable at a higher level is specified first, followed by a colon, and then a predictor variable or a list of predictor variables that have random coefficients. The final part specifies the options following a comma. Several common options include `cov()` for the variance–covariance structure, `mle` for the full maximum likelihood estimation, and `var` for variances instead of standard deviations. For example, the command `xtmixed y x ||schid:, cov(unstructured) mle var` tells Stata to fit a multilevel model to estimate a continuous outcome variable y with a predictor variable x, and `schid:` specifies the random intercepts varying across schools. The `cov(unstructured)` option requests an unstructured variance–covariance matrix for the random effects. The `mle` option requests the full maximum likelihood estimation rather than the restricted likelihood (`reml`). With the `var` option, the output displays variances rather than standard deviations for the variance–covariance matrix.

Unconditional Means Model (Model 1: Null Model)

The unconditional means model or the null model is known as the one-way random-effects ANOVA. Neither level 1 nor level 2 predictor variables are included in the model. This model can be expressed as follows:

$$\text{Level 1: } Y_{ij} = \beta_{0j} + r_{ij}$$

$$\text{Level 2: } \beta_{0j} = \gamma_{00} + u_{0j}$$

The command `xtmixed mathach || SCH_ID: , mle var` is used to fit the unconditional model (model 1). The `xtmixed` command is followed by the fixed part of the model. Because this is the unconditional model without any predictor variables, only the continuous outcome variable `mathach` is specified. The random part of the model is specified after the fixed part by two vertical bars (||).

The command SCH_ID: specifies the random effects at the school level by the identifier variable SCH_ID. No random coefficients for any predictor variables are specified in this model. The mle and var options request the maximum likelihood estimation method and variances for the random effects, respectively. After the model is fitted, the next command estimates store null saves the estimates with a name of null for the future model comparisons using the log likelihood ratio test. The output is displayed as follows.

```
. xtmixed mathach || SCH_ID: , mle var

Performing EM optimization:

Performing gradient-based optimization:

Iteration 0:   log likelihood = -37674.34
Iteration 1:   log likelihood = -37674.34

Computing standard errors:

Mixed-effects ML regression              Number of obs     =       9866
Group variable: SCH_ID                   Number of groups  =        617

                                         Obs per group: min =          1
                                                        avg =       16.0
                                                        max =         45


                                         Wald chi2(0)      =          .
Log likelihood = -37674.34               Prob > chi2       =          .

------------------------------------------------------------------------
     mathach |    Coef.    Std. Err.     z    P>|z|   [95% Conf. Interval]
-------------+----------------------------------------------------------
       _cons |  39.12354  .2537751   154.17   0.000   38.62615   39.62093
------------------------------------------------------------------------


------------------------------------------------------------------------
 Random-effects Parameters  |  Estimate   Std. Err.    [95% Conf. Interval]
----------------------------+-------------------------------------------
SCH_ID: Identity            |
              var(_cons)    |  31.73971   2.306184     27.52677   36.59743
----------------------------+-------------------------------------------
           var(Residual)    |  109.3623   1.610085     106.2517   112.564
------------------------------------------------------------------------
LR test vs. linear regression: chibar2(01) = 1386.75 Prob >= chibar2 = 0.0000

. estimates store null
```

Interpreting the Output

The top of the output displays the steps for the log likelihood estimation. The `xtmixed` program performs the expectation–maximization optimization and gradient-based iterations until they converge. The log likelihood values of two iterations are displayed. The note "`Computing standard errors:`" indicates that the program computes the standard errors for parameter estimates. The title of the regression table "`Mixed-effects ML regression`" tells us the mixed model is fitted with the ML estimation rather than the REML estimation method.

The top right of the fixed effects table displays the number of observations; the number of groups; the minimum, average, and maximum number of observations per group; the Wald chi-square test statistic; and the associated p value. A total of 9,866 observations in level 1 is nested in 617 groups (i.e., schools) in level 2. The range of observations per group is from 1 to 15 with an average of 16.0. The value of the Wald chi-square test (`Wald chi2(0)= .`) is missing since this null model contains no predictor variables.

The first table, which looks the same as other regression tables, reports the fixed effects for the predictor variables. Since no predictors are included in the model, this table displays the estimate for the intercept, its standard error, the Wald z statistic, the associated p value, and the 95% confidence interval. The intercept γ_{00} (labeled `_cons`) is 39.124, which is significant (`P>|z| = .000`). This means that the average math achievement score across all schools is 39.124.

The second table contains the random effects, the variance components of the model. The row labeled `SCH_ID: Identity` reports variances at level 2 (i.e., schools). The row labeled `var(_cons)` reports the between-school variance, and the next row (`var(Residual)`) reports the within-school variance. The between-school variance (τ_{00}) is 31.740, and the within-school variance (σ^2) is 109.362.

The ICC is defined as the proportion of total variance in the outcome variable ($\sigma^2 + \tau_{00}$) explained by the between-group variance (τ_{00}). It is expressed as $ICC = \tau_{00}/(\sigma^2 + \tau_{00})$.

From the earlier output, $ICC = 31.740/(31.740 + 109.362) = .225$, which indicates that 22.5% of the total variance is accounted for by schools in level 2.

At the bottom, the log likelihood ratio test is used to compare the unconditional model and the conventional linear regression model ignoring the nested data structure (`LR test vs. linear regression: chibar2(01) = 1386.75 Prob >= chibar2 = 0.0000`). The null hypothesis is that the between-group variance is equal to zero (i.e., $\tau_{00} = 0$). The significant result of the test shows that we are in favor of the unconditional means model estimating the between-group variance.

Next, the command `estimates store null` is echoed in the output. The estimated results are saved in memory and will be used for model comparisons with the `lrtest` command for the log likelihood ratio test.

Random-Intercept Model (Model 2)

Next, we include the predictor variable gceffic (math self-efficacy) to the level 1 equation, with all other parts of the level 1 equation the same. The model is referred to as the random-intercept model since the intercept is allowed to vary across schools. This model can be expressed as follows:

$$\text{Level 1: } Y_{ij} = \beta_{0j} + \beta_{1j}\text{gceffic}_{ij} + r_{ij}$$

$$\text{Level 2: } \beta_{0j} = \gamma_{00} + u_{0j}$$

$$\beta_{1j} = \gamma_{10}$$

In the level 1 equation, the predictor variable is math self-efficacy (gceffic) and the outcome variable is math achievement (mathach). The predictor variable is grand-mean centered. The level 2 equations express the random intercepts (β_{0j}) and the fixed slopes (β_{1j}).

The command xtmixed mathach gceffic || SCH_ID: , mle var, is used to fit the random-intercept model (model 2) after a predictor variable gceffic is added to the fixed part of the model. Just as in the unconditional model, no random coefficients are specified in the random part of the model. The next command, estimates store ranint, saves the estimates with a name of ranint, which will be used for the model comparison using the log likelihood ratio test. The following output is displayed.

```
. xtmixed mathach gceffic || SCH_ID: , mle var

Performing EM optimization:

Performing gradient-based optimization:

Iteration 0:   log likelihood = -36976.312
Iteration 1:   log likelihood = -36976.312

Computing standard errors:

Mixed-effects ML regression              Number of obs      =       9866
Group variable: SCH_ID                   Number of groups   =        617

                                         Obs per group: min =          1
                                                        avg =       16.0
                                                        max =         45


                                         Wald chi2(1)       =    1499.66
Log likelihood = -36976.312              Prob > chi2        =     0.0000
```

```
--------------------------------------------------------------------
       mathach |      Coef.   Std. Err.       z    P>|z|    [95% Conf. Interval]
---------------+----------------------------------------------------
       gceffic |   4.676707   .1207658    38.73    0.000    4.440011   4.913404
         _cons |   45.75365   .2917481   156.83    0.000    45.18184   46.32547
--------------------------------------------------------------------


--------------------------------------------------------------------
  Random-effects Parameters  |   Estimate   Std. Err.    [95% Conf. Interval]
-----------------------------+--------------------------------------
SCH_ID: Identity             |
               var(_cons)    |   27.49137   1.997849     23.84174   31.69967
-----------------------------+--------------------------------------
             var(Residual)   |   94.9433    1.397775     92.24286   97.7228
--------------------------------------------------------------------
LR test vs. linear regression: chibar2(01) =  1368.36 Prob >= chibar2 = 0.0000

. estimates store ranint
```

Interpreting the Output

In the fixed-effects table, the estimated intercept is 45.754, and the coefficient for gceffic is 4.677. Both estimates are significant ($p < .001$). The intercept can be interpreted as follows: The average math achievement score is 45.754 for students with a value of math self-efficacy at 0. The coefficient for gceffic is 4.677, $z = 38.73$, $p < .001$, which indicates that for a one-unit increase in math self-efficacy, there is an increase of 4.677 points in math achievement scores.

After the level 1 predictor is entered in the model, the variance for the random intercept has decreased to 27.491 compared with the original 31.740 in the unconditional model.

At the bottom of the random-effects table, the log likelihood ratio test is used to test the null hypothesis that the between-group variance is equal to zero (i.e., $\tau_{00} = 0$) by comparing the random-intercept model with the conventional linear regression model (LR test vs. linear regression: chibar2(01) = 1368.36 Prob >= chibar2 = 0.0000). The significant result shows that we are in favor of the random-intercept model with significant between-group variance.

Log Likelihood Ratio Test Comparing
the Unconditional Model and the Random Intercept Model

```
. lrtest null ranint

Likelihood-ratio test                           LR chi2(1)   =    1396.06
(Assumption: null nested in ranint)             Prob > chi2  =     0.0000
```

The log likelihood ratio test or the deviance difference test is used to compare the unconditional model (model 1) and the random-intercept model (model 2). The log likelihood ratio chi-square test $\chi^2_{(1)}$ = 1,396.06, p < .001, which indicates that the random-intercept model fits the data better than the unconditional model.

Random-Coefficient Model: Random-Intercept and Slope Model With Level 1 Variable (Model 3)

In addition to the random intercept, level 1 slopes (i.e., the coefficients of the level 1 predictors) can also be specified to be random. In other words, a predictor may have a random slope across clusters. For example, we may allow the effect of math self-efficacy on math achievement to vary across different schools. This model can be expressed as follows:

$$\text{Level 1: } Y_{ij} = \beta_{0j} + \beta_{1j}\texttt{gceffic}_{ij} + r_{ij}$$

$$\text{Level 2: } \beta_{0j} = \gamma_{00} + u_{0j}$$

$$\beta_{1j} = \gamma_{10} + u_{1j}$$

The level 1 equation for the random-coefficient model is the same as that for the random-intercept model. The math self-efficacy (gceffic) is still the only predictor variable, and math achievement is the outcome variable. Unlike the random-intercept model, in the random-coefficient model, the level 2 equations specify that both the intercept and the coefficient at level 1 are random across schools.

The command xtmixed mathach gceffic || SCH_ID: gceffic, cov(uns) mle var is used to fit the random-coefficient model (model 3) after a predictor variable gceffic is added to the random part of the model. The cov(uns) option specifies that the variance–covariance matrix is unstructured so that the slope variance of gceffic can be estimated. Next, the command estimates store ranslop1 saves the estimates with a name of ranslop1 for the comparison between the nested models.

```
. xtmixed mathach gceffic || SCH_ID: gceffic, cov(uns) mle var

Performing EM optimization:

Performing gradient-based optimization:

Iteration 0:    log likelihood = -36973.093
Iteration 1:    log likelihood = -36970.736
Iteration 2:    log likelihood = -36970.696
Iteration 3:    log likelihood = -36970.696
```

```
Computing standard errors:

Mixed-effects ML regression                Number of obs    =      9866
Group variable: SCH_ID                     Number of groups =       617

                                           Obs per group: min =         1
                                                          avg =      16.0
                                                          max =        45

                                           Wald chi2(1)     =   1317.08
Log likelihood = -36970.696                Prob > chi2      =    0.0000

------------------------------------------------------------------------------
    mathach |      Coef.   Std. Err.      z    P>|z|     [95% Conf. Interval]
------------+-----------------------------------------------------------------
     gceffic |   4.648547   .1280889    36.29   0.000     4.397497    4.899597
       _cons |   45.69043   .3140238   145.50   0.000     45.07495     46.3059
------------------------------------------------------------------------------


------------------------------------------------------------------------------
  Random-effects Parameters  |   Estimate   Std. Err.     [95% Conf. Interval]
-----------------------------+------------------------------------------------
SCH_ID: Unstructured         |
               var(gceffic)  |     1.0053   .5351668      .3541297    2.853835
                 var(_cons)  |   35.23956      3.485      29.03029    42.77694
         cov(gceffic,_cons)  |   3.467283   1.176671       1.16105    5.773515
-----------------------------+------------------------------------------------
               var(Residual) |   94.26357   1.425518       91.5106    97.09935
------------------------------------------------------------------------------
LR test vs. linear regression:        chi2(3) =   1379.59   Prob > chi2 = 0.0000

Note: LR test is conservative and provided only for reference

. estimates store ranslop1
```

Interpreting the Output

The fixed effects in the estimation table look similar to those in the random-intercept model (model 2). The random-effects table reports the unstructured variance–covariance components. The variance for the random coefficient of `gceffic` or the slope variance (labeled `var(gceffic)`) is 1.005, the between-group variance is 35.240, and the covariance between `gceffic` and the intercept is 3.467. You may be surprised to see that the between-school variance increases from the random-intercept model to the random-coefficient model. This is not a mistake. According to Rabe-Hesketh and Skrondal (2012), the total residual variance in the random-coefficient model is not as constant as that in the random-intercept model since the total variance varies across the values of a predictor variable when we have a varying slope in the random-coefficient model.

The within-school variance or the level 1 residual variance (labeled var(Residual)) is 94.264, which is similar to that of the random-intercept model.

Model Comparisons Using the Likelihood Ratio Test

```
. lrtest ranint ranslop1

Likelihood-ratio test                        LR chi2(2)  =      11.23
(Assumption: ranint nested in ranslop1)      Prob > chi2 =     0.0036

Note: LR test is conservative
```

The log likelihood ratio test is used to compare the random-intercept model (model 2) and the random-coefficient model with a level 1 predictor (model 3). The log likelihood ratio chi-square test $\chi^2_{(2)} = 11.23$, $p < .01$, which indicates that the random-coefficient model has a significantly better fit than the random-intercept model. Therefore, allowing a random coefficient in the model is justified.

Contextual Model With Level 1 and Level 2 Variables (Model 4)

The contextual model is a special case of the random-coefficient model when both level 1 and level 2 predictor variables are included. The two level 2 equations specify the intercept and slope from the level 1 equation to be random. In this example, two school-level variables public and csclimat are added to the level 2 equation. This model can be expressed as follows:

Level 1: $Y_{ij} = \beta_{0j} + \beta_{1j}\text{gceffic}_{ij} + r_{ij}$

Level 2: $\beta_{0j} = \gamma_{00} + \gamma_{01}\text{public}_j + \gamma_{02}\text{csclimat}_j + u_{0j}$ $\beta_{1j} = \gamma_{10} + u_{1j}$

With the level 1 equation the same as that for the previous random-coefficient model (model 3), the two school-level predictor variables are added to the equation for the random intercept at level 2. On the other hand, the equation for the random slope β_{1j} contains no predictor variables.

The command xtmixed mathach gceffic public csclimat || SCH_ID: gceffic, cov(uns) mle var is used to fit the contextual model (model 4) after the two level 2 predictor variables public and csclimat are added to the model.

Interpreting the Output

The fixed-effects table displays the intercept and the coefficients for both the level 1 and level 2 predictor variables. The coefficient for gceffic is 4.627, $z = 36.09$, $p < .001$,

```
. xtmixed mathach gceffic public csclimat || SCH_ID: gceffic, cov(uns) mle var

Performing EM optimization:

Performing gradient-based optimization:

Iteration 0:    log likelihood = -36898.012
Iteration 1:    log likelihood = -36895.344
Iteration 2:    log likelihood = -36895.282
Iteration 3:    log likelihood = -36895.281

Computing standard errors:

Mixed-effects ML regression                    Number of obs      =       9866
Group variable: SCH_ID                         Number of groups   =        617

                                               Obs per group: min =          1
                                                              avg =       16.0
                                                              max =         45

                                               Wald chi2(3)       =    1494.79
Log likelihood = -36895.281                    Prob > chi2        =     0.0000

------------------------------------------------------------------------------
     mathach |      Coef.   Std. Err.      z    P>|z|     [95% Conf. Interval]
-------------+----------------------------------------------------------------
     gceffic |   4.626977   .1282168    36.09   0.000     4.375676    4.878277
      public |   -2.89695   .5247261    -5.52   0.000    -3.925395   -1.868506
    csclimat |   3.027466   .3249077     9.32   0.000     2.390658    3.664273
       _cons |   47.95833   .5014002    95.65   0.000      46.9756    48.94105
------------------------------------------------------------------------------

------------------------------------------------------------------------------
  Random-effects Parameters  |   Estimate   Std. Err.     [95% Conf. Interval]
-----------------------------+------------------------------------------------
SCH_ID: Unstructured         |
               var(gceffic)  |     1.0765   .5391603      .4033608    2.872993
                 var(_cons)  |   29.33441   3.166695      23.74045    36.24649
         cov(gceffic,_cons)  |   4.133116   1.153393      1.872508    6.393725
-----------------------------+------------------------------------------------
               var(Residual) |   94.35758   1.427296      91.60119    97.19691
------------------------------------------------------------------------------
LR test vs. linear regression:       chi2(3) =    902.78   Prob > chi2 = 0.0000

Note: LR test is conservative and provided only for reference

. estimates store ranslop2
```

which indicates that students with higher mathematics self-efficacy tend to have higher mathematics achievement when holding other predictors constant. The effects of two school-level predictor variables are significant. The coefficient for public is −2.897,

$z = -5.52$, $p < .001$, which indicates that students' mathematics scores in public schools tend to be lower than those in private schools. The coefficient for csclimat is 3.027, $z = 9.32$, $p < .001$, which indicates that schools with better social climate tend to have higher mathematics scores.

The variance for the random coefficient of gceffic or the slope variance (labeled var(gceffic)) is 1.077, the between-group variance (labeled var(_cons)) is 29.334, and the covariance between gceffic and the intercept is 4.133. Both the slope variance and the covariance in model 4 look similar to those in the random-coefficient model (model 3). After two school-level predictors are included in the model, the between-group variance decreases from 35.240 to 29.334. In other words, the school-level variables explain 16.8% of between-group variance.

The within-school variance or the level 1 residual variance (labeled var(Residual)) is 94.358, which is similar to that of the random-coefficient model.

Model Comparisons Using the Likelihood Ratio Test

```
. lrtest ranslop1 ranslop2

Likelihood-ratio test                              LR chi2(2)  =     150.83
(Assumption: ranslop1 nested in ranslop2)          Prob > chi2 =     0.0000
```

A comparison between the contextual model with both level 1 and level 2 predictor variables (model 4: ranslop2) and the random-intercept and slope model with a level 1 predictor (model 3: ranslop1) yields the value of the log likelihood ratio chi-square test $\chi^2_{(2)} = 150.83$, $p < .001$, which indicates that the contextual model fits the data better.

Contextual Model With Cross-Level Interactions (Model 5)

We can also add cross-level interactions to the model by including level 2 predictor variables in the equation for the random slope (β_{1j}). This model can be expressed as follows:

$$\text{Level 1: } Y_{ij} = \beta_{0j} + \beta_{1j}\,\texttt{gceffic}_{ij} + r_{ij}$$

$$\text{Level 2: } \beta_{0j} = \gamma_{00} + \gamma_{01}\texttt{public}_j + \gamma_{02}\texttt{csclimat}_j + u_{0j}$$

$$\beta_{1j} = \gamma_{10} + \gamma_{11}\texttt{public}_j + \gamma_{12}\texttt{csclimat}_j + u_{1j}$$

In Stata, we need to create two new variables for the interactions before model fitting. The new variable pub_eff is created for the interaction between public and gceffic, and cli_eff is for the interaction between csclimat and gceffic.

```
. generate pub_eff= public*gceffic

. generate cli_eff=csclimat*gceffic

. xtmixed mathach gceffic public csclimat pub_eff cli_eff || SCH_ID: gceffic, c
> ov(uns) mle var

Performing EM optimization:

Performing gradient-based optimization:

Iteration 0:   log likelihood = -36896.227
Iteration 1:   log likelihood = -36893.399
Iteration 2:   log likelihood = -36893.316
Iteration 3:   log likelihood = -36893.315

Computing standard errors:

Mixed-effects ML regression                    Number of obs      =        9866
Group variable: SCH_ID                         Number of groups   =         617

                                               Obs per group: min =           1
                                                              avg =        16.0
                                                              max =          45


                                               Wald chi2(5)       =     1509.15
Log likelihood = -36893.315                    Prob > chi2        =      0.0000


------------------------------------------------------------------------------
     mathach |      Coef.   Std. Err.      z    P>|z|     [95% Conf. Interval]
-------------+----------------------------------------------------------------
     gceffic |   4.227338   .2791439    15.14   0.000     3.680226     4.77445
      public |  -2.032692   .7514439    -2.71   0.007    -3.505495   -.5598894
    csclimat |   2.859086   .4752097     6.02   0.000     1.927692     3.79048
     pub_eff |   .5229215   .3237115     1.62   0.106    -.1115415    1.157384
     cli_eff |  -.1000364   .2066047    -0.48   0.628    -.5049742    .3049014
       _cons |   47.30584   .6473974    73.07   0.000     46.03696    48.57471
------------------------------------------------------------------------------


------------------------------------------------------------------------------
  Random-effects Parameters  |   Estimate   Std. Err.     [95% Conf. Interval]
-----------------------------+------------------------------------------------
SCH_ID: Unstructured         |
               var(gceffic)  |   1.004614   .5339988      .3544429    2.847425
                 var(_cons)  |   29.23593   3.155693      23.66133     36.1239
          cov(gceffic,_cons) |   4.025937   1.145585      1.780631    6.271243
-----------------------------+------------------------------------------------
               var(Residual) |   94.34819   1.426905      91.59255    97.18673
------------------------------------------------------------------------------
LR test vs. linear regression:        chi2(3) =    904.88   Prob > chi2 = 0.0000

Note: LR test is conservative and provided only for reference

. estimates store ranslop3
```

Interpreting the Output

The fixed-effects table displays the intercept and the coefficients for the level 1 and level 2 predictor variables and two interaction terms. Let us take a look at the coefficients for the two interaction terms first since we are interested in the cross-level interactions. The coefficient for `pub_eff` (γ_{11}) = .523, z = 1.62, $p > .05$, and the coefficient for `cli_eff` (γ_{12}) = −.100, z = −.48, $p > .05$, which means both interactions are not significant. What if there is a cross-level effect? How can it be interpreted? For example, if the interaction between `public` and `gceffic` is significant, then this effect can be interpreted as follows: The relationship between mathematics self-efficacy and mathematics achievement varies across school type.

The variance and covariance components for the random effects for model 5 look similar to those for model 4, so they will not be interpreted here.

Model Comparisons Using the Likelihood Ratio Test

```
. lrtest ranslop2 ranslop3

Likelihood-ratio test                          LR chi2(2)  =      3.93
(Assumption: ranslop2 nested in ranslop3)      Prob > chi2 =    0.1399
```

After two interaction terms are added to the model, the log likelihood ratio chi-square test comparing the contextual models with or without interactions (model 5 vs. model 4) $\chi^2_{(2)}$ = 3.93, p = .1399, which indicates that the contextual model with cross-level interactions does not fit the data better than the contextual model without interactions. Therefore, we should remove the interaction terms from the model.

10.2.4 Making Publication-Quality Tables

```
1. .quietly xtmixed mathach || SCH_ID: , mle var
2. .outreg2 using chap10aout, dec(3) long word replace
3. .generate pub_eff= public*gceffic
4. .generate cli_eff=csclimat*gceffic
5. .quietly xtmixed mathach gceffic public csclimat pub_eff cli_eff || SCH_ID:
   gceffic, cov(uns) mle var
6. .outreg2 using chap10aout, dec(3) long word append
```

For illustration purposes, we will make a table containing the results for only two models, the unconditional model and the contextual model with cross-level interactions. The commands for creating this table are explained as follows. By following this example, readers should easily be able to create a table containing the results for all the fitted models.

The first command `quietly xtmixed mathach || SCH_ID: , mle var` estimates the unconditional model without showing the output. The second command `outreg2 using chap10aout, dec(3) long word replace` tells Stata to create a regression table and save the results containing both the fixed and random effects to a file named chap10aout. The `dec(3)` option requests three decimal places for all statistics. The `long` option aligns the fixed and random effects and lists them in one column. Without this option, the results will not be aligned properly. The `word` option requests that the table be created in a Word document with the extension .rtf. The `replace` option asks Stata to replace any files with the same name as chap10aout.

The remaining commands create the two cross-level interactions using the `generate` command, fit the contextual model (model 5), and append the estimated results to the created table.

These commands automatically produce Table 10.1, as shown here in its original format, presenting the results for the unconditional model and the contextual model with cross-level interactions.

In the original table, the fixed effects for both models are properly displayed. However, the variances and the covariance look different from those in previous output since Stata does not estimate them directly. Please note that the estimated within-group and between-group variances by the `xtmixed` command are stored as log standard deviations. In the preceding table, `lns1_1_1` stands for the log standard deviation for `gceffic`, `lnsig_e` for the within-group log standard deviation, and `lns1_1_2` for the log standard deviation for the between-group. In addition, `atr1_1_1_2` stands for the covariance between `gceffic` and the intercept, which is on the arc hyperbolic tangent scale. Although we could transform the log standard deviations to standard deviations and then to variances, it is easier to type the variances and the covariance directly from the output of the unconditional model (model 1) and the contextual model (model 5). The edited table is displayed as Table 10.2.

10.2.5 Reporting the Results

Since multilevel models estimate the fixed effects and random effects, the results of both of them need to be reported. The following are the basic guidelines for reporting. Several common reporting guidelines provided in previous chapters can also be applied to the reporting for the multilevel modeling. Please note that what needs to be reported in the Results section in a research article and the formats for displaying results vary across disciplines and journals.

Table 10.1   Results of the Two Multilevel Models: The Unconditional Model and the Contextual Model (Shown in Original Format Generated by Stata)

Variables	(1) mathach	(2) mathach
mathach		
gceffic		4.227***
		(0.279)
public		−2.033***
		(0.751)
csclimat		2.859***
		(0.475)
pub_eff		0.523
		(0.324)
cli_eff		−0.100
		(0.207)
Constant	39.124***	47.306***
	(0.254)	(0.647)
lns1_1_1		
Constant	1.729***	0.002
	(0.036)	(0.266)
lnsig_e		
Constant	2.347***	2.274***
	(0.007)	(0.008)
lns1_1_2		
Constant		1.688***
		(0.054)
atr1_1_1_2		
Constant		0.957***
		(0.233)
Observations	9,866	9,866
Number of groups	617	617

Note: Standard errors in parentheses.

***$p < 0.01$, **$p < 0.05$, *$p < 0.1$

Table 10.2   Results of the Two Multilevel Models: The Unconditional Model and the Contextual Model (Edited)

Variables	(1) **Unconditional** **Model (model 1)**	(2) **Contextual** **Model (model 5)**
Fixed Effects		
mathach		
gceffic		4.227***
		(0.279)
public		-2.033***
		(0.751)
csclimat		2.859***
		(0.475)
pub_eff		0.523
		(0.324)
cli_eff		-0.100
		(0.207)
Intercept	39.124***	47.306***
	(0.254)	(0.647)
Random Effects (Var. Components)		
Slope variance (gceffic)		
	—	1.005
		(0.534)
Within-school variance (σ^2)	109.362***	94.348***
	(1.610)	(1.427)
Between-group variance (τ_{00})	31.740***	29.236***
	(2.306)	(3.156)
Covariance	—	4.026***
		(1.146)
Observations	9,866	9,866
Number of groups	617	617

Note: Standard errors are shown in parentheses.

***$p < .01$.

First, as with other research examples, describe the purpose of your study and explain why the multilevel modeling is needed for the analysis.

Second, if a series of nested models is fitted, then report model-building steps and briefly describe each model. Report and interpret the intraclass correlation coefficient.

Third, if necessary, report the results of the fitted models in a table including both the parameter estimates for the fixed effects and the variances and covariances for the random effects. If available, include deviance statistics (i.e., −2LL) and AIC and BIC statistics for these models in the table.

Fourth, report and interpret the fixed effects of the predictor variables and variances and covariances in the final model. The following is an example of summarized results for the unconditional model and the contextual model.

Multilevel modeling was used to examine the relationships between high-school students' mathematics achievement and mathematics self-efficacy, school type, and school climate. Five models, from the unconditional (null) model to the contextual model with cross-level interactions, were fitted. Table 10.2 presents the parameter estimates for the fixed effects and random effects for the fitted models. For illustration purposes, the following interpretations focused only on the results of the unconditional model and the final contextual model without cross-level interactions.

Results for the Unconditional Model

The between-school variance (τ_{00}) was 31.740, and the within-school variance (σ^2) was 109.362. The log likelihood ratio test compared the unconditional model and the conventional linear regression model ignoring the nested data structure, $\chi^2_{(1)} = 1386.75$, $p < .001$, which indicated that the between-school variance was significant.

The ICC = 31.740/(31.740 + 109.362) = .225, which indicated that 22.5% of the total variance was explained by schools in level 2. This empirical evidence showed that it was appropriate to use multilevel modeling for data analysis.

Results for the Contextual Model With Cross-Level Interactions

The coefficient for gceffic was 4.227, $z = 15.14$, $p < .001$, which indicated that students with higher mathematics self-efficacy tended to have higher mathematics achievement when holding other predictors constant. The effects of two school-level predictor variables were also significant. The coefficient for public was −2.033, $z = −2.71$, $p < .01$, which indicated that students' mathematics scores in public schools tended to be lower than those in private schools. The coefficient for csclimat was 2.859, $z = 6.02$, $p < .001$, which indicated that schools with better social climate tended to have higher mathematics scores.

With regard to the cross-level interactions, the coefficient for `pub_eff` $(\gamma_{11}) = .523$, $z = 1.62, p > .05$, and the coefficient for `cli_eff` $(\gamma_{12}) = -.100, z = -.48, p > .05$, so both interactions were not significant.

Regarding the random effects, the variance and covariance components are displayed in Table 10.2. After two school-level variables were included in the contextual model (model 5), the between-school variance (τ_{00}) decreased from 35.240 to 29.236 when compared with that for the random coefficient model (model 3): $(35.240 - 29.236)/35.240 = 17\%$, which indicated that there was a decrease of 17% in the between-school variance from the random-coefficient model (model 3) to the contextual model with the cross-level interactions (model 5) after the two school-level variables were included.

10.3 Multilevel Modeling for Binary Outcome Variables

In the previous section, we introduced multilevel models for continuous outcome variables. Multilevel modeling can be extended for categorical response variables. In this section, multilevel models for binary response variables (Goldstein, 2003; Heck, Thomas, & Tabata, 2012; Hedeker & Gibbons, 1994; Hox, 2010; O'Connell, Goldstein, Rogers, & Peng, 2008; Raudenbush & Bryk, 2002) will be introduced.

Recall that in the single-level binary logistic regression model introduced in Chapter 3, the response variable only has two values (i.e., 1 and 0). Rather than estimating the binary outcome variable directly, we estimate the probability of success or of having an event via a logistic transformation, $\text{logit}(\pi)$ or $\ln(\text{odds})$. The logit transformation is defined as follows:

$$\eta = \text{logit}(\pi) = \ln \frac{\pi}{1 - \pi} \qquad 10.3$$

The generic form for the multiple logistic regression model is then expressed as:

$$\eta = \ln\left(\frac{\pi(x)}{1 - \pi(x)}\right) = \beta_0 + \beta_1 X_1 + \beta_2 X_2 + \dots + \beta_p X_p \qquad 10.4$$

where β_0 is the intercept; X_1, X_2, ..., X_p are the predictor variables, and β_1, β_2, ..., β_p are the logit coefficients of these predictors. This model estimates the relationship between a set of independent variables and the binary outcome variable on a scale of the logit or log odds. The link function transforms the original outcome variable Y into the transformed outcome η.

In multilevel modeling for binary response variables, we have a nested data structure that allows us to estimate the relationships between predictor variables from different levels and the binary response variable.

Let us look at a basic two-level logistic regression model with only one predictor variable at each level of the equations. For example, researchers are interested in estimating whether students are proficient in mathematics from a student-level variable, student's major language, and a school-level variable, public or private. The outcome variable is whether students are capable of doing simple arithmetical operations on whole numbers. The level 1 predictor variable is whether student's native language is English (stlang). The level 2 predictor is whether the school is public or private (public). Both the intercept and the slope of the level 1 predictor variable stlang are allowed to vary randomly across schools.

The level 1 equation is similar to that for the single-level logistic regression model, whereas in multilevel modeling, students are nested with schools so we have a subscript i to represent students and a subscript j to represent schools. As with a single-level logistic regression model, no error term is specified in the level 1 equation:

$$\text{Level 1: } \eta_{ij} = \text{logit} \left[\pi(x_{ij})\right] = \ln\left(\frac{\pi(x_{ij})}{1-\pi(x_{ij})}\right) = \beta_{0j} + \beta_{1j}\text{stlang}_{ij} \qquad 10.5$$

$$\text{Level 2: } \beta_{0j} = \gamma_{00} + \gamma_{01}\text{public}_j + u_{0j} \qquad 10.6$$

$$\beta_{1j} = \gamma_{10} + \gamma_{11}\text{public}_j + u_{1j}$$

The equations at level 2 look the same as those for the multilevel models for continuous outcome variables. Both the intercept and the coefficient (i.e., the slope of stlang) are allowed to vary randomly across schools at level 2 equations, where they are specified as the outcome variables.

In the level 1 equation, η_{ij} or $[\pi(x_{ij})]$ is the logit link for the probability of success for the ith student in the jth school. In level 2 equations, γ_{00}, γ_{01}, γ_{10}, and γ_{11} are level 2 coefficients, which are the fixed effects. γ_{00} is the overall average logit or log odds of reaching math proficiency across schools. γ_{01} is the overall average slope for stlang when the school-level predictor public is zero (public = 0). u_{0j} is an error term associated with the random intercept, the random variation for the intercept, and u_{1j} is the error term associated with the random slope, the random variation for the level 1 slope across schools.

Estimation of the level 2 random effects results in the variance and covariance matrix, which includes the intercept variance, the slope variance, and the covariance between the intercept and the slope.

10.3.1 Odds and Odds Ratios in Multilevel Logistic Regression Models

As with the single-level conventional binary logistic regression model, the odds and odds ratios in multilevel models for binary response variables can be interpreted in a similar way. The odds of success or of having an event are still the ratio of the probability of success or of having an event (p) to the probability of failure or of not having that event ($1 - p$).

The odds ratios are the exponentiated logit coefficients. They are interpreted as the change in odds of success or of having an event for a one-unit increase in an independent variable. When an odds ratio (OR) is larger than 1, the odds of success or of having an event increase for a one-unit increase in the predictor variable; when an OR is less than 1, the odds of success or of having an event decrease for a one-unit increase in the predictor variable. In addition, an OR of 1 indicates that there is no relationship between the predictor and the odds.

10.3.2 Research Example and Research Questions

In this research example, using the ELS:2002, researchers are interested in predicting whether high-school students can do simple arithmetical operations on whole numbers from three predictor variables: whether English is the student's native language, school type, and school climate. The first predictor is a student-level variable, and the other two are school-level variables. The following research questions will be addressed:

1. Can the probability of being in mathematics proficient level 1 (i.e., doing simple arithmetical operations on whole numbers) be significantly predicted by whether English is the student's native language?

2. Do school characteristics, such as school type and school climate, significantly impact mathematics proficiency?

3. Does student's mathematics proficiency level 1 vary across schools?

4. Does the relationship between student's native language and the probability of being in mathematics proficient level 1 vary across schools?

5. Are there any interaction effects between the two school-level variables (i.e., school type and school climate) and student language? In other words, does the relationship between student's native language and the probability of being in mathematics proficiency level 1 vary across school type and school climate?

10.3.3 Description of the Data and Sample

The ELS:2002 base-year data are used for the following analyses. The variables are listed as follows:

- Profmath1: mathematics proficiency level 1
- stlang: student language
- public: school type (1 = public, 0 = private and others)
- sclimat: school climate
- csclimat: school climate (grand-mean centered)

For demonstration purposes, the missing values of the variables are deleted. The descriptive statistics are summarized and the school-level variable sclimat is grand-mean centered.

```
. drop if missing(Profmath1, stlang, public, sclimate)
(3459 observations deleted)

. summarize Profmath1 stlang public sclimate

    Variable |       Obs        Mean    Std. Dev.       Min        Max
-------------+--------------------------------------------------------
   Profmath1 |     12793     .9498163    .2183324         0          1
      stlang |     12793     .8447589    .3621486         0          1
      public |     12793     .7788634    .4150285         0          1
    sclimate |     12793     3.922561    .6735278       1.2          5

. generate csclimat = sclimate -r(mean)
```

10.3.4 Multilevel Modeling for Binary Outcome Variables With Stata: Commands and Output

The Stata Command melogit *or* xtmelogit

The Stata command melogit or xtmelogit is used for multilevel models with binary outcome variables or mixed-effects logistic regression models. The command xtmelogit works with older versions from Stata 10 to 12 and has been replaced by the new command melogit in Stata 13 and 14. The syntax structure of melogit for mixed-effects logistic regression models is similar to that for xtmixed for the linear mixed models. The command melogit is followed by the dependent variable with or without the predictor variable(s). Next, two vertical lines (| |) specify the random effects of the model. The options are separated from the command by a comma. For example, the command melogit y x | |schid:, cov(unstructured) tells Stata to fit a multilevel model to estimate a binary outcome variable y on a predictor variable x with random intercepts varying across schools by specifying `||schid:`. The `cov(unstructured)`option requests an unstructured variance–covariance matrix for the random effects. Different from `xtmixed`, the `mle` and `var` options do not work with `melogit`.

Unconditional Model or Null Model (Model 1)

The unconditional model, or null model, for binary response variables includes no predictor variable at any level. This model estimates the overall probability of success or of having an event and the variability in the probability of success between groups (e.g., schools). It can be expressed as follows:

$$\text{Level 1: logit } [\pi(x_{ij})] = \beta_{0j}$$

$$\text{Level 2: } \beta_{0j} = \gamma_{00} + u_{0j}$$

Recall in binary logistic regression that logit(π) is the logistic transformation of the probability of success or an event occurrence. In multilevel logistic regression models, logit(π_{ij}) is the logit link for the probability of success for the ith student in the jth school. It can be expressed as $\ln \dfrac{\pi_{ij}}{1 - \pi_{ij}}$. In the level 2 equation, γ_{00} is the overall average logit or log odds of reaching math proficiency across schools and u_{0j} is an error term at the school level.

The command `melogit Profmath1 || SCH_ID:` is used to fit the unconditional model (model 1). The `melogit Profmath1` command tells Stata to fit a mixed logistic regression model for the binary response variable `Profmath1`. Since this is the unconditional model, no predictor variables are specified. The random part of the model is specified after the fixed part by two vertical bars (`||`). The command `SCH_ID:` specifies the random effects at the school level by the identifier variable `SCH_ID`. No random coefficients for any predictor variables are specified in this model. The next command including the `or` option `melogit, or` requests odds ratios. The third command, `estimates store null`, saves the estimates named null for future use. The output is displayed as follows.

```
. melogit Profmath1 || SCH_ID:

Fitting fixed-effects model:

Iteration 0:    log likelihood = -2764.5155
Iteration 1:    log likelihood = -2546.5935
Iteration 2:    log likelihood = -2546.5204
Iteration 3:    log likelihood = -2546.5204

Refining starting values:

Grid node 0:    log likelihood = -2455.7983
```

```
Fitting full model:

Iteration 0:   log likelihood = -2455.7983
Iteration 1:   log likelihood = -2431.4024
Iteration 2:   log likelihood =  -2429.073
Iteration 3:   log likelihood = -2429.0615
Iteration 4:   log likelihood = -2429.0617

Mixed-effects logistic regression         Number of obs     =       12793
Group variable:         SCH_ID            Number of groups  =         619

                                          Obs per group: min =           2
                                                         avg =        20.7
                                                         max =          50

Integration method: mvaghermite           Integration points =          7

                                          Wald chi2(0)      =           .
Log likelihood = -2429.0617               Prob > chi2       =           .
------------------------------------------------------------------------------
    Profmath1 |     Coef.   Std. Err.      z    P>|z|     [95% Conf. Interval]
--------------+---------------------------------------------------------------
        _cons |  3.471657    .086027    40.36   0.000     3.303048    3.640267
--------------+---------------------------------------------------------------
SCH_ID        |
   var(_cons) |  1.338871    .189729                      1.014183    1.767506
------------------------------------------------------------------------------
LR test vs. logistic regression: chibar2(01) =   234.92 Prob>=chibar2 = 0.0000

. melogit, or

Mixed-effects logistic regression         Number of obs     =       12793
Group variable:         SCH_ID            Number of groups  =         619

                                          Obs per group: min =           2
                                                         avg =        20.7
                                                         max =          50

Integration method: mvaghermite           Integration points =          7

                                          Wald chi2(0)      =           .
Log likelihood = -2429.0617               Prob > chi2       =           .
------------------------------------------------------------------------------
    Profmath1 | Odds Ratio  Std. Err.      z    P>|z|     [95% Conf. Interval]
--------------+---------------------------------------------------------------
        _cons |  32.19005   2.769215    40.36   0.000     27.19539    38.10202
--------------+---------------------------------------------------------------
SCH_ID        |
   var(_cons) |  1.338871    .189729                      1.014183    1.767506
------------------------------------------------------------------------------
LR test vs. logistic regression: chibar2(01) =   234.92 Prob>=chibar2 = 0.0000

. estimates store null
```

Interpreting the Output

The top of the output reports the estimation log for a series of estimation. The `melogit` program starts the estimation process from fitting the fixed-effects model to refining the starting values to fitting the final model.

Just like the output for the linear mixed models with `xtmixed`, the top right of the output for mixed-effects logistic regression displays the number of observations; the number of groups; the minimum, average, and maximum number of observations per group; the Wald chi-square test statistic; and the associated p value. A total of 12,793 observations in level 1 is nested in 619 schools in level 2. The range of observations per group is from 2 to 50 with an average of 20.7.

Stata uses the full maximum likelihood estimation with numerical integration for multilevel logistic regression models. The integration method in the model is the mean and variance adaptive Gauss–Hermite quadrature (`mvaghermite`) with a default of 7 integration points. Other integration methods, such as the mode and curvature adaptive Gauss–Hermite quadrature (`mcaghermite`), nonadaptive Gauss–Hermite quadrature (`ghermite`), or the Laplacian approximation (`laplace`) can be specified in the option. With the increase of the integration points, the estimation gets more accurate, but it takes longer to compute.

The results of the Wald chi-square test for the overall model (`Wald chi2(0)= .`) and the associated p value are missing since this null model contains no predictor variables.

The estimation table has two parts. The first section reports the fixed effects, and the second section at the bottom reports the variance components for the random effects of the model. The output displays the estimate for the intercept, its standard error, the Wald z statistic, the associated p value, and the 95% confidence interval. The intercept γ_{00} (labeled `_cons`) is 3.472, which is significant (z = 40.36, P>|z| = .000). This means that the overall average logit or log odds of being proficient in mathematics across all schools is 3.472. The estimated logit can be transformed to the odds ratio, which will be explained next.

Next, let us look at the variance components. The row (labeled `var(_cons)`) reports that the between-school variance (τ_{00}) is 1.339, which is the variance in the intercepts across all schools. The log likelihood ratio test comparing the null model with the ordinary logistic regression yields a reduction of 234.92 (`chibar2(01)=` 234.92), $p < .001$, which indicates that the between-school variance (τ_{00}) is significant. Another method is to look at the ratio of the variance in the intercept and its standard error, $1.339/.190 = 7.047$, which is larger than 2. This also indicates that the between-school variance is significant.

The ICC in multilevel logistic regression is also defined as the proportion of total variance in the outcome variable explained by the between-group variance. Since the variance of the logistic distribution is $\pi^2/3 = 3.29$, ICC = $\tau_{00}/(\tau_{00} + 3.29) = 1.339/(1.339$

+ 3.29) = .289, which indicates that 28.9% of the total variance is accounted for by schools in level 2. This is a large ICC, so we can justify that using multilevel logistic regression is appropriate for the analysis.

With the command `melogit,  or`, we can get the odds ratios for the predictor variables. Since this is the unconditional model, only the odds ratio for the intercept is reported. OR = 32.19, which is the overall average odds of being proficient in math.

Random-Intercept Model (Model 2)

Next, we include the predictor variable, whether student's native language is English (`stlang`), to the level 1 equation. The model is referred to as the random-intercept model since the intercept is allowed to vary freely across schools. This model can be expressed as follows:

$$\text{Level 1: logit } [\pi(x_{ij})] = \beta_{0j} + \beta_{1j}\texttt{stlang}_{ij}$$

$$\text{Level 2: } \beta_{0j} = \gamma_{00} + u_{0j}$$

$$\beta_{1j} = \gamma_{10}$$

The level 1 equation includes the predictor variable, whether student's native language is English (`stlang`), and the binary response variable, whether students are capable of doing simple arithmetical operations on whole numbers (`Profmath1`). The level 2 equation for the intercept (β_{0j}) is specified to be random, and the slope for `stlang` (β_{1j}) is constrained to be fixed.

The command `melogit Profmath1  stlang || SCH_ID:` is used to fit the random-intercept model (model 2) after a predictor variable `stlang` is added to the fixed part of the model. Just like the unconditional model, no random coefficients are specified in the random part of the model. The next command, `melogit,  or`, requests odds ratios. The third command, `estimates store  ranint`, saves the estimates named "ranint" for the following likelihood ratio test. The following output is displayed.

```
. melogit Profmath1 stlang || SCH_ID:

Fitting fixed-effects model:

Iteration 0:    log likelihood = -2746.7305
Iteration 1:    log likelihood = -2505.2039
Iteration 2:    log likelihood = -2504.5723
Iteration 3:    log likelihood =  -2504.572
Iteration 4:    log likelihood =  -2504.572

Refining starting values:

Grid node 0:    log likelihood = -2429.2466
```

```
Fitting full model:

Iteration 0:    log likelihood = -2429.2466
Iteration 1:    log likelihood = -2407.6286
Iteration 2:    log likelihood = -2404.2853
Iteration 3:    log likelihood = -2404.2468
Iteration 4:    log likelihood = -2404.2472
Iteration 5:    log likelihood = -2404.2472

Mixed-effects logistic regression           Number of obs     =      12793
Group variable:         SCH_ID               Number of groups  =        619

                                             Obs per group: min =          2
                                                            avg =       20.7
                                                            max =         50

Integration method: mvaghermite             Integration points =          7

                                             Wald chi2(1)      =      53.17
Log likelihood = -2404.2472                  Prob > chi2       =     0.0000
-------------------------------------------------------------------------------
    Profmath1 |      Coef.   Std. Err.      z    P>|z|     [95% Conf. Interval]
--------------+----------------------------------------------------------------
       stlang |   .7805386   .1070465     7.29   0.000     .5707312    .990346
        _cons |    2.80956   .1181369    23.78   0.000     2.578016   3.041104
--------------+----------------------------------------------------------------
SCH_ID        |
    var(_cons)|   1.194246   .1752651                      .8957219   1.592262
-------------------------------------------------------------------------------
LR test vs. logistic regression: chibar2(01) =    200.65 Prob>=chibar2 = 0.0000

. melogit, or

Mixed-effects logistic regression           Number of obs     =      12793
Group variable:         SCH_ID               Number of groups  =        619

                                             Obs per group: min =          2
                                                            avg =       20.7
                                                            max =         50

Integration method: mvaghermite             Integration points =          7

                                             Wald chi2(1)      =      53.17
Log likelihood = -2404.2472                  Prob > chi2       =     0.0000
-------------------------------------------------------------------------------
    Profmath1 | Odds Ratio   Std. Err.      z    P>|z|     [95% Conf. Interval]
--------------+----------------------------------------------------------------
       stlang |   2.182648   .2336449     7.29   0.000     1.769561   2.692166
        _cons |   16.60262   1.961382    23.78   0.000     13.17099   20.92835

--------------+----------------------------------------------------------------
SCH_ID        |
    var(_cons)|   1.194246   .1752651                      .8957219   1.592262
-------------------------------------------------------------------------------
LR test vs. logistic regression: chibar2(01) =    200.65 Prob>=chibar2 = 0.0000

. estimates store ranint
```

Interpreting the Output

     The estimation table reports the coefficients for `stlang` and the intercept. The intercept γ_{00} (labeled `_cons`) = 2.810. It is significantly different from 0 (z = `23.78`, `P>|z|` = `.000`), which means that the estimated logit or log odds of being proficient in doing simple arithmetical operations on whole numbers for students whose native language is not English (`stlang` = 0) in a typical school is 2.810. The logit coefficient for `stlang` (γ_{10}) = .781, $z = 7.29$, $p < .001$, which indicates that the logit or log odds of being proficient in math for students with English as their native language is .781 times as great as that for those students whose native language is not English. The odds ratio for `stlang` is reported with the `melogit` command with the `or` option. Interpretation of odds ratios in multilevel logistic regression is the same as that for binary logistic regression. OR = 2.183. It can be interpreted as meaning that the odds of being proficient in math for students whose native language is English are 2.183 times as large as the odds for those whose native language is not English.

     The row labeled `var(_cons)` reports that the intercept variance (τ_{00}) is 1.194, which is significantly different from 0 (`chibar2(01)`= `200.65` `Prob>=chibar2` = `0.0000`).

Log Likelihood Ratio Test Comparing
the Unconditional Model and the Random-Intercept Model

```
. lrtest null ranint

Likelihood-ratio test                                    LR chi2(1)   =      49.63
(Assumption: null nested in ranint)                      Prob > chi2 =     0.0000
```

     The log likelihood ratio test is used to compare the unconditional model and the random-intercept model: $\chi^2_{(1)}$ = 49.63, $p < .001$, which indicates that the random-intercept model fits the data better than the unconditional model, so we are in favor of the former model with a varying intercept.

Random-Coefficient Model With a Level 1 Variable (Model 3)

     Just like the random-coefficient model for linear mixed modeling, we can also fit a random-coefficient logistic regression model. Although the expressions for the level 1 equation are different between the two models, both level 2 equations for the random intercept and random coefficient are the same. This model is referred to as the random-coefficient model since both the intercept and coefficient in the level 1 equation are allowed to vary across schools at level 2. This model also estimates variance–covariance components, including variances in intercepts and coefficients/slopes and the covariance between them. The random coefficient logistic regression model can be expressed as follows:

$$\text{Level 1: logit } [\pi(x_{ij})] = \beta_{0j} + \beta_{1j} \text{stlang}_{ij}$$

$$\text{Level 2: } \beta_{0j} = \gamma_{00} + u_{0j}$$

$$\beta_{1j} = \gamma_{10} + u_{1j}$$

The level 1 equation includes the predictor variable, student's native language (stlang), and the binary response variable (Profmath1). The level 2 equations for the intercept and the slope for stlang are both specified to be random with level 2 residuals u_{0j} and u_{1j}, respectively.

The command melogit Profmath1 stlang || SCH_ID: stlang, cov(uns) is used to fit the random-coefficient model (model 3). The fixed part of the model is the same as that of the random-intercept model. Different from the random-intercept model, the predictor variable stlang is specified in the random part of the model, which indicates that the slope of the level 1 predictor stlang is allowed to vary randomly across schools. The next command, melogit, or, requests the odds ratios. The third command, estimates store ranslop1, saves the estimates in a file named "ranslop1" for the future likelihood ratio test. The following results are produced with the previous command.

```
. melogit Profmath1 stlang || SCH_ID: stlang, cov(uns)

Fitting fixed-effects model:

Iteration 0:   log likelihood = -2746.7305
Iteration 1:   log likelihood = -2505.2039
Iteration 2:   log likelihood = -2504.5723
Iteration 3:   log likelihood =  -2504.572
Iteration 4:   log likelihood =  -2504.572

Refining starting values:

Grid node 0:   log likelihood = -2442.8055

Fitting full model:

Iteration 0:   log likelihood = -2442.8055
Iteration 1:   log likelihood = -2421.9825
Iteration 2:   log likelihood = -2402.2901
Iteration 3:   log likelihood = -2401.2103
Iteration 4:   log likelihood = -2401.2027
Iteration 5:   log likelihood = -2401.2028
```

```
Mixed-effects logistic regression            Number of obs     =      12793
Group variable:        SCH_ID                Number of groups  =        619

                                             Obs per group: min =          2
                                                            avg =       20.7
                                                            max =         50

Integration method: mvaghermite              Integration points =          7

                                             Wald chi2(1)      =      25.43
Log likelihood = -2401.2028                  Prob > chi2       =     0.0000
------------------------------------------------------------------------------
    Profmath1 |      Coef.   Std. Err.      z    P>|z|     [95% Conf. Interval]
--------------+---------------------------------------------------------------
       stlang |   .8620904   .1709481     5.04   0.000     .5270383    1.197143
        _cons |   2.785555   .1545206    18.03   0.000        2.4827    3.08841
--------------+---------------------------------------------------------------
SCH_ID        |
  var(stlang) |   .6392856   .3397445                       .2255948    1.811594
   var(_cons) |   1.293923    .385562                       .7215526    2.320324
--------------+---------------------------------------------------------------
SCH_ID        |
cov(_cons,stlang)| -.2824301   .3107553    -0.91   0.363   -.8914993    .3266392
------------------------------------------------------------------------------
LR test vs. logistic regression:       chi2(3) =     206.74   Prob > chi2 = 0.0000

Note: LR test is conservative and provided only for reference.

. melogit, or

Mixed-effects logistic regression            Number of obs     =      12793
Group variable:        SCH_ID                Number of groups  =        619

                                             Obs per group: min =          2
                                                            avg =       20.7
                                                            max =         50

Integration method: mvaghermite              Integration points =          7

                                             Wald chi2(1)      =      25.43
Log likelihood = -2401.2028                  Prob > chi2       =     0.0000
------------------------------------------------------------------------------
    Profmath1 | Odds Ratio   Std. Err.      z    P>|z|     [95% Conf. Interval]
--------------+---------------------------------------------------------------
       stlang |   2.368106   .4048232     5.04   0.000     1.693908    3.310644
        _cons |   16.20881   2.504594    18.03   0.000     11.97355    21.94215
--------------+---------------------------------------------------------------
SCH_ID        |
  var(stlang) |   .6392856   .3397445                       .2255948    1.811594
   var(_cons) |   1.293923    .385562                       .7215526    2.320324
--------------+---------------------------------------------------------------
SCH_ID        |
cov(_cons,stlang)| -.2824301   .3107553    -0.91   0.363   -.8914993    .3266392
------------------------------------------------------------------------------
LR test vs. logistic regression:       chi2(3) =     206.74   Prob > chi2 = 0.0000

Note: LR test is conservative and provided only for reference.

. estimates store ranslop1
```

Interpreting the Output

The estimated logit coefficients and the corresponding odds ratios for the intercept and `stlang` in the output are similar to those for the previous random-intercept model. They can also be interpreted in a similar way. The odds ratio for `stlang` is 2.368, which indicates that the odds of being proficient in math for students whose native language is English are 2.368 times as great as the odds for those whose native language is not English. In other words, being students whose native language is English increases the odds of being proficient in math by a factor of 2.368.

Regarding the random effects in the model, the intercept variance (τ_{00}) is 1.294, the slope variance for `stlang` is .639, and the covariance between the intercept and the `stlang` slope is –.282.

Model Comparisons Using the Likelihood Ratio Test

```
. lrtest ranint ranslop1

Likelihood-ratio test                          LR chi2(2)    =      6.09
(Assumption: ranint nested in ranslop1)        Prob > chi2 =    0.0476

Note: The reported degrees of freedom assumes the null hypothesis is not on the
boundary of the parameter space.  If this is not true, then the reported test is
conservative.
```

The log likelihood ratio test compares the random-intercept model (model 2) and the random-coefficient model (model 3): $\chi^2_{(2)} = 6.09$, $p < .05$, which indicates that we are in favor of the random-coefficient model, so having a random coefficient in model 3 is justified.

Contextual Model With Level-2 Variables (Model 4)

Just like the contextual model in the linear mixed modeling, the contextual model for the binary response variable also includes both level 1 and level 2 predictor variables. The two level 2 equations specify the random intercept and random slope. In this example, two school-level variables, `public` and `csclimat,` are added to the level 2 equation for the random intercept. In the level 2 equation for the random slope, no school-level variables are specified. This model is expressed as follows:

$$\text{Level 1: logit } [\pi(x_{ij})] = \beta_{0j} + \beta_{1j}\texttt{stlang}_{ij}$$

$$\text{Level 2: } \beta_{0j} = \gamma_{00} + \gamma_{01}\texttt{public}_j + \gamma_{02}\texttt{csclimat}_j + u_{0j}$$

$$\beta_{1j} = \gamma_{10} + u_{1j}$$

The command `melogit Profmath1 stlang public csclimat || SCH_ID:`
`stlang, cov(uns)` is used to fit the contextual model (model 4). With other parts of
the equations the same as those of the random-coefficient model (model 3), the fixed part
of the model includes both the level 1 predictor `stlang` and the two school-level
predictor variables `public` and `csclimat`. Again, the odds ratios are requested using
the command `melogit, or`. The third command, `estimates store ranslop2`,
saves the estimates in a file named "ranslop2" for the future likelihood ratio test.

```
. melogit Profmath1 stlang public csclimat || SCH_ID: stlang, cov(uns)

Fitting fixed-effects model:

Iteration 0:    log likelihood = -2715.8421
Iteration 1:    log likelihood = -2420.1492
Iteration 2:    log likelihood = -2405.9032
Iteration 3:    log likelihood = -2405.7093
Iteration 4:    log likelihood = -2405.7093

Refining starting values:

Grid node 0:    log likelihood = -2389.1995

Fitting full model:

Iteration 0:    log likelihood = -2389.1995   (not concave)
Iteration 1:    log likelihood = -2366.1464
Iteration 2:    log likelihood = -2350.9729
Iteration 3:    log likelihood = -2346.0687
Iteration 4:    log likelihood = -2345.9496
Iteration 5:    log likelihood = -2345.9495

Mixed-effects logistic regression           Number of obs     =      12793
Group variable:           SCH_ID            Number of groups  =        619

                                            Obs per group: min =          2
                                                           avg =       20.7
                                                           max =         50

Integration method: mvaghermite             Integration points =         7

                                            Wald chi2(3)       =     123.70
Log likelihood = -2345.9495                 Prob > chi2        =     0.0000
-----------------------------------------------------------------------------
   Profmath1 |     Coef.   Std. Err.      z    P>|z|    [95% Conf. Interval]
-------------+---------------------------------------------------------------
      stlang |  .6829654   .1574742     4.34   0.000    .3743217    .9916091
      public | -1.288413   .2152774    -5.98   0.000   -1.710349   -.8664767
    csclimat |  .6029364    .094154     6.40   0.000    .418398    .7874748
       _cons |  3.135276   .2804155    11.18   0.000    2.585672    3.68488
-------------+---------------------------------------------------------------
SCH_ID       |
 var(stlang) |  .815965   .3719859                      .3339042   1.993982
 var(_cons)  |  1.18652   .368578                       .6454456   2.181175
-------------+---------------------------------------------------------------
```

```
SCH_ID           |
cov(_cons,stlang)|  -.5661113     .326873    -1.73    0.083    -1.20677      .074548
-----------------------------------------------------------------------------------
LR test vs. logistic regression:      chi2(3) =   119.52   Prob > chi2 = 0.0000

Note: LR test is conservative and provided only for reference.

. melogit, or

Mixed-effects logistic regression             Number of obs      =       12793
Group variable:            SCH_ID              Number of groups   =         619

                                               Obs per group: min =           2
                                                              avg =        20.7
                                                              max =          50

Integration method: mvaghermite                Integration points =           7

                                               Wald chi2(3)       =      123.70
Log likelihood = -2345.9495                    Prob > chi2        =      0.0000
-----------------------------------------------------------------------------------
       Profmath1 | Odds Ratio   Std. Err.      z    P>|z|     [95% Conf. Interval]
-----------------+-----------------------------------------------------------------
          stlang |   1.97974    .3117579     4.34   0.000     1.454005     2.695568
          public |   .2757081   .0593537    -5.98   0.000     .1808028     .4204303
         csclimat |   1.827477   .1720642     6.40   0.000     1.519525     2.197839
           _cons |   22.99498   6.448148    11.18   0.000     13.2722      39.84035
-----------------+-----------------------------------------------------------------
SCH_ID           |
     var(stlang) |   .815965    .3719859                       .3339042     1.993982
     var(_cons)  |   1.18652    .368578                         .6454456     2.181175
-----------------+-----------------------------------------------------------------
SCH_ID           |
cov(_cons,stlang)|  -.5661113    .326873    -1.73    0.083    -1.20677      .074548
-----------------------------------------------------------------------------------
LR test vs. logistic regression:      chi2(3) =   119.52   Prob > chi2 = 0.0000

Note: LR test is conservative and provided only for reference.

. estimates store ranslop2
```

Interpreting the Output

The fixed-effects part in the estimation table reports the estimated logit coefficients for both level 1 and level 2 predictor variables and the intercept. The logit coefficients for all predictor variables are significant. The corresponding odds ratios are reported in the table produced by the command `melogit, or`. The odds ratio for `stlang` is 1.980, which indicates that the odds of being proficient in math for students whose native language is English are 1.980 times as great as the odds for those whose native language is not English.

The odds ratio for `public` is .276, which indicates that the odds of being proficient in math for students in public schools are .276 times the odds for those in private schools. In other words, being students in public schools decreases the odds of being proficient in math by 72.4%.

The odds ratio for csclimat is 1.827, which indicates that a one-unit increase in school climate corresponds to an increase of 1.827 points in the odds of being proficient in math. In other words, being students in schools with a more positive school climate increases the odds of being proficient in math.

The residual variance in the intercept (τ_{00}) is 1.187, the slope variance for stlang is .816, and the covariance between the intercept and the stlang slope is –.566.

Model Comparisons Using the Likelihood Ratio Test

```
. lrtest ranslop1 ranslop2

Likelihood-ratio test                          LR chi2(2)   =    110.51
(Assumption: ranslop1 nested in ranslop2)      Prob > chi2 =    0.0000
```

The log likelihood ratio chi-square test compares the contextual model with both the level 1 and level 2 predictor variables (model 4: ranslop2) with the random-coefficient model with a level 1 predictor only (model 3: ranslop1). It yields a value of $\chi^2_{(2)} = 110.51$, $p < .001$, which indicates that the contextual model fits the data better.

Contextual Model With Cross-Level Interactions (Model 5)

The contextual model with cross-level interactions is used to investigate whether there are interaction effects between level 1 and level 2 predictor variables. For example, if we are interested in whether the two school-level variables public and csclimat interact with the level 1 predictor stlang, we can include two interaction terms in the model. The two interaction terms need to be created before model fitting. One interaction, pub_lang, is created for two categorical variables, stlang and public; the other interaction, cli_lang, is for the categorical variable stlang and the continuous variable csclimat. This model can be expressed as follows:

$$\text{Level 1: logit } [\pi(x_{ij})] = \beta_{0j} + \beta_{1j}\text{stlang}_{ij}$$

$$\text{Level 2: } \beta_{0j} = \gamma_{00} + \gamma_{01}\text{public}_j + \gamma_{02}\text{csclimat}_j + u_{0j}$$

$$\beta_{1j} = \gamma_{10} + \gamma_{11}\text{public}_j + \gamma_{12}\text{csclimat}_j + u_{1j}$$

The command melogit Profmath1 stlang public csclimat pub_lang cli_lang || SCH_ID: stlang, cov(uns) is used to fit the contextual model with cross-level interactions (model 5). With other parts of the equations the same as those of the contextual model (model 4), the fixed part of the model includes two extra variables, pub_lang and cli_lang, which are the created cross-level interactions. The next command, melogit, or, requests the odds ratios, and the third command, estimates store ranslop3, saves the estimates in a file named "ranslop3". The following output is presented.

```
. generate pub_lang= public*stlang

. generate cli_lang=csclimat*stlang

. melogit Profmath1 stlang public csclimat pub_lang cli_lang || SCH_ID: stlang,
cov(uns)

Fitting fixed-effects model:

Iteration 0:    log likelihood = -2714.2983
Iteration 1:    log likelihood = -2420.3437
Iteration 2:    log likelihood = -2405.1987
Iteration 3:    log likelihood = -2404.9167
Iteration 4:    log likelihood = -2404.9166

Refining starting values:

Grid node 0:    log likelihood = -2387.4869

Fitting full model:

Iteration 0:    log likelihood = -2387.4869  (not concave)
Iteration 1:    log likelihood = -2364.5431
Iteration 2:    log likelihood = -2349.3768
Iteration 3:    log likelihood = -2344.9613
Iteration 4:    log likelihood =  -2344.854
Iteration 5:    log likelihood =  -2344.854

Mixed-effects logistic regression          Number of obs      =      12793
Group variable:          SCH_ID             Number of groups   =        619

                                            Obs per group: min =          2
                                                           avg =       20.7
                                                           max =         50

Integration method: mvaghermite               Integration points =        7

                                            Wald chi2(5)       =     121.69
Log likelihood =  -2344.854                 Prob > chi2        =     0.0000
------------------------------------------------------------------------------
   Profmath1 |      Coef.   Std. Err.      z    P>|z|     [95% Conf. Interval]
-------------+----------------------------------------------------------------
      stlang |   .2702196   .5559571     0.49   0.627    -.8194363    1.359876
      public |  -1.405411   .4572263    -3.07   0.002    -2.301558    -.509264
     csclimat |   .4153246   .1530077     2.71   0.007     .1154351    .7152141
     pub_lang |   .1650978   .4985976     0.33   0.741    -.8121356    1.142331
     cli_lang |   .2539083   .1682822     1.51   0.131    -.0759188    .5837354
        _cons |   3.432849   .5095942     6.74   0.000     2.434063    4.431636
-------------+----------------------------------------------------------------
```

```
SCH_ID          |
    var(stlang)|    .6878203    .3621313                         .2450896    1.930301
    var(_cons)|    1.076623    .3539386                         .5652391    2.050667
---------------+--------------------------------------------------------------------
SCH_ID          |
cov(_cons,stlang)|  -.4364309    .3182447    -1.37   0.170   -1.060179    .1873172
---------------+--------------------------------------------------------------------
LR test vs. logistic regression:     chi2(3) =    120.13   Prob > chi2 = 0.0000

Note: LR test is conservative and provided only for reference.

. melogit, or

Mixed-effects logistic regression              Number of obs      =     12793

Group variable:           SCH_ID               Number of groups   =       619

                                               Obs per group: min =         2
                                                              avg =      20.7
                                                              max =        50

Integration method: mvaghermite                Integration points =         7

                                               Wald chi2(5)       =    121.69
Log likelihood = -2344.854                     Prob > chi2        =    0.0000
--------------------------------------------------------------------------------
     Profmath1 | Odds Ratio   Std. Err.      z    P>|z|     [95% Conf. Interval]
---------------+----------------------------------------------------------------
        stlang |   1.310252    .728444     0.49   0.627      .44068    3.895708
        public |   .2452662   .1121422    -3.07   0.002    .1001027    .6009377
       csclimat|   1.514862   .2317855     2.71   0.007    1.122362    2.044624
       pub_lang|   1.179508   .5881001     0.33   0.741    .4439091    3.134066
       cli_lang|   1.289054   .2169248     1.51   0.131    .9268914    1.792722
          _cons|   30.96475   15.77946     6.74   0.000    11.40513    84.06881
---------------+----------------------------------------------------------------
SCH_ID          |
    var(stlang)|    .6878203    .3621313                         .2450896    1.930301
    var(_cons)|    1.076623    .3539386                         .5652391    2.050667
---------------+--------------------------------------------------------------------
SCH_ID          |
cov(_cons,stlang)|  -.4364309    .3182447    -1.37   0.170   -1.060179    .1873172
---------------+--------------------------------------------------------------------
LR test vs. logistic regression:     chi2(3) =    120.13   Prob > chi2 = 0.0000

Note: LR test is conservative and provided only for reference.

. estimates store ranslop3
```

Interpreting the Output

In the output, let us look at the coefficients for these two interaction terms in the fixed-effects table. The coefficient for pub_lang (γ_{11}) = .165 (OR = 1.180), z = .33, $p > .05$, and the coefficient for cli_lang (γ_{12}) = .254 (OR = 1.289), z = 1.51, $p > .05$.

So both interactions are not significant, which indicates that the effect of `stlang` does not differ in public or private schools. In addition, the effect of `stlang` is not influenced by school climate, either. How can a cross-level effect be interpreted if it is identified in the model? For example, if `cli_lang` is significant, then this effect can be interpreted as follows: The relationship between students' native language and the probability of being in mathematics proficiency level 1 varies across school climate.

Since the variance and covariance components for the random effects for model 5 are similar to those for model 4, their interpretation is omitted here.

Model Comparison Using the Likelihood Ratio Test

```
. lrtest ranslop2 ranslop3

Likelihood-ratio test                                  LR chi2(2)  =      2.19
(Assumption: ranslop2 nested in ranslop3)              Prob > chi2 =    0.3344
```

The log likelihood ratio chi-square test is again used to compare the contextual models with or without interactions (model 5 vs. model 4): $\chi^2_{(2)} = 2.19$, $p > .05$, which indicates that we are in favor of the null hypothesis. We can conclude that the contextual model with cross-level interactions does not fit the data better than the contextual model without interactions. Therefore, the interaction terms can be removed from model 5.

Interpreting the Estimated Probabilities With the `margins` *Command in Stata 13*

With the `margins` command, we can estimate the margins or the predicted probabilities for predictor variables at specified values in the multilevel logistic regression model. Since the interaction terms have been removed from model 5, the `margins` command is executed immediately after the contextual model without interactions (model 4).

```
. quietly melogit Profmath1 i.stlang i.public csclimat || SCH_ID: stlang, cov(uns)
. margins public, atmeans predict(mu fixedonly) vsquish

Adjusted predictions                           Number of obs   =      12793
Model VCE    : OIM

Expression   : Predicted mean, fixed portion only, predict(mu fixedonly)
at           : 0.stlang        =      .1552411 (mean)
               1.stlang        =      .8447589 (mean)
               0.public        =      .2211366 (mean)
```

```
            1.public       =    .7788634  (mean)
            csclimat       =   -3.55e-09  (mean)


                        |              Delta-method
                        |     Margin   Std. Err.       z    P>|z|     [95% Conf. Interval]
------------------------+---------------------------------------------------------------
                 public |
catholic or other private |  .9892548   .0022311   443.40   0.000     .984882    .9936277
                 public |  .9621884   .0030146   319.17   0.000     .9562798   .9680969
```

The contextual model without interactions (model 4) is refitted with the `quietly` command so that the output is suppressed. Since the `margins` command only works with factor variables when the predictor variables are categorical, the two categorical variables `stlang` and `public` are specified with the `i.` prefix. The command `margins public, atmeans predict(mu fixedonly) vsquish` tells Stata 13 to compute the estimated probabilities for the binary predictor variable `public` while holding the other predictor variables at their means. The `fixedonly` option needs to be used so that the random effects in the model (i.e., u_{0j} and u_{1j}) are set to zero. Otherwise, Stata will display an error message. The `vsquish` option saves space for the output.

In the output, the estimated margin or the probability for students in Catholic or other private schools is .989 when `public = 0` and other predictor variables are held at their means (`1.stlang` = .845, and `csclimat` = −3.55e−09). The estimated probability for students in public schools is .962 when `public = 1` and the other two predictor variables are held at their means.

To visualize the results, the command `marginsplot` is used next. Figure 10.1 shows the estimated probabilities of being proficient in doing simple arithmetical operations on whole numbers for students in public schools and private schools.

The graph created by `marginsplot` shows that students in Catholic or other private schools are more likely than students in public schools to be proficient in doing simple arithmetical operations on whole numbers.

Computing the Estimated Probabilities
With the Improved `margins` Command in Stata 14

To compute the estimated probabilities, the `margins` command in Stata 14 now can integrate over the random effects in multilevel modeling so that we do not need to set them to zero. The command `margins public, atmeans vsquish` tells Stata 14 to compute the estimated probabilities for the binary predictor variable `public` when holding the other predictor variables at their means. The same contextual model without interactions (model 4) is refitted before running the command.

Figure 10.1   Estimated Probabilities for the Binary Predictor Variable `public`

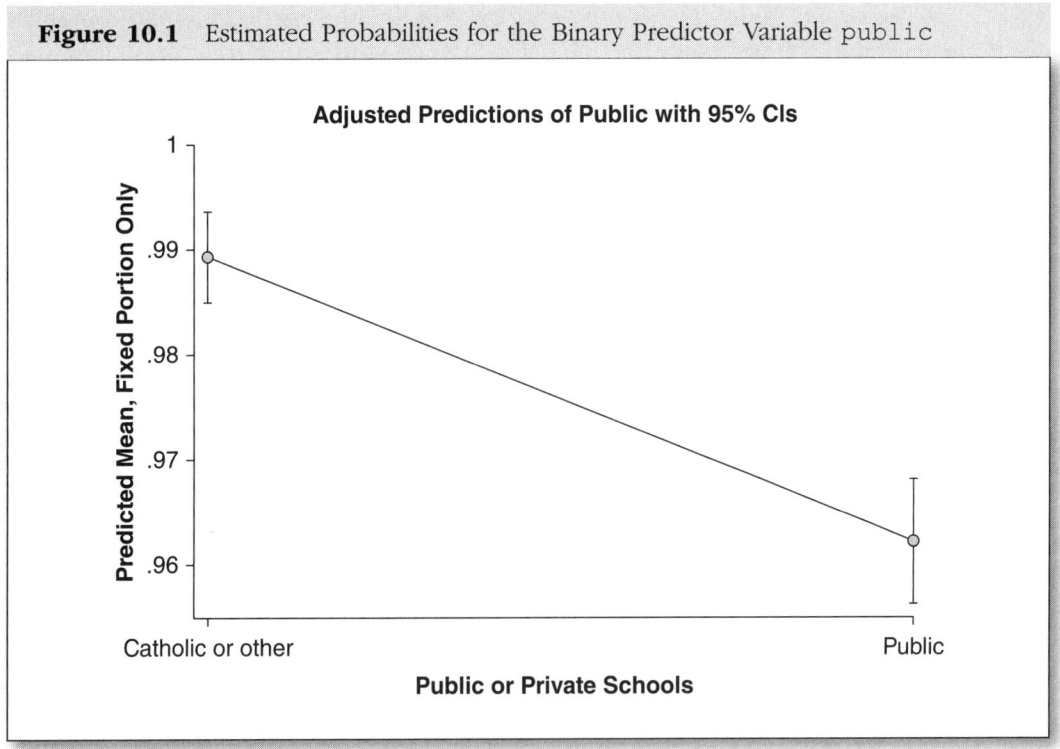

```
. quietly melogit Profmath1 i.stlang i.public csclimat || SCH_ID: stlang, cov(uns)
. margins public, atmeans vsquish

Adjusted predictions                            Number of obs     =      12,793
Model VCE     : OIM

Expression    : Marginal predicted mean, predict()
at            : 0.stlang      =      .1552411 (mean)
                1.stlang      =      .8447589 (mean)
                0.public      =      .2211366 (mean)
                1.public      =      .7788634 (mean)
                csclimat      =     -3.55e-09 (mean)

------------------------------------------------------------------------------
                        |            Delta-method
                        |     Margin   Std. Err.      z    P>|z|     [95% Conf. Interval]
------------------------+-----------------------------------------------------
                 public |
catholic or other private |  .9843652   .0030589   321.80   0.000     .9783699    .9903605
                 public |  .9476904   .0034751   272.71   0.000     .9408793    .9545014
------------------------------------------------------------------------------
```

The results in the output produced by Stata 14 are slightly different from those estimated in the preceding section using Stata 13 since the current version of Stata takes the random effects of the model into account whereas the older version sets them to zero. The `marginsplot` command can be used to visualize the estimated probabilities by following the `margins` command. The graph is omitted to save space.

10.3.5 Making Publication-Quality Tables

```
1.  .quietly melogit Profmath1 || SCH_ID:
2.  .outreg2 using chap10bout, dec(3) long word replace
3.  .outreg2 using chap10bout, eform long word append
4.  *.generate pub_lang= public*stlang
5.  *.generate cli_lang=csclimat*stlang
6.  .quietly melogit Profmath1 stlang public csclimat pub_lang cli_lang || SCH_ID:
    stlang, cov(uns)
7.  .outreg2 using chap10bout,  dec(3) long word append
8.  .outreg2 using chap10bout, eform long word append
```

To save space, we will make a table containing the results for only two models: the unconditional model and the contextual model with cross-level interactions. The commands for creating this table are explained as follows. Following the examples, readers may easily create a table containing the results for all the fitted models:

- The first command, `quietly melogit Profmath1|| SCH_ID`, estimates the unconditional model without showing the output.
- The second command, `outreg2 using chap10bout, dec(3) long word replace`, tells Stata to create a regression table and save the results containing both the fixed and random effects to a file named chap10bout.
- The third command, `outreg2 using chap10bout, eform long word append`, tells Stata to display odds ratios for the fixed effects with the `eform` option. With this option, the command `melogit, or` for the odds ratios can be omitted.
- The next two commands generate two cross-level interactions for the contextual model.
- The commands from six to eight fit the contextual model (model 5) and append the estimated results to the created table.

These commands automatically produce Table 10.3, as shown here in its original format, presenting the results for the unconditional model and the contextual model with cross-level interactions.

Table 10.3 Results of the Two Multilevel Logistic Regression Models: Unconditional Model and Contextual Model (Shown in Original Format Generated by Stata)

Variables	(1) Profmath1	(2) Profmath1	(3) Profmath1	(4) Profmath1
Profmath1				
stlang			0.270	1.310
			(0.556)	(0.728)
public			−1.405***	0.245***
			(0.457)	(0.112)
csclimat			0.415***	1.515***
			(0.153)	(0.232)
pub_lang			0.165	1.180
			(0.499)	(0.588)
cli_lang			0.254	1.289
			(0.168)	(0.217)
Profmath1				
Constant	3.472***	32.19***	3.433***	30.96***
	(0.086)	(2.769)	(0.510)	(15.78)
var(_cons[SCH_ID])				
Constant	1.339***	3.815***	1.077***	2.935***
	(0.190)	(0.724)	(0.354)	(1.039)
var(stlang[SCH_ID])				
Constant			0.688*	1.989*
			(0.362)	(0.720)
cov(_cons[SCH_ID],stlang[SCH_ID])				
Constant			−0.436	0.646
			(0.318)	(0.206)
Observations	12,793	12,793	12,793	12,793
Number of groups	619	619	619	619

Note: Standard errors in parentheses.

***$p < 0.01$, **$p < 0.05$, *$p < 0.1$

Please note that the logit coefficients and the corresponding odds ratios for the predictor variables may be incorrectly listed, so check the accuracy of the table after running the `outreg2` command. After adding the model titles, the labels for logit coefficients and odds ratios, and the names for the variances and the covariance on the original table, the final table is displayed as in Table 10.4.

Table 10.4 Results of the Two Multilevel Logistic Regression Models: Unconditional Model and Contextual Model (Edited)

	Unconditional Model (Model 1)		Contextual Model (Model 5)	
	Coefficient (SE)	Odds Ratio (SE)	Coefficient (SE)	Odds Ratio (SE)
Fixed Effects				
Stlang			0.270	1.310
			(0.556)	(0.728)
Public			−1.405***	0.245***
			(0.457)	(0.112)
Csclimat			0.415***	1.515***
			(0.153)	(0.232)
pub_lang			0.165	1.180
			(0.499)	(0.588)
cli_lang			0.254	1.289
			(0.168)	(0.217)
Intercept	3.472***	32.19***	3.433***	30.96***
	(0.086)	(2.769)	(0.510)	(15.78)
Random Effects				
var(_cons[SCH_ID])				
Between-School Variance	1.339***		1.077***	
	(0.190)		(0.354)	
var(stlang[SCH_ID])				
Slope Variance (stlang)			0.688*	
			(0.362)	

	Unconditional Model (Model 1)		Contextual Model (Model 5)	
	Coefficient (SE)	Odds Ratio (SE)	Coefficient (SE)	Odds Ratio (SE)
cov(_cons[SCH_ID],stlang[SCH_ID])				
Covariance			−0.436	
			(0.318)	
Observations	12,793	12,793	12,793	12,793
Number of groups	619	619	619	619

Note: Standard errors are shown in parentheses.

***$p < .01$.

10.3.6 Reporting the Results

The same reporting guidelines for multilevel modeling for continuous response variables can be applied to the reporting for the multilevel logistic regression modeling. The only difference is that the odds ratios may need to be reported and interpreted in the latter models. The following is an example of summarizing the results of a multilevel contextual logistic regression model with cross-level interactions.

Multilevel logistic regression models were fitted to predict whether high-school students were capable of doing simple arithmetical operations on whole numbers from three predictor variables: whether English is the students' native language, the school type, and the school climate. The first predictor was a student-level variable, and the other two were school-level variables. Multilevel modeling for binary response variables was used since the outcome variable was dichotomous. Five models, from the unconditional (null) model to the contextual model with cross-level interactions, were fitted. To save space, only the results for the final contextual model with cross-level interactions (model 5) were reported as follows.

**Results for the Contextual Model
With Cross-Level Interactions (Model 5)**

The logit coefficient for stlang (γ_{10}) = .270, z = .49, p = .627. The odds ratio for stlang was 1.310, which indicated that the odds of being proficient in math for students whose

(Continued)

(Continued)

native language is English were 1.310 times the odds for those whose native language is not English. However, this effect was not significant.

The logit coefficient for `public` (γ_{01}) = −1.405, z = −3.07, p < .01. The odds ratio for `public` is .245, which indicated that the odds of being proficient in math for students in public schools were .245 times the odds for those in private schools. In other words, being students in public schools decreased the odds of being proficient in math by 75.5%.

The logit coefficient for `csclimat` (γ_{02}) = .415, z = 2.71, p < .01. OR = 1.515, which indicated that for a one-unit increase in school climate, the odds of being above a particular category of math proficiency increased by a factor of 1.515. So the odds of being in higher math proficiency levels were associated with schools with a more positive school climate.

With regard to cross-level interactions, the coefficient for `pub_lang` (γ_{11}) = .165, z = .33, p > .05, and the coefficient for `cli_lang` (γ_{12}) = .254, z = 1.51, p > .05. So both interactions were not significant, which indicated that the effect of `stlang` does not differ in public or private schools. In addition, the effect of `stlang` was not influenced by school climate either.

Regarding the variance and covariance components for the random effects, the between-group variance, the intercept variance (τ_{00}), was 1.077. The slope variance (τ_{10}) for `stlang` was .688. The covariance between the intercept and the slope was −.436.

After two school-level variables and two cross-level interactions were included in the contextual model (model 5), the between-school variance (τ_{00}) decreased from 1.388 to 1.077 compared with that for the random-coefficient model (model 3): (1.388 − 1.077)/ 1.388 = 22%. This indicated that there was a decrease of 22% in the between-school variance from the random coefficient model (model 3) to the contextual model (model 5) after the two school-level variables were included.

The log likelihood ratio chi-square test was used to compare the contextual models with or without interactions (model 5 vs. model 4): $\chi^2_{(2)}$ = 2.19, p >.05, which indicated that there was no improvement in deviance from the contextual model without cross-level interactions to the contextual model without interactions. Therefore, the interaction terms should be removed from model 5.

10.4 Summary of Stata Commands in This Chapter

```
*Multilevel modeling for continuous outcome variables
use chap10-els2002, clear

*Deleting missing values and grand-mean centering sclimate
drop if missing(mathach, efficacy, public, sclimate)
```

```
summarize mathach efficacy public sclimate
generate csclimat = sclimate -r(mean)
gen gceffic = efficacy -r(mean)

*Unconditional means model (Model 1)
xtmixed mathach || SCH_ID: , mle var
estimates store null

*Random-intercept model (Model 2)
xtmixed mathach gceffic || SCH_ID: , mle var
estimates store ranint

*Log likelihood ratio test comparing the unconditional model and random-intercept
model
lrtest null ranint

*Random-coefficient model (Model 3)
xtmixed mathach gceffic || SCH_ID: gceffic, cov(uns) mle var
estimates store ranslop1

*Log likelihood ratio test
lrtest ranint ranslop1

*Contextual Model without Cross-Level Interactions (Model 4)
xtmixed mathach gceffic public csclimat || SCH_ID: gceffic, cov(uns) mle var
estimates store ranslop2
lrtest ranslop1 ranslop2

*Creating interaction terms
generate pub_eff= public*gceffic
generate cli_eff=csclimat*gceffic

*Contextual Model with Cross-Level Interactions (Model 5)
xtmixed mathach gceffic public csclimat pub_eff cli_eff ///
|| SCH_ID: gceffic, cov(uns) mle var
estimates store ranslop3
lrtest ranslop2 ranslop3

*Making publication-quality tables using outreg2
quietly xtmixed mathach || SCH_ID: , mle var
outreg2 using chap10aout, dec(3) long word replace
*generate pub_eff= public*gceffic
*generate cli_eff=csclimat*gceffic
quietly xtmixed mathach gceffic public csclimat pub_eff cli_eff ///
 || SCH_ID: gceffic, cov(uns) mle var
outreg2 using chap10aout, dec(3) long word append
```

*Multilevel modeling for binary response variables

```
use chap10-els2002, clear
drop if missing(Profmath1, stlang, public, sclimate)
summarize Profmath1 stlang public sclimate efficacy
generate csclimat = sclimate -r(mean)
sum csclimat

*Unconditional model (Model 1)
melogit Profmath1 || SCH_ID:
```

```
melogit, or
estimates store null

*Random-intercept model (Model 2)
melogit Profmath1 stlang || SCH_ID:
melogit, or
estimates store ranint

*Log likelihood ratio test
lrtest null ranint

*Random-coefficient model (Model 3)
melogit Profmath1 stlang || SCH_ID: stlang, cov(uns)
melogit, or
estimates store ranslop1

*Log likelihood ratio test
lrtest ranint ranslop1

*Contextual Model without Cross-Level Interactions (Model 4)
melogit Profmath1 stlang public csclimat || SCH_ID: stlang, cov(uns)
melogit, or
estimates store ranslop2

*Log likelihood ratio test
lrtest ranslop1 ranslop2

*Creating interaction terms
generate pub_lang= public*stlang
generate cli_lang=csclimat*stlang

*Contextual Model with Cross-Level Interactions (Model 5)
melogit Profmath1 stlang public csclimat pub_lang cli_lang || SCH_ID: stlang, cov(uns)
melogit, or
estimates store ranslop3

*Log likelihood ratio test
lrtest ranslop2 ranslop3

*Computing estimated probabilities with the margins command in Stata 13
quietly melogit Profmath1 i.stlang i.public csclimat || SCH_ID: stlang, cov(uns)
margins public, atmeans predict(mu fixedonly) vsquish
marginsplot

*margins in Stata 14
quietly melogit Profmath1 i.stlang i.public csclimat || SCH_ID: stlang, cov(uns)
margins public, atmeans vsquish

*Making publication-quality tables using outreg2
quietly melogit Profmath1 || SCH_ID: ,
outreg2 using chap10bout, dec(3) long word replace
outreg2 using chap10bout, eform long word append
*generate pub_lang= public*stlang
*generate cli_lang=csclimat*stlang
quietly melogit Profmath1 stlang public csclimat pub_lang cli_lang ///
 || SCH_ID: stlang, cov(uns)
outreg2 using chap10bout,  dec(3) long word append
outreg2 using chap10bout, eform long word append
```

10.5 EXERCISES

Use the ELS:2002 data for the following problems. The following variables are used for the multilevel modeling.

> mathach: mathematics IRT scores of high-school students
>
> Profmath2: math proficiency level 2 (1 = capable of doing simple operations with decimals, fractions, powers, and roots; 0 = not achieving this level)
>
> gender: gender (1 = female; 0 = male)
>
> byses: socioeconomic status composite for base-year data
>
> usecalc: using calculators
>
> usecompu: using computers
>
> urban: urban schools or not (1 = urban; 0 = suburban or rural areas) (school-level variable)

1. We will conduct a study investigating the relationships between students' math achievement and a set of student-level predictors and a school-level predictor. The outcome variable is mathach, and the predictor variables are gender, byses, usecalc, usecompu, and urban. The multilevel modeling for the continuous response variable will be used for data analysis. Answer the following questions or perform the following analyses:

    a. Fit an unconditional model and obtain the between-group variance. Compute the ICC. What does it tell us?

    b. Fit a two-level, random-intercept model with a random intercept and four student-level predictor variables gender, byses, usecalc, and usecompu. Use the grand-mean centering for byses, usecalc, and usecompu before fitting the model.

    c. Interpret the fixed effects of four predictor variables.

    d. Conduct a likelihood ratio test comparing the random-intercept model and the unconditional model. Interpret the results.

    e. Fit a two-level, contextual model with gender, byses, usecalc, and usecompu as the student-level predictor variables and urban as the school-level predictor variable. Specify a random intercept and a random slope for byses. No cross-level interactions are included.

    f. Write level 1 and level 2 equations for the contextual model presented.

g. Interpret the coefficient of urban in the contextual model.

h. Among the three fitted models, which one fits the data best? Why?

2. Multilevel logistic regression models will be fitted to estimate whether students are capable of doing simple operations with decimals, fractions, powers, and roots. The binary outcome variable is Profmath2, and the predictor variables are gender, byses, usecalc, and urban. Answer the following questions or perform the following analyses:

a. Fit an unconditional model and obtain the between-group variance. Compute the ICC and interpret it.

b. Fit a two-level, random-intercept model with a random intercept and three student-level predictor variables gender, byses, and usecalc. Use the grand-mean centering for byses and usecalc before fitting the model.

c. Interpret the odds ratios of three predictor variables.

d. Conduct a likelihood ratio test comparing the random-intercept model and the unconditional model. Interpret the results.

e. Fit a two-level, contextual model with gender, byses, and usecalc as the student-level predictor variables and urban as the school-level predictor variable. Specify a random intercept and a random slope for usecalc. No cross-level interactions are included.

f. Write level 1 and level 2 equations for the contextual model in Exercise 2e.

g. Interpret the odds ratio of urban in the contextual model.

h. Among the three fitted models, which one fits the data best? Why?

i. Use the margins command to calculate the estimated probabilities of being proficient in math for gender when holding other predictor variables at their means.

j. Make a publication-quality table containing the estimated logit coefficients and odds ratios for the contextual model.

k. Write a report to summarize the results of the contextual model.

CHAPTER 11

Multilevel Modeling for Ordinal Response Variables

Objectives of This Chapter

This chapter introduces multilevel modeling for ordinal response variables. It starts with an introduction of the model followed by a discussion of the model specification, the odds and odds ratios in the model, and the log likelihood ratio test, as well as a description of the research questions and data. Then several models are illustrated using Stata with step-by-step instructions, including the unconditional (null) model, the random-intercept model, the random-coefficient model with a level 1 variable, the contextual model with both level 1 and level 2 variables, and the contextual model with cross-level interactions. Stata commands are explained, and the output is interpreted for each model in detail. This chapter also illustrates how the results are displayed in publication-quality tables using the Stata command and reported in text. It focuses on model fitting with Stata, as well as on interpreting and presenting the results. After reading this chapter, you should be able to

- Determine when multilevel modeling for ordinal response variables is used.
- Formulate multilevel models for ordinal response variables.
- Fit multilevel models for ordinal response variables using `meologit`, `meglm`, or `gllamm`.
- Interpret the output.

(Continued)

(Continued)

- Compute and interpret the intraclass correlation coefficient (ICC).
- Compute and plot the estimated probabilities using the `margins` and `marginsplot` commands, respectively.
- Be familiar with model fitting strategies.
- Compare models using log likelihood ratio tests and AIC and BIC statistics.
- Present results in publication-quality tables using Stata.
- Write the results for publication.

11.1 Multilevel Modeling for Ordinal Response Variables: An Introduction

Multilevel proportional odds (PO) models (Goldstein, 2003; Hedeker, 2007; Hedeker & Gibbons, 1994; Hox, 2010; O'Connell, Goldstein, Rogers, & Peng, 2008; Rabe-Hesketh & Skrondal, 2012; Raudenbush & Bryk, 2002; Snijders & Bosker, 2012) are an extension of multilevel binary logistic regression models when the response variables are ordinal. They are also a generalization of single-level proportional odds models when the data structure is nested or hierarchical. Just like the proportional odds models, they estimate the cumulative odds of being at or below a particular category of the response variable, or the inversed odds, the odds of being above that particular category. The cumulative odds of being at or below a particular ordinal category are the probability of being at or below a category divided by the probability of being above that category. On the other hand, the cumulative odds of being above a particular category compare the probability of being above a category with the probability of being at or below that particular category. Since below and above a particular category are two opposite directions, these two probabilities are complementary, and thus, these two odds are inversed.

With the nested data structure, the multilevel modeling for ordinal response variables allows us to estimate the relationships between predictor variables at different levels and the ordinal response variable. Similar to multilevel logistic regression models, level 2 models can also specify varying intercepts and slopes, so the multilevel proportional odds models also estimate variance–covariance components for the random effects.

11.1.1 Model Specification

The level 1 equation for the multilevel PO model can be expressed as:

$$\text{Level 1: } \text{logit}[\pi_{kij}(Y \le k)] = \ln\left(\frac{\pi\left(Y_{ij} \le k \,|\, x_1, x_2, \ldots, x_p\right)}{\pi\left(Y_{ij} > k \,|\, x_1, x_2, \ldots, x_p\right)}\right) = \alpha_k + (-\beta_{1j}X_{1ij} - \beta_{2j}X_{2ij} - \ldots - \beta_{pj}X_{pij}) \quad (11.1)$$

where α_k are the cut points with $k = 1, 2,..., K - 1$; X_{1ij}, X_{2ij}, ..., X_{pij} are the predictor variables for the ith individual in the jth cluster; and β_{1j}, β_{2j}, ..., β_{pj} are the logit coefficients of these predictors in the jth cluster.

The model specification for multilevel PO models is more complex than those for the multilevel modeling for continuous and binary response variables. People may often skip equations since the details of the equations are not well explained. Do not feel intimidated by the subscripts in the equation. To help you understand the equation (Equation 11.1), detailed explanations are as follows.

First, the subscripts i and j mean the ith individual in the jth cluster. For example, they mean the ith student in the jth school when students are nested within schools. To understand the equation, we can first ignore these subscripts and put them back later. Without the subscripts, this model is just the single-level PO model introduced in Chapter 4. It estimates the relationship between a set of independent variables and the ordinal outcome variable on a scale of the logit or log odds.

Second, let us look at the three parts of the equation with these subscripts. The left side of the equation, $\text{logit}[\pi_{kij}(Y \leq k)]$, is the link function. It transforms the original ordinal response variable Y for an ith individual in the jth cluster into the transformed outcome, logit or log odds of the cumulative probability of being at or below a particular category k. Recall that the odds of being at or below a particular category in proportional odds models equals the ratio of the probability of being at or below a category to the probability of being above that category, which is expressed in the middle of the equation. The right side of the equation describes the linear combination of the predictor variables with each logit coefficient associated with cluster j.

Just as with the single-level PO model, the multilevel PO model also assumes that the logit coefficients for each predictor variable are the same across the categories of the ordinal outcome variable. In other words, the logit coefficients for the underlying binary logistic regression models that dichotomize the ordinal response variable are the same. This is known as the proportional odds or parallel lines assumption. Do not feel confused with the varying coefficients (slopes) across clusters. Although the logit coefficients for each predictor do not vary across the categories of the ordinal outcome variable, they are allowed to vary across clusters in the multilevel data structure.

Unlike the multilevel modeling for binary response variables where an intercept is estimated, the multilevel PO models estimate a series of cut points for the underlying binary logistic models for the dichotomized ordinal response variable. When the ordinal response variable has k levels, $K - 1$ cut points will be estimated. As with the randomly varying intercept in the multilevel modeling for binary response variables, the cut points in the multilevel PO models may also be allowed to vary across clusters.

Since the odds of being above a particular category are the inversed odds of being at or below that particular category of the response variable, to estimate the odds of being above a particular category versus being at or below that category, the level 1 equation can also be expressed as:

$$\text{Level 1: logit}[\pi_{kij}(Y > k)] = \ln\left(\frac{\pi\left(Y_{ij} > k \mid x_1, x_2, \ldots, x_p\right)}{\pi\left(Y_{ij} \le k \mid x_1, x_2, \ldots, x_p\right)}\right) = -\alpha_k + (\beta_{1j}X_{1ij} + \beta_{2j}X_{2ij} + \ldots + \beta_{pj}X_{pij}) \quad (11.2)$$

Please be aware that the signs before the cut points and the coefficient in Equation 11.2 are reversed from those in Equation 11.1.

Model Specification: An Example of a Basic Two-Level PO Model

To understand the specification of multilevel modeling with both level 1 and level 2 equations, let us specify a basic two-level PO model with one predictor variable at each level of the equations. For example, researchers are interested in estimating the relationships between an ordinal response variable, student's mathematics proficiency, and a student-level variable, student's native language, and a school-level variable, public. The level 1 predictor variable is whether student's native language is English (stlang), and the level 2 predictor is whether a school is public or private (public). Both the cut points and the slope of the level 1 predictor variable stlang are allowed to vary randomly across schools.

The level 1 equation is similar to that for the single-level PO model, whereas in multilevel modeling, students are nested with schools so we have a subscript i to represent students and a subscript j to represent schools. As with the single-level PO model, no error term is specified in the level 1 equation:

$$\text{Level 1: } \eta_{ij} = \text{logit}[\pi_{kij}(Y \le k)] = \ln\left(\frac{\pi\left(Y_{ij} \le k\right)}{\pi\left(Y_{ij} > k\right)}\right) = \alpha_k - (\beta_{0j} + \beta_{1j}\,\text{stlang}_{ij}) \quad (11.3)$$

$$\text{Level 2: } \beta_{0j} = \gamma_{00} + \gamma_{01}(\text{public}_j) + u_{0j} \quad (11.4)$$

$$\beta_{1j} = \gamma_{10} + \gamma_{11}(\text{public}_j) + u_{1j} \quad (11.5)$$

The equations at level 2 look the same as those for the multilevel models for continuous outcome variables and binary response variables. Both the intercept and the coefficient (i.e., the slope of stlang) are allowed to vary randomly across schools at level 2 equations, where they are specified as the outcome variables.

In the level 1 equation, η_{ij} or logit(π_{kij}) is the logit link for the cumulative probability of being at or below a particular category k for the ith student in the jth school. It can be expressed as:

$$\ln\left(\frac{\pi\left(Y_{ij} \le k\right)}{\pi\left(Y_{ij} > k\right)}\right)$$

In the level 2 equation, γ_{00}, γ_{01}, γ_{10}, and γ_{11} are level 2 coefficients, which are the fixed effects. γ_{00} is the intercept, the overall logit or log odds of being at or below a particular math proficiency across schools, and u_{0j} is an error term at the school level. In Stata, the overall intercept is fixed to zero for identification purposes and is removed from the model, so u_{0j} is an error term associated with the cut points, the random variation for the cut points; γ_{10} is the overall average slope for `stlang` when the school-level predictor `public` is zero (`public = 0`), and u_{1j} is the error term associated with the random slope, the random variation for the level 1 slope across schools; γ_{01} is the fixed effect of the level 2 predictor `public` on the log odds of being at or below a particular ordinal category; and γ_{11} is the cross-level interaction between `stlang` and `public`.

The level 1 and level 2 equations can be expressed in a combined form as follows:

$$\eta_{ij} = \text{logit}[\pi_{kij}(Y_{ij} \le k)] = \ln \left(\frac{\pi(Y_{ij} \le k)}{\pi(Y_{ij} > k)} \right) = \alpha_k - (\gamma_{00} + \gamma_{01}\,\text{public}_j + \gamma_{10}\,\text{stlang}_{ij} +$$

$$\gamma_{11}\text{public}_j \times \text{stlang}_{ij} + u_{0j} + u_{1j}\text{stlang}_{ij}) \tag{11.6}$$

In Stata, the intercept γ_{00} is set to be zero and the cut points are estimated. Other software packages may estimate the overall intercept and set the first cut point to be zero. After removing the intercept, we can rewrite this composite equation as follows:

$$\eta_{ij} = \text{logit}[\pi_{kij}(Y_{ij} \le k)] = (\alpha_k - u_{0j}) - (\gamma_{01}\text{public}_j + \gamma_{10}\text{stlang}_{ij} +$$

$$\gamma_{11}\text{public}_j \times \text{stlang}_{ij} + u_{1j}\text{stlang}_{ij}) \tag{11.7}$$

We can easily see that α_k are the estimated cut points and that u_{0j} is the random effect associated with the cut points. When the cut points are allowed to vary randomly across schools, u_{0j} represents the deviation from the cut points across schools.

To estimate the cumulative odds of being above a particular ordinal category versus being at or below that category, an alternative form for the level 1 equation is as follows:

$$\text{Level 1: } \eta_{ij} = \text{logit}[\pi_{kij}(Y_{ij} > k)] = \ln \left(\frac{\pi(Y_{ij} > k)}{\pi(Y_{ij} \le k)} \right) = -\alpha_k + (\beta_{0j} + \beta_{1j}\text{stlang}_{ij}) \tag{11.8}$$

The level 2 equations remain the same:

$$\text{Level 2: } \beta_{0j} = \gamma_{00} + \gamma_{01}\text{public}_j + u_{0j} \tag{11.9}$$

$$\beta_{1j} = \gamma_{10} + \gamma_{11}\text{public}_j + u_{1j} \tag{11.10}$$

The combined form can be expressed as:

$$\eta_{ij} = \text{logit}[\pi_{kij}(Y_{ij} > k)] = (-\alpha_k + u_{0j}) + \gamma_{01}\text{public}_j + \gamma_{10}\text{stlang}_{ij} +$$

$$\gamma_{11}\text{public}_j \times \text{stlang}_{ij} + u_{1j}\text{stlang}_{ij} \tag{11.11}$$

11.1.2 Odds and Odds Ratios in Multilevel PO Models

As with the single-level conventional PO model, the odds and odds ratios can be interpreted in a similar way. The cumulative odds of being at or below a category in ordinal logistic regression are the cumulative probability of being at or below a category divided by the probability of being above that category:

$$\text{Odds } (Y_{ij} \leq k) = \frac{p(Y_{ij} \leq k)}{p(Y_{ij} > k)}$$

where the cumulative probability $p(Y_{ij} \leq k)$ equals the sum of the probabilities of all categories at or below that category. The corresponding odds ratio of being above a particular category in the PO model is the exponentiated logit coefficient $\exp(\beta)$. It is interpreted as the change in the odds of being above a particular category for each one-unit change in a predictor variable.

Similarly, the inversed cumulative odds, the odds of being above a particular category versus being at or below that category, are the ratio of the cumulative probability of being above that category $p(Y_{ij} > k)$ to its complementary probability $p(Y_{ij} \leq k)$.

$$\text{Odds } (Y_{ij} > k) = \frac{p(Y_{ij} > k)}{p(Y_{ij} \leq k)}$$

The odds ratio of being at or below a particular category in the PO model is the exponentiated negative coefficient $\exp(-\beta)$. It is interpreted as the change in the odds of being at or below a particular category for each one-unit change in a predictor variable.

11.1.3 Likelihood Ratio Test

As with the multilevel binary logistic regression, the log likelihood ratio test can be used to compare nested models since several types of Gauss–Hermite quadrature can be used as integration methods to estimate log likelihood values. The difference in –2 log likelihood (–2LL) or deviance is expressed as $G = D_{\text{Reduced}} - D_{\text{Full}}$. It is the difference between the deviance for the reduced model and that for the full model. When model A is nested within model B, the log likelihood ratio test $G = D_{\text{ModelA}} - D_{\text{ModelB}}$. The likelihood ratio test statistic follows a chi-square distribution with the degrees of freedom of the distribution equal to the difference in the number of parameters between these two models.

In addition to the log likelihood ratio test, the Akaike information criterion (AIC) and Bayesian information criterion (BIC) statistics can be used to compare nested models. One advantage of these two statistics is that they can also be used to compare non-nested models.

11.2 Research Example: Research Problem and Questions

Using the Educational Longitudinal Study of 2002 (ELS:2002), researchers are interested in investigating the relationships between an ordinal response variable, students' math proficiency, and a student-level variable, student's native language, and two school-level variables, school type and school climate. The research questions are as follows:

1. Can the cumulative odds of being at or below a particular mathematics proficiency level be significantly predicted by whether English is the student's native language?

2. Do school characteristics, such as school type and school climate, significantly impact the odds of being at or below a particular level of mathematics proficiency?

3. Does the relationship between the student's native language and the odds of being at or below a particular level of mathematics proficiency vary across schools?

4. Are there any interaction effects between the two school-level variables (i.e., school type and school climate) and student's native language? In other words, does the relationship between student's native language and the cumulative odds of being at or below a particular mathematics proficiency level vary across school type and school climate?

11.2.1 Description of the Data and Sample

The ELS:2002 base-year data are used for the following analyses. The variables are listed as follows:

- `Profmath`: mathematics proficiency with six levels from 0 to 5
- `stlang`: whether English is the student's native language
- `public`: school type (1 = public, 0 = private and others)
- `sclimat`: school climate
- `csclimat`: school climate (grand-mean centered)

For demonstration purposes, the missing values of the variables are deleted. The descriptive statistics are summarized, and the school-level variable `sclimat` is grand-mean centered. The commands and the output are displayed as follows.

```
. drop if missing(Profmath, stlang, public, sclimate)
(3459 observations deleted)

. tab Profmath

    Profmath |      Freq.      Percent        Cum.
-------------+-----------------------------------
           0 |        642         5.02        5.02
           1 |      2,970        23.22       28.23
           2 |      2,716        21.23       49.46
           3 |      3,683        28.79       78.25
           4 |      2,683        20.97       99.23
           5 |         99         0.77      100.00
-------------+-----------------------------------
       Total |     12,793       100.00

. summarize stlang public sclimate

    Variable |        Obs         Mean    Std. Dev.         Min         Max
-------------+-----------------------------------------------------------
      stlang |      12793     .8447589    .3621486           0           1
      public |      12793     .7788634    .4150285           0           1
    sclimate |      12793     3.922561    .6735278         1.2           5

. generate csclimat = sclimate -r(mean)
```

11.3 Multilevel Modeling for Ordinal Response Variables With Stata: Commands and Output

The Stata Command meologit

The Stata command meologit, which was introduced in Stata 13, is used for mixed-effects ordinal logistic regression models. If you are using an earlier version of Stata, then you can use the user-written command gllamm to estimate these models (Rabe-Hesketh, Skrondal, & Pickles, 2004). The syntax structure of meologit for mixed-effects ordinal logistic regression models is similar to those of xtmixed for the linear mixed models and melogit for the mixed-effects logistic regression models. The command meologit is immediately followed by the dependent variable with or without the predictor variable(s). Next, after two vertical lines (||), we specify the random effects of the model. The options are separated from the command by a comma. For example, the command meologit y x || schid:, cov(unstructured) tells Stata to fit a multilevel model to estimate an ordinal response variable y on a predictor variable x with a randomly varying intercept across schools by specifying ||schid:. The cov(unstructured) option requests an unstructured variance–covariance matrix for the random effects. Just as with melogit, neither the mle nor the var option works with meologit.

The same estimation methods for multilevel modeling for binary response variables can be applied to multilevel modeling for ordinal response variables. Stata uses the full maximum likelihood estimation with numerical integration for both models. This method provides more accurate estimates than the quasi-likelihood methods, such as the marginal quasi-likelihood and the penalized quasi-likelihood methods. However, when using the numerical integration method, you may have converging problems. Sometimes it takes longer to converge or it does not converge at all. The default integration method is the mean and variance adaptive Gauss–Hermite quadrature, which allows for model comparisons using the likelihood ratio test. Discussion of the technical details of the estimation methods is beyond the scope of the text.

11.3.1 Unconditional Model or Null Model (Model 1)

The unconditional model or null model for the ordinal response variable includes no predictor variable at any level. This model estimates the overall cumulative probabilities of being at or below a particular category and the variability in the cumulative probabilities between groups (e.g., schools). It can be expressed as follows:

$$\text{Level 1: logit}[\pi_{kij}(Y \le k)] = \ln\left(\frac{\pi\left(Y_{ij} \le k\right)}{\pi\left(Y_{ij} > k\right)}\right) = \alpha_k - \beta_{0j}$$

$$\text{Level 2: } \beta_{0j} = \gamma_{00} + u_{0j}$$

Since γ_{00} is set to be zero in Stata, the composite model is $\eta_{ij} = \text{logit}[\pi_{kij}(Y_{ij} \le k)] = a_k - u_{0j}$.

In multilevel ordinal logistic regression models, $\text{logit}(\pi_{kij})$ is the logit link for the cumulative probability of being at or below a particular category k for the ith student in the jth school. It can be expressed as:

$$\ln\left(\frac{\pi\left(Y_{ij} \le k\right)}{\pi\left(Y_{ij} > k\right)}\right)$$

In the level 2 equation, γ_{00} is the overall logit or log odds of being at or below a particular math proficiency across schools, and u_{0j} is an error term at the school level. The intercept γ_{00} is set to zero since the cut points α_k are estimated.

Stata Command and Output

The command `meologit Profmath || SCH_ID:` is used to fit the unconditional model (model 1). The `meologit Profmath` command tells Stata to fit a mixed ordinal logistic regression model for the ordinal response variable `Profmath`. Since this is the unconditional model, no predictor variables are specified. The command `SCH_ID:` specifies the random effects at the school level by the identifier variable `SCH_ID`. No random coefficients for any predictor variables are specified in this model. The second command, meologit, or, requests odds ratios. The third command, estimates store null, saves the estimates named "null" for future use. The output is displayed as follows.

```
. meologit Profmath || SCH_ID:

Fitting fixed-effects model:

Iteration 0:   log likelihood = -19725.193
Iteration 1:   log likelihood = -19725.193

Refining starting values:

Grid node 0:   log likelihood = -18938.381

Fitting full model:

Iteration 0:   log likelihood = -18938.381
Iteration 1:   log likelihood = -18819.038
Iteration 2:   log likelihood = -18809.513
Iteration 3:   log likelihood = -18809.263
Iteration 4:   log likelihood = -18809.263

Mixed-effects ologit regression          Number of obs    =      12793
Group variable:            SCH_ID         Number of groups =        619

                                          Obs per group: min =          2
                                                         avg =       20.7
                                                         max =         50

Integration method: mvaghermite          Integration points =          7

                                          chi2()           =          .
Log likelihood = -18809.263               Prob > chi2      =          .
-------------------------------------------------------------------------
   Profmath |    Coef.   Std. Err.      z    P>|z|    [95% Conf. Interval]
------------+------------------------------------------------------------
      /cut1 |  -3.29173   .0575999   -57.15  0.000   -3.404623   -3.178836
      /cut2 | -1.069975   .0438133   -24.42  0.000   -1.155848   -.9841025
      /cut3 | -.0002538   .0427716    -0.01  0.995   -.0840846    .0835771
      /cut4 |  1.516327   .0450073    33.69  0.000    1.428115     1.60454
      /cut5 |  5.319195   .1104609    48.15  0.000    5.102696    5.535694
------------+------------------------------------------------------------
SCH_ID      |
  var(_cons)|  .8882411   .0637512                    .7716814    1.022407
-------------------------------------------------------------------------
LR test vs. ologit regression:   chibar2(01) =  1831.86 Prob>=chibar2 = 0.0000

. meologit, or

Mixed-effects ologit regression          Number of obs    =      12793
Group variable:            SCH_ID         Number of groups =        619
```

```
                                        Obs per group: min =           2
                                                        avg =        20.7
                                                        max =          50

Integration method: mvaghermite          Integration points =           7

                                          chi2()              =           .
Log likelihood = -18809.263               Prob > chi2         =           .
------------------------------------------------------------------------------
   Profmath | Odds Ratio   Std. Err.      z    P>|z|     [95% Conf. Interval]
------------+-----------------------------------------------------------------
      /cut1 |   .0371895    .0021421   -57.15   0.000     .0332193    .0416341
      /cut2 |  -1.069975    .0438133   -24.42   0.000    -1.155848   -.9841025
      /cut3 |  -.0002538    .0427716    -0.01   0.995    -.0840846    .0835771
      /cut4 |   1.516327    .0450073    33.69   0.000     1.428115     1.60454
      /cut5 |   5.319195    .1104609    48.15   0.000     5.102696    5.535694
------------+-----------------------------------------------------------------
SCH_ID      |
  var(_cons)|   .8882411    .0637512                      .7716814    1.022407
------------------------------------------------------------------------------
LR test vs. ologit regression:   chibar2(01) =  1831.86 Prob>=chibar2 = 0.0000

. estimates store null
```

Interpreting the Output

The output for the `meologit` command shown earlier is similar to that for the mixed-effects logistic regression model using the `melogit` command. The estimation log is displayed at the top. After the log likelihood estimation for the full model is the estimation table for the mixed-effects ordinal logistic regression, including both fixed effects and random effects.

A total of 12,793 observations in level 1 are nested in 619 schools in level 2. The range of observations per group is from 2 to 50 with an average of 20.7. The integration method is the mean and variance adaptive Gauss–Hermite quadrature (`mvaghermite`), and the default number of integration points is 7. As with the `melogit` command, other types of integration methods, such as the mode and curvature adaptive Gauss–Hermite quadrature (`mcaghermite`), nonadaptive Gauss–Hermite quadrature (`ghermite`), or the Laplacian approximation (`laplace`) can be specified in the option.

Just as with the other mixed-effects unconditional models, the results of the Wald chi-square test and the associated p value are reported as missing since no predictor variables are included in the null model.

The estimation table has two parts. The first section reports the fixed effects, and the second section at the bottom reports the variance components for the random effects of the model. The output displays the estimates for the cut points, their standard errors, the Wald z statistics, the associated p values, and the 95% confidence intervals.

Five cut points (labeled /cut1, /cut2, /cut3, /cut4, and /cut5) are reported. As with the single-level proportional odds models, the cut points are used to differentiate the adjacent levels of the latent variable Y^* (i.e., math proficiency). The latent variable falls at or below the first cut point α_1 when the response category is 0. It falls between the first cut point α_1 and the second cut point α_2 when the response category is 1. By using the same method to divide the regions by other cut points, the response category reaches 5 if the latent variable is at or beyond the fifth cut point α_5.

In this example, the observed mathematics proficiency is an ordinal response variable with six levels ranging from 0 to 5. It is defined as follows:

$$y = \begin{cases} 0 & \text{if } y^* \leq \alpha_1 \\ 1 & \text{if } \alpha_1 < y^* \leq \alpha_2 \\ 2 & \text{if } \alpha_2 < y^* \leq \alpha_3 \\ 3 & \text{if } \alpha_3 < y^* \leq \alpha_4 \\ 4 & \text{if } \alpha_4 < y^* \leq \alpha_5 \\ 5 & \text{if } \alpha_5 < y^* \leq \infty \end{cases} \tag{11.12}$$

The variance components for the random effects are separated from the estimated cut points. Since no random variables are specified in the model, only the between-school variance (τ_{00}) is reported. The row labeled var(_cons) reports that the between-school variance (τ_{00}) is .888, which is the intercept variance across all schools. The ratio of the variance in the intercept and its standard error is 13.875 (.888/.064 = 13.875). The value is larger than 2, which indicates that the between-school variance is significantly different from zero.

The log likelihood ratio test comparing the unconditional model with the single-level ordinal logistic regression suggests that the unconditional model be preferred (chibar2(01)= 1831.86, Prob>=chibar2 = 0.0000). The results also suggest that the between-school variance (τ_{00}) is significant.

The intraclass correlation coefficient (ICC) in the multilevel ordinal logistic regression can be computed in the same way as that for the multilevel logistic regression. Since the variance of the logistic distribution is $\pi^2/3 = 3.29$, ICC = $\tau_{00}/(\tau_{00} + 3.29) = .888/(.888 + 3.29) = .213$, which indicates that 21.3% of the total variance is accounted for by schools in level 2. Therefore, using multilevel ordinal logistic regression is appropriate for the analysis.

With the command meologit, or, we can obtain odds ratios for the fixed effects. Since this is the unconditional model, only the odds ratio for the intercept is reported.

11.3.2 Random-Intercept Model (Model 2)

In the random-intercept model, we include the predictor variable stlang (whether a student's native language is English) in the level 1 equation. The model is referred to as the random-intercept model since the intercept is allowed to vary randomly across schools. This model can be expressed as follows:

$$\text{Level 1: logit}[\pi_{kij}(Y \leq k)] = \ln\left(\frac{\pi(Y_{ij} \leq k)}{\pi(Y_{ij} > k)}\right) = \alpha_k - (\beta_{0j} + \beta_{1j}\text{stlang}_{ij})$$

Level 2: $\beta_{0j} = \gamma_{00} + u_{0j}$

$\beta_{1j} = \gamma_{10}$

Substituting the level 2 equations into the level 1 equation, the composite model is expressed as follows:

$$\eta_{ij} = \text{logit}[\pi_{kij}(Y_{ij} \leq k)] = (\alpha_k - u_{0j}) - (\gamma_{10}\text{stlang}_{ij})$$

The level 1 equation includes the predictor variable, student's native language (stlang), and the ordinal response variable (Profmath). The level 2 equation for the intercept is specified to be random, and the slope for stlang is constrained to be fixed.

Stata Command and Output

The command meologit Profmath stlang || SCH_ID: is used to fit the random-intercept model (model 2). The fixed part of the model includes a predictor variable stlang immediately after the ordinal response variable. Just as with the unconditional model, no random coefficients are specified in the random part of the model. The second command, meologit, or, requests the odds ratios for the fixed effects. The third command, estimates store ranint, saves the estimates in a file named "ranint" for the following likelihood ratio test. The following output is displayed.

```
. meologit Profmath stlang || SCH_ID:

Fitting fixed-effects model:

Iteration 0:    log likelihood = -19725.193
Iteration 1:    log likelihood = -19665.556
Iteration 2:    log likelihood = -19665.521
Iteration 3:    log likelihood = -19665.521

Refining starting values:

Grid node 0:    log likelihood = -18911.751

Fitting full model:

Iteration 0:    log likelihood = -18911.751
Iteration 1:    log likelihood = -18799.543
```

```
Iteration 2:   log likelihood = -18783.827
Iteration 3:   log likelihood = -18783.193
Iteration 4:   log likelihood = -18783.191
Iteration 5:   log likelihood = -18783.191

Mixed-effects ologit regression            Number of obs     =      12793
Group variable:          SCH_ID            Number of groups  =        619

                                           Obs per group: min =          2
                                                          avg =       20.7
                                                          max =         50

Integration method: mvaghermite            Integration points =         7

                                           Wald chi2(1)      =      52.19
Log likelihood = -18783.191                Prob > chi2       =     0.0000
-----------------------------------------------------------------------------
    Profmath |    Coef.   Std. Err.      z    P>|z|    [95% Conf. Interval]
-------------+---------------------------------------------------------------
      stlang |  .3787448   .0524265    7.22   0.000    .2759908    .4814989
-------------+---------------------------------------------------------------
       /cut1 |  -2.97676   .0716262  -41.56   0.000   -3.117144   -2.836375
       /cut2 | -.7488928   .0620082  -12.08   0.000   -.8704266    -.627359
       /cut3 |  .3235997   .0616847    5.25   0.000    .2026999    .4444996
       /cut4 |  1.841916   .0635884   28.97   0.000    1.717285    1.966547
       /cut5 |  5.643419   .1192678   47.32   0.000    5.409659     5.87718
-------------+---------------------------------------------------------------
SCH_ID       |
  var(_cons) |   .866438    .062509                    .7521903    .9980384
-----------------------------------------------------------------------------
LR test vs. ologit regression:   chibar2(01) =  1764.66 Prob>=chibar2 = 0.0000

. meologit, or

Mixed-effects ologit regression            Number of obs     =      12793
Group variable:          SCH_ID            Number of groups  =        619

                                           Obs per group: min =          2
                                                          avg =       20.7
                                                          max =         50

Integration method: mvaghermite            Integration points =         7

                                           Wald chi2(1)      =      52.19
Log likelihood = -18783.191                Prob > chi2       =     0.0000
-----------------------------------------------------------------------------
    Profmath | Odds Ratio Std. Err.      z    P>|z|    [95% Conf. Interval]
-------------+---------------------------------------------------------------
      stlang |   1.46045   .0765663    7.22   0.000    1.317836    1.618498
-------------+---------------------------------------------------------------
       /cut1 |  -2.97676   .0716262  -41.56   0.000   -3.117144   -2.836375
       /cut2 | -.7488928   .0620082  -12.08   0.000   -.8704266    -.627359
       /cut3 |  .3235997   .0616847    5.25   0.000    .2026999    .4444996
       /cut4 |  1.841916   .0635884   28.97   0.000    1.717285    1.966547
       /cut5 |  5.643419   .1192678   47.32   0.000    5.409659     5.87718
-------------+---------------------------------------------------------------
```

```
SCH_ID        |
   var(_cons)|    .866438     .062509                    .7521903    .9980384
-----------------------------------------------------------------------------
LR test vs. ologit regression:   chibar2(01) =   1764.66 Prob>=chibar2 = 0.0000

. estimates store ranint
```

Interpreting the Output

The estimation table reports the coefficient for stlang and five cut points or thresholds. The logit coefficient for stlang (γ_{10}) = .379, z = 7.22, p < .001, which indicates that the cumulative logit or log odds of being above a particular math proficiency level for students with English as their native language is .379 times as large as that for those whose native language is not English. The odds ratio for stlang is reported with the meologit command with the or option. Interpretation of odds ratios in multilevel ordinal logistic regression is the same as that for the single-level ordinal logistic regression. Odds ratio (OR) = 1.460, which means that the odds of being above a particular math proficiency level for students whose native language is English are 1.460 times as great as the odds for those whose native language is not English.

The row labeled var(_cons) reports that the intercept variance (τ_{00}) is .866, which is significantly different from 0 (chibar2(01)= 1764.66). The intercept variance here is the variance associated with the cut points.

Log Likelihood Ratio Test Comparing the
Unconditional Model and Random-Intercept Model

```
. lrtest null ranint

Likelihood-ratio test                        LR chi2(1)   =      52.14
(Assumption: null nested in ranint)          Prob > chi2 =      0.0000
```

We can compare the random-intercept model (model 2) and the unconditional model (model 1) using the log likelihood ratio test: $\chi^2_{(1)}$ = 52.14, p < .001, which indicates that the random-intercept model fits the data better than the unconditional model, so we are in favor of the random-intercept model.

11.3.3 Random-Coefficient Model With a Level 1 Variable (Model 3)

Both intercepts and coefficients at level 1 are allowed to vary across schools at level 2 so the model estimates variance–covariance components. This model is referred to as the random-coefficient model since both the level 1 intercept and slope are allowed to vary across schools. When the variance–covariance components are specified as unstructured, the variances in intercepts and coefficients/slopes and the covariance between them will be estimated. This model can be expressed as follows:

$$\text{Level 1: logit}[\pi_{kij}(Y \le k)] = \ln\left(\frac{\pi\left(Y_{ij} \le k\right)}{\pi\left(Y_{ij} > k\right)}\right) = \alpha_k - (\beta_{0j} + \beta_{1j}\text{stlang}_{ij})$$

$$\text{Level 2: } \beta_{0j} = \gamma_{00} + u_{0j}$$
$$\beta_{1j} = \gamma_{10} + u_{1j}$$

Since Stata set γ_{00} to zero, the composite model is expressed as follows:

$$\eta_{ij} = \text{logit}[\pi_{kij}(Y_{ij} \le k)] = (\alpha_k - u_{0j}) - (\gamma_{10}\text{stlang}_{ij} + u_{1j}\text{stlang}_{ij})$$

The level 1 equation includes the predictor variable, student's native language (stlang), and the ordinal response variable (Profmath). In the level 2 equations, the level 1 intercept and the slope for stlang are both specified to be random with level 2 residuals u_{0j} and u_{1j}, respectively.

Stata Command and Output

The command meologit Profmath stlang || SCH_ID: , stlang, cov(uns) is used to fit the random-coefficient model (model 3). The fixed part of the model is the same as that of the random-intercept model. Unlike the random-intercept model, the predictor variable stlang is specified in the random part of the model, which indicates that the slope of the level 1 predictor stlang is allowed to vary randomly across schools. The next command, meologit, or, requests the odds ratios. The third command, estimates store ranslop1, saves the estimates in a file named "ranslop1" for the future likelihood ratio test. The following results are produced with the earlier command.

```
. meologit Profmath stlang || SCH_ID: stlang, cov(uns)

Fitting fixed-effects model:

Iteration 0:   log likelihood = -19725.193
Iteration 1:   log likelihood = -19665.556
Iteration 2:   log likelihood = -19665.521
Iteration 3:   log likelihood = -19665.521

Refining starting values:

Grid node 0:   log likelihood = -19016.483
```

```
Fitting full model:

Iteration 0:   log likelihood = -19016.483  (not concave)
Iteration 1:   log likelihood = -18916.991
Iteration 2:   log likelihood = -18874.162
Iteration 3:   log likelihood = -18773.027
Iteration 4:   log likelihood = -18770.622
Iteration 5:   log likelihood =   -18770.6
Iteration 6:   log likelihood =   -18770.6

Mixed-effects ologit regression          Number of obs     =      12793
Group variable:          SCH_ID           Number of groups  =        619

                                          Obs per group: min =          2
                                                         avg =       20.7
                                                         max =         50

Integration method: mvaghermite           Integration points =          7

                                          Wald chi2(1)      =      41.88
Log likelihood =    -18770.6              Prob > chi2       =     0.0000
------------------------------------------------------------------------------
     Profmath |      Coef.   Std. Err.      z    P>|z|     [95% Conf. Interval]
--------------+---------------------------------------------------------------
       stlang |   .4173941   .0645001     6.47   0.000     .2909762    .5438121
--------------+---------------------------------------------------------------
        /cut1 |  -2.963759   .0816494   -36.30   0.000    -3.123789   -2.803729
        /cut2 |  -.7158377   .0741185    -9.66   0.000    -.8611073   -.5705681
        /cut3 |   .3628601   .0741764     4.89   0.000     .217477     .5082431
        /cut4 |   1.887071   .0760913    24.80   0.000    1.737935    2.036207
        /cut5 |   5.700663   .1266798    45.00   0.000    5.452376    5.948951
--------------+---------------------------------------------------------------
SCH_ID        |
  var(stlang) |   .3726675   .1217428                      .1964497    .7069547
   var(_cons) |    1.38836   .1658741                      1.098512    1.754685
--------------+---------------------------------------------------------------
SCH_ID        |
cov(_cons,stlang)| -.4717146   .1277357    -3.69   0.000   -.7220719   -.2213572
------------------------------------------------------------------------------
LR test vs. ologit regression:        chi2(3) =  1789.84   Prob > chi2 = 0.0000

Note: LR test is conservative and provided only for reference.

. meologit, or

Mixed-effects ologit regression          Number of obs     =      12793
Group variable:          SCH_ID           Number of groups  =        619

                                          Obs per group: min =          2
                                                         avg =       20.7
                                                         max =         50

Integration method: mvaghermite           Integration points =          7

                                          Wald chi2(1)      =      41.88
Log likelihood =    -18770.6              Prob > chi2       =     0.0000
```

```
                                                      max =           50

Integration method: mvaghermite              Integration points =       7

                                              Wald chi2(1)      =      41.88
Log likelihood =    -18770.6                  Prob > chi2       =     0.0000
-----------------------------------------------------------------------------
      Profmath | Odds Ratio  Std. Err.     z    P>|z|    [95% Conf. Interval]
---------------+-------------------------------------------------------------
        stlang |   1.518001  .0979113    6.47   0.000    1.337733    1.722561
---------------+-------------------------------------------------------------
         /cut1 |  -2.963759  .0816494  -36.30   0.000   -3.123789   -2.803729
         /cut2 |  -.7158377  .0741185   -9.66   0.000   -.8611073   -.5705681
         /cut3 |   .3628601  .0741764    4.89   0.000     .217477    .5082431
         /cut4 |   1.887071  .0760913   24.80   0.000    1.737935    2.036207
         /cut5 |   5.700663  .1266798   45.00   0.000    5.452376    5.948951
---------------+-------------------------------------------------------------
SCH_ID         |
   var(stlang) |   .3726675  .1217428                     .1964497    .7069547
    var(_cons) |    1.38836  .1658741                     1.098512    1.754685
---------------+-------------------------------------------------------------
SCH_ID         |
cov(_cons,stlang)| -.4717146  .1277357   -3.69   0.000   -.7220719   -.2213572
-----------------------------------------------------------------------------
LR test vs. ologit regression:        chi2(3) =   1789.84   Prob > chi2 = 0.0000

Note: LR test is conservative and provided only for reference.

. estimates store ranslop1
```

Interpreting the Output

In the estimation table, the coefficient for stlang and five cut points or thresholds are similar to those for the random-intercept model (model 2). The logit coefficient for stlang (γ_{10}) = .417, z = 6.47, p < .001, which indicates that the cumulative logit or log odds of being above a particular category of math proficiency for students with English as their native language is .417 times as great as that for those whose native language is not English. The odds ratio for stlang is reported with the meologit command with the or option: OR = 1.518, which indicates that being students whose native language is English increases the odds of being above a particular math proficiency level by a factor of 1.460.

We are interested in the variance and covariance components for the random effects since the slope of stlang is random across schools. The row labeled var(_cons) reports that the intercept variance (τ_{00}) is 1.388, which is significantly different from 0 (1.388/.166 = 8.361, which is larger than 2). The row labeled var(stlang) reports that the slope variance (τ_{10}) for stlang is .373, which is also significant (.373/.122 = 3.057 > 2). The covariance between the intercept and the slope (labeled cov(_cons,stlang)) is −.472, p < .001.

Model Comparison Using the Likelihood Ratio Test

```
. lrtest ranint ranslop1

Likelihood-ratio test                              LR chi2(2)    =      25.18
(Assumption: ranint nested in ranslop1)            Prob > chi2 =     0.0000

Note: The reported degrees of freedom assumes the null hypothesis is not on the
boundary of the parameter space.  If this is not true, then the reported test is
conservative.
```

The log likelihood ratio test is used to compare the random-intercept model (model 2) and random-coefficient model with a level 1 predictor (model 3). The log likelihood ratio chi-square test $\chi^2_{(2)} = 25.18$, $p < .01$, which indicates that the random-coefficient model has a significantly better fit than the random-intercept model.

11.3.4 Contextual Model With Both Level 1 and Level 2 Variables (Model 4)

The contextual model is a random-coefficient model with both level 1 and level 2 predictor variables. The level 1 predictor stlang remains unchanged. Two school-level variables, public and csclimat, are added to the level 2 equation:

$$\text{Level 1: logit}[\pi_{kij}(Y \leq k)] = \ln\left(\frac{\pi(Y_{ij} \leq k)}{\pi(Y_{ij} > k)}\right) = \alpha_k - (\beta_{0j} + \beta_{1j}\text{stlang}_{ij})$$

$$\text{Level 2: } \beta_{0j} = \gamma_{00} + \gamma_{01}\text{public}_j + \gamma_{02}\text{csclimat}_j + u_{0j}$$
$$\beta_{1j} = \gamma_{10} + u_{1j}$$

The level 1 equation includes the predictor variable, the student's native language (stlang), and the ordinal response variable (Profmath). The level 2 equations for the level 1 intercept and the slope for stlang are both specified as random with level 2 residuals u_{0j} and u_{1j}, respectively. In the level 2 equations, two school-level predictor variables are added to the equation for the level 1 random intercept, whereas the equation for the random slope contains no predictor variables. The composite model is expressed as follows:

$$\eta_{ij} = \text{logit}[\pi_{kij}(Y_{ij} \leq k)] = (\alpha_k - u_{0j}) - (\gamma_{01}\text{public}_j + \gamma_{02}\text{csclimat}_j + \gamma_{10}\text{stlang}_{ij} + u_{1j}\text{stlang}_{ij})$$

Stata Command and Output

The command meologit Profmath stlang public csclimat || SCH_ID: , stlang, cov(uns) is used to fit the contextual model (model 4). The fixed part of the model includes one level 1 predictor, stlang, and two level 2 predictor variables, public and csclimat. The syntax itself does not distinguish between the level 1 and level 2 predictor variables. The predictor variable stlang is specified in the random part of the model, which indicates that the slope of the level 1 predictor stlang is allowed to vary randomly across schools. The cov(uns) option tells Stata that the variance–covariance matrix is unstructured. To request the odds ratios for the fixed effects, we use the command meologit, or. The last command estimates store ranslop2 saves the estimates in a file named "ranslop2" for the future likelihood ratio test. The following output is presented.

```
. meologit Profmath stlang public csclimat || SCH_ID: stlang, cov(uns)

Fitting fixed-effects model:

Iteration 0:    log likelihood = -19725.193
Iteration 1:    log likelihood = -19241.474
Iteration 2:    log likelihood = -19239.506
Iteration 3:    log likelihood = -19239.505

Refining starting values:

Grid node 0:    log likelihood = -18905.481

Fitting full model:

Iteration 0:    log likelihood = -18905.481   (not concave)
Iteration 1:    log likelihood = -18798.539   (not concave)
Iteration 2:    log likelihood = -18755.066
Iteration 3:    log likelihood =  -18687.45
Iteration 4:    log likelihood = -18684.234
Iteration 5:    log likelihood =  -18684.18
Iteration 6:    log likelihood =  -18684.18

Mixed-effects ologit regression         Number of obs      =      12793
Group variable:         SCH_ID          Number of groups   =        619

                                        Obs per group: min =          2
                                                       avg =       20.7
                                                       max =         50

Integration method: mvaghermite         Integration points =          7

                                        Wald chi2(3)       =     242.82
Log likelihood =  -18684.18             Prob > chi2        =     0.0000
-----------------------------------------------------------------------
     Profmath |    Coef.    Std. Err.     z    P>|z|   [95% Conf. Interval]
--------------+--------------------------------------------------------
       stlang |  .4067733    .06491     6.27   0.000    .2795521    .5339945
       public | -.5721059   .0889826   -6.43   0.000   -.7465085   -.3977033
     csclimat |  .5351121    .05594     9.57   0.000    .4254718    .6447524
--------------+--------------------------------------------------------
```

```
            /cut1 |  -3.410053    .1053497   -32.37   0.000   -3.616534   -3.203571
            /cut2 |  -1.165185    .0994702   -11.71   0.000   -1.360143    -.9702267
            /cut3 |  -.0865876    .0991816    -0.87   0.383    -.28098     .1078049
            /cut4 |   1.438011    .1000437    14.37   0.000    1.241929    1.634093
            /cut5 |   5.246318    .1418344    36.99   0.000    4.968328    5.524308
-----------------+----------------------------------------------------------------
SCH_ID           |
      var(stlang)|   .420955     .1262007                      .2339099    .7575699
      var(_cons) |  1.097754     .1439413                      .8489703    1.419442
-----------------+----------------------------------------------------------------
SCH_ID           |
cov(_cons,stlang)|  -.4853097    .1226906    -3.96   0.000    -.7257789   -.2448405
----------------------------------------------------------------------------------
LR test vs. ologit regression:        chi2(3) =  1110.65   Prob > chi2 = 0.0000

Note: LR test is conservative and provided only for reference.

. meologit, or

Mixed-effects ologit regression              Number of obs       =     12793
Group variable:          SCH_ID               Number of groups    =       619

                                              Obs per group: min =         2
                                                             avg =      20.7
                                                             max =        50

Integration method: mvaghermite               Integration points =         7

                                              Wald chi2(3)        =    242.82
Log likelihood = -18684.18                    Prob > chi2         =    0.0000
----------------------------------------------------------------------------------
       Profmath | Odds Ratio   Std. Err.     z    P>|z|    [95% Conf. Interval]
-----------------+----------------------------------------------------------------
          stlang |  1.501964    .0974924     6.27   0.000    1.322537    1.705732
          public |   .5643357   .050216      -6.43   0.000    .4740187    .6718613
         csclimat|  1.70764     .0955253     9.57   0.000    1.530312    1.905515
-----------------+----------------------------------------------------------------
            /cut1 |  -3.410053    .1053497   -32.37   0.000   -3.616534   -3.203571
            /cut2 |  -1.165185    .0994702   -11.71   0.000   -1.360143    -.9702267
            /cut3 |  -.0865876    .0991816    -0.87   0.383    -.28098     .1078049
            /cut4 |   1.438011    .1000437    14.37   0.000    1.241929    1.634093
            /cut5 |   5.246318    .1418344    36.99   0.000    4.968328    5.524308
-----------------+----------------------------------------------------------------
SCH_ID           |
      var(stlang)|   .420955     .1262007                      .2339099    .7575699
      var(_cons) |  1.097754     .1439413                      .8489703    1.419442
-----------------+----------------------------------------------------------------
SCH_ID           |
cov(_cons,stlang)|  -.4853097    .1226906    -3.96   0.000    -.7257789   -.2448405
----------------------------------------------------------------------------------
LR test vs. ologit regression:        chi2(3) =  1110.65   Prob > chi2 = 0.0000

Note: LR test is conservative and provided only for reference.

. estimates store ranslop2
```

Interpreting the Output

The fixed-effects part of the estimation table reports the coefficient for stlang, two new school-level predictors public and csclimat, and five cut points or thresholds. The logit coefficient for stlang (γ_{10}) = .407, z = 6.27, p < .001. The result is similar to that in the random-coefficient table (model 3), so the interpretation for this variable is omitted here. Our interpretation focuses on the other two newly added level 2 predictor variables.

The logit coefficient for public (γ_{01}) = −.572, z = −6.43, p < .001, which indicates that being in public schools decreases the logit of being above a particular category of math proficiency by a factor of .572. The corresponding odds ratio can be obtained by exponentiating the coefficient. The odds ratio is reported in the output by the meologit command with the or option. OR = .564, which indicates that a one-unit increase in public is associated with a decrease of .572 points in the odds of being above a particular category of math proficiency. In other words, the odds of being above a particular category for students in a public school are .564 times as great as the odds for students in a private school when holding all the other predictors constant.

The logit coefficient for csclimat (γ_{02}) = .535, z = 9.57, p < .001. OR = 1.708, which indicates that for a one-unit increase in school climate, the odds of being above a particular category of math proficiency increase by a factor of 1.708. So the odds of being in higher math proficiency levels are associated with schools with a more positive school climate.

Regarding the variance and covariance components for the random effects, the row labeled var(_cons) reports that the intercept variance (τ_{00}) is 1.098, which is significantly different from 0 (1.098/.144 = 7.625, which is larger than 2). The row labeled var(stlang) reports that the slope variance (τ_{10}) for stlang is .421, which is also significant (.421/.126 = 3.341). The covariance between the intercept and the slope (labeled cov(_cons,stlang)) is −.485, p < .001.

After two school-level variables are included in the contextual model (model 4), the between-school variance (τ_{00}) decreases from 1.388 to 1.098 when compared with that for the random-coefficient model (model 3): (1.388 − 1.098)/1.388 = 21%. This indicates that there is a decrease of 21% in the between-school variance from the random-coefficient model (model 3) to the contextual model (model 4) after the two school-level variables are included.

Model Comparisons Using the Likelihood Ratio Test

```
. lrtest ranslop1 ranslop2

Likelihood-ratio test                             LR chi2(2)  =    172.84
(Assumption: ranslop1 nested in ranslop2)         Prob > chi2 =    0.0000
```

The log likelihood ratio test is still used to compare the contextual model (model 4) and random-coefficient model (model 3). The log likelihood ratio chi-square test $\chi^2_{(2)}$ = 172.84, $p < .01$, which indicates that we are in favor of the contextual model (model 4).

11.3.5 Contextual Model With Cross-Level Interactions (Model 5)

The contextual model with cross-level interactions is used to investigate whether there are interaction effects between level 1 and level 2 predictor variables. For example, if we are interested in whether the two school-level variables public and csclimat interact with the level 1 predictor stlang, we can include two interaction terms in the model. The two interaction terms need to be created before model fitting. One interaction term, pub_lang, is created for two categorical variables, stlang and public; and the other interaction term, cli_lang, is for the categorical variable stlang and the continuous variable csclimat. This model can be expressed as follows:

$$\text{Level 1: logit}[\pi_{kij}(Y \le k)] = \ln\left(\frac{\pi(Y_{ij} \le k)}{\pi(Y_{ij} > k)}\right) = \alpha_k - (\beta_{0j} + \beta_{1j}\text{stlang}_{ij})$$

$$\text{Level 2: } \beta_{0j} = \gamma_{00} + \gamma_{01}\text{public}_j + \gamma_{02}\text{csclimat}_j + u_{0j}$$
$$\beta_{1j} = \gamma_{10} + \gamma_{11}\text{public}_j + \gamma_{12}\text{csclimat}_j + u_{1j}$$

The expression of this model (model 5) is similar to that of the previous contextual model (model 4). The only difference is that the level 2 equation for the slope of stlang now includes two predictor variables, public and csclimat.

The composite model is expressed as follows:

$$\eta_{ij} = \text{logit}[\pi_{kij}(Y_{ij} \le k)] = (\alpha_k - u_{0j}) - (\gamma_{01}\text{public}_j + \gamma_{02}\text{csclimat}_j + \gamma_{10}\text{stlang}_{ij} +$$
$$\gamma_{11}\text{public}_j \times \text{stlang}_{ij} + \gamma_{12}\text{csclimat}_j \times \text{stlang}_{ij} + u_{1j}\text{stlang}_{ij})$$

Stata Command and Output

The command meologit Profmath stlang public csclimat pub_lang cli_lang || SCH_ID: , stlang, cov(uns) is used to fit the contextual model with cross-level interactions (model 5). The fixed part of the model includes one level 1 predictor, stlang, two level 2 predictor variables, public and csclimat, and two interaction terms, pub_lang and cli_lang. The random part of the model and the options are the same as those for model 4. The command meologit, or is still used to request the odds ratios for the fixed effects. The last command estimates store ranslop3 saves the estimates in a file named "ranslop3" for the model comparison using the likelihood ratio test. The following output is presented.

```
. generate pub_lang= public*stlang
. generate cli_lang=csclimat*stlang
. meologit Profmath stlang public csclimat pub_lang cli_lang || SCH_ID: stlang,
cov(uns)

Fitting fixed-effects model:

Iteration 0:   log likelihood = -19725.193
Iteration 1:   log likelihood = -19238.627
Iteration 2:   log likelihood = -19236.559
Iteration 3:   log likelihood = -19236.558

Refining starting values:

Grid node 0:   log likelihood = -18905.072

Fitting full model:

Iteration 0:   log likelihood = -18905.072  (not concave)
Iteration 1:   log likelihood = -18798.185  (not concave)
Iteration 2:   log likelihood = -18754.802
Iteration 3:   log likelihood = -18686.569
Iteration 4:   log likelihood = -18683.247
Iteration 5:   log likelihood = -18683.189
Iteration 6:   log likelihood = -18683.188
```

```
Mixed-effects ologit regression            Number of obs     =     12793
Group variable:          SCH_ID            Number of groups  =       619

                                           Obs per group: min =         2
                                                          avg =      20.7
                                                          max =        50

Integration method: mvaghermite            Integration points =        7

                                           Wald chi2(5)      =    245.13
Log likelihood = -18683.188                Prob > chi2       =    0.0000
```

Profmath	Coef.	Std. Err.	z	P>|z|	[95% Conf. Interval]	
stlang	.23695	.1606481	1.47	0.140	-.0779144	.5518145
public	-.7640119	.1916095	-3.99	0.000	-1.13956	-.388464
csclimat	.5826373	.1051742	5.54	0.000	.3764997	.7887749
pub_lang	.2005767	.1796246	1.12	0.264	-.1514811	.5526345
cli_lang	-.0516191	.0981976	-0.53	0.599	-.2440829	.1408448
/cut1	-3.575194	.1757905	-20.34	0.000	-3.919737	-3.230651
/cut2	-1.328874	.1714471	-7.75	0.000	-1.664905	-.9928442
/cut3	-.249987	.1710504	-1.46	0.144	-.5852396	.0852657
/cut4	1.274624	.1714537	7.43	0.000	.9385808	1.610667
/cut5	5.083578	.1984976	25.61	0.000	4.69453	5.472626

SCH_ID						
var(stlang)	.4355385	.1276881			.2451756	.7737057
var(_cons)	1.103144	.1443305			.8536207	1.425607

SCH_ID						
cov(_cons,stlang)	-.4950848	.1235309	-4.01	0.000	-.737201	-.2529687

```
LR test vs. ologit regression:        chi2(3) =  1106.74   Prob > chi2 = 0.0000

Note: LR test is conservative and provided only for reference.

. meologit, or

Mixed-effects ologit regression                Number of obs      =      12793
Group variable:          SCH_ID                 Number of groups   =        619

                                                Obs per group: min =          2
                                                               avg =       20.7
                                                               max =         50

Integration method: mvaghermite                 Integration points =          7

                                                Wald chi2(5)       =     245.13
Log likelihood = -18683.188                     Prob > chi2        =     0.0000
------------------------------------------------------------------------------
     Profmath | Odds Ratio  Std. Err.      z    P>|z|     [95% Conf. Interval]
--------------+---------------------------------------------------------------
       stlang |   1.267378   .2036018     1.47   0.140     .9250436    1.736401
       public |    .465794   .0892506    -3.99   0.000     .3199599     .6780976
     csclimat |   1.790755   .1883412     5.54   0.000     1.457175    2.200699
     pub_lang |   1.222107   .2195206     1.12   0.264     .8594341    1.737825
     cli_lang |   .9496906   .0932574    -0.53   0.599     .7834227    1.151246
--------------+---------------------------------------------------------------
        /cut1 |  -3.575194   .1757905   -20.34   0.000    -3.919737   -3.230651
        /cut2 |  -1.328874   .1714471    -7.75   0.000    -1.664905    -.9928442
        /cut3 |   -.249987   .1710504    -1.46   0.144    -.5852396     .0852657
        /cut4 |   1.274624   .1714537     7.43   0.000     .9385808    1.610667
        /cut5 |   5.083578   .1984976    25.61   0.000      4.69453    5.472626
--------------+---------------------------------------------------------------
SCH_ID        |
  var(stlang) |   .4355385   .1276881                      .2451756     .7737057
   var(_cons) |   1.103144   .1443305                      .8536207    1.425607
--------------+---------------------------------------------------------------
SCH_ID        |
cov(_cons,stlang)| -.4950848 .1235309    -4.01   0.000     -.737201    -.2529687
------------------------------------------------------------------------------
LR test vs. ologit regression:        chi2(3) =  1106.74   Prob > chi2 = 0.0000

Note: LR test is conservative and provided only for reference.

. estimates store ranslop3
```

Interpreting the Output

The estimation table displays the logit coefficients for the level 1 and level 2 predictor variables and two interaction terms. The logit coefficient for pub_lang (γ_{11}) = .201, $z = 1.12$, $p > .05$, and the logit coefficient for cli_lang (γ_{12}) = −.052, $z = −.53$, $p > .05$. So both interactions are not significant. Since the variance and covariance components for model 5 look similar to those for model 4, the interpretation is omitted here.

Model Comparisons Using the Likelihood Ratio Test

```
. lrtest ranslop2 ranslop3

Likelihood-ratio test                              LR chi2(2)  =      1.98
(Assumption: ranslop2 nested in ranslop3)          Prob > chi2 =    0.3711
```

We again use the log likelihood ratio test to compare the contextual model with cross-level interactions (model 5) and the contextual model without interactions (model 4). The log likelihood ratio chi-square test $\chi^2_{(2)} = 1.98$, $p = .37$, which indicates that there is no significant improvement in deviance from model 4 to model 5. Therefore, the two interaction terms need to be removed from the model.

11.3.6 Model Comparisons Using the AIC and BIC Statistics

The command `estimates stats` can be used to display the AIC and BIC statistics for all the fitted models listed earlier following the `estimates store` command. The strength of the `estimates stats` command is that it can provide the AIC and BIC statistics for more than one model. The command is followed immediately by the saved estimates for five models. Recall that `null`, `ranint`, `ranslop1`, `ranslop2`, and `ranslop3` are the names for the stored estimates for models 1 to 5, respectively.

```
. estimates stats null ranint ranslop1 ranslop2 ranslop3

Akaike's information criterion and Bayesian information criterion

-------------------------------------------------------------------------------
    Model |     Obs    ll(null)   ll(model)      df         AIC          BIC
----------+--------------------------------------------------------------------
     null |   12793           .   -18809.26       6    37630.53     37675.27
   ranint |   12793           .   -18783.19       7    37580.38     37632.58
 ranslop1 |   12793           .    -18770.6       9     37559.2     37626.31
 ranslop2 |   12793           .   -18684.18      11    37390.36     37472.38
 ranslop3 |   12793           .   -18683.19      13    37392.38     37489.31
-------------------------------------------------------------------------------
           Note:  N=Obs used in calculating BIC; see [R] BIC note
```

In the output, we can see that the AIC and BIC statistics decrease dramatically from the null model (model 1) to the contextual model without interactions (model 4: ranslop2), whereas both statistics increase slightly from model 4 (labeled `ranslop2`) to model 5 (labeled `ranslop3`). For example, the AIC for model 4 is 37,390.36 and that for model 5 is 37,392.38. The difference in AIC between these two models is only 2.02, which is small. Although there is no test to compare the difference in the AIC statistic, this provides additional information to confirm with the likelihood ratio test that there is no significant improvement from model 4 to model 5.

11.3.7 Computing the Estimated Probabilities With the `margins` Command

Using the `margins` *and* `marginsplot` *Commands in Stata 13*

As with the single-level PO model, the `margins` command can also be used to compute the margins or estimated probabilities of being in each category of the ordinal response variable. In Stata 13, we need to run the `margins` command separately for each category of an ordinal response variable. In the following example, the ordinal response variable, mathematics proficiency (`Profmath`), has six levels from 0 to 5, so we need to execute the program six times with the `predict(fixedonly outcome())` option. For illustration purposes, we only compute the estimated probabilities for mathematics proficiency levels 1 and 4.

The contextual model without interactions (model 4) is refitted with the `quietly` command so that the output is not shown. The `i.` prefix needs to be specified for the two categorical variables `stlang` and `public` since the `margins` command only works with factor variables when categorical predictor variables are included in the model. The command `margins public, atmeans predict(fixedonly outcome(1)) vsquish` tells Stata to compute the estimated probabilities when the ordinal response variable $Y = 1$, for the binary predictor variable `public`, while holding the other predictor variables at their means. The `fixedonly` option is specified so that the random effects in the model (i.e., u_{0j} and u_{1j}) are set to zero.

```
. quietly meologit Profmath i.stlang i.public csclimat || SCH_ID: stlang, cov(ur

. margins public, atmeans predict(fixedonly outcome(1)) vsquish

Adjusted predictions                              Number of obs   =      12793
Model VCE    : OIM

Expression   : Predicted mean (1.Profmath), fixed portion only, predict(fixedonl
outcome(1))
at           : 0.stlang        =      .1552411  (mean)
               1.stlang        =      .8447589  (mean)
               0.public        =      .2211366  (mean)
               1.public        =      .7788634  (mean)
               csclimat        =     -3.55e-09  (mean)

-----------------------------------------------------------------------------
```

```
                            |          Delta-method
                            |   Margin    Std. Err.      z     P>|z|     [95% Conf. Interva
----------------------------+-----------------------------------------------------------
                   public   |
catholic or other private   |  .1590029   .0100498    15.82   0.000     .1393058    .1787C
                   public   |  .2421911   .0073028    33.16   0.000     .2278778    .2565C
----------------------------+-----------------------------------------------------------

. marginsplot

  Variables that uniquely identify margins: public
```

In the output, the estimated margin or the probability of being in math proficiency level 1 (i.e., when $Y = 1$) for students in Catholic or other private schools is .159 when `public = 0` and other predictor variables are held at their means (`1.stlang = .845`, and `csclimat = -3.55e-09`). The estimated probability for students in public schools is .242 when `public = 1` and the other two predictor variables are held at their means.

Similarly, we can also compute the estimated probability of being in the other categories. To save space, the output for the expected probabilities for categories 0, 2, 3, and 5 is omitted. With the `predict(fixedonly outcome(4))` option, we can compute the estimated probability of being in category 4 (i.e., $Y = 4$). The output is displayed as follows.

```
. margins public, atmeans predict(fixedonly outcome(4)) vsquish

Adjusted predictions                            Number of obs   =       12793
Model VCE    : OIM

Expression   : Predicted mean (4.Profmath), fixed portion only, predict(fixedonl
outcome(4))
at           : 0.stlang        =     .1552411  (mean)
               1.stlang        =     .8447589  (mean)
               0.public        =     .2211366  (mean)
               1.public        =     .7788634  (mean)
               csclimat        =    -3.55e-09  (mean)

-----------------------------------------------------------------------------
                            |          Delta-method
                            |   Margin    Std. Err.      z     P>|z|     [95% Conf. Interva
----------------------------+-----------------------------------------------------------
                   public   |
catholic or other private   |   .24438    .0142158    17.19   0.000     .2165175    .27224
                   public   |  .155636    .0058018    26.83   0.000     .1442647    .1670C
----------------------------+-----------------------------------------------------------

. marginsplot

  Variables that uniquely identify margins: public
```

In the output, the estimated margin or the probability of being in math proficiency level 4 (i.e., when $Y = 4$) for students in Catholic or other private schools is .244 when public = 0 and other predictor variables are held at their means (1.stlang = .845, and csclimat = -3.55e-09). The estimated probability for students in public schools is .155 when public = 1 and the other two predictor variables are held at their means. Figure 11.1 shows the estimated probabilities when $Y = 4$.

The graph created by marginsplot shows that students in Catholic or other private schools are more likely to be in category 4 of the math proficiency than students in public schools.

Computing the Estimated Probabilities With the
Improved margins *and* marginsplot *Commands in Stata 14*

The improved margins command in Stata 14 can simultaneously provide the results for all categories of an ordinal response variable, so we do not need to run it multiple times with the predict() option. The command margins public,

Figure 11.1 Estimated Probabilities When $Y = 4$ for public (1 = public; 0 = Catholic or Other Private)

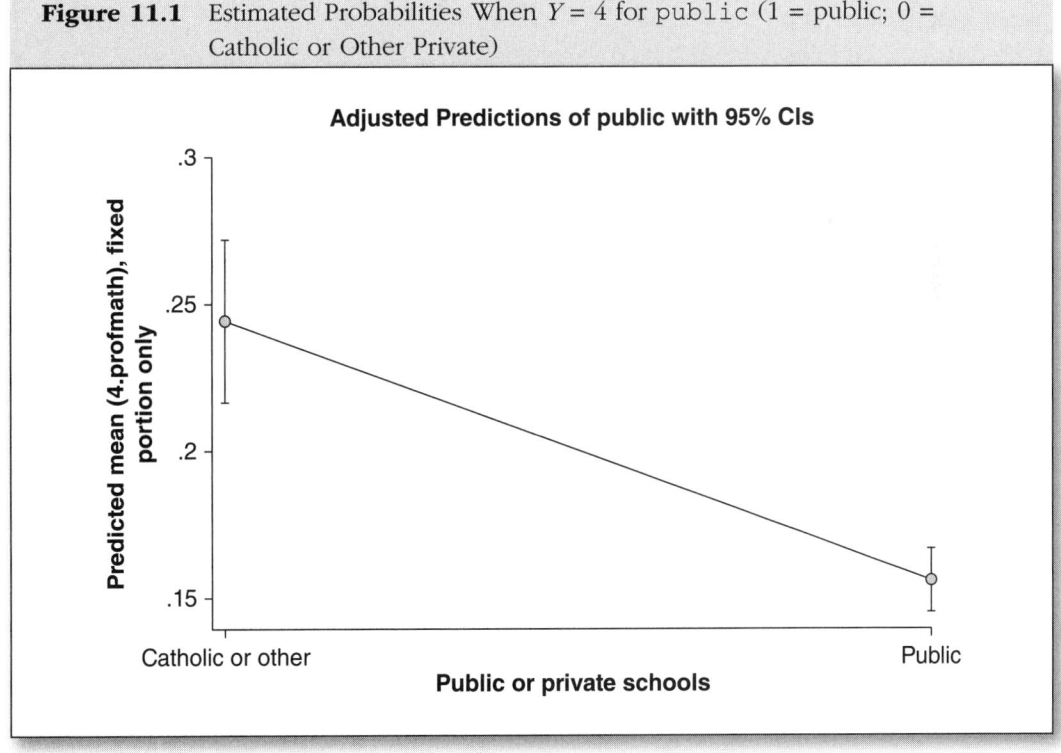

Note: CI = confidence interval.

atmeans vsquish tells Stata 14 to compute the estimated probabilities for all six categories of the ordinal response variable when $Y = 0, 1, 2, 3, 4$, and 5 for the binary predictor variable public when holding the other predictor variables at their means. Please note that the fixedonly option has been removed from the command syntax in the preceding section since the margins command in Stata 14 now integrates the random effects in the model instead of setting them to zero and provides more accurate results.

```
. *magins and marginsplot in Stata 14
. quietly meologit Profmath i.stlang i.public csclimat || SCH_ID: stlang, cov(uns)

. margins public, atmeans vsquish

Adjusted predictions                          Number of obs      =     12,793
Model VCE      : OIM

1._predict   : Marginal predicted mean (0.Profmath), predict(pr outcome(0))
2._predict   : Marginal predicted mean (1.Profmath), predict(pr outcome(1))
3._predict   : Marginal predicted mean (2.Profmath), predict(pr outcome(2))
4._predict   : Marginal predicted mean (3.Profmath), predict(pr outcome(3))
5._predict   : Marginal predicted mean (4.Profmath), predict(pr outcome(4))
6._predict   : Marginal predicted mean (5.Profmath), predict(pr outcome(5))
at           : 0.stlang        =    .1552411  (mean)
               1.stlang        =    .8447589  (mean)
               0.public        =    .2211366  (mean)
               1.public        =    .7788634  (mean)
               csclimat        =   -3.55e-09  (mean)

------------------------------------------------------------------------------
                          |            Delta-method
                          |    Margin   Std. Err.      z    P>|z|   [95% Conf. Interval]
--------------------------+---------------------------------------------------
          _predict#public |
1#catholic or other private |  .0293077    .002495   11.75   0.000    .0244176    .0341979
                 1#public |  .0501258   .0026914   18.62   0.000    .0448507    .0554008
2#catholic or other private |  .1750853   .0098036   17.86   0.000    .1558707    .1942999
                 2#public |  .2518104   .0066195   38.04   0.000    .2388365    .2647844
3#catholic or other private |  .2008829   .0060717   33.08   0.000    .1889826    .2127833
                 3#public |  .2296174   .0042763   53.70   0.000    .2212361    .2379988
4#catholic or other private |  .3219814   .0055394   58.13   0.000    .3111243    .3328385
                 4#public |  .2866787   .0051819   55.32   0.000    .2765223    .2968351
5#catholic or other private |  .2631351   .0134544   19.56   0.000     .236765    .2895052
                 5#public |   .176307   .0060374   29.20   0.000    .1644738    .1881402
6#catholic or other private |  .0096075   .0012086    7.95   0.000    .0072386    .0119764
                 6#public |  .0054607   .0006021    9.07   0.000    .0042807    .0066407
------------------------------------------------------------------------------
```

The margins command executed immediately after the contextual model without interactions (model 4) is refitted with the quietly command. In the output, the notes for the predicted probabilities for each of the six ordinal categories (labeled from 1._predict to 6._predict) and the means of the predictor variables (labeled at:) are listed first. Because category 0 is included in the ordinal response variable, the first prediction is for this category when $Y = 0$. The table for the margins or estimated probabilities is displayed at the bottom. As we can easily see, the estimated probabilities for all six categories of the ordinal response variable for the binary predictor variable `public` are displayed in the output. Please note that the results are slightly different from those estimated by separate commands in the previous section using Stata 13 since the random effects are treated differently by the `margins` command in Stata 13 and 14.

The `marginsplot` command in Stata 14 now automatically plots the estimated probabilities of being in each category of an ordinal response variable in a single graph. Figure 11.2 shows the estimated probabilities when $Y = 0, 1, 2, 3, 4,$ and 5.

11.3.8 Fitting Multilevel PO Models Using the `meglm` Command

In addition to the `ologit` command, the `glm` command for generalized linear models can be used to fit the single-level conventional PO model. Correspondingly, in the multilevel framework, in addition to the `meologit` command introduced earlier, we can use the `meglm` command, which was also newly introduced in Stata 13 for multilevel mixed-effects generalized linear models, to fit the multilevel PO models. The `meglm` command allows us to estimate mixed-effects models for various types of

Figure 11.2 Estimated Probabilities of Being in Categories 0 through 5 for `educ` With Stata 14

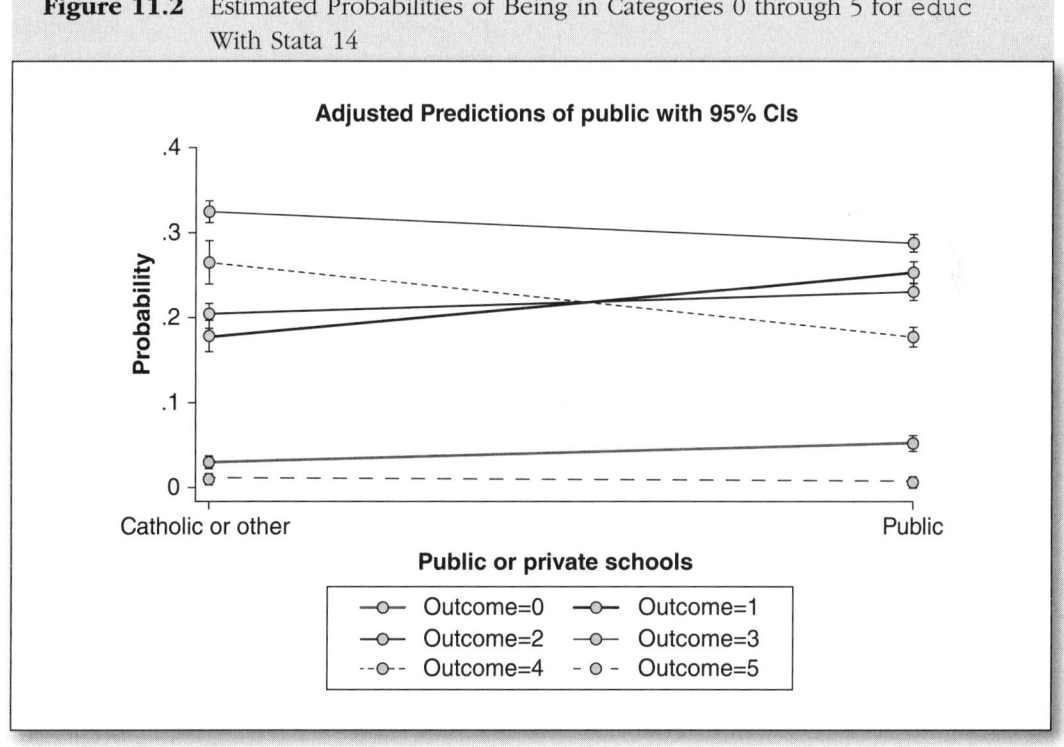

Note: CI = confidence interval.

response variables with different distributions, for example, the continuous, binary, ordinal, and count data. The syntax of the command is similar to that of the `meologit` command except that the distribution of the outcome variable and the link function need to be specified in the `family()` and `link()` options, respectively. You can specify the Gaussian (`gaussian`), Bernoulli (`bernoulli`), binomial (`binomial`), gamma (`gamma`), negative binomial (`nbinomial`), ordinal (`ordinal`), or Poisson (`poisson`) distribution for the `family()` option depending on the distribution of your response variable. For ordinal response variables, we specify the `family(ordinal)` option. The link functions include `identify`, `log`, `logit`, `probit`, and `cloglog`. The `link(logit)`, `link(probit)`, and `link(cloglog)` can be used for the ordinal response variables with the first one as the default.

The following is the syntax to fit the contextual model (model 4) with a single level 1 predictor variable `stlang` and level 2 predictor variables `public` and `csclimat`:

```
meglm Profmath stlang public csclimat || SCH_ID: stlang,
family(ordinal) link(logit) cov(uns)
```

The command `meglm` is followed by the ordinal response variable `Profmath` and three predictor variables. The fixed effects and the random effects of the model in the syntax are separated by two vertical lines (||). In the random effects part, `SCH_ID` is the cluster identifier variable. Following a colon is the predictor variable `stlang`, which has a random coefficient across schools. The `family(ordinal)` option specifies that the response variable is ordinal, and the `link(logit)` specifies the logit link function, which is the default. The `cov(uns)` specifies that the covariance matrix is unstructured so that all variances and covariances for the random effects will be uniquely estimated.

The `meglm` command produces the identical output to that obtained by the `meologit` command. The results from the contextual model (model 4) are displayed as follows.

```
. meglm Profmath stlang public csclimat || SCH_ID: stlang, family(ordinal) link(logit)
cov(uns)

Fitting fixed-effects model:

Iteration 0:   log likelihood = -19725.193
Iteration 1:   log likelihood = -19241.474
Iteration 2:   log likelihood = -19239.506
Iteration 3:   log likelihood = -19239.505

Refining starting values:

Grid node 0:   log likelihood = -18905.481

Fitting full model:

Iteration 0:   log likelihood = -18905.481   (not concave)
Iteration 1:   log likelihood = -18798.539   (not concave)
```

```
Iteration 2:    log likelihood = -18755.066
Iteration 3:    log likelihood =  -18687.45
Iteration 4:    log likelihood = -18684.234
Iteration 5:    log likelihood =  -18684.18
Iteration 6:    log likelihood =  -18684.18

Mixed-effects GLM                          Number of obs      =      12793
Family:                 ordinal
Link:                     logit
Group variable:          SCH_ID             Number of groups   =        619

                                            Obs per group: min =          2
                                                           avg =       20.7
                                                           max =         50

Integration method: mvaghermite             Integration points =          7

                                            Wald chi2(3)       =     242.82
Log likelihood = -18684.18                  Prob > chi2        =     0.0000
------------------------------------------------------------------------------
     Profmath |     Coef.   Std. Err.      z    P>|z|    [95% Conf. Interval]
--------------+---------------------------------------------------------------
       stlang |  .4067733     .06491     6.27   0.000     .2795521    .5339945
       public | -.5721059   .0889826    -6.43   0.000    -.7465085   -.3977033
      csclimat |  .5351121     .05594     9.57   0.000     .4254718    .6447524
--------------+---------------------------------------------------------------
        /cut1 | -3.410053   .1053497   -32.37   0.000    -3.616534   -3.203571
        /cut2 | -1.165185   .0994702   -11.71   0.000    -1.360143   -.9702267
        /cut3 | -.0865876   .0991816    -0.87   0.383     -.28098    .1078049
        /cut4 |  1.438011   .1000437    14.37   0.000     1.241929    1.634093
        /cut5 |  5.246318   .1418344    36.99   0.000     4.968328    5.524308
--------------+---------------------------------------------------------------
SCH_ID        |
  var(stlang) |   .420955   .1262007                      .2339099    .7575699
   var(_cons) |  1.097754   .1439413                      .8489703    1.419442
--------------+---------------------------------------------------------------
SCH_ID        |
cov(_cons,stlang) | -.4853097   .1226906    -3.96   0.000    -.7257789   -.2448405
------------------------------------------------------------------------------
LR test vs. ologit regression:        chi2(3) =   1110.65   Prob > chi2 = 0.0000

Note: LR test is conservative and provided only for reference.
```

The odds ratios for the three predictor variables can be requested using the meglm, or command. The results are the same as those obtained by the meologit, or command. The output is omitted here.

11.3.9 Fitting Multilevel PO Models Using the gllamm Command

In addition to the meologit and meglm commands, the user-written program gllamm can be used to fit the multilevel PO models. In fact, before the official release of the meologit command in Stata 13, the available program for Stata users to fit these multilevel models was gllamm. This program can be used to estimate various generalized linear mixed models, such as the mixed models for binary, ordinal, multinomial response variables; discrete-time survival data; and count data (Rabe-Heskteh, Skrondal, & Pickles, 2004). Since this is a user-written program, you need to install it first by typing ssc install gllamm before fitting the model. If you have a previous version installed and would like to update it, then type ssc install gllamm, replace.

The syntax of gllamm for a multilevel PO model is similar to that of the Stata ologit command except that the random effects of the model need to be specified in the option. Following the gllamm command are the outcome variable and the predictor variables. As with a regular Stata command, the options are separated by a comma. We specify the cluster identifier variable by using the i() option, the link function by using the link() option, and the adaptive quadrature integration method to estimate log likelihood by using the adapt option. To define the random effects of the model, we need to specify the number of random effects by using the nrf() option and the equations for the random effects by using the eqs() option.

For example, to fit the contextual model (model 4) with one student-level variable, stlang, and two school-level predictor variables, public and csclimat, we use the following gllamm syntax:

```
gllamm Profmath stlang public csclimat, i(SCH_ID) nrf(2)
eqs( inter slope) link(ologit) adapt
```

In the syntax, gllamm is followed by the ordinal response variable Profmath and three predictor variables, stlang, public, and csclimat. The i(SCH_ID) option specifies the cluster identifier variable SCH_ID, the link(ologit) specifies the link function, and the adapt option specifies that the numerical integration method is adaptive quadrature. The option nrf(2) tells us there are two random effects in the model since the intercept and the slope of stlang in the level 1 equation are allowed to vary randomly across schools. The eqs (inter slope) option specifies two equations for random effects, one for the intercept and one for the slope. Please note that we need to use the eq command to define these two equations before using the gllamm command. Without specifying the eqs() option, gllamm assumes that we only estimate one random effect, which is the random intercept.

The eq command is used to specify a variable or several variables multiplied by the random effect. Since the random cut points are not associated with any variables, they are multiplied by a constant 1. We first generate a constant of 1 by using the command generate cons=1, and then we specify the equation for the intercept by using the command eq inter: cons. Since we have one random slope in the model, we define the equation for the slope by using the command eq slope: stlang. Novice users might feel confused when they use the eq command to set up equations for the random effects and then when they use the nrf() option and the eqs() to specify the number of random effects and the equations for them. The following output is produced after executing this command.

```
. generate cons=1

. eq inter: cons

. eq slope: stlang

. gllamm Profmath stlang public csclimat, i(SCH_ID) nrf(2) eqs( inter slope)
link(ologit) ad
> apt

Running adaptive quadrature
Iteration 0:     log likelihood = -18786.298
Iteration 1:     log likelihood = -18691.056
Iteration 2:     log likelihood = -18685.151
Iteration 3:     log likelihood = -18684.794
Iteration 4:     log likelihood = -18684.794


Adaptive quadrature has converged, running Newton-Raphson
Iteration 0:     log likelihood = -18684.794
Iteration 1:     log likelihood = -18684.794   (backed up)
Iteration 2:     log likelihood = -18684.184
Iteration 3:     log likelihood =  -18684.18
Iteration 4:     log likelihood =  -18684.18

number of level 1 units = 12793
number of level 2 units = 619

Condition Number = 19.359103

gllamm model

log likelihood = -18684.18


--------------------------------------------------------------------------------
    Profmath |      Coef.    Std. Err.      z     P>|z|     [95% Conf. Interval]
-------------+------------------------------------------------------------------
Profmath     |
      stlang |   .4067739    .0649101     6.27    0.000     .2795525    .5339954
      public |  -.5721051    .0889828    -6.43    0.000    -.7465082   -.3977021
    csclimat |   .5351123     .05594      9.57    0.000     .4254718    .6447527
-------------+------------------------------------------------------------------
_cut11       |
        cons |  -3.410052    .1053501   -32.37    0.000    -3.616535    -3.20357
-------------+------------------------------------------------------------------
_cut12       |
        cons |  -1.165184    .0994706   -11.71    0.000    -1.360142   -.9702247
-------------+------------------------------------------------------------------
_cut13       |
        cons |  -.0865862    .099182     -0.87    0.383    -.2809794    .107807
-------------+------------------------------------------------------------------
```

```
_cut14     |
     _cons |   1.438013    .100044    14.37   0.000     1.24193    1.634095
-----------+----------------------------------------------------------------
_cut15     |
     _cons |   5.246321   .1418346    36.99   0.000     4.96833    5.524312
-----------------------------------------------------------------------------


Variances and covariances of random effects
-----------------------------------------------------------------------------

***level 2 (SCH_ID)

    var(1): 1.0977583 (.14394091)
    cov(2,1): -.48531214 (.12269012) cor(2,1): -.71391944

    var(2): .42095644 (.12620055)
-----------------------------------------------------------------------------
```

The output produced by the gllamm command is almost the same as the one by the meologit command. Please note that the estimation time for using gllamm is much longer when there are many observations and a large number of parameters since adaptive quadrature is used for numerical integration and numerical derivatives are used to maximize the log likelihood value. The first part of the output is for the fixed effects, which are exactly the same as those produced by meologit. The second part is the variance and covariance for the random effects. The output displays var(1), cov(2,1), and var(2), which represent the variance for the intercept, the covariance between intercept and slope, and the variance for the slope, respectively. The intercept variance, the between-school variance, is 1.098 with a standard error of .144; the slope variance is .421 with a standard error of .126; and the covariance is −.485 with a standard error of .123. These results are the same as those estimated by the meologit command. The output also reports the correlation between the intercept and the slope, which equals −.714.

To request the odds ratios, following the original command immediately, we use gllamm with the eform option.

```
. gllamm, eform

number of level 1 units = 12793
number of level 2 units = 619

Condition Number = 19.359103

gllamm model

log likelihood = -18684.18
```

```
------------------------------------------------------------------------
     Profmath |    exp(b)    Std. Err.      z    P>|z|   [95% Conf. Interval]
--------------+---------------------------------------------------------
Profmath      |
       stlang |   1.501965   .0974927    6.27   0.000    1.322538   1.705734
       public |   .5643362   .0502162   -6.43   0.000    .4740189   .6718622
      csclimat |   1.70764   .0955254    9.57   0.000    1.530312   1.905516
--------------+---------------------------------------------------------
_cut11        |
        _cons |  -3.410052   .1053501  -32.37   0.000   -3.616535   -3.20357
--------------+---------------------------------------------------------
_cut12        |
        _cons |  -1.165184   .0994706  -11.71   0.000   -1.360142  -.9702247
--------------+---------------------------------------------------------
_cut13        |
        _cons |  -.0865862    .099182   -0.87   0.383   -.2809794    .107807
--------------+---------------------------------------------------------
_cut14        |
        _cons |   1.438013    .100044   14.37   0.000    1.24193   1.634095
--------------+---------------------------------------------------------
_cut15        |
        _cons |   5.246321   .1418346   36.99   0.000    4.96833   5.524312
------------------------------------------------------------------------


Variances and covariances of random effects
------------------------------------------------------------------------


***level 2 (SCH_ID)

    var(1): 1.0977583 (.14394091)
    cov(2,1): -.48531214 (.12269012) cor(2,1): -.71391944

    var(2): .42095644 (.12620055)
------------------------------------------------------------------------
```

With others (i.e., the cut points and variance and covariance components) the same as in the previous output, the earlier output displays the odds ratios for the fixed effects of the predictor variables (labeled exp(b)). The results are almost the same as those produced by the meologit command with the or option.

11.4 Making Publication-Quality Tables

```
1.  . quietly meologit Profmath || SCH_ID:
2.  . outreg2 using chap11out, e(ll chi2) dec(3) long word replace
3.  . quietly meologit Profmath stlang || SCH_ID:
4.  . outreg2 using chap11out, e(ll chi2)  dec(3) long word append
```

```
5. . quietly meologit Profmath stlang || SCH_ID: stlang, cov(uns)
6. . outreg2 using chap11out, e(ll chi2)  dec(3) long word append
7. . quietly meologit Profmath stlang public csclimat || SCH_ID: stlang, cov(uns)
8. . outreg2 using chap11out, e(ll chi2)  dec(3) long word append
```

The first command, `quietly meologit Profmath || SCH_ID`, estimates the unconditional model without providing the output. The second command, `outreg2 using chap11out, e(ll chi2) dec(3) long word replace`, tells Stata to create a regression table for the estimated fixed and random effects and to save the results to a file named chap11out. The option `e(ll chi2)` requests the estimated log likelihood value and the chi-square test statistic. The `dec(3)` option requests three decimal places for all statistics. The `word` option requests that the table be created in a Word document with the extension .rtf. The `long` option aligns the fixed and random effects and lists them in one column. The `replace` option asks Stata to replace any files with the same name as chap11out.

The remaining commands repeat the same process for the random-intercept model, the random-coefficient model, and the contextual model. The estimated fixed effects and random effects for these three models are appended to the table.

These commands automatically produce Table 11.1, as shown here in its original format, presenting the results for all four multilevel PO models.

Table 11.1   Results of the Multilevel PO Models: From Unconditional Model to the Contextual Model (Shown in Original Format Generated by Stata)

Variables	(1) Profmath	(2) Profmath	(3) Profmath	(4) Profmath
cut1				
Constant	−3.292***	−2.977***	−2.964***	−3.410***
	(0.058)	(0.072)	(0.082)	(0.105)
cut2				
Constant	−1.070***	−0.749***	−0.716***	−1.165***
	(0.044)	(0.062)	(0.074)	(0.099)

	(1)	(2)	(3)	(4)
Variables	**Profmath**	**Profmath**	**Profmath**	**Profmath**
cut3				
Constant	−0.000	0.324***	0.363***	−0.087
	(0.043)	(0.062)	(0.074)	(0.099)
cut4				
Constant	1.516***	1.842***	1.887***	1.438***
	(0.045)	(0.064)	(0.076)	(0.100)
cut5				
Constant	5.319***	5.643***	5.701***	5.246***
	(0.110)	(0.119)	(0.127)	(0.142)
var(_cons[SCH_ID])				
Constant	0.888***	0.866***	1.388***	1.098***
	(0.064)	(0.063)	(0.166)	(0.144)
Profmath				
Stlang		0.379***	0.417***	0.407***
		(0.052)	(0.065)	(0.065)
Public				−0.572***
				(0.089)
Csclimat				0.535***
				(0.056)
var(stlang[SCH_ID])				
Constant			0.373***	0.421***
			(0.122)	(0.126)
cov(_cons[SCH_ID],stlang[SCH_ID])				

(Continued)

Table 11.1   (Continued)

Variables	(1) Profmath	(2) Profmath	(3) Profmath	(4) Profmath
Constant			−0.472***	−0.485***
			(0.128)	(0.123)
Observations	12,793	12,793	12,793	12,793
Number of groups	619	619	619	619
Ll	−18,809	−18,783	−18,771	−18,684
chi2	—	52.19	41.88	242.8

Note: Standard errors in parentheses.

***$p < 0.01$, **$p < 0.05$, *$p < 0.1$

The between-group variance in the table is separated from the slope variance by three predictor variables, so the table needs to be edited. After adding the model titles, the labels for variances and covariances, and the names for the cut points on the original table and grouping variances and covariances together, the final table is displayed as in Table 11.2.

Table 11.2   Results of the Multilevel PO Models: From Unconditional Model to the Contextual Model (Edited)

Variables	Model (1) Unconditional Model	Model (2) Random- Intercept Model	Model (3) Random- Coefficient Model	Model (4) Contextual Model
Fixed Effects				
cut1				
α_1	−3.292***	−2.977***	−2.964***	−3.410***
	(0.058)	(0.072)	(0.082)	(0.105)
cut2				
α_2	−1.070***	−0.749***	−0.716***	−1.165***
	(0.044)	(0.062)	(0.074)	(0.099)

Variables	Model (1) Unconditional Model	Model (2) Random-Intercept Model	Model (3) Random-Coefficient Model	Model (4) Contextual Model
cut3				
α_3	−0.000	0.324***	0.363***	−0.087
	(0.043)	(0.062)	(0.074)	(0.099)
cut4				
α_4	1.516***	1.842***	1.887***	1.438***
	(0.045)	(0.064)	(0.076)	(0.100)
cut5				
α_5	5.319***	5.643***	5.701***	5.246***
	(0.110)	(0.119)	(0.127)	(0.142)
stlang (γ_{10})		0.379***	0.417***	0.407***
		(0.052)	(0.065)	(0.065)
public (γ_{01})				−0.572***
				(0.089)
csclimat (γ_{02})				0.535***
				(0.056)
Random Effects (Variance Components)				
var(_cons[SCH_ID])				
Var. in intercepts	0.888***	0.866***	1.388***	1.098***
	(0.064)	(0.063)	(0.166)	(0.144)
var(stlang[SCH_ID])				
Var. in stlang slope			0.373***	0.421***
			(0.122)	(0.126)

(Continued)

Table 11.2   (Continued)

Variables	Model (1) Unconditional Model	Model (2) Random- Intercept Model	Model (3) Random- Coefficient Model	Model (4) Contextual Model
cov(_cons[SCH_ ID],stlang[SCH_ ID])				
Covariance			−0.472***	−0.485***
			(0.128)	(0.123)
Observations	12,793	12,793	12,793	12,793
Number of groups	619	619	619	619
Ll	−18,809	−18,783	−18,771	−18,684
chi2	—	52.19	41.88	242.8

Note: Standard errors in parentheses.

***$p < .01$.

11.5 Reporting the Results

Writing the results of the multilevel PO models is similar to that of the multilevel models for the continuous and binary response variables.

First, describe the purpose of your study and explain why the multilevel PO models are needed for the analysis.

Second, report the model-building steps and briefly describe each model. Report and interpret the intraclass correlation coefficient, which is computed from the between-group variance in the unconditional model.

Third, report the results of a series of fitted models in a table including both the parameter estimates for the fixed effects and the variances and covariances for the random effects. If available, include deviance statistics (i.e., −2LL) and the AIC and BIC statistics for these models in the table.

Fourth, as with the conventional PO model, report the parameter estimates for the fixed effects of the predictor variables in the final contextual model and interpret the corresponding odds ratios. In addition, report the variances and covariances for the random effects in text. The following is an example of summarizing the results of the unconditional model and the contextual model fitted earlier.

Multilevel modeling for the ordinal response variable was used to estimate the relationships between an ordinal response variable, students' math proficiency, and a student-level variable, the student's native language, and two school-level variables, the school type and the school climate. Five models, from the unconditional (null) model to the contextual model with cross-level interactions, were fitted. Table 11.2 presents the parameter estimates for the fixed effects and random effects for the fitted models. Since no cross-level interactions were identified in the contextual model, the interaction terms were removed from the model. For illustration purposes, the following presentations focused only on the results of the unconditional model and the final contextual model without cross-level interactions.

Results for the Unconditional Model (Model 1)

There was a significant between-group variance: τ_{00} = .888, which is the intercept variance across all schools. The log likelihood ratio test comparing the unconditional model with the single-level ordinal logistic regression model suggested that the unconditional model that estimated the between-group variance was preferred ($\chi^2_{(1)}$ = 1,831.86, p < .001).

The ICC in the unconditional model was .213, which indicated that 21.3% of the total variance was accounted for by schools in level 2. Therefore, using multilevel ordinal logistic regression was justified.

Results for the Contextual Model Without Cross-Level Interactions (Model 4)

The logit coefficient for `stlang` (γ_{10}) = .407, z = 6.27, p < .001. OR = 1.502, which indicated that the odds of being above a particular mathematics proficiency level for students whose native language was English were 1.460 times as great as the odds for those whose native language was not English.

The logit coefficient for `public` (γ_{01}) = −.572, z = −6.43, p < .001, which indicated that being in public schools decreased the logit of being above a particular category of mathematics proficiency by a factor of .572. The corresponding odds ratio was .564, which indicated that the odds of being above a particular category for students in a public school were .572 times as great as the odds for students in a private school when holding all the other predictors constant.

The logit coefficient for `csclimat` (γ_{02}) = .535, z = 9.57, p < .001. OR = 1.708, which indicated that for a one-unit increase in the school climate the odds of being above a particular category of mathematics proficiency increased by a factor of 1.708. So the odds of being in higher math proficiency levels were associated with schools with a more positive school climate.

Regarding the variance and covariance components for the random effects, the between-group variance, the intercept variance (τ_{00}), was 1.098, which was significantly different from zero (p < .001). The slope variance (τ_{10}) for `stlang` was .421, which was also significant (.421/.126 = 3.341). The covariance between the intercept and the slope was −.485, p < .001.

After two school-level variables were included in the contextual model (model 4), the between-school variance (τ_{00}) decreased from 1.388 to 1.098 compared with that for the random-coefficient model (model 3): (1.388 − 1.098)/1.388 = 21%, which indicated that there was a decrease of 21% in the between-school variance from the random-coefficient model (model 3) to the contextual model (model 4) after the two school-level variables were included.

11.6 Summary of Stata Commands in This Chapter

```
use chap11-els2002, clear

*Deleting missing values and grand-mean center sclimate
drop if missing(Profmath, stlang, public, sclimate)
tab Profmath
summarize stlang public sclimate
generate csclimat = sclimate -r(mean)

*Unconditional model
meologit Profmath || SCH_ID: ,
meologit, or
estimates store null

*Random-intercept model
meologit Profmath stlang || SCH_ID:
meologit, or
estimates store ranint

*Log likelihood ratio test comparing the unconditional model and random-intercept
model
lrtest null ranint

*Random-coefficient model
meologit Profmath stlang || SCH_ID: stlang, cov(uns)
meologit, or
estimates store ranslop1

*Contextual model with level 1 and level 2 variables
meologit Profmath stlang public csclimat || SCH_ID: stlang, cov(uns)
meologit, or
estimates store ranslop2

*Log likelihood ratio test
lrtest ranint ranslop1

*Log likelihood ratio test
lrtest ranslop1 ranslop2

*Create cross-level interactions
generate pub_lang= public*stlang
generate cli_lang=csclimat*stlang

*Contextual model with cross-level interactions
meologit Profmath stlang public csclimat pub_lang cli_lang || SCH_ID: stlang, ///
    cov(uns)
meologit, or
estimates store ranslop3
lrtest ranslop2 ranslop3
```

```
*Model comparisons using AIC and BIC statistics
estimates stats null ranint ranslop1 ranslop2 ranslop3

*margins and marginsplot in Stata 13
quietly meologit Profmath i.stlang i.public csclimat || SCH_ID: stlang, cov(uns)
margins public, atmeans predict(fixedonly outcome(1)) vsquish
margins public, atmeans predict(fixedonly outcome(4)) vsquish
marginsplot
*Fitting a Multilevel PO model using meglm
meglm Profmath stlang public csclimat || SCH_ID: stlang, family(ordinal) ///
  link(logit) cov(uns)

*Fitting a Multilevel PO model using gllamm
generate cons=1
eq inter: cons
eq slope: stlang
gllamm Profmath stlang public csclimat, i(SCH_ID) nrf(2) eqs( inter slope) ///
link(ologit) adapt
gllamm, eform

*Making publication-quality tables using outreg2
quietly meologit Profmath || SCH_ID:
outreg2 using chap11out, e(ll chi2) dec(3) long word replace
quietly meologit Profmath stlang || SCH_ID:
outreg2 using chap11out, e(ll chi2)  dec(3) long word append
quietly meologit Profmath stlang || SCH_ID: stlang, cov(uns)
outreg2 using chap11out, e(ll chi2)  dec(3) long word append
quietly meologit Profmath stlang public csclimat || SCH_ID: stlang, cov(uns)
outreg2 using chap11out, e(ll chi2)  dec(3) long word append
```

11.7 EXERCISES

Use the ELS:2002 data with the following variables for the multilevel ordinal logistic regression models.

Profmath: mathematics proficiency level with six ordinal categories from 0 to 5

gender: gender (1 = female; 0 = male)

byses: socioeconomic status composite for base-year data

usecalc: using calculators

urban: urban schools or not (1 = urban; 0 = suburban or rural areas) (school-level variable)

This study investigates how student-level and school-level predictors influence students' mathematics proficiency. The ordinal outcome variable is Profmath with six levels, and the predictor variables are `gender, byses, usecalc,` and `urban`. Conduct the multilevel modeling for ordinal response variables and then answer the following questions or perform the following analyses:

1. Fit an unconditional model and obtain the between-group variance. Compute the ICC and interpret it.

2. Fit a two-level, random-intercept model with a random intercept and three student-level predictor variables, `gender, byses,` and `usecalc`. Use the grand-mean centering for `byses` and `usecalc` before fitting the model.

3. Interpret the odds ratios of three predictor variables.

4. Conduct a likelihood ratio test comparing the random-intercept model and the unconditional model. Interpret the results.

5. Fit a two-level, contextual model with `gender, byses,` and `usecalc` as the student-level predictor variables and `urban` as the school-level predictor variable. Specify a random intercept and a random slope for `usecalc`. No cross-level interactions are included.

6. Write level 1 and level 2 equations for the contextual model shown in Exercise 5.

7. Interpret the odds ratio of `urban` in the contextual model.

8. Conduct log likelihood ratio tests among the three fitted models. Which one fits the data best? Why?

9. Use the `estimates stats` command to compute the AIC and BIC statistics for the three fitted models in Exercises 1, 2, and 5. Interpret the results. Do they support the finding from the log likelihood ratio tests?

10. Use the `margins` command to calculate the estimated probabilities of being in category 2 (i.e., $Y = 2$) for the predictor variable `gender` when holding other predictor variables at their means.

11. Use the `margins` command to calculate the estimated probabilities of being in category 5 (i.e., $Y = 5$) for `gender` when holding other predictor variables at their means.

12. Plot the estimated probabilities for category 5 obtained in Exercise 11 by using the `marginsplot` command.

13. Use the `meglm` command with the `logit` option to fit the same contextual model shown earlier.

14. Present a table containing the estimated logit coefficients and odds ratios for the contextual model.

15. Write a report to summarize the results of the contextual model.

CHAPTER **12**

Beyond Ordinal Logistic Regression Models: Ordinal Probit Regression Models and Multinomial Logistic Regression Models

Objectives of This Chapter

This chapter introduces ordinal probit regression models and multinomial logistic regression models. The first section starts with an introduction of the probit regression model followed by a discussion of how to interpret parameter estimates. After a description of the research example, the data, and the sample, a multiple-predictor ordinal probit regression model using Stata is illustrated with step-by-step instructions. Stata commands and output are explained in detail. The second section starts with an introduction of the multinomial logistic regression model followed by a discussion of the odds and odds ratios or relative risk ratios in the model. After a description of the research example, the data, and the sample, a multiple-predictor multinomial logistic regression model is fitted with Stata. This chapter focuses on fitting the ordinal probit models and multinomial logistic regression models with

(Continued)

(Continued)

Stata, as well as on interpreting and presenting the results. After reading this chapter, you should be able to

- Identify when an ordinal probit model and multinomial logistic regression model are used.
- Fit both models using Stata.
- Interpret the output.
- Compute and interpret marginal effects and estimated probabilities for ordinal probit models using the `margins` command.
- Compute and interpret estimated probabilities for multinomial logistic regression models using the `margins` command.
- Compare models using the likelihood ratio test.
- Present results in publication-quality tables using Stata.
- Write the results for publication.

12.1 Ordinal Probit Regression Models

12.1.1 Ordinal Probit Regression Models: An Introduction

In addition to the logistic regression models, which are the main focuses of this book, the probit models are also well known for analyzing ordinal response variables (Agresti, 2010; Liao, 1994; Long, 1997; Long & Freese, 2006, 2014; Powers & Xie, 2000). These two models were developed from different origins. The logistic regression models, which assume that the error term follows a logistic distribution, were developed by McCullagh (1980), whereas the probit regression models, which assume a standard normal distribution, were developed by McKelvey and Zavoina (1975). When fitting these two models, the logit link function is used for the logistic regression models, whereas the probit link function is used for the probit models. Although they follow different distributions, the results estimated from both models are similar. The choice of these two models depends on the preference of the researchers, the ease of interpretation of parameter estimates, and the fields to which the researchers belong. Specifically, researchers often choose models they are more familiar with, or prefer models when the results are more interpretative. In addition, some disciplines may use one model more frequently than the other.

As with the logistic regression models, the probit regression models belong to the family of generalized linear models. They can also be expressed as a linear function of a set of predictor variables. Assuming we estimate a latent ordinal variable Y^*, we can define Y^* as a function of a set of predictor variables and a random error. Let Y^* be divided by some cut points (thresholds): α_1, α_2, α_3, ..., α_j and $\alpha_1 < \alpha_2 < \alpha_3 < ... < \alpha_j$. The values of the observed ordinal variable Y fall within the regions divided by these cut points (thresholds). For example, $Y = 0$, if $Y^* \leq \alpha_1$. Considering the observed health status is an ordinal outcome y ranging from 1 to 4, it is defined as follows:

$$y = \begin{cases} 1 & \textit{if } y^* \leq \alpha_1 \\ 2 & \textit{if } \alpha_1 < y^* \leq \alpha_2 \\ 3 & \textit{if } \alpha_2 < y^* \leq \alpha_3 \\ 4 & \textit{if } \alpha_3 < y^* \leq \infty \end{cases}$$

Therefore, we can predict the probability of being in a particular health status level and the cumulative probabilities. $P(Y \leq j) = F(\alpha_j - \mathbf{x}\beta)$, where $j = 1, 2, ..., J - 1$. Recall that the logit form for the proportional odds models is as follows:

$$\ln(Y_j') = \text{logit}[\pi_j(x)] = \alpha_j + (-\beta_1 X_1 - \beta_2 X_2 - ... - \beta_p X_p) \tag{12.1}$$

where $\pi_j(x) = \pi(Y \leq j \mid x_1, x_2, ..., x_p)$, which is the probability of being at or below category j, given a set of predictors: $j = 1, 2, ..., J - 1$. α_j are the cut points, and β_1, β_2, ..., β_p are the logit coefficients. The equation can be also rewritten as:

$$\text{logit}[\pi(Y \leq j \mid x_1, x_2, ..., x_p)] = \alpha_j + (-\beta_1 X_1 - \beta_2 X_2 - ... - \beta_p X_p) \tag{12.2}$$

Similarly, the ordinal probit regression model can be expressed in the following form:

$$\text{probit}[\pi(Y \leq j \mid x_1, x_2, ..., x_p)] = \alpha_j + (-\beta_1 X_1 - \beta_2 X_2 - ... - \beta_p X_p) \tag{12.3}$$

where the left side of the equation is the probit link function. The probit model follows a standard cumulative normal distribution with a mean of 0 and the standard deviation set to 1. The probit link, or probit transformation, is often expressed as the inverse of the cumulative density function (cdf) of the standard normal distribution $\Phi^{-1}(\pi)$, where Φ is defined as the cdf for the standard normal. The ordinal probit model can be also expressed as:

$$\Phi^{-1}(\pi) = \text{probit}[\pi(Y \leq j \mid x_1, x_2, ..., x_p)] = \alpha_j + (-\beta_1 X_1 - \beta_2 X_2 - ... - \beta_p X_p) \tag{12.4}$$

where α_j are still the cut points and β_1, β_2, ..., β_p are the probit coefficients. As with the proportional odds models, the coefficients for each predictor variable in the probit models are the same across the ordinal categories. In other words, the slopes are also parallel. This assumption is not called the proportional odds assumption since the probit models do not estimate odds ratios.

The cumulative probability of being at or below a particular category can be expressed as the cumulative standard normal distribution function. By taking the inverse on both sides of the equation, we get the following equation:

$$\pi(Y \le j \mid x_1, x_2, ..., x_p) = \Phi \left[\alpha_j + (-\beta_1 X_1 - \beta_2 X_2 - ... -\beta_p X_p) \right] \quad (12.5)$$

where Φ (.) denotes the cdf for the standard normal and a linear combination of the cut points and a set of predictors is specified within the parentheses. The probit coefficients are z scores or standard normal scores. With the estimated probit coefficients and the values of predictor variables, we can compute the z scores or standard scores and then obtain the estimated probabilities from the table of standard normal distributions.

Interpretation of Model Parameter Estimates

Although ordinal logistic models and ordinal probit models produce similar regression coefficients and model fit statistics, the interpretations for the coefficients in these two models are not identical. The estimated logit coefficients in logistic regression can be transformed into odds ratios, whereas the probit coefficients in probit regression models cannot be interpreted in the same way.

The parameter estimates in ordinal probit models can be interpreted in four ways. First, in ordinal probit regression, the regression coefficient is the probit coefficient. It can be interpreted as the change in the predicted probit of being above a particular category for a one-unit increase in the predictor variable. For example, when the probit coefficient of a predictor variable educ is .706, it can be interpreted that for a one-unit increase in the predictor variable, there is an increase of .706 in the probit of being beyond a particular category of the ordinal response variable, health status.

As with the logit in logistic regression, the probit itself in probit regression is difficult to interpret. However, this interpretation on the scale of probit gives us a sense of the relationship between the predictor variable and the probit function of the probability. Specifically, when the probit coefficient is positive, it indicates the relationship between the predictor variable and the probit function of the probability is positive. In other words, a positive coefficient increases the probability of being above a category and decreases the probability of being at or below that category for an increase in a predictor variable. When the probit coefficient is negative, it indicates that the relationship between the predictor variable and the probit function is negative. A negative coefficient decreases the probability of being above a category and increases the probability of being at or below that category. When the probit coefficient equals zero, there is no relationship between the predictor and the probit function, so there is no change in the latent variable when the values of the predictor variable change.

Second, a probit coefficient can also be interpreted as the change in the mean of a latent variable for each one-unit increase in the predictor variable. It is the change in z scores, which is the number of standard deviations from the mean of a latent variable. Here, the standard deviations are conditional on the predictor variables in the model, so they are also called conditional standard deviations (Agresti, 2010).

Third, a common practice is to interpret the marginal effects of predictor variables in ordinal probit models. A marginal effect can be interpreted as the change in the probability of being in a particular category of the ordinal response variable for a one-unit change in a predictor variable.

Fourth, as with logistic regression models, the estimated probabilities of being in a particular category can be computed for predictor variables at specified values. The marginal effects and estimated probabilities or margins are more interpretable than the probits, but they need complex computations. The examples on how to compute marginal effects and estimated probabilities will be provided after an ordinal probit model is fitted in Section 12.1.3.

12.1.2 Description of the Research Example, Data, and Sample

We will still investigate the relationships between the ordinal response variable, health status, and four predictor variables. Unlike in other chapters, however, here the research interest will focus on using the ordinal probit regression. The General Social Survey 2012 (GSS 2012) data are used for the following analyses. The same ordinal response variable and the predictor variables for the proportional odds models used in Chapter 4 are used to fit the probit models.

- `healthre`: the recoded variable of health (health status) with four ordinal categories (1 = poor health, 2 = fair health, 3 = good health, and 4 = excellent health)
- `maritals`: the recoded variable of marital (marital status) with 1 = currently married and 0 = not currently married
- `educ`: the highest education
- `age`: respondent's age
- `male`: recoded variable of sex with 1 = male and 0 = female

12.1.3 Ordinal Probit Models With Stata: Commands and Output

The Stata Command `oprobit`

The Stata command `oprobit` is used for the ordinal probit regression analysis. The syntax structure for ordinal probit models is the same as that for proportional odds models. The command `oprobit` is followed by the ordinal response variable and the predictor variable(s). For more details on how to use this command, type the `help oprobit` command.

The Ordinal Probit Model: Multiple-Predictor Model

The command `xi: oprobit healthre i.maritals educ age i.male` tells Stata to conduct the ordinal probit regression to estimate the ordinal response variable `healthre` by using four predictor variables: `maritals`, `educ`, `age`, and `male`. In the command syntax, `xi` is the prefix command for the categorical variables `maritals` and `male`. The prefix `xi` can be omitted if you use Stata 11 or later versions since the new coding for factor variables is used. This prefix is still used throughout the book since user-written programs may not work with factor variables. The following output is displayed.

```
. xi: oprobit healthre i.maritals educ age i.male
i.maritals          _Imaritals_0-1      (naturally coded; _Imaritals_0 omitted)
i.male              _Imale_0-1          (naturally coded; _Imale_0 omitted)

Iteration 0:   log likelihood = -1579.4068
Iteration 1:   log likelihood = -1513.2572
Iteration 2:   log likelihood = -1513.2025
Iteration 3:   log likelihood = -1513.2025

Ordered probit regression                     Number of obs   =       1300
                                               LR chi2(4)      =     132.41
                                               Prob > chi2     =     0.0000
Log likelihood = -1513.2025                    Pseudo R2       =     0.0419

------------------------------------------------------------------------------
    healthre |      Coef.   Std. Err.      z    P>|z|     [95% Conf. Interval]
-------------+----------------------------------------------------------------
_Imaritals_1 |   .3006869   .0613369     4.90   0.000     .1804688    .4209049
        educ |   .0757582   .0098139     7.72   0.000     .0565233    .0949932
         age |  -.0111655   .0017531    -6.37   0.000    -.0146015   -.0077294
    _Imale_1 |   .0347169   .0609249     0.57   0.569    -.0846938    .1541276
-------------+----------------------------------------------------------------
       /cut1 |  -.985996    .1693112                     -1.31784    -.6541522
       /cut2 |   .0121435   .1661625                     -.313529     .337816
       /cut3 |   1.30246    .1686844                      .9718445    1.633075
------------------------------------------------------------------------------
```

12.1.4 Interpreting the Output

The output for the ordinal probit model looks similar to that for the proportional odds model. The log likelihood ratio chi-square test, LR $\chi^2_{(4)} = 132.41$, $p < .001$. This indicates that the probit model with four predictors provides a better fit than the null model with no independent variables in predicting the ordinal response variable. The maximum likelihood value, the log likelihood ratio chi-square test statistic, and the likelihood ratio R^2_L for the ordinal probit model are almost identical to those estimated for the proportional odds model (full model) introduced in Chapter 4.

The ordinal probit regression table displays the parameter estimates for the predictor variables and the cut points or intercepts, their standard errors, the Wald z statistics, the associated p values, and the 95% confident intervals of the parameter estimates.

Like ordinal logistic regression models, the Wald test is used to test whether the probit coefficient is statistically significant. It equals the probit coefficient divided by its standard error. The null hypothesis for the Wald test is that the coefficient of the predictor variable is zero, and the alternative hypothesis is that the coefficient of the predictor variable is significantly different from zero. Among the four predictor variables, the probit coefficients for `maritals`, `educ`, and `age` are significant, whereas the coefficient for `male` is not significant.

For the `maritals` predictor, the probit coefficient $\beta = .301$, which is positive. It indicates that being married increases the probability of being beyond a particular category of health status (better health status) when holding all the other predictors constant. It can also be interpreted that being married increases the mean of the latent variable health status by .301 standard deviations.

For the `educ` predictor, $\beta = .076$, which is positive. It indicates that an increase in the predictor `educ` corresponds to the increase in the probability of being beyond a particular category of health status (better health status) when holding all the other predictors constant. For each one-unit increase in education, there is an increase of .076 standard deviations in the mean of the latent health status.

For the `age` predictor, $\beta = -.011$, which is negative. It indicates that the increase in the predictor `age` is associated with a decrease in the probability of being beyond a particular category of health status (better health status) when all the other predictors remain constant. For a one-unit increase in age, the z scores of being healthier decrease by .011.

For the `male` predictor, $\beta = .035$, $z = .57$, $p = .569$, which is not significantly different from 0. It indicates that there is no relationship between being a male and the probability of being in better health status.

The probit coefficients can also be computed using the `listcoef` command (Long & Freese, 2014). The output is displayed as follows.

```
. listcoef

 oprobit (N=1300): Unstandardized and standardized estimates

   Observed SD:   0.8544
     Latent SD:   1.0614
```

	b	z	P>|z|	bStdX	bStdY	bStdXY	SDofX
_Imaritals_1	0.3007	4.902	0.000	0.150	0.283	0.141	0.499
educ	0.0758	7.719	0.000	0.239	0.071	0.225	3.151
age	-0.0112	-6.369	0.000	-0.195	-0.011	-0.183	17.432
_Imale_1	0.0347	0.570	0.569	0.017	0.033	0.016	0.497

The ordinal probit models and the proportional odds models have almost identical goodness-of-fit statistics. All measures of fit statistics in proportional odds models, such as the deviance, log likelihood ratio test, pseudo R^2, and Akaike information criterion (AIC) and Bayesian information criterion (BIC) statistics, can be applied to ordinal probit models. The `fitstat` command computes various fit statistics for the probit model. Interested readers can examine the model fit statistics for both models. The output is displayed as follows.

```
. fitstat

                            |       oprobit
----------------------------+---------------
Log-likelihood              |
                   Model    |    -1513.203
         Intercept-only     |    -1579.407
----------------------------+---------------
Chi-square                  |
       Deviance (df=1293)   |     3026.405
              LR (df=4)     |      132.409
                p-value     |        0.000
----------------------------+---------------
R2                          |
               McFadden     |        0.042
      McFadden (adjusted)   |        0.037
      McKelvey & Zavoina    |        0.112
           Cox-Snell/ML     |        0.097
  Cragg-Uhler/Nagelkerke    |        0.106
                  Count     |        0.464
          Count (adjusted)  |        0.013
----------------------------+---------------
IC                          |
                    AIC     |     3040.405
         AIC divided by N   |        2.339
              BIC (df=7)    |     3076.596
----------------------------+---------------
Variance of                 |
                      e     |        1.000
                 y-star     |        1.127
```

In the output, the fit statistics, such as the deviance, the log likelihood ratio test, various pseudo R^2 measures, and the AIC and BIC statistics are similar to those for the proportional odds models with the logit link in Chapter 4. The variance of error in the output is reported to be one since the error assumes to have a normal distribution with a mean of zero and variance of one in the probit model.

12.1.5 Interpreting the Marginal Effects With the `margins` Command

Marginal effects are more interpretable than the original probit coefficients in ordinal probit models. In a binary logistic regression model, a marginal effect of a predictor variable is the change in the probability of success or of having an event associated with a one-unit change in the predictor variable. In an ordinal probit model, a marginal effect is the change in the probability of being in a particular category of the ordinal response variable for a one-unit change in the value of the predictor variable. It is the instantaneous rate of change for a continuous predictor variable or the discrete change for a binary variable (Long & Freese, 2014). If the ordinal response variables have J categories, then there will be J marginal effects for each predictor variable. A significantly positive marginal effect indicates there is an increase in the probability of being in a particular category for a one-unit change in the predictor variable, whereas a negative marginal effect corresponds to a decrease in the probability. When the marginal effect is not significantly different from zero, it indicates that there is no relationship between the predictor variable and the probability of being in that category.

Using the `margins` Command in Stata 13

The command `margins, dydx(*)` tells Stata 13 to compute the marginal effects for all predictor variables when the ordinal response variable $Y = 1$. This command is equivalent to `margins, dydx(*) predict (outcome(1))` since the default in Stata 13 is to predict the change in the probability of being the lowest category when the first category of the ordinal response variable $Y = 1$. The output is displayed as follows.

```
. margins, dydx(*)

Average marginal effects                      Number of obs   =       1300
Model VCE      : OIM

Expression   : Pr(healthre==1), predict()
dy/dx w.r.t. : 1.maritals educ age 1.male
```

```
-----------------------------------------------------------------------
             |             Delta-method
             |    dy/dx     Std. Err.       z     P>|z|    [95% Conf. Interval]
-------------+---------------------------------------------------------
  1.maritals |  -.0338846    .0072624    -4.67    0.000    -.0481187    -.0196506
        educ |  -.0087562    .0013206    -6.63    0.000    -.0113446    -.0061678
         age |   .0012905    .0002263     5.70    0.000     .0008471     .001734
     1.male  |  -.0040009    .0070084    -0.57    0.568    -.0177371     .0097354
-----------------------------------------------------------------------
Note: dy/dx for factor levels is the discrete change from the base level.
```

The beginning of the output indicates that the average marginal effects are estimated and that the number of observations is 1,300. The next note, Model VCE: OIM, shows the method for variance estimation. The third note, Expression: Pr(healthre==1), predict(), indicates that we estimate the change in the probability of being in category 1 of the ordinal response variable. Finally, dy/dx w.r.t.: 1.maritals educ age 1.male indicates that the marginal effects are estimated for the four predictor variables.

The marginal effects table displays the average marginal effects, Delta-method standard errors, z tests, associated p values, and 95% confidence intervals. A positive marginal effect indicates that there is an increase in the probability of being in the lowest category (i.e., category 1) for a one-unit increase in the predictor variable, whereas a negative marginal effect suggests that there is a decrease in the probability of being in category one.

- The marginal effect for maritals (labeled dy/dx) = −.034, which indicates that the probability of being in category 1 (i.e., poor health) decreases by .034 when the marital status goes from unmarried to married when holding other predictor variables constant.
- The marginal effect for educ = −.009, which indicates that the probability of being in category 1 (i.e., poor health) decreases by .009 for a one-unit increase in education.
- The marginal effect for age = .001, which indicates that the probability of being in category 1 (i.e., poor health) increases by .001 for a one-unit increase in age. In other words, people are more likely to be in poor health with an increase of age.
- The marginal effect for male = −.004, z = −.57, p =. 568, which indicates that there is no change in the probability of being in category 1 (i.e., poor health) for being male. In other words, being male does not impact the likelihood of being in poor health.

To predict the change in the probabilities of being in the other three categories, we can use the predict() option. For example, with the predict(outcome(2)) or predict(outcome(3)) option, the margins command estimates the marginal effects on the probability of being in categories 2 or 3, respectively. To save space, the output for the margins effects with respect to all predictor variables when the ordinal categories 2 and 3 are omitted. The following output shows the results for the marginal effects on the probability for category 4 by using the command margins, predict(outcome(4)) dydx(*).

```
. margins, predict(outcome(4)) dydx(*)

Average marginal effects                        Number of obs  =       1300
Model VCE    : OIM

Expression   : Pr(healthre==4), predict(outcome(4))
dy/dx w.r.t. : 1.maritals educ age 1.male

------------------------------------------------------------------------------
             |            Delta-method
             |     dy/dx   Std. Err.      z    P>|z|     [95% Conf. Interval]
-------------+----------------------------------------------------------------
 1.maritals  |   .0945033   .0193388     4.89   0.000     .0565999    .1324067
       educ  |   .0235549   .0029804     7.90   0.000     .0177134    .0293965
        age  |  -.0034716   .0005385    -6.45   0.000    -.0045271   -.0024161
     1.male  |   .0108066   .0189858     0.57   0.569    -.0264049    .0480182
------------------------------------------------------------------------------
Note: dy/dx for factor levels is the discrete change from the base level.
```

The margins command in the preceding output computes the change in the probability of being in category 4 of the ordinal response variable given a one-unit change in each predictor variable.

- The marginal effect for maritals is .095, which indicates that the probability of being in category 4 (i.e., excellent health) increases by .095 when the marital status goes from unmarried to married when holding other predictor variables constant.
- The marginal effect for educ = .024, which indicates that the probability of being in excellent health increases by .024 for a one-unit increase in education.
- The marginal effect for age = −.003, which indicates that the probability of being in excellent health decreases by .003 for a one-unit increase in age. In other words, people are less likely to be in excellent health with an increase of age.

- The marginal effect for `male` = .011, z = .57, p = .569, which indicates that being male does not impact the probability of being in excellent health.

12.1.6 Computing the Marginal Effects With the Improved `margins` Command in Stata 14

Without computing the marginal effects for each category of an ordinal response variable separately, the improved `margins` command in Stata 14 can simultaneously provide the results of marginal effects for all categories. The command `margins, dydx(*)` tells Stata 14 to compute the marginal effects for all predictor variables when the ordinal response variable Y = 1, 2, 3, and 4. Recall that the same command in Stata 13 in the preceding section only computes the marginal effects for the first category when Y = 1. The output is displayed as follows.

```
. *marginal effects
.
. quietly oprobit healthre i.maritals educ age i.male

. margins, dydx(*)

Average marginal effects                        Number of obs    =       1,300
Model VCE      : OIM

dy/dx w.r.t. : 1.maritals educ age 1.male
1._predict   : Pr(healthre==1), predict(pr outcome(1))
2._predict   : Pr(healthre==2), predict(pr outcome(2))
3._predict   : Pr(healthre==3), predict(pr outcome(3))
4._predict   : Pr(healthre==4), predict(pr outcome(4))

------------------------------------------------------------------------------
             |            Delta-method
             |      dy/dx   Std. Err.      z    P>|z|     [95% Conf. Interval]
-------------+----------------------------------------------------------------
1.maritals   |
   _predict  |
          1  |  -.0338846   .0072624    -4.67   0.000    -.0481187   -.0196506
          2  |  -.0605102    .012602    -4.80   0.000    -.0852098   -.0358107
          3  |  -.0001085   .0034464    -0.03   0.975    -.0068633    .0066464
          4  |   .0945033   .0193388     4.89   0.000     .0565999    .1324067
-------------+----------------------------------------------------------------
educ         |
   _predict  |
          1  |  -.0087562   .0013206    -6.63   0.000    -.0113446   -.0061678
          2  |  -.0150384   .0019613    -7.67   0.000    -.0188825   -.0111942
          3  |   .0002396   .0008478     0.28   0.777    -.0014219    .0019012
          4  |   .0235549   .0029804     7.90   0.000     .0177134    .0293965
-------------+----------------------------------------------------------------
```

```
age          |
    _predict |
           1 |    .0012905    .0002263     5.70   0.000     .0008471      .001734
           2 |    .0022164    .0003471     6.39   0.000     .0015361     .0028967
           3 |   -.0000353    .0001248    -0.28   0.777    -.0002799     .0002093
           4 |   -.0034716    .0005385    -6.45   0.000    -.0045271    -.0024161
-------------+--------------------------------------------------------------------
1.male       |
    _predict |
           1 |   -.0040009    .0070084    -0.57   0.568    -.0177371     .0097354
           2 |   -.0068932    .0120983    -0.57   0.569    -.0306053      .016819
           3 |    .0000874    .0004041     0.22   0.829    -.0007046     .0008793
           4 |    .0108066    .0189858     0.57   0.569    -.0264049     .0480182
-------------------------------------------------------------------------------
Note: dy/dx for factor levels is the discrete change from the base level.
```

The first part of the output lists the names of four predictor variables for which the marginal effects are estimated and the number for the predicted probabilities for each of the four ordinal categories (labeled from 1._predict to 4._predict). The second part of the output displays the marginal effects table for all the predictor variables. The results of the marginal effects are the same as those estimated in Stata 13 if the marginal effects for each category in the previous section are combined. The interpretation of the results is omitted to avoid redundancy.

12.1.7 Interpreting the Estimated Probabilities With the `margins` Command

Using the `margins` Command in Stata 13

As with the proportional odds model, after fitting the probit model, the `margins` command can also be used to compute the estimated probabilities of being in a particular category of the ordinal response variable at specified values of predictor variables. The same `margins` command for the proportional odds model in Chapter 4 is used as follows. The command `margins, predict (outcome(1)) at(educ = (8 13 16)) atmeans vsquish` tells Stata 13 to compute the estimated probabilities when the ordinal response variable $Y = 1$, for the predictor variable `educ`, at the values of 8, 13, and 16, when holding the other predictor variables at their means.

```
. margins, predict (outcome(1)) at(educ = (8 13 16)) atmeans vsquish

Adjusted predictions                              Number of obs   =       1300
Model VCE    : OIM

Expression   : Pr(healthre==1), predict(outcome(1))
1._at        : _Imaritals_1    =    .4684615 (mean)
```

```
              educ           =           8
              age            =        48.2  (mean)
              _Imale_1       =    .4438462  (mean)
  2._at      : _Imaritals_1  =    .4684615  (mean)
              educ           =          13
              age            =        48.2  (mean)
              _Imale_1       =    .4438462  (mean)
  3._at      : _Imaritals_1  =    .4684615  (mean)
              educ           =          16
              age            =        48.2  (mean)
              _Imale_1       =    .4438462  (mean)
```

		Margin	Delta-method Std. Err.	z	P>|z|	[95% Conf.	Interval]
_at							
1		.1131095	.0141324	8.00	0.000	.0854106	.1408085
2		.0560362	.0064299	8.71	0.000	.0434338	.0686385
3		.0346682	.0049953	6.94	0.000	.0248776	.0444587

With the same margins command, the results of the estimated probabilities for the proportional odds model and the probit model are very similar. The margins table at the bottom of the output displays the margins, their Delta-method standard errors, the z statistics and the associated p values, and the 95% confidence intervals.

In the output, the first combination (labeled 1._at) is for the situation where educ = 8 and other predictor variables are held at their means (1.maritals = .468, age = 48.2, and 1.male = .444). The estimated margin or the probability of being in category 1 (i.e., $Y = 1$) given these specified values of the four predictor variables is .113.

Next, the estimated probability of being in category 1 for the second combination (labeled 2._at), where educ = 13 and the other three predictor variables are held at their means, is .056.

Finally, the estimated probability of being in category 1 is .035, when educ = 16 and other predictor variables are held at their means (labeled 3._at). All three predicted probabilities are significantly different from zero ($p < .001$).

The margins or expected probabilities for the other three categories (i.e., 2, 3, and 4) of the ordinal response variable can be computed in a similar way by specifying the option predict (outcome(2)), predict (outcome(3)), or predict (outcome(4)), respectively. To save space, only the output for the expected probabilities for category 4 is displayed as follows.

```
. margins, predict (outcome(4)) at(educ = (8 13 16)) atmeans vsquish

Adjusted predictions                              Number of obs    =       1300
Model VCE      : OIM

Expression   : Pr(healthre==4), predict(outcome(4))
1._at        : _Imaritals_1    =     .4684615  (mean)
               educ            =           8
               age             =        48.2   (mean)
               _Imale_1        =     .4438462  (mean)
2._at        : _Imaritals_1    =     .4684615  (mean)
               educ            =          13
               age             =        48.2   (mean)
               _Imale_1        =     .4438462  (mean)
3._at        : _Imaritals_1    =     .4684615  (mean)
               educ            =          16
               age             =        48.2   (mean)
               _Imale_1        =     .4438462  (mean)

-------------------------------------------------------------------------------
             |            Delta-method
             |    Margin   Std. Err.      z    P>|z|     [95% Conf. Interval]
-------------+-----------------------------------------------------------------
        _at  |
          1  |    .14045   .0152515     9.21   0.000    .1105575    .1703424
          2  |  .2421171   .0121561    19.92   0.000    .2182916    .2659426
          3  |  .3183799   .0156762    20.31   0.000    .2876552    .3491046
-------------------------------------------------------------------------------
```

Again, the results in the preceding output for the ordinal probit model are similar to those for the proportional odds model fitted in Chapter 4. When educ equals 8, 13, and 16, and other predictor variables are held at their means, the margins or estimated probabilities for being in excellent health (i.e., category 4) are .140, .242, and .318, respectively. They are all statistically significant ($p < .001$). Following the margins command, we can visualize the estimated probabilities by using the marginsplot command. Figure 12.1 shows the estimated probabilities of being in excellent health condition (i.e., $Y = 4$).

Figure 12.1 shows that with the increase of the highest year of school completed from 8 to 13 to 16 years, the probability of being in excellent health (i.e., $Y = 4$) increases.

The estimated probabilities of being in each category (i.e., $Y = 1, 2, 3$, and 4) can be plotted in a single graph by using the graph combine command after each separate graph is drawn. Please note that we need to name each plot first when using the marginsplot command. The following output provides the complete commands for Figure 12.2 using Stata 13.

Figure 12.1   Estimated Probabilities of Being in Category 4 for educ at 8, 13, and 16

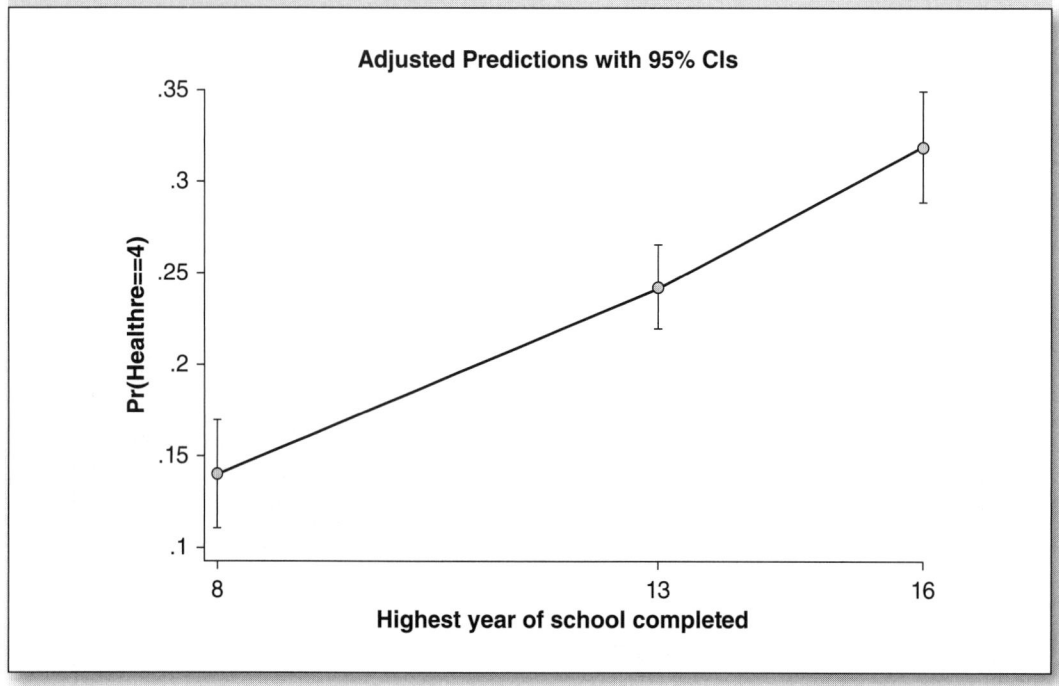

Note: CI = confidence interval.

```
margins, predict (outcome(1)) at(educ = (8 13 16)) atmeans vsquish
marginsplot, name(plot1)
margins, predict (outcome(2)) at(educ = (8 13 16)) atmeans vsquish
marginsplot, name(plot2)
margins, predict (outcome(3)) at(educ = (8 13 16)) atmeans vsquish
marginsplot, name(plot3)
margins, predict (outcome(4)) at(educ = (8 13 16)) atmeans vsquish
marginsplot, name(plot4)
graph combine plot1 plot2 plot3 plot4, ycommon
```

Using the Improved `margins` and `marginsplot` Commands in Stata 14

Using the improved `margins` and `marginsplot` commands for ordinal response variables in Stata 14, we can simultaneously obtain and plot the results for all categories of the ordinal response variable without multiple commands.

Figure 12.2 Estimated Probabilities of Being in Categories 1, 2, 3, and 4 for educ

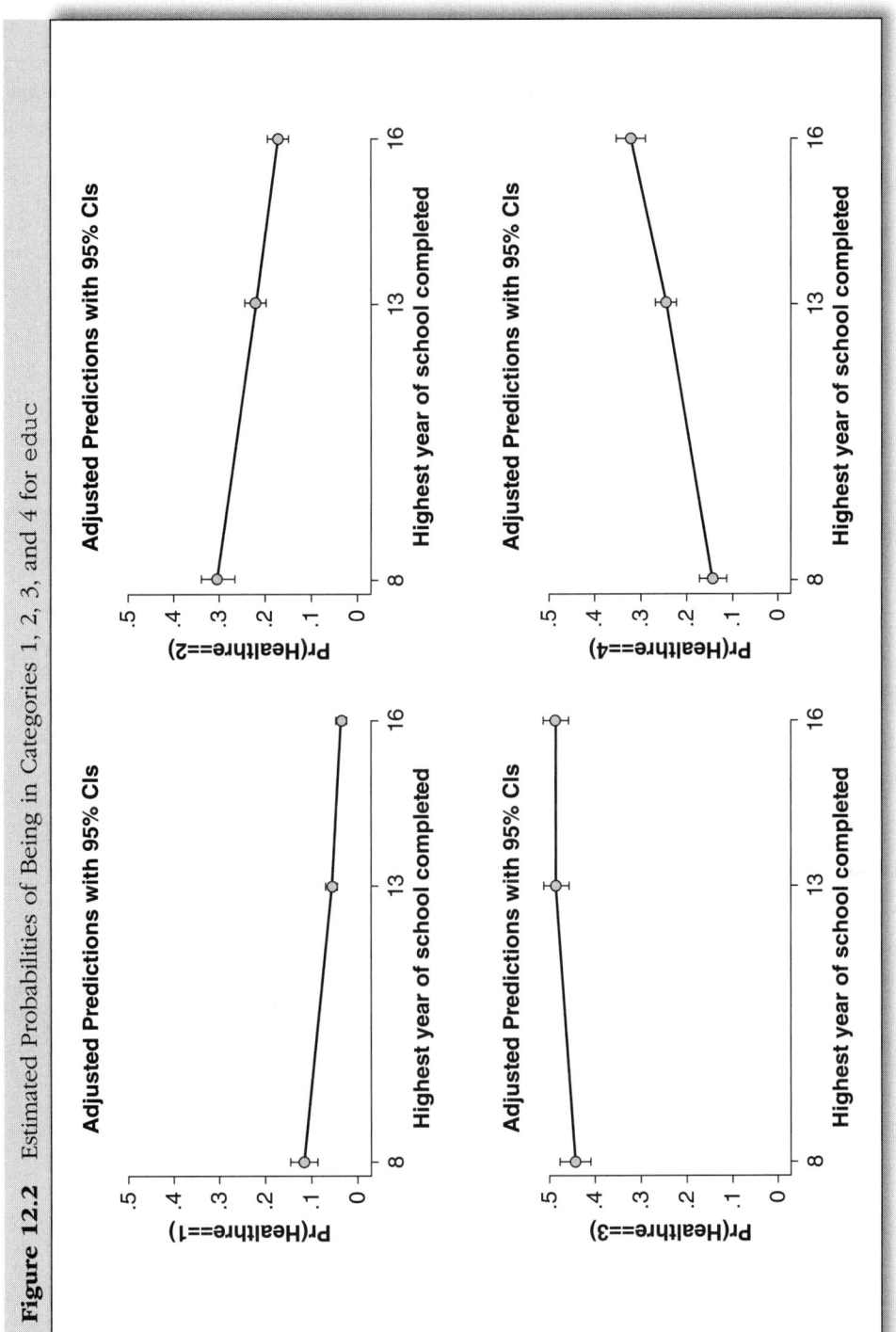

Note: CI = confidence interval.

To replicate the results in the preceding section, the command `margins, at(educ = (8 13 16)) atmeans vsquish` tells Stata 14 to compute the estimated probabilities for all four categories of the ordinal response variable when $Y = 1$, 2, 3, and 4 for the predictor variable `educ` at the values of 8, 13, and 16 when holding the other predictor variables at their means. The output is shown as follows.

```
. *margins and marginsplot in Stata 14
. quietly oprobit healthre i.maritals edu age i.male

. margins, at(educ = (8 13 16)) atmeans vsquish

Adjusted predictions                              Number of obs    =      1,300
Model VCE    : OIM

1._predict   : Pr(healthre==1), predict(pr outcome(1))
2._predict   : Pr(healthre==2), predict(pr outcome(2))
3._predict   : Pr(healthre==3), predict(pr outcome(3))
4._predict   : Pr(healthre==4), predict(pr outcome(4))
1._at        : 0.maritals       =      .5315385  (mean)
               1.maritals       =      .4684615  (mean)
               educ             =             8
               age              =          48.2  (mean)
               0.male           =      .5561538  (mean)
               1.male           =      .4438462  (mean)
2._at        : 0.maritals       =      .5315385  (mean)
               1.maritals       =      .4684615  (mean)
               educ             =            13
               age              =          48.2  (mean)
               0.male           =      .5561538  (mean)
               1.male           =      .4438462  (mean)
3._at        : 0.maritals       =      .5315385  (mean)
               1.maritals       =      .4684615  (mean)
               educ             =            16
               age              =          48.2  (mean)
               0.male           =      .5561538  (mean)
               1.male           =      .4438462  (mean)

------------------------------------------------------------------------------
             |            Delta-method
             |    Margin   Std. Err.      z    P>|z|    [95% Conf. Interval]
-------------+----------------------------------------------------------------
_predict#_at |
         1 1 |  .1131095   .0141324    8.00   0.000    .0854106    .1408085
         1 2 |  .0560362   .0064299    8.71   0.000    .0434338    .0686385
         1 3 |  .0346682   .0049953    6.94   0.000    .0248776    .0444587
         2 1 |  .3029376   .0184358   16.43   0.000     .266804    .3390711
         2 2 |  .2212885   .0118967   18.60   0.000    .1979714    .2446056
         2 3 |  .1719869   .0112651   15.27   0.000    .1499077    .1940661
         3 1 |  .4435029   .0163581   27.11   0.000    .4114417    .4755642
         3 2 |  .4805582   .0143568   33.47   0.000    .4524194     .508697
         3 3 |   .474965    .014327   33.15   0.000    .4468845    .5030455
         4 1 |    .14045   .0152515    9.21   0.000    .1105575    .1703424
         4 2 |  .2421171   .0121561   19.92   0.000    .2182916    .2659426
         4 3 |  .3183799   .0156762   20.31   0.000    .2876552    .3491046
------------------------------------------------------------------------------
```

As we can easily see, the estimated probabilities for all four categories of the ordinal response variable are displayed in the output. The results are the same as the combined results estimated from each separate command in the previous section using Stata 13.

Without plotting each estimated probability separately, the improved `marginsplot` command in Stata 14 can automatically plot the estimated probabilities of being in each category of an ordinal response variable in a single graph. Figure 12.3 shows the estimated probabilities when $Y = 1, 2, 3$, and 4.

12.1.8 Model Comparison Using the Log Likelihood Ratio Test

The log likelihood ratio test or the deviance difference test can be used for model comparisons. As with the model comparison in the proportional odds models, we first fit the single-predictor ordinal probit model and save the estimated log likelihood value, and then we fit the full model with all four predictor variables and save the estimated log likelihood value. The `estimates store` command is used to save

Figure 12.3 Estimated Probabilities of Being in Categories 1, 2, 3, and 4 for educ With Stata 14

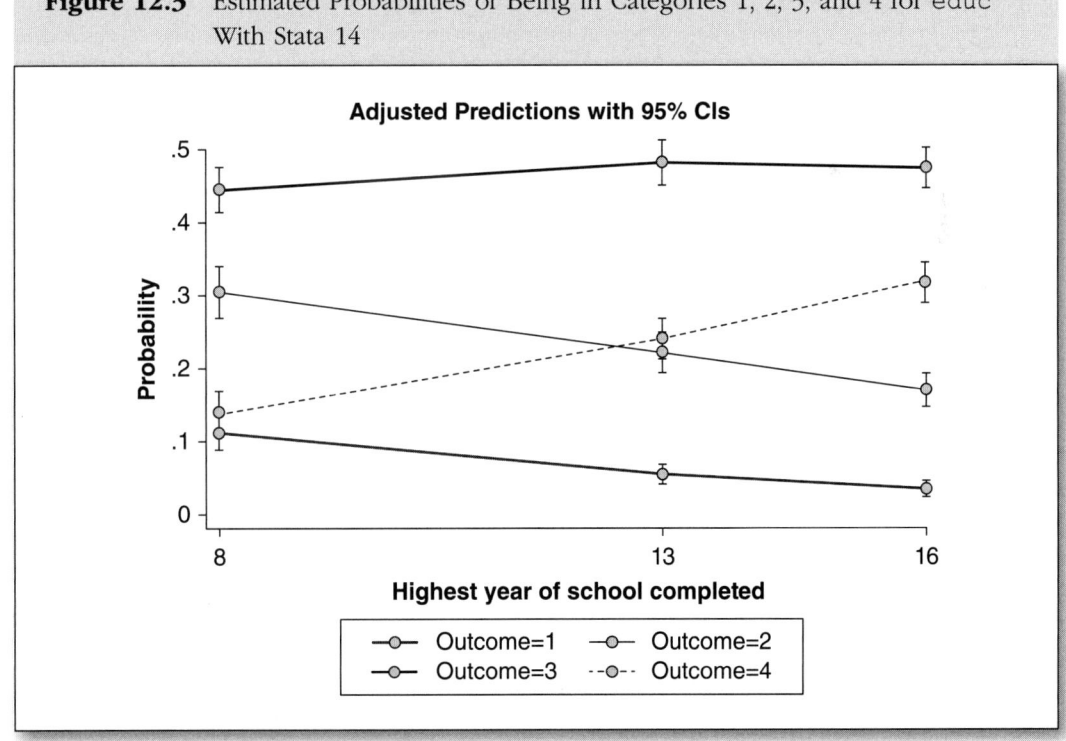

Note: CI = confidence interval.

the estimated log likelihood values. Finally, we conduct the log likelihood ratio test to compare these two models using the `lrtest` command. The following output is displayed.

```
. quietly xi: oprobit healthre i.maritals
. estimates store model1
. quietly xi: oprobit healthre i.maritals age educ i.male
. estimates store fullmodel
. lrtest model1 fullmodel
Likelihood-ratio test                              LR chi2(3)  =     106.49
(Assumption: model1 nested in fullmodel)           Prob > chi2 =     0.0000
```

The log likelihood ratio chi-square test is reported in the output $\chi^2_{(3)} = 106.49$, $p < .001$. This result suggests that the full model with all four predictor variables fits the data better than the single-predictor model.

12.1.9 Making Publication-Quality Tables
Comparing the Probit Model and Proportional Odds Model

Both proportional odds models and ordinal probit models can be used to estimate ordinal response variables. Since both models are only different in terms of error distributions, the estimated coefficients may look similar. We can compare the coefficients for both models.

```
1.  . quietly xi: oprobit healthre i.maritals age educ i.male
2.  . estimates store probit
3.  . quietly xi: ologit healthre i.maritals age educ i.male
4.  . estimates store logit
5.  . outreg2 [probit logit] using probitout, e(ll df_m chi2) addstat
       (Pseudo R-squared, `e(r2_p)') dec(3) word replace
probitout.rtf
dir : seeout
```

The first command fits the ordinal probit model without displaying the output. The second command, `estimates store`, is then used to save the estimates with a name of "probit". The third command fits a proportional odds model with the same predictor variables, and the estimates are saved using the fourth command, `estimates store logit`. Finally, we use the `outreg2` command to display the estimated coefficients and model fit statistics for these two models in a table. In the syntax, `outreg2` is the command for making a regression table. The square bracket [probit logit] contains the saved estimates from the two models. The `using probitout` command saves the final table in a file named probitout. The e(ll df_m chi2) option adds the log likelihood ratio test statistic (ll), the degrees of freedom (df_m), and the chi square test statistic (chi2), and the addstat () option adds the pseudo R^2 to the table. The option dec(3) reports three decimal places. The word option requests that the file probitout be saved as a Word document. The replace option will replace previous files that have the same name.

These commands automatically produce Table 12.1, as shown here in its original format, presenting the results of both the ordinal probit model and the proportional odds model.

The probit coefficient for each predictor variable in the ordinal probit model looks similar to the logit coefficient in the proportional odds model. In fact, the logit coefficients are approximately 1.7 times as large as the probit coefficients. For example, the probit coefficient for maritals is .301, its corresponding logit coefficient is .514, and the ratio of the latter to the former is 1.708. The model fit statistics between these two models, such as the log-likelihood ratio chi-square test statistic and the pseudo R^2, are almost the same.

12.1.10 Reporting the Results

Writing the results of the ordinal probit regression models is similar to that of the ordinal logistic regression models. Most common reporting guidelines for the ordinal logistic regression models can be applied to the reporting for the probit models. The only difference is that no odds ratios are reported in the probit models. The following are the generic guidelines for reporting the results. You may need to adjust your writing when submitting your manuscript for publication since your journals may have different requirements.

First, describe the statistical method you used for data analysis, the dependent variable and the independent variables in the models, and your research hypothesis or the purpose of your study.

Second, report model fit statistics, including but not limited to the log likelihood ratio statistic, the associated p value, and the pseudo R^2, followed by a concise statement of interpretation on whether the fitted model is better than the null model. If more fit statistics, such as various pseudo R^2 values, deviance statistics, and AIC and BIC statistics, are available, then include them in a table.

Third, report the parameter estimates for the predictor variables, their standard errors, and the associated p values either in a table or in the text. A table is preferable for models with multiple predictors. If more than one model is fitted, then the results of all the competing models from the simple model to the full model can be presented in a table. The following is an example of summarizing the results from the ordinal probit regression model.

Table 12.1 Results of the Ordinal Probit Model and the Proportional Odds Model (Shown in Original Format Generated by Stata)

Variables	(1) probit healthre	(2) logit healthre
_Imaritals_1	0.301***	0.514***
	(0.061)	(0.106)
Age	−0.011***	−0.019***
	(0.002)	(0.003)
Educ	0.076***	0.134***
	(0.010)	(0.017)
_Imale_1	0.035	0.053
	(0.061)	(0.105)
Constant cut1	−0.986***	−1.701***
	(0.169)	(0.298)
Constant cut2	0.012	0.125
	(0.166)	(0.286)
Constant cut3	1.302***	2.258***
	(0.169)	(0.294)
Observations	1,300	1,300
Pseudo R	0.0424	0.0424
Ll	−1,513	−1,512
df_m	4	4
chi2	132.4	133.9

Note: Standard errors in parentheses.

***$p < 0.01$, **$p < 0.05$, *$p < 0.1$

The ordinal probit regression analysis was conducted to examine the relationships between the ordinal outcome variable, health status, and a set of predictor variables, such as marital status, years of education, age, and gender. A single-predictor model with marital status as the predictor was fitted first, and then the full model with all predictors was fitted. The log likelihood ratio test or the deviance difference test is used to compare the two models, $\chi^2_{(3)}$ = 106.49, p < .001. The result indicated that the full model fitted data better than the single predictor model. The results for the full model are presented in Table 12.1.

For the full model, $LR\ \chi^2_{(4)}$ = 132.40, p < .001, which indicated that the full probit model with four predictors provided a better fit than the null model with no independent variables in predicting the ordinal response variable. The likelihood ratio R^2_L = .042 was larger than that of the one-predictor model but still small, which suggested that the relationship between the response variable, health status, and four predictors was still small.

For the `maritals` predictor, the probit coefficient β = .301, which was positive. It indicated that being married increased the probability of being beyond a particular category of health status (better health status) when holding all the other predictors constant. It could also be interpreted that being married increased the z scores by .301.

For the `educ` predictor, β = .076, which was positive. It indicated that an increase in the predictor `educ` corresponded to the increase in the probability of being beyond a particular category of health status (better health status) when holding all the other predictors constant. For each one-unit increase in education, there was an increase in z scores by .076.

For the `age` predictor, β = −.011, which was negative. It indicated that the increase in the predictor `age` was associated with a decrease in the probability of being beyond a particular category of health status (better health status) when all the other predictors remain constant. For a one-unit increase in age, the z scores of being healthier decreased by .011.

For the `male` predictor, β = .035, z = .57, p = .569, which was not significantly different from 0. It indicated that there was no relationship between being a male and the probability of being in a better health status.

12.2 Multinomial Logistic Regression Models

12.2.1 Multinomial Logistic Regression Models: An Introduction

The multinomial logistic regression model is used to estimate nominal response variables that have multiple unordered categories. This model can be used for an ordinal response variable when the proportional odds assumption does not hold. It estimates the odds of being in a category versus the base category of a nominal variable. Although the proportional odds model compares the cumulative probabilities of being at or below a particular category and the probabilities of being above that category, the multinomial logistic regression model compares a particular category with the base category. If a nominal response variable has J levels, there are $J − 1$ comparisons between any other categories to the base category. For example, if we disregard the ordinal nature of the ordinal response variable, health status, and treat it as a nominal response variable with four categories, then we compare category 2 and category 1, category 3 and category 1, and category 4 and category 1 in the multinomial logistic model where the base category is set to be one.

The multinomial logistic model can be expressed as follows:

$$\ln\left(\frac{\pi\left(Y = j \mid x_1, x_2, \dots, x_p\right)}{\pi\left(Y = J \mid x_1, x_2, \dots, x_p\right)}\right) = \alpha_j + \beta_{j1}X_1 + \beta_{j2}X_2 + \dots + \beta_{jp}X_p \tag{12.6}$$

where $j = 1, 2, \dots, J-1$; J is the base category, which can be any category but is generally the highest one; α_j are the intercepts; and $\beta_{j1}, \beta_{j2}, \dots, \beta_{jp}$ are the logit coefficients for each comparison. The model estimates $J - 1$ logit coefficients for each predictor. If we set the base category to be category 1, the lowest category, then the model can be rewritten as follows:

$$\ln\left(\frac{\pi\left(Y = j \mid x_1, x_2, \dots, x_p\right)}{\pi\left(Y = 1 \mid x_1, x_2, \dots, x_p\right)}\right) = \alpha_j + \beta_{j1}X_1 + \beta_{j2}X_2 + \dots + \beta_{jp}X_p \tag{12.7}$$

It can be treated as a simultaneous estimation of a series of binary logistic regression models comparing a particular category and the base category. In each binary model, being in a particular category is coded as the binary outcome of 1 and being in the base category is coded as 0.

12.2.2 Odds in Multinomial Logistic Models

The multinomial logistic model estimates the logit or log odds of being in a particular category relative to the baseline category. The odds in the multinomial logistic model can be defined as the ratio of the probability of being in a particular category to the probability of being in the base category. It is expressed as:

$$\text{Odds } (Y = j \text{ vs. } Y = J) = \frac{p(Y = j)}{p(Y = J)} \tag{12.8}$$

where j can be any categories from 1 to $J - 1$ categories.

For example, if we treat the ordinal response variable health status as nominal with four categories from 1 to 4, with 1 = poor, 2 = fair, 3 = good, and 4 = excellent, then we estimate three odds with category 1 as the base category: The odds of being in category 2 versus category 1, the odds of being in category 3 versus category 1, and the odds of being in category 4 versus category 1.

Specifically, odds $(Y = 2 \text{ vs. } Y = 1)$ equal the ratio of the probability of being in category 2 to the probability of being in category 1:

$$\text{Odds } (Y = 2 \text{ vs. } 1) = \frac{p(Y = 2)}{p(Y = 1)} = \frac{p(2)}{p(1)}$$

The other two odds, odds $(Y = 3$ vs. $Y = 1)$ and odds $(Y = 4$ vs. $Y = 1)$, are expressed as follows:

$$\text{Odds } (Y = 3 \text{ vs. } 1) = \frac{p(Y = 3)}{p(Y = 1)} = \frac{p(3)}{p(1)}$$

$$\text{Odds } (Y = 4 \text{ vs. } 1) = \frac{p(Y = 4)}{p(Y = 1)} = \frac{p(4)}{p(1)}$$

Table 12.2 presents the logits, odds, and category comparisons for the multinomial logistic regression model for the nominal response variable with four levels.

12.2.3 Odds Ratios or Relative Risk Ratios in Multinomial Logistic Regression Models

Since the multinomial logistic model can be treated as a series of binary logistic regression models estimated simultaneously with the comparison of any other categories to the base category, the logit coefficients can be interpreted in a similar way as that for the binary logistic regression. The odds ratio of being in a category j versus the baseline category J is obtained by taking the exponential of the logit coefficient β. The odds ratio in the multinomial logistic regression is also called the relative risk ratio. Hilbe (2009) preferred the use of the relative risk ratio rather than the odds ratio since the categories of the nominal response variable are independent of each other. The odds ratio or relative risk ratio in multinomial logistic regression can be interpreted as the change in the odds or relative risk for a one-unit change in a predictor variable when holding other predictor variables constant. To obtain the inversed odds, the odds of being in the base category versus a particular category, we exponentiate the logit coefficient with a negative sign $\exp(-\beta)$.

Table 12.2 Category Comparisons for the Multinomial Logistic Regression Model With Four Levels of Health Status $(j = 1, 2, 3, 4)$

Equation	Logit $P(Y = j$ vs. $J)$	Odds	Probability Comparisons
1	logit $P(Y = 2$ vs. 1)	$\dfrac{P(Y = 2)}{P(Y = 1)}$	Category 2 vs. category 1
2	logit $P(Y = 3$ vs. 1)	$\dfrac{P(Y = 3)}{P(Y = 1)}$	Category 3 vs. category 1
3	logit $P(Y = 4$ vs. 1)	$\dfrac{P(Y = 4)}{P(Y = 1)}$	Category 4 vs. category 1

12.2.4 Description of the Research Example, Data, and Sample

We will investigate the relationships between the nominal response variable, health status, and four predictor variables. Unlike in other chapters, however, here the research interest focuses on using the multinomial logistic regression to predict the nominal response variable. The GSS 2012 data are used for the following analyses. The same outcome variable and the predictor variables used in the first section of this chapter are used to fit the following multinomial logistic regression models.

12.2.5 Multinomial Logistic Regression Models With Stata: Commands and Output

The Stata Command mlogit

The command mlogit is used for multinomial logistic regression models. In the syntax, the command mlogit is followed by the dependent variable and the independent variable(s). Since the single-predictor multinomial logistic regression model has been estimated in Chapter 7 (Adjacent Categories Logistic Regression Models), it will not be discussed here. Instead, we will fit a multinomial logistic model with multiple predictor variables. For more details on how to use this command, type the help mlogit command.

The xi: mlogit healthre i.maritals educ age i.male, baseoutcome(1) tells Stata to conduct the multinomial logistic regression to estimate the outcome healthre from the four predictor variables. The option baseoutcome(1) specifies the lowest category 1 as the base category or the reference group. The default is the category with the highest value (i.e., category 4 in this example) if you do not specify this option. As noted, the prefix xi can be omitted if you use Stata 11 or later versions since the new coding for factor variables is used. The output is displayed as follows.

```
. xi: mlogit healthre i.maritals educ age i.male, baseoutcome(1)
i.maritals         _Imaritals_0-1      (naturally coded; _Imaritals_0 omitted)
i.male             _Imale_0-1          (naturally coded; _Imale_0 omitted)

Iteration 0:   log likelihood = -1579.4068
Iteration 1:   log likelihood = -1506.8551
Iteration 2:   log likelihood = -1503.7817
Iteration 3:   log likelihood = -1503.7655
Iteration 4:   log likelihood = -1503.7655

Multinomial logistic regression              Number of obs   =       1300
                                              LR chi2(12)     =     151.28
                                              Prob > chi2     =     0.0000
Log likelihood = -1503.7655                   Pseudo R2       =     0.0479
```

```
------------------------------------------------------------------------------
    healthre |      Coef.   Std. Err.       z    P>|z|     [95% Conf. Interval]
-------------+----------------------------------------------------------------
1            |  (base outcome)
-------------+----------------------------------------------------------------
2            |
 _Imaritals_1 |  .1515596   .2618387     0.58   0.563    -.3616348     .664754
        educ |  .0572264   .0370078     1.55   0.122    -.0153075    .1297603
         age | -.0136654   .0068847    -1.98   0.047    -.0271592   -.0001716
   _Imale_1 | -.2468661   .2535898    -0.97   0.330    -.7438931    .2501608
       _cons |  1.303891   .6486351     2.01   0.044     .0325897    2.575193
-------------+----------------------------------------------------------------
3            |
 _Imaritals_1 |  .4265435   .2499166     1.71   0.088    -.0632839     .916371
        educ |  .2063935   .0368721     5.60   0.000     .1341255    .2786615
         age | -.0285959   .0066426    -4.30   0.000    -.0416152   -.0155767
   _Imale_1 | -.2744464   .2421955    -1.13   0.257    -.7491408     .200248
       _cons |  .7407376   .6317226     1.17   0.241     -.497416    1.978891
-------------+----------------------------------------------------------------
4            |
 _Imaritals_1 |  .9181182   .2617502     3.51   0.000     .4050972    1.431139
        educ |   .233781   .0392539     5.96   0.000     .1568448    .3107172
         age | -.0376157   .0071261    -5.28   0.000    -.0515826   -.0236488
   _Imale_1 | -.0521476   .2540155    -0.21   0.837    -.5500088    .4457137
       _cons | -.1170394   .6686723    -0.18   0.861    -1.427613    1.193534
------------------------------------------------------------------------------
```

12.2.6 Interpreting the Output

The output for the multinomial logistic regression model looks like that for binary logistic regression models except that there are multiple binary comparisons with category 1 as the base outcome. The beginning of the output displays the iterations for the maximum likelihood estimation.

The log likelihood ratio test statistic LR chi2 (12) = 151.28 and the associated p value Prob > chi2 = 0.0000. It indicates that the four-predictor model provides a better fit than the null model with no independent variables in predicting the logit of being in any other category of health status compared with being in the base category (i.e., poor health).

The multinomial logistic regression table displays the parameter estimates for the predictor variables and the intercepts, their standard errors, the Wald z statistics, the associated p values, and the 95% confident intervals. Since the multinomial logistic regression model is treated as a series of binary logistic regression models, the table displays the parameter estimates for three binary logistic models comparing each category versus the base category. Only three binary models are estimated since the base outcome is category 1. The three models are numbered 2, 3, and 4. These three equations compare categories 2 with 1, 3 with 1, and 4 with 1, respectively.

Based on the parameter estimates in the output, the three equations can be expressed as:

$$\ln\left(\frac{\pi(Y=2)}{\pi(Y=1)}\right) = 1.304 + .152\text{maritals} + .057\text{educ} - .014\text{age} - .247\text{male}$$

$$\ln\left(\frac{\pi(Y=3)}{\pi(Y=1)}\right) = .741 + .427\text{maritals} + .206\text{educ} - .028\text{age} - .274\text{male}$$

$$\ln\left(\frac{\pi(Y=4)}{\pi(Y=1)}\right) = -.117 + .918\text{maritals} + .234\text{educ} - .038\text{age} - .052\text{male}$$

The Wald test is used to test whether the coefficient of each predictor variable is significantly different from zero.

The first equation, labeled "2" in the output, compares category 2 and category 1. Among four predictor variables, only age is significant (β = −.123, Wald z = −1.98, p = .047), whereas the other three predictor variables maritals, educ, and age are not significant in predicting the log odds of being in category 2 versus 1. The 95% confidence interval of the regression coefficient for age is [−.0271592, −.0001716], which also indicates the coefficient is significant since it does not contain the value of zero.

The second equation, labeled "3" in the output, compares category 3 and category 1. The predictor variables educ and age are significant. For educ, β = .206, Wald z = 5.60, p < .001; for age, β = −.029, Wald z = −4.30, p < .001. The other two predictor variables, maritals and male, are not significant.

The third equation, labeled "4" in the output, compares category 4 and category 1. The three predictor variables maritals, educ, and age are significant, whereas male is not significant. For the predictor variable maritals, β = .918, Wald z = 3.51. The associated p value, P>|z|=.000, so we reject the null hypothesis; for educ, β = .234, Wald z = 5.96, p < .001; for age, β = −.038, Wald z = −5.28, p < .001; however, for male, β = −.052, Wald z = −.21, p = .837, so we fail to reject the null hypothesis that the coefficient for male is zero.

As with ordinal logistic regression models, a better way to interpret the multinomial logistic regression model is to transform the logit coefficients into odds ratios. Exponentiating the logit coefficients gives us the odds ratios or the relative risk ratios in multinomial logistic regression. Either the command mlogit with the rrr option or the command listcoef (Long & Freese, 2014) immediately following the original model reports the odds ratios or the relative risk ratios. The mologit, rrr command reports the relative risk ratios of being in any other category versus the base category, whereas the listcoef command of the SPost13 package (Long & Freese, 2014) computes the odds comparing any two combinations of all the categories. The listcoef command produces the following output for the odds comparing any two of the four categories.

```
. listcoef

mlogit (N=1300): Factor change in the odds of healthre

Variable: _Imaritals_1 (sd=0.499)
------------------------------------------------------------------------
                       |      b        z      P>|z|      e^b     e^bStdX
-----------------------+------------------------------------------------
1        vs 2          |  -0.1516   -0.579    0.563    0.859    0.927
1        vs 3          |  -0.4265   -1.707    0.088    0.653    0.808
1        vs 4          |  -0.9181   -3.508    0.000    0.399    0.632
2        vs 1          |   0.1516    0.579    0.563    1.164    1.079
2        vs 3          |  -0.2750   -1.802    0.071    0.760    0.872
2        vs 4          |  -0.7666   -4.492    0.000    0.465    0.682
3        vs 1          |   0.4265    1.707    0.088    1.532    1.237
3        vs 2          |   0.2750    1.802    0.071    1.317    1.147
3        vs 4          |  -0.4916   -3.541    0.000    0.612    0.782
4        vs 1          |   0.9181    3.508    0.000    2.505    1.581
4        vs 2          |   0.7666    4.492    0.000    2.152    1.466
4        vs 3          |   0.4916    3.541    0.000    1.635    1.278
------------------------------------------------------------------------

Variable: educ (sd=3.151)
------------------------------------------------------------------------
                       |      b        z      P>|z|      e^b     e^bStdX
-----------------------+------------------------------------------------
1        vs 2          |  -0.0572   -1.546    0.122    0.944    0.835
1        vs 3          |  -0.2064   -5.598    0.000    0.814    0.522
1        vs 4          |  -0.2338   -5.956    0.000    0.792    0.479
2        vs 1          |   0.0572    1.546    0.122    1.059    1.198
2        vs 3          |  -0.1492   -6.025    0.000    0.861    0.625
2        vs 4          |  -0.1766   -6.300    0.000    0.838    0.573
3        vs 1          |   0.2064    5.598    0.000    1.229    1.916
3        vs 2          |   0.1492    6.025    0.000    1.161    1.600
3        vs 4          |  -0.0274   -1.176    0.240    0.973    0.917
4        vs 1          |   0.2338    5.956    0.000    1.263    2.089
4        vs 2          |   0.1766    6.300    0.000    1.193    1.744
4        vs 3          |   0.0274    1.176    0.240    1.028    1.090
------------------------------------------------------------------------

Variable: age (sd=17.432)
------------------------------------------------------------------------
                       |      b        z      P>|z|      e^b     e^bStdX
-----------------------+------------------------------------------------
1        vs 2          |   0.0137    1.985    0.047    1.014    1.269
1        vs 3          |   0.0286    4.305    0.000    1.029    1.646
1        vs 4          |   0.0376    5.279    0.000    1.038    1.927
2        vs 1          |  -0.0137   -1.985    0.047    0.986    0.788
2        vs 3          |   0.0149    3.549    0.000    1.015    1.297
2        vs 4          |   0.0240    4.874    0.000    1.024    1.518
3        vs 1          |  -0.0286   -4.305    0.000    0.972    0.607
3        vs 2          |  -0.0149   -3.549    0.000    0.985    0.771
3        vs 4          |   0.0090    2.144    0.032    1.009    1.170
4        vs 1          |  -0.0376   -5.279    0.000    0.963    0.519
4        vs 2          |  -0.0240   -4.874    0.000    0.976    0.659
4        vs 3          |  -0.0090   -2.144    0.032    0.991    0.855
------------------------------------------------------------------------
```

```
Variable: _Imale_1 (sd=0.497)
--------------------------------------------------------------------------------
                              |       b       z    P>|z|       e^b   e^bStdX
------------------------------+-------------------------------------------------
1            vs 2             |  0.2469   0.973   0.330     1.280     1.131
1            vs 3             |  0.2744   1.133   0.257     1.316     1.146
1            vs 4             |  0.0521   0.205   0.837     1.054     1.026
2            vs 1             | -0.2469  -0.973   0.330     0.781     0.885
2            vs 3             |  0.0276   0.182   0.855     1.028     1.014
2            vs 4             | -0.1947  -1.156   0.248     0.823     0.908
3            vs 1             | -0.2744  -1.133   0.257     0.760     0.872
3            vs 2             | -0.0276  -0.182   0.855     0.973     0.986
3            vs 4             | -0.2223  -1.626   0.104     0.801     0.895
4            vs 1             | -0.0521  -0.205   0.837     0.949     0.974
4            vs 2             |  0.1947   1.156   0.248     1.215     1.102
4            vs 3             |  0.2223   1.626   0.104     1.249     1.117
--------------------------------------------------------------------------------
```

The output contains one table for each predictor variable. In each table, the first column displays all possible comparisons for two of the four categories. The columns labeled "b", "z", and "P>|z|" display the logit coefficients, z statistics, and associated p values for each comparison. The column labeled e^b lists the odds ratios.

The next four tables display the odds ratios comparing any of the two categories for the four predictor variables with one table for each of them. We are interested in the odds ratios comparing a particular category (2, 3, or 4) with the base category 1, or the odds ratios comparing the base category 1 with any other category. The results of the odds ratios across three binary comparisons are summarized in Table 12.3.

12.2.7 Interpreting the Odds Ratios of Being in a Category j Versus the Base Category 1

The odds ratios in the multinomial model can be interpreted in a similar way to other logistic regression models except that the former model compares any category with the base category. As with binary logistic regression models, a positive logit regression coefficient in the multinomial model corresponds to an odds ratio greater than 1, a negative coefficient is associated with an odds ratio less than 1, and a coefficient of 0 corresponds to an odds ratio of 1.

The odds ratio of being in a particular category compared with the base category can be interpreted as the change in the odds of being in that category versus the base category for a one-unit increase in the predictor variable when holding other predictors constant. Recall there are $J - 1$ binary comparisons for a nominal response variable with J categories.

Table 12.3   Odds Ratios for All Four Predictor Variables Across Three Comparisons $(Y = j$ vs. $Y = 1)$

Category Comparisons Variables	$Y = 2$ vs. $Y = 1$ OR	$Y = 3$ vs. $Y = 1$ OR	$Y = 4$ vs. $Y = 1$ OR
maritals	1.164	1.532	2.505**
Educ	1.059	1.229**	1.63**
Age	.986*	.972**	.963**
Male	.781	.860	.949

$*p < .05, **p < .01.$

The odds ratio for each predictor needs to be interpreted across three comparisons. For maritals, the odds ratios of being in category 2 versus category 1, category 3 versus category 1, and category 4 versus category 1 are 1.164, 1.532, and 2.505, respectively. Among them, only the odds ratio of being in category 4 versus the base category is significant, $z = 3.508$, $p < .001$, OR(4,1) = 2.505, which indicates that the odds of being in category 4 versus the base category for the married are 2.505 times as large as the odds for the unmarried when holding other predictors constant. The odds ratios for the other three variables can be interpreted in a similar way.

For educ, the odds ratios for three binary comparisons (i.e., categories 2 vs. 1, 3 vs. 1, and 4 vs. 1) are 1.059, 1.229, and 1.263, respectively. OR(2,1) = 1.059, $z = 1.546$, $p > .05$, which indicates that there is no relationship between educ and the odds of being in category 2 versus the base category 1. The odds ratios for the other two comparisons are significant, OR(3,1) = 1.229, OR(4,1) = 1.263, which indicates that the odds of being in category 3 versus the base category and category 4 versus the base category increase by 1.229 and 1.263, respectively, for a one-unit increase in the educ predictor when holding all the other predictors constant.

With regard to age, the odds ratios for three binary comparisons (i.e., categories 2 vs. 1, 3 vs. 1, and 4 vs. 1) are .986, .972, and .963, respectively, and they are all significant. It indicates that the odds of being in categories 2, 3, and 4 versus the base category decrease by .986, .972, and .963, respectively, for a one-unit increase in the age predictor when holding all the other predictors constant.

For the male predictor, none of the odds ratios for the binary comparisons are significant. It indicates that being male does not impact the odds of being in any particular category versus the base category when holding other predictors constant.

The fitstat command, which is also part of the SPost package (Long & Freese, 2014), produces various model fit statistics, such as the log likelihood statistic, deviance statistic, log likelihood ratio test, six types of pseudo R^2 statistics, and AIC and BIC statistics.

```
. fitstat

                         |       mlogit
-------------------------+-------------
Log-likelihood           |
                   Model |   -1503.765
          Intercept-only |   -1579.407
-------------------------+-------------
Chi-square               |
         Deviance (df=1285) |    3007.531
             LR (df=12) |     151.283
                 p-value |       0.000
-------------------------+-------------
R2                       |
                McFadden |       0.048
     McFadden (adjusted) |       0.038
            Cox-Snell/ML |       0.110
Cragg-Uhler/Nagelkerke   |       0.120
                   Count |       0.466
         Count (adjusted) |      0.017
-------------------------+-------------
IC                       |
                     AIC |    3037.531
         AIC divided by N |       2.337
             BIC (df=15) |    3115.083
```

12.2.8 Interpreting the Estimated Probabilities With the margins Command

Using the margins and marginsplot Commands in Stata 13

The margins command can also be used to compute the estimated probabilities of being in a particular category of the nominal response variable at specified values of predictor variables. The same margins command for the proportional odds model and the ordinal probit model is used for the following analysis after the multinomial logistic regression model is fitted. The command margins, predict (outcome(1)) at(educ = (8 13 16)) atmeans vsquish tells Stata 13 to compute the estimated probabilities when the nominal response variable $Y = 1$, for the predictor variable educ at the values of 8, 13, and 16, when holding the other predictor variables at their means.

```
. margins, predict (outcome(1)) at(educ = (8 13 16)) atmeans vsquish

Adjusted predictions                              Number of obs    =       1300
Model VCE      : OIM

Expression   : Pr(healthre==1), predict(outcome(1))
1._at        : _Imaritals_1    =    .4684615 (mean)
               educ            =           8
               age             =        48.2 (mean)
               _Imale_1        =    .4438462 (mean)
2._at        : _Imaritals_1    =    .4684615 (mean)
               educ            =          13
               age             =        48.2 (mean)
               _Imale_1        =    .4438462 (mean)
3._at        : _Imaritals_1    =    .4684615 (mean)
               educ            =          16
               age             =        48.2 (mean)
               _Imale_1        =    .4438462 (mean)

------------------------------------------------------------------------------
             |            Delta-method
             |    Margin   Std. Err.      z    P>|z|     [95% Conf. Interval]
-------------+----------------------------------------------------------------
         _at |
          1  |  .1205813   .0200847     6.00   0.000     .081216     .1599467
          2  |   .056261   .007065      7.96   0.000     .0424139    .0701081
          3  |  .0330092   .0060246     5.48   0.000     .0212012    .0448172
------------------------------------------------------------------------------
```

When educ = 8 and the other predictor variables are held at their means (1.mar-itals = .468, age = 48.2, and 1.male = .444), the estimated margin or the probability of being in category 1 (i.e., $Y = 1$) is .121.

Next, when educ = 13 and the other three predictor variables are held at their means (labeled 2._at), the estimated probability of being in category 1 is .056.

Finally, when educ = 16 and the other predictor variables are held at their means (labeled 3._at), the estimated probability of being in category 1 is .033. All three predicted probabilities are significantly different from zero ($p < .001$).

Immediately following the margins command, we can plot the estimated probabilities at the specified values of the predictor variables using the marginsplot command. Figure 12.4 shows the estimated probabilities of being in poor health condition (i.e., $Y = 1$).

The margins plot shows that when other predictor variables are held at their means, the increase of the highest year of school completed from 8 to 13 to 16 years corresponds to the decrease of the probability of being in poor health (i.e., $Y = 1$).

Figure 12.4   Estimated Probabilities of Being in Category 1 for educ at 8, 13, and 16

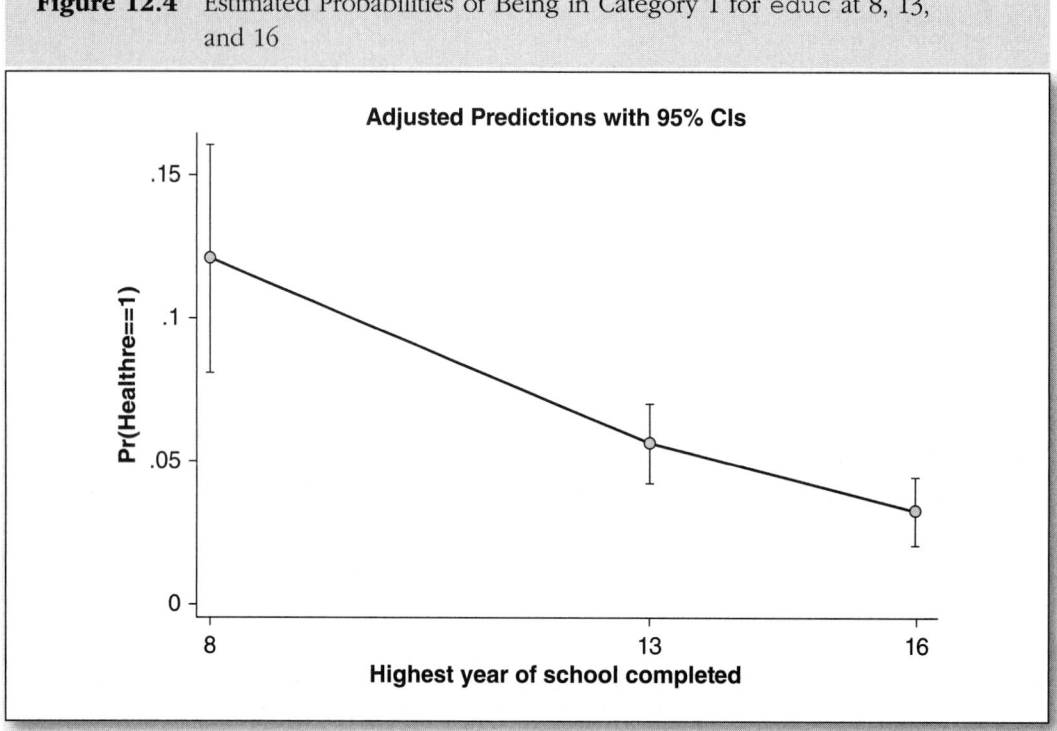

Note: CI = confidence interval.

We can also compute the margins or expected probabilities for the other three categories (i.e., 2, 3, and 4) of the nominal response variable by specifying the option `predict()` option. With the `predict (outcome(4))` option, we can compute the expected probabilities for category 4. The output is displayed as follows.

```
. margins, predict (outcome(4)) at(educ = (8 13 16)) atmeans vsquish

Adjusted predictions                              Number of obs    =      1300
Model VCE    : OIM

Expression   : Pr(healthre==4), predict(outcome(4))
1._at        : _Imaritals_1     =     .4684615   (mean)
               educ             =            8
               age              =         48.2   (mean)
               _Imale_1         =     .4438462   (mean)
2._at        : _Imaritals_1     =     .4684615   (mean)
               educ             =           13
               age              =         48.2   (mean)
               Imale 1          =     .4438462   (mean)
```

```
3._at          : _Imaritals_1  =    .4684615 (mean)
                 educ          =          16
                 age           =        48.2 (mean)
                 _Imale_1      =    .4438462 (mean)
```

	Margin	Delta-method Std. Err.	z	P>|z|	[95% Conf. Interval]
_at						
1	.1706155	.0206975	8.24	0.000	.1300492	.2111819
2	.2562093	.0131674	19.46	0.000	.2304016	.282017
3	.3031182	.0168223	18.02	0.000	.2701471	.3360892

In the output, when educ equals 8, 13, and 16, and other predictor variables are held at their means, the margins or estimated probabilities for being in excellent health (i.e., category 4) are .171, .256, and .303, respectively. They are all statistically significant ($p < .001$). Again, the marginsplot command is used to visualize the results from the output. Figure 12.5 shows the estimated probabilities of being in excellent health condition (i.e., $Y = 4$).

Figure 12.5   Estimated Probabilities of Being in Category 4 for educ at 8, 13, and 16

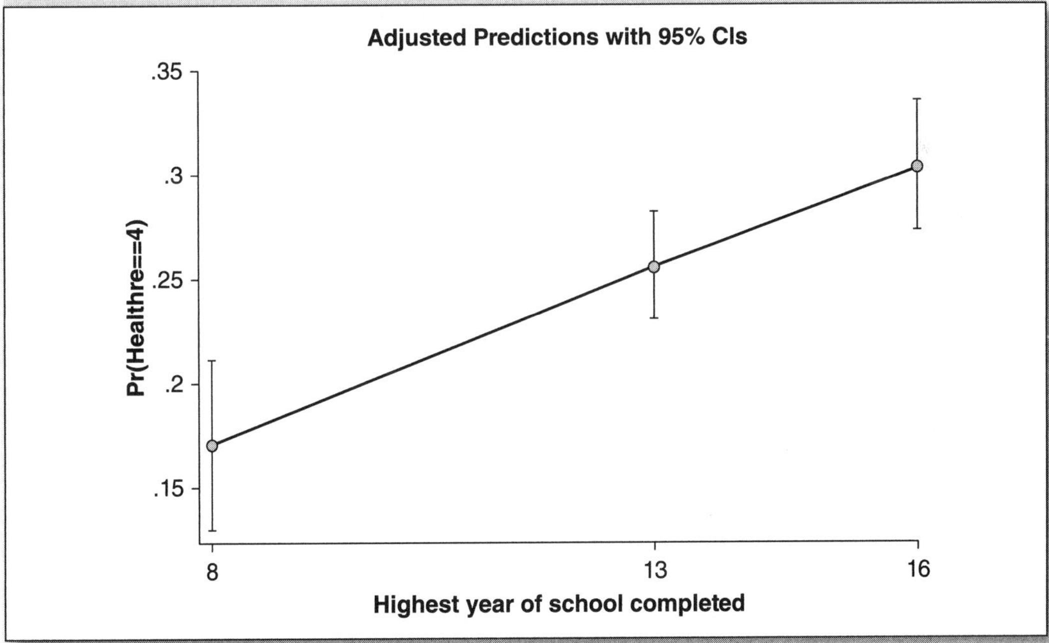

Note: CI = confidence interval.

Figure 12.5 shows that when the other three predictor variables are held at their means, the increase of years of education is associated with an increase in the probability of being in excellent health. The two plots for the margins or estimated probabilities tell us that the increase of years of education decreases the probability of being in poor health while increasing the probability of being in excellent health.

By using the graph combine command after each separate graph is drawn, we can plot the estimated probabilities of being in each category (i.e., Y = 1, 2, 3, and 4) in a single graph (Figure 12.6). The command graph combine plot1 plot2 plot3 plot4, ycommon combines all four plots, named plot1 to plot4, which are created by the marginsplot command.

Computing the Estimated Probabilities With the Improved margins and marginsplot Commands in Stata 14

To replicate the results in the preceding section, the same command, margins, at(educ = (8 13 16)) atmeans vsquish, tells Stata 14 to compute simultaneously the estimated probabilities for all four categories of the ordinal response variable for the predictor variable educ at the values of 8, 13, and 16 when holding the other predictor variables at their means. The output is shown as follows.

```
. *margins and marginsplot in Stata 14
. margins, at(educ = (8 13 16)) atmeans vsquish

Adjusted predictions                          Number of obs    =     1,300
Model VCE      : OIM

1._predict     : Pr(healthre==1), predict(pr outcome(1))
2._predict     : Pr(healthre==2), predict(pr outcome(2))
3._predict     : Pr(healthre==3), predict(pr outcome(3))
4._predict     : Pr(healthre==4), predict(pr outcome(4))
1._at          : _Imaritals_1    =     .4684615 (mean)
                 educ            =            8
                 age             =         48.2 (mean)
                 _Imale_1        =     .4438462 (mean)
2._at          : _Imaritals_1    =     .4684615 (mean)
                 educ            =           13
                 age             =         48.2 (mean)
                 _Imale_1        =     .4438462 (mean)
3._at          : _Imaritals_1    =     .4684615 (mean)
                 educ            =           16
```

```
                  age           =       48.2 (mean)
                  _Imale_1      =   .4438462 (mean)


-----------------------------------------------------------------------------
               |             Delta-method
               |    Margin    Std. Err.      z    P>|z|    [95% Conf. Interval]
---------------+-------------------------------------------------------------
_predict#_at   |
         1  1  |  .1205813    .0200847     6.00   0.000    .081216    .1599467
         1  2  |   .056261    .007065      7.96   0.000    .0424139   .0701081
         1  3  |  .0330092    .0060246     5.48   0.000    .0212012   .0448172
         2  1  |  .3495982    .0294041    11.89   0.000    .2919672   .4072292
         2  2  |  .2171511    .0121735    17.84   0.000    .1932914   .2410108
         2  3  |  .1512687    .0130285    11.61   0.000    .1257333   .1768042
         3  1  |   .359205    .0281752    12.75   0.000    .3039825   .4144274
         3  2  |  .4703786    .0148527    31.67   0.000    .4412679   .4994892
         3  3  |  .5126039    .0181685    28.21   0.000    .4769942   .5482136
         4  1  |  .1706155    .0206975     8.24   0.000    .1300492   .2111819
         4  2  |  .2562093    .0131674    19.46   0.000    .2304016   .282017
         4  3  |  .3031182    .0168223    18.02   0.000    .2701471   .3360892
-----------------------------------------------------------------------------
```

The results in the output are the same as the combined results for each of the four categories estimated from separate commands in the previous section using Stata 13.

Using the improved `marginsplot` command in Stata 14, we now can simultaneously plot the estimated probabilities of being in each category of an ordinal response variable in a single graph instead of plotting them separately and then combining them. Figure 12.7 shows the estimated probabilities when $Y = 1, 2, 3,$ and 4.

12.2.9 Independence of Irrelevant Alternatives (IIA) Tests

Multinomial logistic regression models assume that nominal outcome categories have the property of IIA. Simply put, the IIA assumes that the inclusion of one category does not impact the effects of the predictor variables in other categories. For example, if a nominal response variable has three categories, categories 1, 2, and 3, then it assumes that the inclusion of category 3 does not influence the odds ratios or relative risks of being in a particular category versus the base category for a predicator variable in categories 1 and 2. The command `mlogtest, iia`, which is part of the SPost13 package (Long & Freese, 2014), can be used to test the IIA assumption. If the test results are statistically significant, then we conclude the IIA assumption is violated.

Figure 12.6 Estimated Probabilities of Being in Categories 1, 2, 3, and 4 for educ

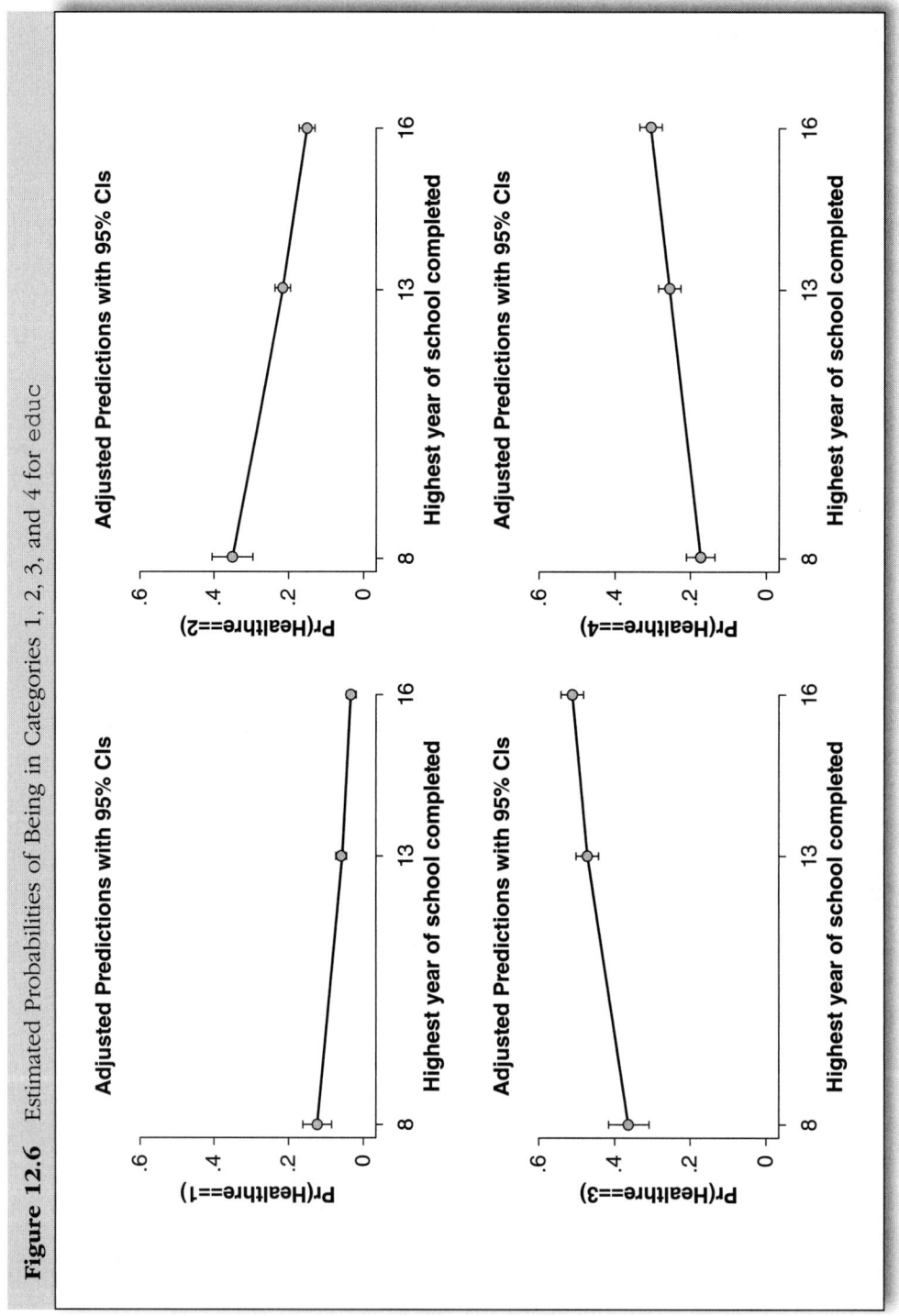

Note: CI = confidence interval.

Figure 12.7 Estimated Probabilities of Being in Categories 1, 2, 3, and 4 for educ With Stata 14

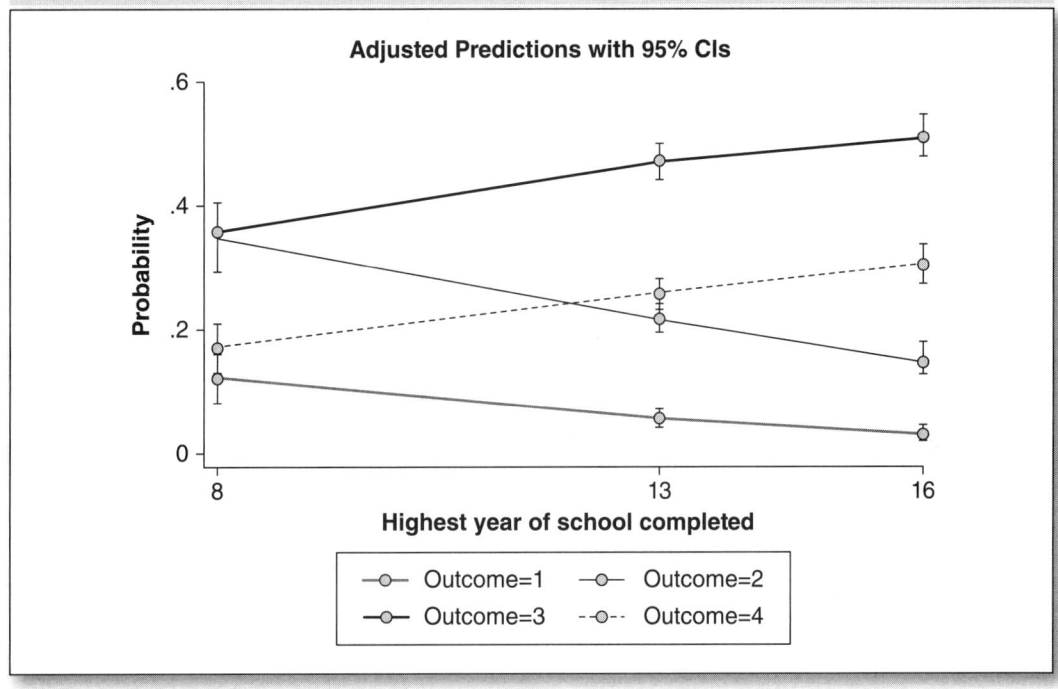

Note: CI = confidence interval.

```
. mlogtest, iia

Hausman tests of IIA assumption (N=1300)

  Ho: Odds(Outcome-J vs Outcome-K) are independent of other alternatives

                    |    chi2    df    P>chi2
  ------------------+----------------------------
                 1  |  23.029    10    0.011
                 2  |   2.128     9    0.989
                 3  |   8.581    10    0.572
                 4  |  -9.089    10      .

  Note: A significant test is evidence against Ho.
  Note: If chi2<0, the estimated model does not meet asymptotic assumptions.

suest-based Hausman tests of IIA assumption (N=1300)

  Ho: Odds(Outcome-J vs Outcome-K) are independent of other alternatives
```

```
               |      chi2     df    P>chi2
---------------+---------------------------
           1 |    10.134     10     0.429
           2 |     8.731     10     0.558
           3 |     7.243     10     0.702
           4 |    13.496     10     0.197
```

Note: A significant test is evidence against Ho.

Small-Hsiao tests of IIA assumption (N=1300)

Ho: Odds(Outcome-J vs Outcome-K) are independent of other alternatives

```
               | lnL(full)  lnL(omit)       chi2     df    P>chi2
---------------+--------------------------------------------------
           1 |  -623.892   -618.062     11.661     10     0.308
           2 |  -445.921   -439.728     12.386     10     0.260
           3 |  -319.044   -314.377      9.334     10     0.501
           4 |  -405.663   -403.376      4.572     10     0.918
```

Note: A significant test is evidence against Ho.

The output produced by the mlogtest, iia command displays the results of three tests of the IIA assumption: the Hausman test, suest-based Hausman test, and Small-Hsiao test. All the results indicate that the IIA assumption is not violated for the model. Hilbe (2009) and Long and Freese (2006, 2014) pointed out that the three tests presented previously could produce conflicting results, so researchers should conduct the IIA test with caution.

12.2.10 Making Publication-Quality Tables

```
1.  . quietly xi: mlogit healthre i.maritals educ age i.male, baseoutcome(1)

2.  . outreg2 using multinomout, e(ll chi2) dec(3) long word replace
    multinomout.rtf
    dir : seeout

3.  . outreg2 using multinomout, eform dec(3) long word append
    multinomout.rtf
    dir : seeout
```

The first command fits the multinomial logistic regression model without displaying the output. The second command, outreg2 using multinomout, e(ll chi2), dec(3), word long replace, tells Stata to create a regression table for the estimated model and to save it to a Word file named "multinomout". With the option eform,

the odds ratios for the estimation can be requested using the following command:
`outreg2 using multinomout, eform dec(3) long word append.`

These commands automatically produce Table 12.4, as shown here in its original format, presenting the results for the multinomial logistic regression model.

Table 12.4 Results of the Multinomial Logistic Regression Model: Four-Predictor Model (Shown in Original Format Generated by Stata)

Variables	(1) healthre	(2) Healthre
1		
o._Imaritals_1	—	—
o.educ	—	—
o.age	—	—
o._Imale_1	—	—
o._cons	0.000	1.000
	(0.000)	(0.000)
2		
_Imaritals_1	0.152	1.164
	(0.262)	(0.305)
Educ	0.057	1.059
	(0.037)	(0.039)
Age	−0.014**	0.986**
	(0.007)	(0.007)
_Imale_1	−0.247	0.781
	(0.254)	(0.198)
Constant	1.304**	3.684**
	(0.649)	(2.389)
3		
_Imaritals_1	0.427*	1.532*
	(0.250)	(0.383)

(Continued)

Table 12.4   (Continued)

Variables	(1) healthre	(2) Healthre
Educ	0.206***	1.229***
	(0.037)	(0.045)
Age	−0.029***	0.972***
	(0.007)	(0.006)
_Imale_1	−0.274	0.760
	(0.242)	(0.184)
Constant	0.741	2.097
	(0.632)	(1.325)
4		
_Imaritals_1	0.918***	2.505***
	(0.262)	(0.656)
Educ	0.234***	1.263***
	(0.039)	(0.050)
Age	−0.038***	0.963***
	(0.007)	(0.007)
_Imale_1	−0.052	0.949
	(0.254)	(0.241)
Constant	−0.117	0.890
	(0.669)	(0.595)
healthre		
healthre		.
		(.)
Observations	1,300	1,300
LL	−1,504	
chi2	151.3	

Note: Standard errors in parentheses.

***$p < 0.01$, **$p < 0.05$, *$p < 0.1$

After deleting the empty cells at the beginning of the table, editing the intercept names, adding the labels for category comparisons and for the logit coefficients and odds ratios, and adding model fit statistics, the final table is displayed as in Table 12.5.

Table 12.5 Results of the Multinomial Logistic Regression Model: Four-Predictor Model (Edited)

Variables	b (SE(b))	OR
Model 1 ($Y = 2$ vs. $Y = 1$)		
_Imaritals_1	0.152	1.164
	(0.262)	(0.305)
Educ	0.057	1.059
	(0.037)	(0.039)
Age	−0.014**	0.986**
	(0.007)	(0.007)
_Imale_1	−0.247	0.781
	(0.254)	(0.198)
α_1	1.304**	3.684**
	(0.649)	(2.389)
Model 2 ($Y = 3$ vs. $Y = 1$)		
_Imaritals_1	0.427*	1.532*
	(0.250)	(0.383)
Educ	0.206***	1.229***
	(0.037)	(0.045)
Age	−0.029***	0.972***
	(0.007)	(0.006)
_Imale_1	−0.274	0.760
	(0.242)	(0.184)
α_2	0.741	2.097

(Continued)

Table 12.5 (Continued)

Variables	b (SE(b))	OR
	(0.632)	(1.325)
Model 3 ($Y = 4$ vs. $Y = 1$)		
_Imaritals_1	0.918***	2.505***
	(0.262)	(0.656)
Educ	0.234***	1.263***
	(0.039)	(0.050)
Age	−0.038***	0.963***
	(0.007)	(0.007)
_Imale_1	−0.052	0.949
	(0.254)	(0.241)
α_3	−0.117	0.890
	(0.669)	(0.595)
Observations	1,300	1,300
LR R^2	.048	
Log likelihood	−1,504	
LR χ^2_{12}	151.3	

Note: Standard errors are shown in parentheses.

$*p < .10.$ $**p < .05.$ $***p < .01.$

12.2.11 Reporting the Results

Reporting the results for multinomial logistic regression is similar to that used for binary logistic regression. The following are the generic guidelines for reporting the results. You may need to adjust your writing since your discipline or journals may have different requirements.

First, describe the multinomial logistic regression model, the nominal response variable and the independent variables, and your research hypothesis or the purpose of your study. Include a couple of sentences explaining why this model is appropriate for the analysis.

Second, report the Wald chi-square statistic for the model and the associated p value, followed by the interpretation of whether the fitted model is better than the null model. If more than one model is developed, then compare models using likelihood ratio test statistics and/or the AIC and BIC statistics.

Third, report the parameter estimates for the predictor variables, their standard errors, the associated p values, and the odds ratios in a table. Since a multinomial logistic regression model includes $J - 1$ binary comparisons, label them in the table. In addition, report the odds ratios or relative risk ratios for each predictor in the table, and interpret the results in the text. If more than one model is fitted, then the results of all the competing models need to be presented in a table. The following is an example of summarizing the results for the multinomial logistic regression model illustrated previously.

The multinomial logistic regression analysis was conducted to predict the ordinal outcome variable, health status, from a set of predictor variables, such as marital status, years of education, age, and gender. Although the multinomial logistic regression model is normally used to estimate the nominal response variables, it is an alternative to estimate ordinal response variables when the proportional odds assumption is violated in the proportional odds models.

The log likelihood ratio test for the fitted model $\chi^2_{(12)} = 151.28$, $p < .001$, which indicated that the four-predictor model provided a better fit than the null model with no independent variables in predicting the logit of being in any other category of health status compared with being in the base category (i.e., poor health).

Table 12.5 displays the parameter estimates and corresponding odds ratios for three binary logistic models comparing each category versus the base category since the multinomial logistic regression model is treated as a series of binary logistic regression models. These three equations compare categories 2 to 1, 3 to 1, and 4 to 1, respectively. Table 12.2 also summarizes the odds ratios across three binary comparisons for all four predictor variables in the model.

For `maritals`, the odds ratios of being in category 2 versus category 1, category 3 versus category 1, and category 4 versus category 1 were 1.164, 1.532, and 2.505, respectively. Among them, only the odds ratio of being in category 4 versus the base category was significant, $z = 3.508$, $p < .001$. OR(4,1) = 2.505, which indicated that the odds of being in category 4 versus the base category for the married were 2.505 times as large as the odds for the unmarried when holding all the other predictors constant.

For `educ`, the odds ratios for three binary comparisons (i.e., categories 2 vs. 1, 3 vs. 1, and 4 vs. 1) were 1.059, 1.229, and 1.263, respectively. OR(2,1) = 1.059, $z = 1.546$, $p > .05$, which indicated that there was no relationship between `educ` and the odds of being in category 2 versus the base category 1. The odds ratios for the other two comparisons were significant, OR(3,1) = 1.229, and

(Continued)

(Continued)

OR(4,1) = 1.263. They indicated that the odds of being in category 3 versus the base category and category 4 versus the base category increased by 1.229 and 1.263, respectively, for a one-unit increase in the predictor educ when holding other predictors constant.

With regard to age, the odds ratios for three binary comparisons (i.e., categories 2 vs. 1, 3 vs. 1, and 4 vs. 1) were .986, .972, and .963, respectively, and they were all significant. It indicated that the odds of being in categories 2, 3, and 4 versus the base category decreased by .986, .972, and .963, respectively, for a one-unit increase in the predictor age when holding all the other predictors constant.

For the male predictor, none of the odds ratios for the binary comparisons were significant. It indicated that being male did not influence the odds of being in any particular category versus the base category when holding other predictors constant.

12.3 Summary of Stata Commands in This Chapter

```
use chap12-gss2012, clear

*Ordinal Probit Models

*The multiple-predictor ordinal probit model
xi: oprobit healthre i.maritals educ age i.male
listcoef
fitstat

*marginal effects in Stata 13
quietly oprobit healthre i.maritals educ age i.male
margins, dydx(*)
margins, predict(outcome(4)) dydx(*)

*marginal effects in Stata 14
quietly oprobit healthre i.maritals educ age i.male
margins, dydx(*)

*margins and marginsplot in Stata 13
quietly oprobit healthre i.maritals educ age i.male
margins, predict (outcome(1)) at(educ = (8 13 16)) atmeans vsquish
margins, predict (outcome(4)) at(educ = (8 13 16)) atmeans vsquish
marginsplot

*margins plots combined in Stata 13
margins, predict (outcome(1)) at(educ = (8 13 16)) atmeans vsquish
marginsplot, name(pplot1)
margins, predict (outcome(2)) at(educ = (8 13 16)) atmeans vsquish
```

```
margins, predict (outcome(3)) at(educ = (8 13 16)) atmeans vsquish
marginsplot, name(pplot3)
margins, predict (outcome(4)) at(educ = (8 13 16)) atmeans vsquish
marginsplot, name(pplot4)
graph combine pplot1 pplot2 pplot3 pplot4, ycommon

*margins and marginsplot in Stata 14
quietly oprobit healthre i.maritals educ age i.male
margins, at(educ = (8 13 16)) atmeans vsquish
marginsplot

*Model comparison using the log likelihood ratio test
quietly xi: oprobit healthre i.maritals
estimates store model1
quietly xi: oprobit healthre i.maritals age educ i.male
estimates store fullmodel
lrtest model1 fullmodel

*Comparing the Probit Model and Proportional Odds Model
quietly xi: oprobit healthre i.maritals age educ i.male
estimates store probit
quietly xi: ologit healthre i.maritals age educ i.male
estimates store logit

*Making publication-quality tables using outreg2
outreg2 [probit logit] using probitout, e(ll df_m chi2) ///
 addstat(Pseudo R-squared, `e(r2_p)') dec(3) word replace

*Multinomial Logistic Regression Models

*The multiple-predictor multinomial logistic regression model
xi: mlogit healthre i.maritals educ age i.male, baseoutcome(1)
listcoef
fitstat

*margins and marginsplot in Stata 13
margins, predict (outcome(1)) at(educ = (8 13 16)) atmeans vsquish
marginsplot, name(plot1)
margins, predict (outcome(2)) at(educ = (8 13 16)) atmeans vsquish
marginsplot, name(plot2)
margins, predict (outcome(3)) at(educ = (8 13 16)) atmeans vsquish
marginsplot, name(plot3)
margins, predict (outcome(4)) at(educ = (8 13 16)) atmeans vsquish
marginsplot, name(plot4)

*margins plots combined in Stata 13
graph combine plot1 plot2 plot3 plot4, ycommon

*margins and marginsplot in Stata 14
quietly mlogit healthre i.maritals educ age i.male, baseoutcome(1)
margins, at(educ = (8 13 16)) atmeans vsquish
marginsplot
```

```
*IIA tests
mlogtest, iia

*Making publication-quality tables using outreg2
quietly xi: mlogit healthre i.maritals educ age i.male, baseoutcome(1)
outreg2 using mulnomout, e(ll chi2) dec(3) long word replace
outreg2 using mulnomout, eform dec(3) long word append
```

12.4 EXERCISES

Use the GSS 2012 data for the following problems.

1. Fit an ordinal probit model to estimate the ordinal response variable fechld from the five predictor variables sex, educ, age, kidjob, and sibs using the oprobit command. Answer the following questions or perform the following analyses:

    a. Identify the likelihood ratio test and the p value of the overall model. What do they tell us?

    b. Compute the deviance statistic for the model.

    c. In the regression table, identify the probit coefficients for the predictor variables sex and educ. Are they statistically significant?

    d. Interpret the probit coefficients of sex and educ.

    e. Use the margins command to compute the estimated probability of being in category 4 at 30, 40, and 50 years of age when holding other predictor variables at their means.

    f. Use the margins command to compute the estimated probability of being in category 1 for the binary categorical predictor variable sex when holding other predictor variables at their means.

    g. Use the outreg2 command to make a table containing the estimated probit coefficients.

2. Conduct an analysis for a multinomial logistic regression model, and estimate the ordinal response variable happy from the four predictor variables, sex, educ, age, and satfin. Choose category 3 (i.e., not too happy) as the referent category. Answer the following questions or perform the following analyses:

    a. Identify the likelihood ratio test of the model, and interpret it.

    b. In the regression table, identify the logit coefficients for the predictor variable educ across two binary comparisons. Are they both statistically significant? What categories are they comparing?

c. Interpret the odds ratios for the predictor variable `educ` across two binary comparisons.

d. Interpret the relative risk ratio/odds ratios of `satfin`.

e. Based on the parameter estimates in the output, write the two equations for the model.

f. Use the `listcoef` command to display the odds ratios. List the odds ratios for `satfin` comparing the other two categories to the base category, and interpret them.

g. Make a publication-quality table containing the estimated logit coefficients and odds ratios.

h. Write a report to summarize the results from the output.

Key Formulas
for Statistical Models

Chapter 3 Logistic Regression for Binary Data

$$\ln\left(\frac{\pi(x)}{1-\pi(x)}\right) = \alpha + \beta_1 X_1 + \beta_2 X_2 + \ldots + \beta_p X_p$$

Chapter 4 Proportional Odds Models for Ordinal Response Variables

$$\ln(Y_j') = \text{logit}[\pi(x)] = \ln\left(\frac{\pi_j(x)}{1-\pi_j(x)}\right) = \alpha_j + (-\beta_1 X_1 - \beta_2 X_2 - \ldots - \beta_p X_p)$$

$$\text{logit}[\pi(Y \le j \mid x_1, x_2, \ldots, x_p)] = \ln\left(\frac{\pi(Y \le j \mid x_1, x_2, \ldots, x_p)}{\pi(Y > j \mid x_1, x_2, \ldots, x_p)}\right) = \alpha_j + (-\beta_1 X_1 - \beta_2 X_2 - \ldots - \beta_p X_p)$$

Chapter 5 Partial Proportional Odds Models and Generalized Ordinal Logistic Regression Models

$$\text{logit}[\pi(Y > j \mid x_1, x_2, \ldots, x_p)] = \ln\left(\frac{\pi(Y > j \mid x_1, x_2, \ldots, x_p)}{\pi(Y \le j \mid x_1, x_2, \ldots, x_p)}\right) = \alpha_j + \left(\beta_{1j} X_1 + \beta_{2j} X_2 + \ldots + \beta_{pj} X_p\right)$$

Chapter 6 Continuation Ratio Models

$$\ln\left(\frac{\pi(Y = j \mid x_1, x_2, \ldots, x_p)}{\pi(Y > j \mid x_1, x_2, \ldots, x_p)}\right) = \alpha_j + (-\beta_1 X_1 - \beta_2 X_2 - \ldots - \beta_p X_p)$$

Chapter 7 Adjacent Categories Logistic Regression Models

$$\text{logit}[\pi(Y = j+1 \mid x_1, x_2, \ldots, x_p)] = \ln\left(\frac{\pi(Y = j+1 \mid x_1, x_2, \ldots, x_p)}{\pi(Y = j \mid x_1, x_2, \ldots, x_p)}\right) = \alpha_j + \beta_1 X_1 + \beta_2 X_2 + \ldots + \beta_p X_p$$

$$\ln\left(\frac{\pi\left(Y = j|x_1, x_2,...,x_p\right)}{\pi\left(Y = 1|x_1, x_2,...,x_p\right)}\right) = (\alpha_1 + \alpha_2 + ... + \alpha_{j-1}) + (J-1)\beta_1 X_1 + (J-1)\beta_2 X_2 + ... + (J-1)\beta_p X_p$$

Chapter 8 Stereotype Logistic Regression Models

$$\text{logit}[\pi(j, J)] = \ln\left(\frac{\pi\left(Y = j|x_1, x_2,...,x_p\right)}{\pi\left(Y = J|x_1, x_2,...,x_p\right)}\right) = \alpha_j - \phi_j(\beta_1 X_1 + \beta_2 X_2 + ... + \beta_p X_p)$$

where $1 = \phi_1 > \phi_2 > \phi_3 > ... \phi_{J-1} > \phi_J = 0$

Chapter 10 Multilevel Modeling for Continuous and Binary Response Variables

10.2 Multilevel Modeling for Continuous Outcome Variables

Level 1: $Y_{ij} = \beta_{0j} + \beta_{1j}\texttt{gceffic}_{ij} + r_{ij}$
Level 2: $\beta_{0j} = \gamma_{00} + \gamma_{01}\texttt{public}_j + \gamma_{02}\texttt{csclimat}_j + u_{0j}$
$\beta_{1j} = \gamma_{10} + \gamma_{11}\texttt{public}_j + \gamma_{12}\texttt{csclimat}_j + u_{1j}$

10.3 Multilevel Modeling for Binary Outcome Variables

Level 1: $\text{logit}[\pi(x_{ij})] = \beta_{0j} + \beta_{1j}\texttt{stlang}_{ij}$
Level 2: $\beta_{0j} = \gamma_{00} + \gamma_{01}\texttt{public}_j + \gamma_{02}\texttt{csclimat}_j + u_{0j}$
$\beta_{1j} = \gamma_{10} + \gamma_{11}\texttt{public}_j + \gamma_{12}\texttt{csclimat}_j + u_{1j}$

Chapter 11 Multilevel Modeling for Ordinal Response Variables

Level 1: $\text{logit}[\pi_{kij}(Y \leq k)] = \ln\left(\frac{\pi\left(Y_{ij} \leq k\right)}{\pi\left(Y_{ij} > k\right)}\right) = \alpha_k - (\beta_{0j} + \beta_{1j}\texttt{stlang}_{ij})$

Level 2: $\beta_{0j} = \gamma_{00} + \gamma_{01}\texttt{public}_j + \gamma_{02}\texttt{csclimat}_j + u_{0j}$
$\beta_{1j} = \gamma_{10} + \gamma_{11}\texttt{public}_j + \gamma_{12}\texttt{csclimat}_j + u_{1j}$

Chapter 12 Beyond Ordinal Logistic Regression Models: Ordinal Probit Regression Models and Multinomial Logistic Regression Models

Ordinal Probit Regression Models

$$\Phi^{-1}(\pi) = \text{probit}[\pi(Y \leq j \mid x_1, x_2, ..., x_p)] = \alpha_j + (-\beta_1 X_1 - \beta_2 X_2 - ... - \beta_p X_p)$$

Multinomial Logistic Regression Models

$$\ln\left(\frac{\pi\left(Y = j|x_1, x_2,...,x_p\right)}{\pi\left(Y = J|x_1, x_2,...,x_p\right)}\right) = \alpha_j + \beta_{j1} X_1 + \beta_{j2} X_2 + ... + \beta_{jp} X_p$$

Appendix: List of Stata User-Written Commands

User-Written Command	First Appearing in Chapter and Section Number
outreg2	2.11
esttab	2.11
fitstat	3.3.1
listcoef	4.3.1
brant	4.3.1
mtable	4.3.2
gologit	5.1
gologit2	5.1
ocratio	6.3
aic	6.3.1
seqlogit	6.3.3
gllamm	11.3.9
mlogtest	12.2.9

References

Acock, A. C. (2014). *A gentle introduction to Stata* (4th ed.). College Station, TX: Stata Press.

Agresti, A. (1996). *An introduction to categorical data analysis.* New York, NY: John Wiley & Sons.

Agresti, A. (2002). *Categorical data analysis* (2nd ed.). New York, NY: John Wiley & Sons.

Agresti, A. (2007). *An introduction to categorical data analysis* (2nd ed.). New York: John Wiley & Sons.

Agresti, A. (2010). *Analysis of ordinal categorical data* (2nd ed.). Hoboken, NJ: John Wiley & Sons.

Akaike, H. (1974). A new look at the statistical model identification. *IEEE Transactions on Automatic Control, AC-19*, 716–723.

Allison, P. D. (1999). *Logistic regression using the SAS system: Theory and application.* Cary, NC: SAS Institute.

Allison, P. D. (2012). *Logistic regression using the SAS system: Theory and application* (2nd ed.). Cary, NC: SAS Institute.

American Psychological Association (APA). (2010). *Publication manual of the American Psychological Association* (6th ed.). Washington, DC: Author.

Ananth, C. V., & Kleinbaum, D. G. (1997). Regression models for ordinal responses: A review of methods and applications. *International Journal of Epidemiology, 26*, 1323–1333.

Anderson, J. A. (1984). Regression and ordered categorical variables. *Journal of the Royal Statistical Society, Series B, 46*, 1–30.

Armstrong, B. B., & Sloan, M. (1989). Ordinal regression models for epidemiological data. *American Journal of Epidemiology, 129*(1), 191–204.

Azen, R., & Walker, C. M. (2011). *Categorical data analysis for behavioral and social sciences.* New York, NY: Routledge.

Baum, C. F. (2006). *An introduction to modern econometrics using Stata.* College Station, TX: Stata Press.

Baum, C. F. (2009). *An introduction to Stata programming.* College Station, TX: Stata Press.

Bender, R., & Benner, A. (2000). Calculating ordinal regression models in SAS and S-Plus. *Biometrical Journal, 42*(6), 677–699.

Bender, R., & Grouven, U. (1998). Using binary logistic regression models for ordinal data with non-proportional odds. *Journal of Clinical Epidemiology, 51*(10), 809–816.

Binder, D. A. (1983). On the variances of asymptotically normal estimators from complex surveys. *International Statistical Review, 51*, 279–292.

Borooah, V. K. (2002). *Logit and probit: Ordered and multinomial models*. Thousand Oaks, CA: Sage.

Brant, R. (1990). Assessing proportionality in the proportional odds model for ordinal logistic regression. *Biometrics, 46*, 1171–1178.

Buis, M. L. (2007). *seqlogit: Stata module to fit a sequential logit model*, Statistical Software Components, Boston College Department of Economics. Retrieved from http://econpapers .repec.org/software/bocbocode/s456843.htm

Cameron, A. C., & Trivedi, P. K. (2010). *Microeconometrics using Stata* (Rev. ed.). College Station, TX: Stata Press.

Cargill, M., & O'Connor, P. (2013). *Writing scientific research articles: Strategy and steps* (2nd ed.). Hoboken, NJ: Wiley-Blackwell.

Clogg, C. C., & Shihadeh, E. S. (1994). *Statistical models for ordinal variables*. Thousand Oaks, CA: Sage.

Cole, S. R., & Ananth, C. V. (2001). Regression models for unconstrained, partially or fully constrained continuation odds ratios. *International Journal of Epidemiology, 30*, 1379–1382.

Collett, D. (2003). *Modelling binary data* (2nd ed.). Boca Raton, FL: Chapman & Hall/CRC.

Cramer, J. S. (2003). *Logit models from economics and other fields*. New York, NY: Cambridge University Press.

Demaris, A. (1992). *Logit modeling*. Newbury Park, CA: Sage.

Dunn, D. S. (2012). *The practical researcher: A student guide to conducting psychological research*. Malden, MA: Wiley-Blackwell.

Dupont, W. D. (2009). *Statistical modeling for biomedical researchers: A simple introduction to the analysis of complex data* (2nd ed.). New York, NY: Cambridge University Press.

Enders, C. K., & Tofighi, D. (2007). Centering predictor variables in cross-sectional multilevel models: A new look at an old issue. *Psychological Methods, 12*, 121–138.

Fienberg, S. E. (1980). *The analysis of cross-classified categorical data*. Cambridge, MA: The MIT Press.

Fox, J. (2008). *Applied regression analysis and generalized linear models* (2nd ed.). Thousand Oaks, CA: Sage.

Fu, V. (1998). Estimating generalized ordered logit models. *Stata Technical Bulletin, 44*, 27–30.

Fullerton, A. S. (2009). A conceptual framework for ordered logistic regression models. *Sociological Methods and Research, 38*, 306–347.

Gallup, J. L. (1998). Formatting regression output for published tables. *Stata Technical Bulletin, 46*, 28–30.

Gallup, J. L. (1999). Revision of outreg. *Stata Technical Bulletin, 49*, 23.

Gallup, J. L. (2000). Update to formatting regression output. *Stata Technical Bulletin, 58*, 9–13.

Garson, G. D. (Ed.). (2013). *Hierarchical linear modeling: Guide and applications*. Thousand Oaks, CA: Sage.

Goldstein, H. (2003). Multilevel *statistical models* (3rd ed.). London, England: Edward Arnold.

Goodman, L. A. (1983). The analysis of dependence in cross-classifications having ordered categories, using log-linear models for dependencies and log-linear models for odds. *Biometrics, 39*, 149–160.

Gould, W. (2000). Interpreting logistic regression in all its forms. *Stata Technical Bulletin, 53*, 19–29.

Greene, W. H. (1990). *Econometric analysis.* New York, NY: Macmillan.

Greenland, S. (1994). Alternative models for ordinal logistic regression. *Statistics in Medicine, 13*(16), 1665–1677.

Hahs-Vaughn, D. L. (2005). A primer for understanding and using weights with national datasets. *Journal of Experimental Education, 73*(3), 221–240.

Hahs-Vaughn, D. L. (2006). Analysis of data from complex samples. *International Journal of Research and Method in Education, 29*(2), 163–181.

Hahs-Vaughn, D. L., & Lomax, R. G. (2006). Utilization of sample weights in single level structural equation modeling. *Journal of Experimental Education, 74*(2), 163–190.

Hamilton, L. C. (2012). *Statistics with Stata: Updated for version 12* (8th ed.). Boston, MA: Brooks/Cole, Cengage Learning.

Hardin, J. W., & Hilbe, J. M. (2007). *Generalized linear models and extensions* (2nd ed.). College Station, TX: Stata Press.

Hardin, J. W., & Hilbe, J. M. (2012). *Generalized linear models and extensions* (3rd ed.). College Station, TX: Stata Press.

Heck, R. H., & Thomas, S. L. (2009). *An introduction to multilevel modeling techniques* (2nd ed.). New York, NY: Routledge.

Heck, R. H., Thomas, S. L., & Tabata, L. N. (2010). *Multilevel and longitudinal modeling with IBM SPSS.* New York, NY: Routledge.

Heck, R. H., Thomas, S. L., & Tabata, L. N. (2012). *Multilevel modeling of categorical outcomes using IBM SPSS.* New York, NY: Routledge.

Hedeker, D. (2007). Multilevel models for ordinal and nominal variables. In J. de Leeuw & E. Meijer (Eds.), *Handbook of multilevel analysis* (pp. 237–274). New York, NY: Springer.

Hedeker, D., & Gibbons, R. D. (1994). A random-effects ordinal regression model for multilevel analysis. *Biometrics, 50*(4), 933–944.

Heeringa, S. G., West, B. T., & Berglund, P. A. (2010). *Applied survey data analysis.* Boca Raton, FL: Chapman & Hall/CRC.

Hilbe, J. M. (2009). *Logistic regression models.* Boca Raton, FL: Chapman & Hall/CRC.

Hofmann, D. A., & Gavin, M. B. (1998). Centering decisions in hierarchical linear models: Implications for research in organizations. *Journal of Management, 24*(5), 623–641.

Hosmer, D. W., & Lemeshow, S. (2000). *Applied logistic regression* (2nd ed.). New York, NY: John Wiley & Sons.

Hosmer, D. W., Lemeshow, S., & Sturdivant, R. X. (2013). *Applied logistic regression* (3rd ed.). New York, NY: John Wiley & Sons.

Hox, J. J. (2010). *Multilevel analysis: Techniques and applications* (2nd ed.). New York, NY: Routledge.

Huck, S. W. (2012). *Reading statistics and research* (6th ed.). Boston, MA: Pearson.

Ingels, S. J., Pratt, D. J., Roger, J., Siegel, P. H., & Stutts, E. (2004). *ELS: 2002 base year data file user's manual.* Washington, DC: NCES (NCES 2004-405).

Ingels, S. J., Pratt, D. J., Roger, J., Siegel, P. H., & Stutts, E. (2005). *Education Longitudinal Study: 2002/04 public use base-year to first follow-up data files and electronic codebook system.* Washington, DC: NCES (NCES 2006-346).

Jann, B. (2005). Making regression tables from stored estimates. *Stata Journal, 5,* 288–308.

Jann, B. (2007). Making regression tables simplified. *Stata Journal, 7,* 227–244.

Juul, S. (2014). *An introduction to Stata for health researchers* (4th ed.). College Station, TX: Stata Press.

Kalton, G. (1983). *Introduction to survey sampling*. Beverly Hills, CA: Sage.

Kleinbaum, D. G., & Klein, M. (2010). *Logistic regression: A self-learning text* (3rd ed.). New York, NY: Springer.

Kline, R. B. (2013). *Beyond significance testing: Statistics reform in the behavioral sciences* (2nd ed.). Washington, DC: American Psychological Association.

Kohler, U., & Kreuter, F. (2012). *Data analysis using Stata* (3rd ed.). College Station, TX: Stata Press.

Kreft, I. G. G., & de Leeuw, J. (1998). *Introducing multilevel modeling*. London, England: Sage Ltd.

Kreft, I. G. G., de Leeuw, J., & Aiken, L. (1995). The effect of different forms of centering in hierarchical linear models. *Multivariate Behavioral Research, 30*, 1–22.

Kuss, O. (2006). On the estimation of the stereotype regression model. *Computational Statistics & Data Analysis, 50*, 1877–1890.

Lee, E. S., & Forthofer, R. N. (2006). *Analyzing complex survey data* (2nd ed.). Thousand Oaks, CA: Sage.

Levy, P. S., & Lemeshow, S. (2008). *Sampling of populations: Methods and application* (4th ed.). New York, NY: John Wiley & Sons.

Liao, T. F. (1994). *Interpreting probability models: Logit, probit, and other generalized linear models*. Thousand Oaks, CA: Sage.

Liu, X. (2009). Ordinal regression analysis: Fitting the proportional odds model using Stata, SAS and SPSS. *Journal of Modern Applied Statistical Methods, 8*(2), 632–645.

Liu, X. (2014). Fitting stereotype logistic regression models for ordinal response variables in educational research. *Journal of Modern Applied Statistical Methods, 13*(2), 528–543.

Liu, X., O'Connell, A. A., & Koirala, H. (2011). Ordinal regression analysis: Predicting mathematics proficiency using the continuation ratio model. *Journal of Modern Applied Statistical Methods, 10*(2), 513–527.

Liu, X., & Koirala, H. (2012). Ordinal regression analysis: Using generalized ordinal logistic regression models to estimate educational data. *Journal of Modern Applied Statistical Methods, 11*(1), 242–254.

Liu, X., & Koirala, H. (2013). Fitting proportional odds models to educational data with complex sampling designs in ordinal logistic regression. *Journal of Modern Applied Statistical Methods, 12*(1), 235–248.

Lohr, S. L. (1999). *Sampling: Design and analysis*. Pacific Grove, CA: Duxbury Press.

Lokshin, M., & Sajaia, Z. (2008). Creating print-ready tables in Stata. *Stata Journal, 8*(3), 374–389.

Long, J. S. (1997). *Regression models for categorical and limited dependent variables*. Thousand Oaks, CA: Sage.

Long, J. S. (2009). *The workflow of data analysis using Stata*. College Station, TX: Stata Press.

Long, J. S., & Freese, J. (2006). *Regression models for categorical dependent variables using Stata* (2nd ed.). College Station, TX: Stata Press.

Long, J. S., & Freese, J. (2014). *Regression models for categorical dependent variables using Stata* (3rd ed.). College Station, TX: Stata Press.

Longest, K. C. (2015). *Using Stata for quantitative analysis* (2nd ed.). Thousand Oaks, CA: Sage.

Luke, D. A. (2004). *Multilevel modeling.* Thousand Oaks, CA: Sage.

Lunt, M. (2001). Stereotype ordinal regression. *Stata Technical Bulletin, 61*, 12–18.

Ma, X., Ma, L., & Bradley, K. D. (2008). Using multilevel modeling to investigate school effects. In A. A. O'Connell & D. B. McCoach (Eds.), *Multilevel modeling of educational data* (pp. 59–110). Charlotte, NC: Information Age.

McCoach, D. B. (2010). Hierarchical linear modeling. In G. R. Hancock & R. O. Mueller (Eds.), *The reviewers' guide to quantitative methods in the social sciences* (pp. 123–140). New York, NY: Routledge.

McCullagh, P. (1980). Regression models for ordinal data [with discussion]. *Journal of the Royal Statistical Society Ser. B, 42,* 109–142.

McCullagh, P., & Nelder, J. A. (1989). *Generalized linear models* (2nd ed.). London, England: Chapman and Hall.

McInerney, D. M. (2002). *Publishing your psychology research: A guide to writing for journals in psychology and related fields.* London, England: Sage Ltd.

McKelvey, R. D., & Zavoina, W. (1975). A statistical model for the analysis of ordinal level dependent variables. *Journal of Mathematical Sociology, 4,* 103–120.

Menard, S. (2000). Coefficients of determination for multiple logistic regression analysis. *The American Statistician, 54*(1), 17–24.

Menard, S. (2002). *Applied logistic regression analysis* (2nd ed.). Thousand Oaks, CA: Sage.

Menard, S. (2010). *Logistic regression: From introductory to advanced concepts and applications.* Thousand Oaks, CA: Sage.

Mills, M. (2011). *Introducing survival and event history analysis.* London, England: Sage Ltd.

Mitchell, M. N. (2010). *Data management using Stata: A practical handbook.* College Station, TX: Stata Press.

Mitchell, M. N. (2012a). *A visual guide to Stata graphics* (3rd ed.). College Station, TX: Stata Press.

Mitchell, M. N. (2012b). *Interpreting and visualizing regression models using Stata.* College Station, TX: Stata Press.

Muthén, B. O., & Satorra, A. (1995). Complex sample data in structural equation modeling. *Sociological Methodology, 25,* 267–316.

Nagler, J. (1994). Scobit: An alternative estimator to logit and probit. *American Journal of Political Science, 38*(1), 230–255.

O'Connell, A. A. (2000). Methods for modeling ordinal outcome variables. *Measurement and Evaluation in Counseling and Development, 33*(3), 170–193.

O'Connell, A. A. (2006). *Logistic regression models for ordinal response variables.* Thousand Oaks, CA: Sage.

O'Connell, A. A., & Amico, K. R. (2010). Logistic regression. In G. R. Hancock & R. O. Mueller (Eds.), *The reviewers' guide to quantitative methods in the social sciences* (pp. 221–239). New York, NY: Routledge.

O'Connell, A. A., Goldstein, J., Rogers, H. J., & Peng, C. Y. J. (2008). Multilevel logistic models for dichotomous and ordinal data. In A. A. O'Connell & D. B. McCoach (Eds.), *Multilevel modeling of educational data* (pp. 199–242). Charlotte, NC: Information Age.

O'Connell, A. A., & Liu, X. (2011). Model diagnostics for proportional and partial proportional odds models. *Journal of Modern Applied Statistical Methods, 10*(1), 139–175.

Paccagnella, O. (2006). Centering or not centering in multilevel models? *Evaluation Review, 30,* 66–85.

Peng, C. Y. J., Lee, K. L., & Ingersoll, G. M. (2002). An introduction to logistic regression and reporting. *The Journal of Educational Research, 96*(1), 3–14.

Peterson, B., & Harrell, F. E. (1990). Partial proportional odds models for ordinal response variables. *Applied Statistics, 39*(2), 205–217.

Powers, D. A., & Xie, Y. (2000). *Statistical models for categorical data analysis.* San Diego, CA: Academic Press.

Pregibon, D. (1981). Logistic regression diagnostics. *Annals of Statistics, 9*, 705–724.

Pulkstenis, E., & Robinson, T. J. (2004). Goodness-of-fit tests for ordinal response regression models. *Statistics in Medicine, 23*(6), 999–1014.

Rabe-Hesketh, S., & Everitt, B. S. (2007). *A handbook of statistical analysis using Stata* (4th ed.). Boca Raton, FL: Chapman & Hall/CRC.

Rabe-Hesketh, S., & Skrondal, A. (2008). *Multilevel and longitudinal modeling using Stata* (2nd ed.). College Station, TX: Stata Press.

Rabe-Hesketh, S., & Skrondal, A. (2012). *Multilevel and longitudinal modeling using Stata* (3rd ed.). College Station, TX: Stata Press.

Rabe-Hesketh, S., Skrondal, A., & Pickles, A. (2004). *GLLAMM manual.* U.C. Berkeley Division of Biostatistics Working Paper Series. Retrieved from http://biostats.bepress.com/ucbbiostat/paper160

Raudenbush, S. W., & Bryk, A. S. (2002). *Hierarchical linear models: Applications and data analysis methods* (2nd ed.). Thousand Oaks, CA: Sage.

Schwarz, G. (1978). Estimating the dimension of a model. *Annals of Statistics, 6*, 461–464.

Singer, J. D., & Willett, J. B. (2003). *Applied longitudinal data analysis: Modeling change and event occurrence.* New York, NY: Oxford University Press.

Snijders, T. A., & Bosker, R. (2012). *Multilevel analysis: An introduction to basic and advanced multilevel modeling* (2nd ed.). London, England: Sage Ltd.

Stapleton, L. M. (2002). The incorporation of sample weights into multilevel structural equation models. *Structural Equation Modeling, 9*(4), 475–502.

Stapleton, L. M. (2006). An assessment of practical solutions for structural equation modeling with complex sample data. *Structural Equation Modeling, 13*(1), 28–58.

Stapleton, L. M. (2008). Variance estimation using replication methods in structural equation modeling with complex sample data. *Structural Equation Modeling, 15*(2), 183–210.

Sternberg, R. J., & Sternberg, K. (2010). *The psychologist's companion: A guide to writing scientific papers for students and researchers* (5th ed.). New York, NY: Cambridge University Press.

Stokes, M. E., Davis, C. S., & Koch, G. G. (2000). *Categorical data analysis using the SAS system* (2nd ed.). Cary, NC: SAS Institute.

Thody, A. (2006). *Writing and presenting research.* London, England: Sage Ltd.

Thomas, S. L., & Heck, R. H. (2001). Analysis of large-scale secondary data in higher education research: Potential perils associated with complex sampling designs. *Research in Higher Education, 42*(5), 517–540.

Thyer, B. A. (2008). *Preparing research articles.* New York, NY: Oxford University Press.

Tomz, M., King, G., & Wittenberg, J. (2003). Clarify: Software for interpreting and presenting statistical results. *Journal of Statistical Software, 8*(1), 1–30.

Tutz, G. (1991). Sequential models in categorical regression. *Computational Statistics & Data Analysis, 11*(3), 275–295.

Tutz, G. (2012). *Regression for categorical data*. New York, NY: Cambridge University Press.

Vittinghoff, E., Shiboski, S. C., Glidden, D. V., & McCulloch, C. E. (2011). *Regression methods in biostatistics: Linear, logistic, survival, and repeated measures models* (2nd ed.). New York, NY: Springer.

Wada, R. (2008). *outreg2: Stata module to arrange regression outputs into an illustrative table*. Statistical Software Components, Boston College Department of Economics. Retrieved from http://fmwww.bc.edu/repec/bocode/o

Watson, I. (2007). *tabout: Stata module to export publication quality cross-tabulations*. Statistical Software Components, Boston College Department of Economics. Retrieved from http://fmwww.bc.edu/repec/bocode/t

West, B. T., Welch, K. B., & Gałecki, A. T. (2014). *Linear mixed models: A practical guide using statistical software* (2nd ed.). Boca Raton, FL: Chapman & Hall/CRC.

Williams, R. (2006). Generalized ordered logit/partial proportional odds models for ordinal dependent variables. *The Stata Journal, 6*(1), 58–82.

Winter, N. (2005). *mktab: Stata module to print table of estimates in delimited or screen presentation format*. Statistical Software Components, Boston College Department of Economics. Retrieved from http://fmwww.bc.edu/repec/bocode/m

Wolfe, R. (1998). Continuation-ratio models for ordinal response data. *Stata Technical Bulletin, 44*, 18–21.

Index